Wavelets in Electromagnetics and Device Modeling

MILLIMETER WAVE OPTICAL DIELECTRIC INTEGRATED GUIDES AND CIRCUITS • *Shiban K. Koul*

MICROWAVE DEVICES, CIRCUITS AND THEIR INTERACTION • *Charles A. Lee and G. Conrad Dalman*

ADVANCES IN MICROSTRIP AND PRINTED ANTENNAS • *Kai-Fong Lee and Wei Chen (eds.)*

SPHEROIDAL WAVE FUNCTIONS IN ELECTROMAGNETIC THEORY • *Le-Wei Li, Xiao-Kang Kang, and Mook-Seng Leong*

OPTICAL FILTER DESIGN AND ANALYSIS: A SIGNAL PROCESSING APPROACH • *Christi K. Madsen and Jian H. Zhao*

THEORY AND PRACTICE OF INFRARED TECHNOLOGY FOR NONDESTRUCTIVE TESTING • *Xavier P. V. Maldague*

OPTOELECTRONIC PACKAGING • *A. R. Mickelson, N. R. Basavanhally, and Y. C. Lee (eds.)*

OPTICAL CHARACTER RECOGNITION • *Shunji Mori, Hirobumi Nishida, and Hiromitsu Yamada*

ANTENNAS FOR RADAR AND COMMUNICATIONS: A POLARIMETRIC APPROACH • *Harold Mott*

INTEGRATED ACTIVE ANTENNAS AND SPATIAL POWER COMBINING • *Julio A. Navarro and Kai Chang*

ANALYSIS METHODS FOR RF, MICROWAVE, AND MILLIMETER-WAVE PLANAR TRANSMISSION LINE STRUCTURES • *Cam Nguyen*

FREQUENCY CONTROL OF SEMICONDUCTOR LASERS • *Motoichi Ohtsu (ed.)*

WAVELETS IN ELECTROMAGNETICS AND DEVICE MODELING • *George W. Pan*

SOLAR CELLS AND THEIR APPLICATIONS • *Larry D. Partain (ed.)*

ANALYSIS OF MULTICONDUCTOR TRANSMISSION LINES • *Clayton R. Paul*

INTRODUCTION TO ELECTROMAGNETIC COMPATIBILITY • *Clayton R. Paul*

ELECTROMAGNETIC OPTIMIZATION BY GENETIC ALGORITHMS • *Yahya Rahmat-Samii and Eric Michielssen (eds.)*

INTRODUCTION TO HIGH-SPEED ELECTRONICS AND OPTOELECTRONICS • *Leonard M. Riaziat*

NEW FRONTIERS IN MEDICAL DEVICE TECHNOLOGY • *Arye Rosen and Harel Rosen (eds.)*

ELECTROMAGNETIC PROPAGATION IN MULTI-MODE RANDOM MEDIA • *Harrison E. Rowe*

ELECTROMAGNETIC PROPAGATION IN ONE-DIMENSIONAL RANDOM MEDIA • *Harrison E. Rowe*

NONLINEAR OPTICS • *E. G. Sauter*

COPLANAR WAVEGUIDE CIRCUITS, COMPONENTS, AND SYSTEMS • *Rainee N. Simons*

ELECTROMAGNETIC FIELDS IN UNCONVENTIONAL MATERIALS AND STRUCTURES • *Onkar N. Singh and Akhlesh Lakhtakia (eds.)*

FUNDAMENTALS OF GLOBAL POSITIONING SYSTEM RECEIVERS: A SOFTWARE APPROACH • *James Bao-yen Tsui*

InP-BASED MATERIALS AND DEVICES: PHYSICS AND TECHNOLOGY • *Osamu Wada and Hideki Hasegawa (eds.)*

COMPACT AND BROADBAND MICROSTRIP ANTENNAS • *Kin-Lu Wong*

DESIGN OF NONPLANAR MICROSTRIP ANTENNAS AND TRANSMISSION LINES • *Kin-Lu Wong*

PLANAR ANTENNAS FOR WIRELESS COMMUNICATIONS • *Kin-Lu Wong*

FREQUENCY SELECTIVE SURFACE AND GRID ARRAY • *T. K. Wu (ed.)*

ACTIVE AND QUASI-OPTICAL ARRAYS FOR SOLID-STATE POWER COMBINING • *Robert A. York and Zoya B. Popović (eds.)*

OPTICAL SIGNAL PROCESSING, COMPUTING AND NEURAL NETWORKS • *Francis T. S. Yu and Suganda Jutamulia*

SiGe, GaAs, AND InP HETEROJUNCTION BIPOLAR TRANSISTORS • *Jiann Yuan*

ELECTRODYNAMICS OF SOLIDS AND MICROWAVE SUPERCONDUCTIVITY • *Shu-Ang Zhou*

Wavelets in Electromagnetics and Device Modeling

GEORGE W. PAN
Arizona State University
Tempe, Arizona

WILEY-INTERSCIENCE
A JOHN WILEY & SONS PUBLICATION

Published by John Wiley & Sons, Inc., Hoboken, New Jersey.
Published simultaneously in Canada.

Library of Congress Cataloging-in-Publication Data:

Pan, George W., 1944–

Wavelets in electromagnetics & device modeling / George W. Pan.

p. cm. — (Wiley series in microwave and optical engineering)

Includes index.

ISBN 0-471-41901-X (cloth : alk. paper)

1. Integrated circuits—Mathematical models. 2. Wavelets (Mathematics) 3. Electromagnetism—Mathematical models. 4. Electromagnetic theory. I. Title: Wavelets in electromagnetics and device modeling. II. Title. III. Series

TK7874.P3475 2002
621.3815—dc21 2002027207

Printed in the United States of America

10 9 8 7 6 5 4 3 2 1

Dedicated to my father Pan Zhen and mother Lei Tian-Lu

Contents

Preface

Applied mathematics has made considerable progress in wavelets. In recent years interest in wavelets has grown at a steady rate, and applications of wavelets are expanding rapidly. A virtual flood of engineers, with little mathematical sophistication, is about to enter the field of wavelets. Although more than 100 books on wavelets have been published since 1992, there is still a large gap between the mathematician's rigor and the engineer's interest. The present book is intended to bridge this gap between mathematical theory and engineering applications.

In an attempt to exploit the advantages of wavelets, the book covers basic wavelet principles from an engineer's point of view. With a minimum number of theorems and proofs, the book focuses on providing physical insight rather than rigorous mathematical presentations. As a result the subject matter is developed and presented in a more basic and familiar way for engineers with a background in electromagnetics, including linear algebra, Fourier analysis, sampling function of $\sin \pi x/\pi x$, Dirac δ function, Green's functions, and so on. The multiresolution analysis (MRA) is naturally delivered in Chapter 2 as a basic introduction that shows a signal decomposed into several resolution levels. Each level can be processed according to the requirement of the application. The application of MRA lies within the Mallat decomposition and reconstruction algorithm. MRA is further explained in a fast wavelet transform section with an example of frequency-dependent transmission lines. Mathematically elegant proofs and derivations are presented in a smaller font if their content is beyond the engineering requirement. Readers with no time or interest in this depth of mathematics may always skip the paragraphs or sections written in smaller font without jeopardizing their understanding of the main subjects.

The main body of the book came from conference presentations, including the IEEE Microwave Theory and Techniques Symposium (IEEE-MTT), IEEE Antennas and Propagation (IEEE-AP), Radio Science (URSI), IEEE Magnetics, Progress in Electromagnetic Research Symposium (PIERS), Electromagnetic and Light Scattering (ELS), COMPUMAG, Conference on Electromagnetic Field Computation (CEFC), Association for Computational Electromagnetic Society (ACES), International Conference on Microwave and Millimeter Wave Technology (ICMTT), and

International Conference on Computational Electromagnetics and its Applications (ICCEA). The book has evolved from curricula taught at the graduate level in the Department of Electronic Engineering at Canterbury University (Christchurch, New Zealand) and Arizona State University. The material was taught as short courses at Moscow State University, CSIRO (Sydney, Australia), IEEE Microwave Theory and Techniques Symposium, Beijing University, Aerospace 207 Institute, and the 3rd Institute of China. The participants in these courses were electrical engineering and computer science students as well as practicing engineers in industry. These people had little or no prior knowledge of wavelets.

The book may serve as a reference book for engineers, practicing scientists, and other professionals. Real-world state-of-the-art issues are extensively discussed, including full-wave modeling of coupled lossy and dispersive transmission lines, scattering of electromagnetic waves from 2D/3D bodies and from randomly rough surfaces, radiation from linear and patch antennas, and modeling of 2D semiconductor devices. The book can also be used as a textbook, as it contains questions, working examples, and 11 exercise assignments with a solution manual. It has been used several times in teaching a one-semester graduate course in electrical engineering.

The book consists of 10 chapters. The first six chapters are dedicated to basic theory and training, followed by four chapters in real-world applications. Chapter 1 summarizes mathematical preliminaries, which may be skipped on the first reading. Chapter 2 provides some background and theoretical insights. Chapter 3 covers the basic orthogonal wavelet theory. Other wavelet topics are discussed in Chapters 4 through 10, including biorthogonal wavelets, weighted wavelets, interpolating wavelets, Green's wavelets, and multiwavelets. Chapter 4 presents applications of wavelets in solving integral equations. Special treatments of edges are discussed here, including periodic wavelets and intervallic wavelets. Chapter 5 derives the positive sampling functions and their biorthogonal counterparts using Daubechies wavelets. Many advantages derive from the use of the sampling biorthogonal time domain (SBTD) method to replace the finite difference time domain (FDTD) scheme. Chapter 6 studies multiwavelet theory, including biorthogonal and orthogonal multiwavelets with applications in the edge-based finite element method (EEM). Advanced topics are presented in Chapter 7, 8, and 9, respectively, for scattering and radiation, 3D rough surface scattering, packaging and interconnects. Chapter 10 is devoted to semiconductor device modeling using the aforementioned knowledge of wavelets. Numerical procedures are fully detailed so as to help interested readers develop their own algorithms and computer codes.

ACKNOWLEDGMENTS

I am indebted to all colleagues who have made contributions in this area. I am grateful to my former colleague, Professor Gilbert G. Walter, a pioneering mathematician, who introduced wavelets to me.

The grants for my research, from the Mayo Foundation, Boeing Aerospace Co., DARPA University Research Initiative, Cray Research, DARPA/MTO, DARPA/DSO, Oregon Medical Systems, W. L. Gore Associates, and the sabbatical funds of Arizona State University are greatly acknowledged. I am particularly grateful to Dr. Barry Gilbert, who has funded my research for nearly two decades. My appreciation is also due to Dr. James Murphy and Dr. Douglas Cochran for their long-term support of the research grants that have made long lasting research possible. The results of the research have formed the trunk of this book.

I wish to thank my former students, Dr. Mikhail Toupikov, Dr. Jilin Tan, Dr. Gaofeng Wang, Dr. Youri Tretiakov, Dr. Allen Zhu, Mr. Pierre Piel, Mr. Janyuan Du, and my students Ke Wang, Stanislav Ogurtsov, and Zhichao Zhang for their contributions. Finally, thanks are due to Professor Kai Chang for his invitation of the book proposal, and to my editor Mr. George Telecki for his patience and timely supervision during the course of publishing the book.

GEORGE W. PAN

Tempe, Arizona

Wavelets in Electromagnetics and Device Modeling

CHAPTER ONE

Notations and Mathematical Preliminaries

1.1 NOTATIONS AND ABBREVIATIONS

The notations and abbreviations used in the book are summarized here for ease of reference.

$D^{(\alpha)} f = f^{\alpha}(t) := df^{\alpha}(t)/dt^{\alpha}$
$\bar{f}$—complex conjugate of f
$\hat{f} := \int_{-\infty}^{\infty} f(t)e^{-i\omega t}\, dt$, Fourier transform of $f(t)$
$f(t) := \frac{1}{2\pi}\int_{-\infty}^{\infty} \hat{f}(\omega)e^{i\omega t}\, d\omega$, inverse Fourier transform of $\hat{f}(\omega)$
$\|f\|$—norm of a function
$f * g$—convolution
$\langle f, h\rangle := \int \overline{f(t)}h(t)\, dt$, inner product
$f_n = O(n)$-order of n, $\exists C$ such that $f_n \leq Cn$
$\mathcal{C}$—complex
N—nonnegative integers
R—real number
R^n—real numbers of size n
Z—integers
Z^+—positive integers
$L^2(R)$—functional space consisting finite energy functions $\int |f(t)|^2\, dt < +\infty$
$L^p(R)$—function space that $\int |f(t)|^p\, dt < +\infty$
$l^2(Z)$—finite energy series $\sum_{n=-\infty}^{\infty} |a_n|^2 < +\infty$
Ω—set
$H^s(\Omega) := W^{s,2}(\Omega)$-Sobolev space equipped with inner product of
$\quad \langle u, v\rangle_{s,2} := \sum_{|\alpha|\leq s} \int_{\Omega} \overline{D^{\alpha}u} D^{\alpha} v d\Omega$

$V \oplus W$—direct sum
$V \otimes W$—tensor product
∇f—gradient
$\vec{H}, \vec{E}$—vector fields
$\nabla \times \vec{H}$—curl
$\nabla \cdot \vec{E}$—divergence
$\lfloor \alpha \rfloor$—largest integer $m \leq \alpha$
$\delta_{m,n}$—Kronecker delta
$\delta(t)$—Dirac delta
$\chi[a, b]$—characteristic function, which is 1 in $[a, b]$ and zero outside
□—end of proof
∃—exist
∀—any
iff—if and only if
a.e.—almost everywhere
d.c.—direct current
o.n.—orthonormal
o.w.—otherwise

1.2 MATHEMATICAL PRELIMINARIES

This chapter is arranged here to familiarize the reader with the mathematical notation, definitions and theorems that are used in wavelet literature and in this book. Important mathematical concepts are briefly reviewed. In most cases no proof is given. For more detailed discussions or in depth studies, readers are referred to the corresponding references [1–5].

Readers are suggested to skip this chapter in their first reading. They may then return to the relevant sections of this chapter if unfamiliar mathematical concepts present themselves during the course of the book.

1.2.1 Functions and Integration

A function $f(t)$ is called integrable if

$$\int_{-\infty}^{\infty} |f(t)| \, dt < +\infty, \tag{1.2.1}$$

and we say that $f \in L^1(R)$.

Two functions $f_1(t)$ and $f_2(t)$ are equal in $L^1(R)$ if

$$\int_{-\infty}^{\infty} |f_1(t) - f_2(t)| \, dt = 0.$$

This implies that $f_1(t)$ and $f_2(t)$ may differ only on a set of points of zero measure. The two functions f_1 and f_2 are almost everywhere (a.e.) equal.

Fatou Lemma. Let $\{f_n\}_{n\in N}$ be a set of positive functions. If

$$\lim_{n\to\infty} f_n(t) = f(t)$$

almost everywhere, then

$$\int_{-\infty}^{\infty} f(t)\,dt \le \lim_{n\to\infty} \int_{-\infty}^{\infty} f_n(t)\,dt.$$

This lemma provides an inequality when taking a limit under the Lebesgue integral for positive functions.

Lebesgue Dominated Convergence Theorem. Let $f_k(t) \in L(E)$ for $k = 1, 2, \ldots,$ and

$$\lim_{k\to\infty} f_k(t) = f(t) \quad \text{a.e.}$$

If there exists an integrable function $F(t)$ such that

$$|f_k(t)| \le F(t) \quad \text{a.e.,} \quad k = 1, 2, \ldots,$$

then

$$\lim_{k\to\infty} \int_E f_k(t)\,dt = \int_E f(t)\,dt.$$

This theorem allows us to exchange the limit with integration.

Fubini Theorem. If

$$\int_{-\infty}^{\infty} \left(\int_{-\infty}^{\infty} f(t_1, t_2)\,dt_1 \right) dt_2 < \infty,$$

then

$$\begin{aligned}\int_{-\infty}^{\infty}\int_{-\infty}^{\infty} f(t_1, t_2)\,dt_1\,dt_2 &= \int_{-\infty}^{\infty} dt_2 \int_{-\infty}^{\infty} f(t_1, t_2)\,dt_1 \\ &= \int_{-\infty}^{\infty} dt_1 \int_{-\infty}^{\infty} f(t_1, t_2)\,dt_2.\end{aligned}$$

This theorem provides a sufficient condition for commuting the order of the multiple integration.

1.2.2 The Fourier Transform

The Fourier transform pair is defined as

$$\hat{f}(\omega) = \int_{-\infty}^{\infty} f(t)e^{-i\omega t}\,dt,$$

$$f(t) = \frac{1}{2\pi}\int_{-\infty}^{\infty} \hat{f}(\omega)e^{i\omega t}\,d\omega.$$

Rigorously speaking, the Fourier transform of $f(t)$ exists if the Dirichlet conditions are satisfied, that is,

(1) $\int_{-\infty}^{\infty} |f(t)|\,dt < +\infty$, as in (1.2.1).
(2) $f(t)$ has a finite number of maxima and minima within any finite interval, and any discontinuities of $f(t)$ are finite. There are only a finite number of such discontinuities in any finite interval.

All functions satisfying (1.2.1) form a functional space L^1. A weaker condition for the existence of the Fourier transform of $f(t)$, in replace of (1.2.1), is given as

$$\int_{-\infty}^{\infty} |f(t)|^2\,dt < +\infty. \tag{1.2.2}$$

All functions satisfying (1.2.2) form a functional space L^2.

When the Dirichlet conditions are satisfied, the inverse Fourier transform converges to $f(t)$ if $f(t)$ is continuous at t, or to

$$\frac{f(t^+) + f(t^-)}{2}$$

if $f(t)$is discontinuous at t. When $f(t)$ has infinite energy, its Fourier transform may be defined by incorporating generalized functions. The resultant is called the generalized Fourier transform of the original function.

1.2.3 Regularity

Lipschitz Regularity. If a function $f(t)$ has a singularity at $t = v$, this implies that $f(t)$ is not differentiable at v. Lipschitz exponent at v characterizes the singularity behavior.

The Taylor expansion relates the differentiability of a function to a local polynomial approximation. Suppose that f is m times differentiable in $[v - h, v + h]$. Let p_v be the Taylor polynomial in the neighborhood of v:

$$p_v(t) = \sum_{k=0}^{m-1} \frac{f^{(k)}(v)}{k!}(t - v)^k.$$

Then the error

$$|\varepsilon_v(t)| \le \frac{|t-v|^m}{m!} \sup_{u\in[v-h,v+h]} |f^{(m)}(u)|$$

where

$$t \in [v-h, v+h], \quad \varepsilon_v(t) := f(t) - p_v(t).$$

The Lipschitz regularity refines the upper bound on the error $\varepsilon_v(t)$ with noninteger exponents. Lipschitz exponents are also referred to as Hölder exponents.

Definition 1 (Lipschitz). A function $f(t)$ is pointwise Lipschitz $\alpha \ge 0$ at $t = v$, if there exist $M > 0$ and a polynomial $p_v(t)$ of degree $m = \lfloor \alpha \rfloor$ such that

$$\forall t \in R, \quad |f(t) - p_v(t)| \le M|t-v|^\alpha. \tag{1.2.3}$$

Definition 2. A function $f(t)$ is uniformly Lipschitz α over $[a, b]$ if it satisfies (1.2.3) for all $v \in [a, b]$ with a constant M independent of v.

Definition 3. The Lipschitz regularity of $f(t)$ at v or over $[a, b]$ is the sup of the α such that $f(t)$ is Lipschitz α.

Theorem 1. A function $f(t)$ is bounded and uniform Lipschitz α over R if

$$\int_{-\infty}^{\infty} |\hat{f}(\omega)|(1+|\omega|^\alpha)\, d\omega < +\infty. \tag{1.2.4}$$

If $0 \le \alpha < 1$, then $p_v(t) = f(v)$ and the Lipschitz condition reduces to

$$\forall t \in R, \quad |f(t) - f(v)| \le M|t-v|^\alpha.$$

Here the function is bounded but discontinuous at v, and we say that the function is Lipschitz 0 at v.

Proof. When $0 \le \alpha < 1$, it follows $m := \lfloor \alpha \rfloor = 0$, and $p_v(t) = f(v)$.
The uniform Lipschitz regularity implies that $\exists M > 0$ such that

$$\forall (t, v) \in R^2.$$

We need to have

$$\frac{|f(t) - f(v)|}{|t-v|^\alpha} \le M.$$

Since

$$f(t) = \frac{1}{2\pi}\int_{-\infty}^{\infty} \hat{f}(\omega)e^{i\omega t}\, d\omega,$$

$$\frac{|f(t) - f(v)|}{|t - v|^{\alpha}} = \frac{1}{2\pi}\left|\int_{-\infty}^{\infty} \hat{f}(\omega)\left[\frac{e^{i\omega t}}{|t - v|^{\alpha}} - \frac{e^{i\omega v}}{|t - v|^{\alpha}}\right] d\omega\right|$$

$$\leq \frac{1}{2\pi}\int_{-\infty}^{\infty} |\hat{f}(\omega)|\frac{|e^{i\omega t} - e^{i\omega v}|}{|t - v|^{\alpha}}\, d\omega.$$

(1) For $|t - v|^{-1} \leq |\omega|$,

$$\frac{|e^{i\omega t} - e^{i\omega v}|}{|t - v|^{\alpha}} \leq \frac{2}{|t - v|^{\alpha}} \leq 2|\omega|^{\alpha}.$$

(2) For $|t - v|^{-1} \geq |\omega|$,

$$|e^{i\omega t} - e^{i\omega v}| = \left|i\omega(t - v) - \frac{\omega^2}{2!}(t - v)^2 - i\frac{(t - v)^3}{3!} + - \cdots\right|.$$

On the right-hand side of the equation above, the imaginary part

$$I = \omega(t - v) - \frac{[\omega(t - v)]^3}{3!} + \frac{[\omega(t - v)]^5}{5!} - + \cdots \leq \omega(t - v),$$

and the magnitude of the real part

$$R = \left\{\frac{[\omega(t - v)]^2}{2!} - \frac{[\omega(t - v)]^4}{4!} + - \cdots\right\} \leq \frac{[\omega(t - v)]^2}{2!}.$$

Thus

$$|(t - v)\omega| \leq 1 \quad \text{and} \quad [(t - v)\omega]^2 \leq |(t - v)\omega|$$

and

$$|e^{i\omega t} - e^{i\omega v}| \leq \left|i\omega(t - v) + \frac{[\omega(t - v)]^2}{2!}\right|$$

$$= \sqrt{[\omega(t - v)]^2 + \frac{\omega^4(t - v)^4}{4}}$$

$$\leq |2\omega(t - v)|.$$

Hence

$$\frac{|e^{i\omega t} - e^{i\omega v}|}{|t - v|^{\alpha}} \leq \frac{2|\omega||t - v|}{|t - v|^{\alpha}} \leq 2|\omega|^{\alpha}.$$

Combining (1) and (2) yields

$$\frac{|f(t) - f(v)|^2}{|t - v|^\alpha} \leq \frac{1}{2\pi} \int_{-\infty}^{\infty} 2|\hat{f}(\omega)|\, |\omega|^\alpha \, d\omega := M.$$

It can be verified that if

$$\int_{-\infty}^{\infty} |\hat{f}(\omega)|[1 + |\omega|^p]\, d\omega < \infty,$$

then $f(t)$ is p times continuously differentiable. Therefore, if

$$\int_{-\infty}^{\infty} \hat{f}(\omega)[1 + |\omega|^\alpha]\, d\omega < \infty,$$

then $f^{(m)}(t)$ is uniformly Lipschitz $\alpha - m$, and hence $f(t)$ is uniformly Lipschitz α, where $m = \lfloor \alpha \rfloor$. □

1.2.4 Linear Spaces

Linear Space. A linear space H is a nonempty set. Let $\mathcal{C}$ be complex. H is called a complex linear space if

(1) $x + y = y + x$.
(2) $(x + y) + z = x + (y + z)$.
(3) There exists a unique element $\theta \in H$ such that for $\forall x \in H, x + \theta = \theta + x$.
(4) For $\forall x \in H$, there exists a unique $-x$ such that $x + (-x) = \theta$.

In addition we define scalar multiplication $\forall(\alpha, x) \in \mathcal{C} \times H$ such that

(1) $\alpha(\beta x) = (\alpha\beta)x, \forall \alpha, \beta \in \mathcal{C}, \forall x \in H$.
(2) $1x = x$.
(3) $(\alpha + \beta)x = \alpha x + \beta x, \forall \alpha, \beta \in \mathcal{C}, \forall x \in H$.
$\quad\;\alpha(x + y) = \alpha x + \alpha y, \forall \alpha \in \mathcal{C}, \forall x, y \in H$.

Norm of a Vector

Definition. Mapping of $\| x \|: R^n \to R$ is called the norm of x on R^n iff

(1) $\| x \| \geq 0, \forall x \in R^n$.
(2) $\| \alpha x \| = |\alpha| \, \| x \|, \forall \alpha \in R, x \in R^n$.
(3) $\| x + y \| \leq \| x \| + \| y \|, \forall x, y \in R^n$.
(4) $\| x \| = 0 \Longleftrightarrow x = 0$.

Let $x = (x_1, x_2, \ldots, x_n)^T \in R^n$. The following are commonly used norms:

$$\| x \|_\infty = \max_i |x_i|, \qquad \ell^\infty \text{ norm},$$

$$\| x \|_1 = \sum_{i=1}^{n} |x_i|, \qquad \ell^1 \text{ norm},$$

$$\| x \|_2 = \left(\sum_{i=1}^{n} x_i^2 \right)^{1/2}, \qquad \ell^2 \text{ norm},$$

$$\| x \|_p = \left(\sum_{i=1}^{n} |x_i|^p \right)^{1/p}, \qquad \ell^p \text{ norm}.$$

1.2.5 Functional Spaces

Metric, Banach, Hilbert, and Sobolev spaces are functional spaces. A functional space is a collection of functions that possess a certain mathematical structure pattern.

Metric Space. A metric space H is a nonempty set that defines the distance of a real-valued function $\rho(x, y)$ that satisfies:

(1) $\rho(x, y) \geq 0$ and $\rho(x, y) = 0$ iff $x = y$.
(2) $\rho(x, y) = \rho(y, x)$.
(3) $\rho(x, y) \leq \rho(x, z) + \rho(z, y), \forall x, y, z \in H$.

Banach Space. Banach space is a vector space H that admits a norm, $\| \cdot \|$, that satisfies:

(1) $\forall f \in H, \| f \| \geq 0$ and $\| f \| = 0$ iff $f = 0$.
(2) $\forall \alpha \in C, \| \alpha f \| = |\alpha| \, \| f \|$.
(3) $\| f + g \| \leq \| f \| + \| g \|, \forall f, g \in H$.

These properties of norms are similar to those of distance, except the homogeneity of (2) is not required in defining a distance. The convergence of $\{f_n\}_{n \in N}$ to $f \in H$ implies that $\lim_{n\to\infty} \| f_n - f \| = 0$ and is denoted as $\lim_{n\to\infty} f_n = f$.

To guarantee that we remain in H when taking the limits, we define the Cauchy sequences. A sequence $\{f_n\}_{n \in N}$ is a Cauchy sequence if for $\forall \varepsilon > 0$, there exist n and m large enough such that $\| f_m - f_n \| < \varepsilon$. The space H is said to be complete if every Cauchy sequence in H converges to an element of H. A complete linear space equipped with norm is called the Banach space.

Example 1 Let S be a collection of sequences $x = (x_1, x_2, \ldots, x_n, \ldots)$. We define addition and multiplication naturally as

$$x + y = (x_1 + y_1, x_2 + y_2, \ldots, x_n + y_n, \ldots),$$
$$\alpha x = (\alpha x_1, \alpha x_2, \ldots, \alpha x_n, \ldots),$$

and define distance as

$$\rho(x, y) = \sum \frac{1}{2^n} \frac{|x_n - y_n|}{1 + |x_n - y_n|}.$$

It can be verified that such a space S is not a Banach space, because $\rho(x, y)$ does not satisfy the homogeneous condition of the norm.

Example 2 For any integer p we define over discrete sequence f_n the norm

$$\|f\|_p = \left[\sum_{n=-\infty}^{\infty} |f_n|^p \right]^{1/p}.$$

The space $\ell^p = \{f : \|f\|_p < \infty\}$ is a Banach space with norm $\|f\|_p$.

Example 3 The space $L^p(R)$ is composed of measurable functions f on R that

$$\| f \|_p = \left\{ \int_{-\infty}^{\infty} |f(t)|^p \right\}^{1/p} < \infty.$$

The space $L^p(R) = \{f : \| f \|_p < \infty\}$ is a Banach space.

Hilbert Space. A Hilbert space is an inner product space that is complete. The inner product satisfies:

(1) $\langle \alpha f + \beta g, h \rangle = \overline{\alpha} \langle f, g \rangle + \overline{\beta} \langle g, h \rangle$ for $\alpha, \beta, \in C$ and $f, g, h \in H$.
(2) $\langle f, g \rangle = \overline{\langle g, f \rangle}$.
(3) $\langle f, f \rangle \geq 0$ and $\langle f, f \rangle = 0$ iff $f = 0$. One may verify that

$$\| f \| = \langle f, f \rangle^{1/2}$$

is a norm.

(4) The Cauchy–Schwarz inequality states that

$$|\langle f, g \rangle| \leq \| f \| \| g \|,$$

where the equality is held iff f and g are linearly dependent.

In a Banach space the norm is defined, which allows us to discuss the convergence. However, the angles and orthogonality are lacking. A Hilbert space is a Banach space equipped with an inner product.

1.2.6 Sobolev Spaces

The Sobolev space is a functional space, and it could have been listed in the previous subsection. However, we have placed it in a separate subsection because of its contents and role in the text.

On many occasions involving differential operators, it is convenient to incorporate the L^p norms of the derivative of a function into a Banach norm. Consider the functions in the class $C^\infty(\Omega)$. For any number $p \geq 1$ and number $s \geq 0$, let us take the closure of $C^\infty(\Omega)$ with respect to the norm

$$\|u\|_{s,p} = \left\{ \sum_{|\alpha| \leq s} \|D^\alpha u\|_{L^p}^p \right\}^{1/p}. \tag{1.2.5}$$

The resulting Banach space is called the Sobolev space $W^{s,p}(\Omega)$. For $p = 2$ we denote $W^s(\Omega) = W^{s,2}(\Omega)$, which is a Hilbert space with respect to the inner product

$$\langle u, v \rangle_{s,2} = \sum_{|\alpha| \leq s} \int_\Omega \overline{D^\alpha u} \cdot D^\alpha v \, dx.$$

Sometimes $W^s(R)$ is also denoted as $H^s(R)$. Note that the differentiation in (1.2.5) can be of a noninteger.

Recall that the Fourier transform of the derivative $f'(t)$ is $i\omega\hat{f}(\omega)$. The Plancherel–Parseval formula proves that $f'(t) \in L^2(R)$ if

$$\int_{-\infty}^{\infty} |f'(t)|^2 \, dt = \frac{1}{2\pi} \int_{-\infty}^{\infty} |\omega|^2 |\hat{f}(\omega)|^2 < +\infty.$$

This expression can be generalized for any $s > 0$,

$$\int_{-\infty}^{\infty} |\omega|^{2s} |\hat{f}(\omega)|^2 \, d\omega < +\infty$$

if $f \in L^2(R)$ is s times differentiable.

Considering the summation nature of (1.2.5), we can write the more precise expression of Sobolev space in the Fourier domain as

$$\int_{-\infty}^{\infty} (1 + \omega^2)^s |f(\omega)|^2 \, d\omega < +\infty.$$

For $s > n + \frac{1}{2}$, f is n times continuously differentiable. The Sobolev space H^α, $\alpha \in R$ consists of functions $f(t) \in S'$ such that

$$\int_{-\infty}^{\infty} \hat{f}(\omega)(1 + \omega^2)^\alpha \, d\omega < \infty.$$

For $\alpha = 0$, the H^α reduces to $L^2(R)$. For $\alpha = 1, 2, \ldots$, H^α is composed of ordinary $L^2(R)$ functions that are $(\alpha - 1)$ times differentiable and whose αth derivative are in $L^2(R)$. For $\alpha = -1, -2, \ldots,$ H^α contains the $-\alpha$th derivatives of $L^2(R)$ and all distributions with point support of order $< \alpha$.

It can be seen $H^\alpha \supset H^\beta$ when $\alpha > \beta$. The inner product of $f, g \in H^\alpha$ is

$$\langle f, g \rangle_\alpha = \frac{1}{2\pi} \int \overline{\hat{f}(\omega)} \hat{g}(\omega)(1 + \omega^2)^\alpha \, d\omega$$

and is complete with respect to this inner product. Therefore it is a Hilbert space.

1.2.7 Bases in Hilbert Space H

Orthonormal Basis. A sequence $\{f_n\}_{n\in N}$ in a Hilbert space H is orthonormal if

$$\langle f_m, f_n \rangle = \delta_{m,n}.$$

If for $f \in H$ there exist α_n such that

$$\lim_{N\to\infty} \| f - \sum_{n=0}^{N} \alpha_n f_n \| = 0,$$

then $\{f_n\}_{n\in N}$ is called an orthogonal basis of H.

For an orthonormal basis we require $\|f_n\| = 1$. A Hilbert space that admits an orthogonal basis is said to be separable. The norm of $f \in H$ is

$$\|f\|^2 = \sum_{n=0}^{\infty} |\langle f, f_n \rangle|^2$$

Riesz Basis. Let $\{f_n\}$ be linear independent and complete in $L^2(a, b)$, meaning that the closed linear span of $\{f_n\}$ is $L^2(a, b)$. The set is called a Riesz basis if there exist $A > 0$ and $B > 0$ such that

$$A \sum_i |c_i|^2 \leq \| \sum_i c_i f_i \|^2 \leq B \sum_i |c_i|^2 \tag{1.2.6}$$

for each sequence $\{c_i\}$ of complex numbers. The Riesz representation theorem guarantees the existence of the dual $\{\tilde{f}_n\}$ in $L^2(a, b)$ such that:

(1) $\{\tilde{f}_n\}$ is the unique biorthogonal sequence to $\{f_n\}$; namely $\langle f_m, \tilde{f}_n \rangle = \delta_{m,n}$.
(2) If $\{c_n\} \in \ell^2$, then $\sum_n c_n f_n$ converges in $L^2(a, b)$.
(3) For each $f \in L^2(a, b)$, $\{\langle f, \tilde{f}_n \rangle\} \in \ell^2$.
(4) For each $f \in L^2(a, b)$,

$$f = \sum_{i=0}^{\infty} \langle f, \tilde{f}_i \rangle f_i = \sum_{i=0}^{\infty} \langle f, f_i \rangle \tilde{f}_i.$$

A Riesz basis of a separable Hilbert space H is a basis that is close to being orthogonal. The right inequality in (1.2.6) is essential. It prevents the expansion from blowing up. The left inequality in (1.2.6) is important too, since it ensures the existence of the inverse.

1.2.8 Linear Operators

In computational electromagnetics, the method of moments and finite element method are based on linear operations. An operator T from a Hilbert space H_1 to another Hilbert space H_2 is linear if

$$\forall \alpha_1, \alpha_2 \in C,\ \forall f_1, f_2 \in H_1, \quad T(\alpha_1 f_1 + \alpha_2 f_2) = \alpha_1 T(f_1) + \alpha_2 T(f_2).$$

Sup Norm. The sup operator norm of T is defined as

$$\|T\|_S = \sup_{f \in H_1} \frac{\|Tf\|}{\|f\|}. \tag{1.2.7}$$

If this norm is finite, then T is continuous; namely $\|Tf_1 - Tf_2\|$ becomes arbitrarily small if $\|f_1 - f_2\|$ is sufficiently small.

Adjoint. The adjoint of T is the operator T^a from H_2 to H_1 such that for any $f_1 \in H_1$ and $f_2 \in H_2$

$$\langle Tf_1, f_2\rangle = \langle f_1, T^a f_2\rangle.$$

When T is defined from H into itself, it is self-adjoint if $T = T^a$. A nonzero vector $f \in H$ is a called an *eigenvector* if there exists an *eigenvalue* $\lambda \in C$ such that

$$Tf = \lambda f.$$

In a finite-dimensional Hilbert space, meaning that Euclidean space, a self-adjoint operator is always diagonalized by an orthogonal basis $\{e_n\}_{0 \le n < N}$ of eigenvectors

$$Te_n = \lambda_n e_n.$$

For a self-adjoint operator T, the eigenvalues λ_n are real, and for any $f \in H$

$$Tf = \sum_{n=0}^{N-1} \langle Tf, e_n\rangle e_n = \sum_{n=0}^{N-1} \lambda_n \langle f, e_n\rangle e_n.$$

In an infinite-dimensional Hilbert space, the previous result can be generalized in terms of the spectrum of the operator, which must be manipulated with caution.

Orthogonal Projector. Let V be a subspace of H. A *projector* P_V on V is a linear operator that satisfies $\forall f \in H, \quad P_V f \in V$ and $\forall f \in V, \quad P_V u = f$.

The projector P_V is *orthogonal* if

$$\forall f \in H, \forall g \in V, \quad \langle f - P_V f, g\rangle = 0.$$

The following properties are often used in the text:

Property 1. If P_V is *a projector* on V, then the following statements are equivalent:

(1) P_V is *orthogonal.*
(2) P_V is *self-adjoint.*
(3) $\|P_V\|_S = 1.$
(4) $\forall f \in H, \|f - P_V f\| = \min_{g\in \mathbf{V}} \|f - g\|.$

If $\{e_n\}_{n\in N}$ is an orthogonal basis of V, then

$$P_V f = \sum_{n=0}^{+\infty} \frac{\langle f, e_n\rangle}{\|e_n\|^2} e_n.$$

If $\{e_n\}_{n\in N}$ is a Riesz basis of V and $\{\tilde{e}_n\}_{n\in N}$ is the biorthogonal basis, then

$$P_V f = \sum_{n=0}^{+\infty} \langle f, e_n\rangle \tilde{e}_n = \sum_{n=0}^{+\infty} \langle f, \tilde{e}_n\rangle e_n.$$

Density and Limit. A space V is *dense* in H if for any $f \in H$ there exist $\{f_m\}_{m\in N}$ with $f_m \in V$ such that

$$\lim_{m\to+\infty} \|f - f_m\| = 0.$$

Let $\{T_n\}_{n\in N}$ be a sequence of linear operators from H to H. Such a sequence *converges* weakly to a linear operator T_∞ if

$$\forall f \in H, \quad \lim_{n\to+\infty} \|T_n f - T_\infty f\| = 0.$$

To find the limit of operators it is preferable to work in a well chosen subspace $V \subset H$ which is dense. The density and limit are justified by the property below.

Property 2 (Density). Let V be a *dense subspace* of H. Suppose that there exists C such that $\|T_n\|_S \le C$ for all $n \in N$. If

$$\forall f \in V, \quad \lim_{n\to+\infty} \|T_n f - T_\infty f\| = 0,$$

then

$$\forall f \in H, \quad \lim_{n\to+\infty} \|T_n f - T_\infty f\| = 0.$$

For numerical computations, an operator is often discretized into a matrix. Only then digital computers can be utilized.

Norm of a Matrix. For a matrix $A \in R^{n\times n}$, the norm of A is defined, similarly to (1.2.7), as

$$\| A \| = \max_{x\neq 0} \left\{ \frac{\| Ax \|}{\| x \|} \right\}.$$

In particular, the commonly used norms are as follows:

(1) The column norm (ℓ^1 norm)

$$\| A \|_1 = \max_j \sum_{i=1}^{n} |a_{ij}|.$$

(2) The row norm (ℓ^∞ norm)

$$\| A \|_\infty = \max_i \{\|a_{i\cdot}\|_1\} = \max_i \sum_{j=1}^{n} |a_{i,j}|.$$

(3) The spectral norm (ℓ^2 norm)

$$\|A\|_2 = (\lambda_{A^TA})^{1/2},$$

where λ_{A^TA} is the maximum eigenvalue of A^TA.

(4) The Frobenius norm

$$\|A\|_F = \left(\sum_{j=1}^{n} \sum_{i=1}^{n} |a_{i,j}|^2 \right)^{1/2} = [\mathrm{tr}\,\{A^TA\}]^{1/2}.$$

BIBLIOGRAPHY

[1] R. L. Wheedan and A. Zygmund, *Measure and Integration*, Marcel Dekker, New York, 1977.

[2] A.N. Kolmogorov and S.V. Fomin, *Introductory Real Analysis*, Dover, New York, 1970.

[3] F. Riesz and B. Sz.-Nagy, *Functional Analysis*, Dover Publication, Inc. New York, 1990.

[4] M. S. Berger, *Nonlinear and Functional Analysis*, Academic Press, New York, 1977.

[5] S. Mallat, *A Wavelet Tour of Signal Processing*, Academic Press, New York, 1998.

CHAPTER TWO

Intuitive Introduction to Wavelets

2.1 TECHNICAL HISTORY AND BACKGROUND

The first questions from those curious about wavelets are: What is a wavelet? Who invented wavelets? What can one gain by using wavelets?

2.1.1 Historical Development

Wavelets are sometimes referred to as the twentieth-century Fourier analysis. Wavelets exploit the multiresolution analysis just like microscopes do in microbiology. The genesis of wavelets began in 1910 when A. Haar proposed the staircase approximation to approximate a function, using the piecewise constants now called the Haar wavelets [1]. Afterward many mathematicians, physicists, and engineers made contributions to the development of wavelets:

- Paley–Littlewood proposed dyadic frequency grouping in 1938 [2].
- Shannon derived sampling theory in 1948 [3].
- Calderon employed atomic decomposition of distributions in parabolic H^p spaces in 1977 [4].
- Stromberg improved the Haar systems in 1981 [5].
- Grossman and Morlet decomposed the Hardy functions into square integrable wavelets for seismic signal analysis in 1984 [6].
- Meyer constructed orthogonal basis in L^2 with dilation and translation of a smooth function in 1986 [7].

- Mallat introduced the multiresolution analysis (MRA) in 1988 and unified the individual constructions of wavelets by Stromberg, Battle–Lemarie, and Meyer [8].
- Daubechies first constructed compactly supported orthogonal wavelet systems in 1987 [9].

2.1.2 When Do Wavelets Work?

Most of the data representing physical problems that we are modeling are not totally random but have a certain correlation structure. The correlation is local in time (spatial domain) and frequency (spectral domain). We should approximate these data sets with building blocks that possess both time and frequency localization. Such building blocks will be able to reveal the intrinsic correlation structure of the data, resulting in powerful approximation qualities: only a small number of building blocks can accurately represent the data. In electromagnetics the compactly supported (strictly localized in space) wavelets may be used as basis functions. These wavelets, by the Heisenberg uncertainty principle (or by Fourier analysis), cannot have strictly finite spectrum, but they can be approximately localized in spectrum. If most of their spectral components are beyond the visible region, for example, $\kappa_x > k_0$, they will produce little radiation, resulting in a sparse impedance matrix in the method of moments.

The previous observations may be generalized and described more precisely:

(1) Wavelets and their duals are local in space and spectrum. Some wavelets are even compactly supported, meaning strictly local in space (e.g., Daubechies and Coifman wavelets) or strictly local in spectrum (e.g., Meyer wavelets). Spatial localization implies that most of the energy of a wavelet is confined to a finite interval. In general, we prefer fast (exponential or inverse polynomial) decay away from the center of mass of the function. The frequency localization means band limit. The decay toward high frequencies corresponds to the smoothness of the wavelets; the smoother the function is, the faster the decay. If the decay is exponential, the function is infinitely many times differentiable. The decay toward low frequencies corresponds to the number of vanishing polynomial moments of the wavelet. Because of the time-frequency localization of wavelets, efficient representation can be obtained. The idea of frequency localization in terms of smoothness and vanishing moments may generalize the concept of "frequency localization" to a manifold, where the Fourier transform is not available.

(2) Wavelet series converge uniformly for all continuous functions, while Fourier series do not. In electromagnetics, the fields are often discontinuous across material boundaries. For piecewise smooth functions, Fourier-based methods give very slow convergence, for example, $\alpha = 1$, while nonlinear (i.e. with truncation) wavelet-based methods, exhibit fast convergence [10], for example, $\alpha \geq 2$, where α is the convergence rate defined by $\| f - f_M \| = O(M^{-\alpha})$

and the M-term approximate of f is given by

$$f_M = \sum_{\lambda \in \Lambda_M} c_\lambda \psi_\lambda. \tag{2.1.1}$$

(3) Wavelets belong to the class of orthogonal bases that are continuous and problem independent. As such, they are more suitable for developing systematic algorithms for general purpose computations. In contrast, the pulse bases, although orthogonal and compact in space, are not smooth. Indeed, they are discontinuous and are not localized in the spectral domain. On the other hand, Chebyshev, Hermite, Legendre, and Bessel polynomials are orthogonal but not localized in space within the domain (in comparison with intervallic and periodic wavelets). Shannon's *sinc* functions are localized in the transform domain but not in the original domain. The eigenmode expansion method is based on orthogonal expansion, but is problem dependent and works only for limited specific cases (e.g., rectangular, circular waveguides) [11].

(4) Wavelets decompose and reconstruct functions effectively due to the multiresolution analysis (MRA), that is, the passing from one scale to either a coarser or a finer scale efficiently. The MRA provides the fast wavelet transform, which allows conversion between a function f and its wavelet coefficients c with linear or linear-logarithmic complexity.

2.1.3 A Wave Is a Wave but What Is a Wavelet?

The title of this section is a note in the June 1994 issue of *IEEE Antennas and Propagation Magazine* from Professor Leopold B. Felson. Wavelet is literally translated from the French word *ondelette*, meaning small wave.

Wavelets are a topic of considerable interest in applied mathematics. One may use wavelets to decompose data, functions, and operators into different frequency components, and then study each component with a "resolution" level that matches the "scale" of the particular component. This "multiresolution" technique outperforms the Fourier analysis in such a way that both time domain and frequency domain information can be preserved. In a loose sense, one may say that the wavelet transform performs the optimized sampling. In contrast to the wavelet transform, the windowed Fourier transform oversamples the object under investigation, with respect to the Nyquist sampling criterion. Again, in a loose sense, one can say that wavelets decompose and compress data, images, and functions with good basis systems to reach high efficiency or sparseness. A key point to understand about wavelets is the introduction of both the dilation (frequency information) and translation (local time information).

Wavelets have been applied with great success to engineering problems, including signal processing, data compression, pattern recognition, target identification, computational graphics, and fluid dynamics. Recently wavelets have also been used in boundary value problems because they permit the accurate representation of a variety of operators without redundancy.

2.2 WHAT CAN WAVELETS DO IN ELECTROMAGNETICS AND DEVICE MODELING?

2.2.1 Potential Benefits of Using Wavelets

Owing to their ability to represent local high-frequency components with local basis elements, wavelets can be employed in a consistent and straightforward way. It is well known to the electromagnetic modeling community that the finite element method (FEM) is a technique that results in sparse matrices amenable to efficient numerical solutions. For the FEM the solution times tend to increase by $n \log(n)$, where $n \sim N^3$, with N being the number of points in one dimension. In using surface integral equations, implemented by the method of moments (MoM), the solution times have been demonstrated to increase by M^3, where $M \sim N^2$. It is obvious that N^2 is much smaller than N^3, and that therefore the MoM deals with many fewer unknowns than the FEM. Unfortunately, the matrix from the MoM is dense. The corresponding computational cost, using the direct solver, is on the order of $O(n^3)$, where $n \sim N^2$. It is clear that the solution of dense complex matrices is prohibitively expensive, especially for electrically large problems.

Integral operators are represented in a classical basis as a dense matrix. In contrast, wavelets can be seen as a quasi-diagonalizing basis for a wider class of integral operators. The "quasi" is necessary because the resulting wavelet expansion of integral operators is not truly diagonal. Instead, it has a peculiar palm pattern. This palm-type sparse structure represents an approximation of the original integral operator to arbitrary precision. It was reported that wavelet-based impedance matrices contain 90 to 99% zero entries. It has been shown by mathematicians that the solution of a wide range of integral equations can be transformed, using wavelets, from a direct procedure requiring order $O(n^3)$ operations to that requiring only order $O(n)$ [12]. In recent years, wavelets have been applied to electromagnetics and semiconductor device modeling for several purposes:

(1) To solve surface integral equations (SIE) originating from scattering, antenna, packaging and EMC (electromagnetic compatibility) problems, where very sparse impedance matrices have been obtained. It was reported that the wavelet scheme reduces the two-norm condition number of the MoM matrix by almost one order of magnitude [13].

(2) To improve the finite difference time domain (FDTD) algorithms in terms of convergence and numerical dispersion using Daubechies sampling biorthogonal time domain method (SBTD).

(3) To improve the convergence of the finite element method (FEM) using multiwavelets as basis functions.

(4) To solve nonlinear partial differential equations (PDEs) via the collocation method, in which the nonlinear terms in the PDEs are treated in the physical space while the derivatives are computed in the wavelet space [14].

(5) To model nonlinear semiconductor devices, where the finite difference method is implemented on the adaptive mesh, based on the interpolating wavelets and sparse point representation.

Some fascinating features of wavelets in the aforementioned applications are as follows:

(1) For the finite difference time domain (FDTD) method, numerical dispersion has been improved greatly. By imposing the Daubechies wavelet-based sampling function and its dual reproducing kernel, the SBTD requires much coarser mesh size in comparison with the Yee-FDTD while achieving the same precision. For a 3D resonator problem, the SBTD improves the CPU time by a factor of 13, and memory by 64. Material inhomogeneity and boundary conditions can be easily incorporated [15].

(2) For the finite element method (FEM), the multiwavelet basis functions are in C^1. At the node/edge, they can match not only the function but also its derivatives, yielding faster convergence than the traditional high-order FEM. For a partially loaded waveguide, the improvement of multiwavelet FEM over linear basis EEM exceeds 435 in CPU time reduction [16].

(3) For packaging and interconnects, the wavelet-based MoM speeds up parasitic parameter extraction by 1000 [17].

(4) Often in semiconductor device modeling, a small part of the computational interval or domain contains most of the activity, and the representation must have high resolution there. In the rest of the domain such high resolution is a high-cost waste. Various adaptive mesh techniques have been developed to address this issue. However, they often suffer accuracy problems in the application of operators, multiplication of functions, and so on. Wavelets offer promise in providing a systematic, consistent and simple adaptive framework. In the simulation of a 2D abrupt diode, the potential distribution was computed using wavelets to achieve a precision of 1.6% with 423 nodes. The same structure was simulated by a commercial package ATLAS, and 1756 triangles were used to reach a 5% precision [18].

(5) Coifman wavelets allow the derivation of a single-point quadrature of precision $O(h^5)$, which reduces the impedance filling process from $O(n^2)$ to $O(n)$.

2.2.2 Limitations and Future Direction of Wavelets

Wavelets are relatively new and are still in their infancy. Despite the advantages and beneficial features mentioned above, there are difficulties and problems associated in using wavelets for EM modeling.

Classical wavelets are defined on the real line, while many real world problems are in the finite domain. Periodic and intervallic wavelets have provided part of the solution, but they have also increased the complexity of the algorithm. Multiwavelets seem to be very promising in solving problems on intervals because of their orthogonality and interpolating properties.

The problems and difficulties encountered in practical fields have stimulated the interest of mathematicians. In recent years mathematicians have constructed wavelets on closed sets of the real line, satisfying certain types of boundary conditions. They have also studied wavelets of increasing order in arbitrary dimensions [19], wavelets on irregular point sets [20], and wavelets on curved surfaces as in the case of spherical wavelets [21].

2.3 THE HAAR WAVELETS AND MULTIRESOLUTION ANALYSIS

One of the most important properties of wavelets is the multiresolution analysis (MRA). Without losing generality, we discuss the MRA through the Haar wavelets. The Haar is the simplest wavelet system that can be studied immediately without any prerequisite. Later we will pass these conclusions on to other orthogonal wavelets. Therefore mathematical proofs are bypassed.

The Haar scaling functions (or scalets) are defined as

$$\varphi(x) = \begin{cases} 1 & \text{if } 0 < x < 1 \\ 0 & \text{otherwise.} \end{cases} \tag{2.3.1}$$

The Haar mother wavelets (or wavelets) are defined as

$$\psi(x) = \begin{cases} 1 & 0 \le x < \frac{1}{2} \\ -1, & \frac{1}{2} \le x < 1 \\ 0 & \text{otherwise.} \end{cases} \tag{2.3.2}$$

These two functions are sketched in Fig. 2.1. In the rest of the book, we will refer to mother wavelets as wavelets and scaling functions as scalets, in order to emphasize their roles as counterparts of wavelets. Notice that the term "wavelets" has a dual meaning. Depending on the context, wavelet can mean the wavelet or both the scalet and wavelet.

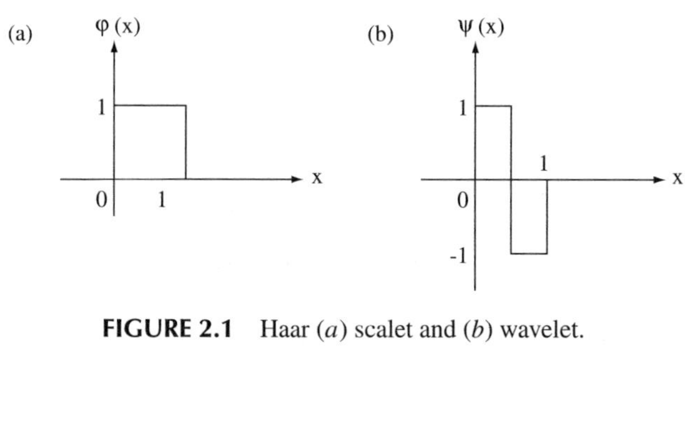

FIGURE 2.1 Haar (*a*) scalet and (*b*) wavelet.

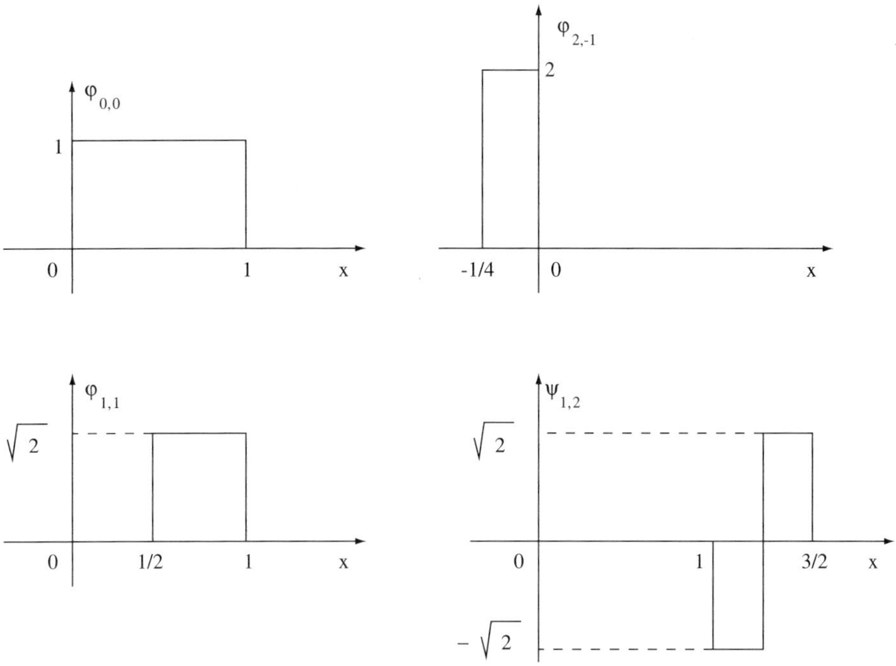

FIGURE 2.2 Dilation and translation.

It is easy to verify that the scalets and wavelets are orthogonal, namely

$$\begin{aligned}\langle \varphi(x), \psi(x) \rangle &= \int \varphi^*(x)\psi(x)\,dx \\ &= 0,\end{aligned}$$

where the asterisk denotes the complex conjugate. Higher-resolution scalets and wavelets are

$$\varphi_{m,n}(x) = 2^{m/2}\varphi(2^m x - n) \tag{2.3.3}$$

and

$$\psi_{m,n}(x) = 2^{m/2}\psi(2^m x - n), \tag{2.3.4}$$

where m denotes the "scale" or "level" and n the "translation" or "shift." As will be seen, the scale represents the frequency information while the translation contains the time (local) information. For instance, in Fig. 2.2, we give $\varphi_{0,0}(x)$, $\varphi_{1,1}(x)$, $\varphi_{2,-1}(x)$, and $-\psi_{1,2}(x)$:

$$\begin{aligned}\varphi_{0,0}(x) &= \varphi(x), \\ \varphi_{1,1}(x) &= \sqrt{2}\varphi(2x-1),\end{aligned}$$

$$\varphi_{2,-1}(x) = 2\varphi(4x+1),$$
$$-\psi_{1,2}(x) = -\sqrt{2}\psi(2x-2).$$

We can verify the following properties:

$$\int \varphi_{1,m}(x)\varphi_{1,n}(x)\,dx = \delta_{m,n},$$
$$\int \psi_{1,m}(x)\psi_{1,n}(x)\,dx = \delta_{m,n},$$
$$\int \varphi_{0,m}(x)\psi_{0,n}(x)\,dx = 0,$$
$$\int \varphi_{0,m}(x)\psi_{1,n}(x)\,dx = 0,$$
$$\int \varphi_{1,m}(x)\psi_{2,n}(x)\,dx = 0,$$

where $\delta_{m,n}$ is the Kronecker delta.

From the previous discussion, it appears that:

(1) The scalets on the same level form an orthonormal system.
(2) The wavelets on the same level form an orthonormal system.
(3) The scalets are orthogonal to all wavelets of the same or higher levels regardless of the translation of wavelets.
(4) Wavelets on different levels are orthogonal regardless of the translations.

These properties originate from the subspace decomposition of the wavelets. For any function $\varphi_{m,n}(x)$ in subspace V_m, namely

$$\varphi_{m,n}(x) \in V_m$$

and

$$\psi_{m,n}(x) \in W_m,$$

we have

$$\begin{aligned} V_m &= W_{m-1} \oplus V_{m-1} \\ &= W_{m-1} \oplus W_{m-2} \oplus V_{m-2} \\ &= W_{m-1} \oplus W_{m-2} \oplus \cdots \oplus W_0 \oplus V_0, \end{aligned} \tag{2.3.5}$$

where $\oplus$ denotes the direct sum. These properties apply not only to the Haar wavelets, but also to all orthogonal wavelets (Battle–Lemarie, Meyer, Daubechies, Coifman, etc.).

Next let us concentrate on how an arbitrary finite energy function $f \in L^2(R)$ is approximated by linear combinations of Haar wavelets. The notation $f \in L^2(R)$, or f is in $L^2(R)$ space implies that

$$\int f^*(x) f(x)\, dx < +\infty, \tag{2.3.6}$$

as discussed in (1.2.2).

2.4 HOW DO WAVELETS WORK?

We concentrate now on how an arbitrary function f can be approximated by linear combinations of Haar wavelets.

Figure 2.3*a* depicts a staircase signal $P_{V1}f$ or f^1, which is a digitized signal coming from the detected voltage f (where f is a continuous function) after conversion by an analog to digital (A/D) converter. The notation indicates that a function $f \in L^2$ is projected on the subspace V_1. In this case the sampling interval (step width) is a half-grid. We call f^1 the original signal with the highest resolution. This resolution depends on the sensitivity and physical parameters of the device and system.

Let us average the signal on the first and second intervals, the third and fourth, and so on. The resultant signal is shown in Fig. 2.3*b*, which is a "blurred" version with resolution twice as coarse as the original, and we denote it as $P_{V0}f$ or f^0. The detailed information is stored in Fig. 2.3*c* as δ^0. Adding Fig. 2.3*c* to Fig. 2.3*b* restores Fig. 2.3*a*, the original signal. The previous decomposition procedure that applied to f^1 may be applied to f^0 and the resultant, f^{-1} and δ^{-1}, are plotted in Fig. 2.4.

Formally, we may obtain the following mathematical description: any f in $L^2(R)$ can be approximated to an arbitrary precision by a function that is piecewise constant on its support (interval) and identically zero beyond the support of $[l2^{-j}, (l+1)2^{-j})$ (it suffices to take the support and j large enough). We can therefore restrict ourselves only to such piecewise constant functions. Assume that f is supported on $[0, 2^{J_1}]$ and is piecewise constant on $[l2^{-J_0}, (l+1)2^{-J_0}]$, where J_1 and J_0 can both be arbitrarily large. In Fig. 2.3 we selected $J_1 = 3$ and $J_0 = 1$ for ease of description. Let us denote the constant value of $f^1 = f^0 + \delta^0$ where f^0 is an approximation to f^1, which is piecewise constant over intervals twice as large as the original, namely, $f^0|_{[k2^{-J_0+1},(k+1)2^{-J_0+1})} \equiv \text{constant}: = f_k^0$. The values f_k^0 are given by the averages of the two corresponding constant values for f^1, $f_k^0 = \frac{1}{2}(f_{2k}^1 + f_{2k+1}^1)$. The function δ^0 is piecewise constant with the same step width as f^1. Hence one immediately has

$$\delta_{2l}^0 = f_{2l}^1 - f_l^0 = \tfrac{1}{2}(f_{2l}^1 - f_{2l+1}^1)$$

and

$$\delta_{2l+1}^0 = f_{2l+1}^1 - f_l^0 = \tfrac{1}{2}(f_{2l+1}^1 - f_{2l}^1) = -\delta_{2l}^0.$$

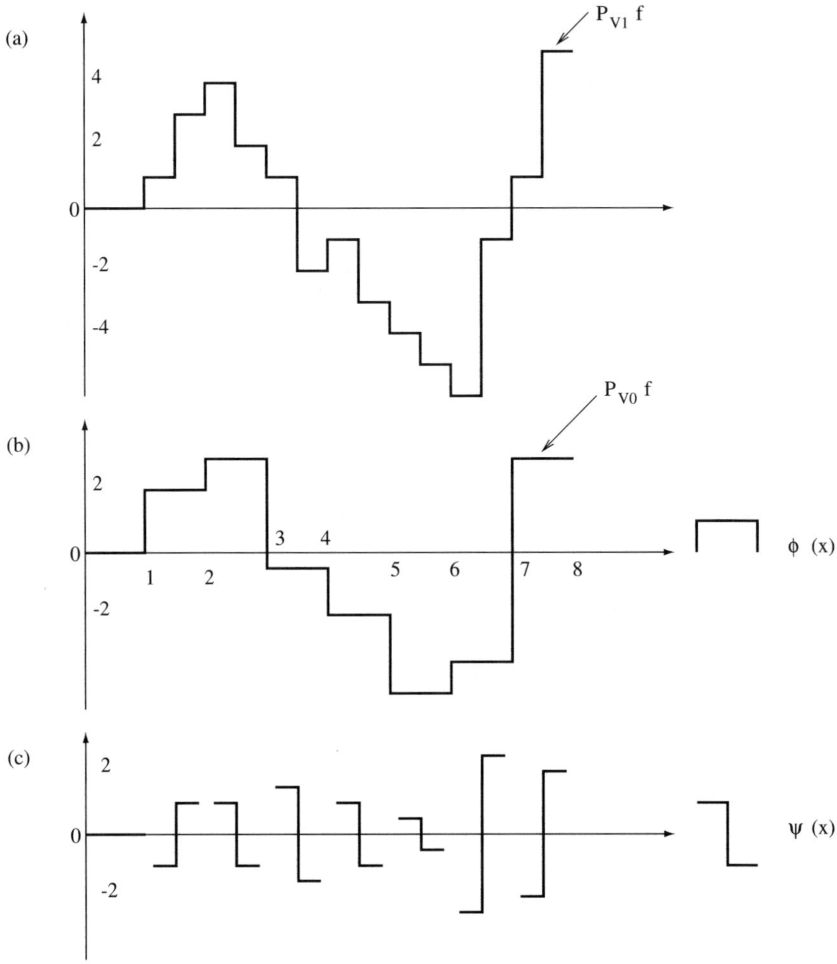

FIGURE 2.3 Decomposition of a signal f^1 into f^0 and δ^0.

Notice that δ^0 is piecewise constant with the same step width as f^1. It follows that δ^0 is a linear combination of scaled and shifted Haar functions. For this example we have

$$\delta^0(x) = 0\psi(x) + (-1)\psi(x-1) + 1\psi(x-2) + 1.5\psi(x-3) \\ + (-1)\psi(x-4) + (-0.5)\psi(x-5) + (-2.5)\psi(x-6) + (-2)\psi(x-7).$$

In general,

$$\delta^0(x) = \sum_{l=0}^{2^{J_1+J_0-1}-1} g_{J_0-1,l}\,\psi(2^{J_0-1}x - l),$$

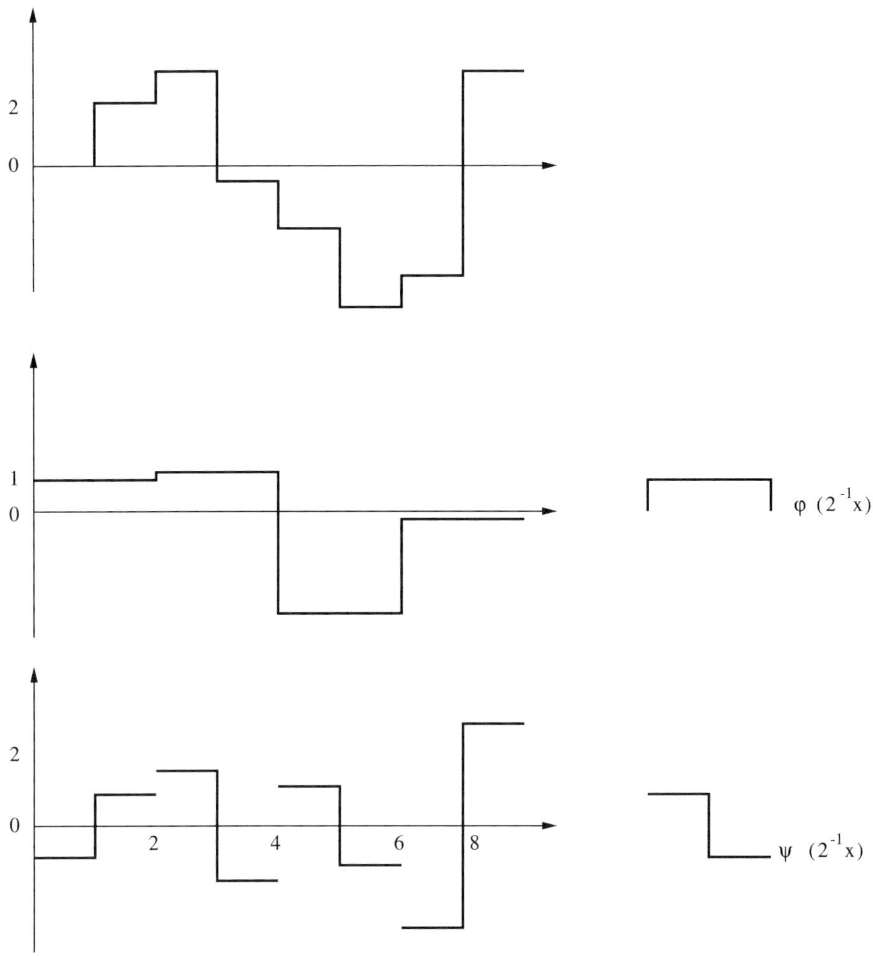

FIGURE 2.4 Further decomposition of f^0 into f^{-1} and δ^{-1}.

where $g_{J_0-1,l} = \langle f, \psi(2^{J_0-1}x - l)\rangle$. One can verify the coefficients in the summation. For instance, the coefficient of the second term is

$$g_{0,1} = \langle f, \psi(x-1)\rangle = 1 \times 1 \times (0.5) + 3 \times (-1) \times (0.5) = -1.$$

We have therefore written f as

$$f := f^1 = f^0 + \sum_l g_{J_0-1,l}\psi_{J_0-1,l} = f^0 + \delta^0,$$

where f^0 is of the same type as f^1, but with step width twice as large or resolution twice as coarse. We can apply the same procedure to f^0 so that

$$f^0 = f^{-1} + \sum_l g_{J_0-2,l}\psi(2^{J_0-2}x - l),$$

with f^{-1} still supported on $[0, 2^{J_1}]$, but piecewise constant on even larger intervals $[k2^{-J_0+2}, (k+1)2^{-J_0+2}]$. We continue this decomposition in Figs. 2.5 and 2.6 until the step width occupies the whole support. Hence we have

$$f^1 = f^{1-(J_0+J_1)} + \sum_{m=-J_1}^{J_0-1} \sum_l g_{m,l}\psi_{m,l}.$$

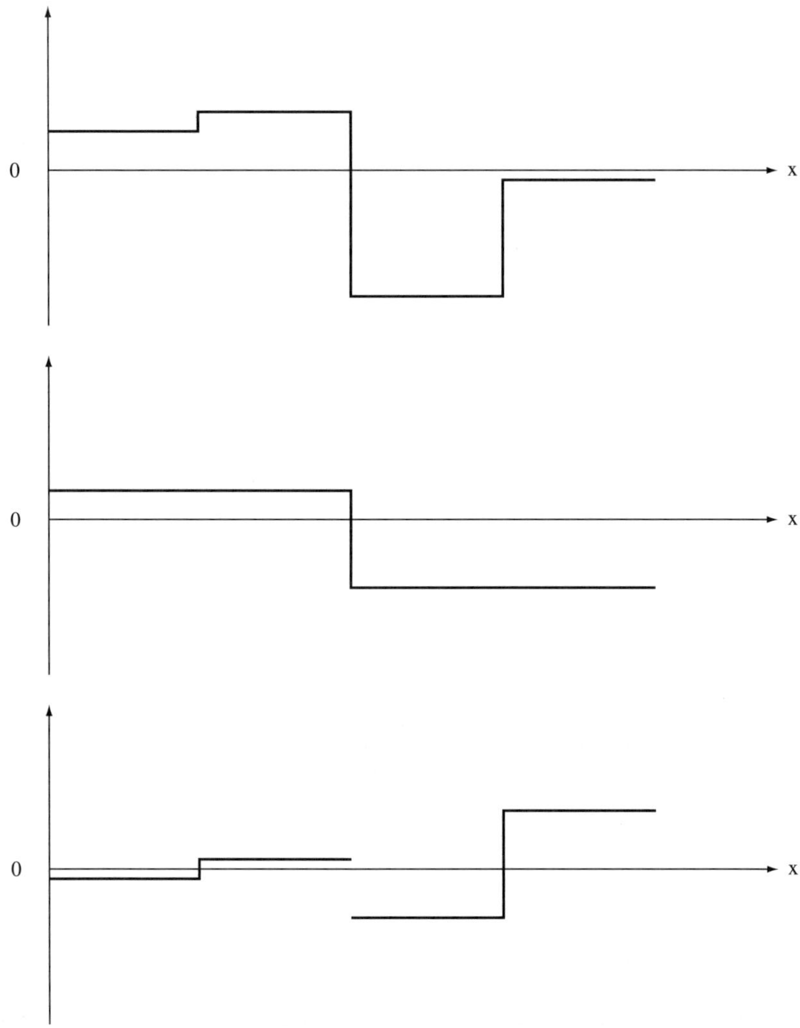

FIGURE 2.5 Decomposition of f^{-1} into f^{-2} and δ^{-2}.

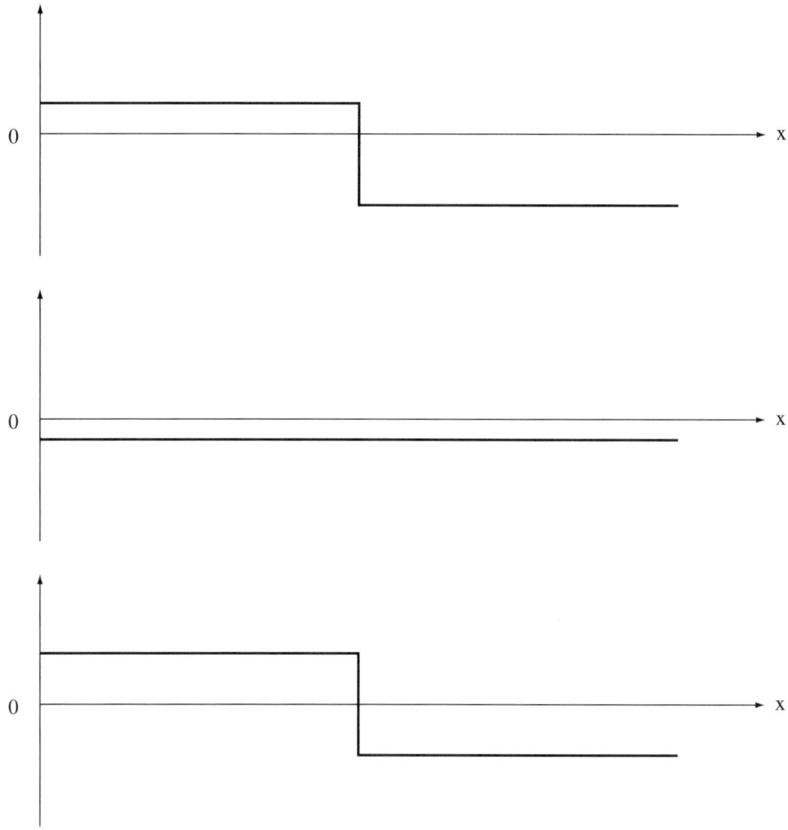

FIGURE 2.6 Decomposition of f^{-2} into f^{-3} and δ^{-3}.

For the numerical example in the figure, the final decomposed multiscale expression is

$$f^1 = f^{-3} + \delta^{-3} + \delta^{-2} + \delta^{-1} + \delta^0.$$

It is worth recognizing the orthogonality of the decomposed signals. For instance, one may verify from the figures that f^0 is orthogonal to δ^0, and that f^{-1} is orthogonal to δ^{-1} and δ^0. This is due to the fact that $\delta^0 \in W_0$, $\delta^{-1} \in W_{-1}$, $f^0 \in V_0$, $f^{-1} \in V_{-1}$ and

$$\begin{aligned}
V_1 &= V_0 \oplus W_0 \\
&= V_{-1} \oplus W_{-1} \oplus W_0 \\
&= V_{-2} \oplus W_{-2} \oplus W_{-1} \oplus W_0 \\
&= V_{-3} \oplus W_{-3} \oplus W_{-2} \oplus W_{-1} \oplus W_0.
\end{aligned}$$

Finally, all the decomposed signals in the highest hierarchical structure, f^{-3}, δ^{-3}, δ^{-2}, δ^{-1}, and δ^{0}, are mutually orthogonal as depicted in Fig. 2.3 to Fig. 2.6.

It can be proved mathematically that f may therefore be approximated to arbitrary precision by a finite linear combination of Haar wavelets. Readers interested in the mathematical proofs are referred to [22].

BIBLIOGRAPHY

[1] A. Haar, "Zur Theorie der orthogonalen Funktionensysteme," *Math. Annal.*, **69**, 331–371, 1910.

[2] R. E. A. C. Paley and J. E. Littlewood, "Theorems on Fourier series and power series," (I). *J.L.M.S.*, **6**, 230–233, 1931; (II). *P.L.M.S.*, **42**, 52–89, 1936; (III). ibid, **43**, 105–126, 1937.

[3] C. E. Shannon, "Mathematical theory of communication," *Bell Syst. Tech. J.*, **27**, 379–423, 623–656, 1948.

[4] A. P. Calderon, "An atomic decomposition of distributions in parabolic H^p spaces," *Adv. Math.*, **25**, 216–255, 1977.

[5] J. Stromberg, "A modified Franklin system and higher-order systems of R^n as unconditional bases for Hardy spaces," *Conf. Harmonic Analysis in Honor of A. Zygmund*, W. Beckner et al., ed., 475–493 *Wadsworth Math Series*, Wadworth, Belmont, CA, 1981.

[6] A. Grossman and J. Morlet, "Decomposition of Hardy functions into square integrable wavelets of constant shape," *SIAM J. Math. Anal.*, **15**, 723–736, 1984.

[7] Y. Meyer, Principle d'incertitude, basis Hilbertiennes et algebras d'operateurs, *Bourbaki Seminar*, no. 662, 1985–1986.

[8] S. Mallat, "Multiresolution approximation and wavelet orthogonal bases of L^2," *Trans. AMS,* **315**, 69–87, 1989.

[9] I. Daubechies, "Orthogonal bases of compactly supported wavelets," *Comm. Pure Appl. Math.*, **41**, 909–996, 1988.

[10] W. Sweldens, "Wavelets: What next?" *Proc. IEEE,* **84**(4), 680–685, 1996.

[11] G. Wang and G. Pan, "Full-Wave analysis of microstrip floating line structures by wavelet expansion method," *IEEE Trans. Microw. Theory Tech.*, **43**, 131–142, Jan. 1995.

[12] D. Gines, G. Beylkin, and J. Dunn, "LU factorization of non-standard forms and direct multiresolution solvers," *App. Comput. Harmon. Anal.*, **5**, 156–201, 1998.

[13] P. Pirinoli, G. Vecchi, and L. Matekovits, "Multiresolution analysis of printed antennas and circuits: a dual-isoscalar approach," *IEEE Trans. Ant. Propg.*, **49**(6), 858–874, June 2001.

[14] W. Cai and J. Wang, "Adaptive multiresolution collocation methods for initial boundary value problems of nonlinear PDEs," *SIAM J. Numer. Anal.*, **33**(3), 937–970, June 1996.

[15] Y. Tretiakov and G. Pan, "Sampling biorthogonal time domain scheme based on Daubechies biorthonormal sampling systems," *IEEE Antennas Propg. Int'l Symposium*, **4**, 810–813, 2001.

[16] G. Pan, K. Wang, and B. Gilbert, "Multiwavelet based finite element method," to appear in *IEEE Trans. Microw. Theory Tech.*, Jan. 2003.

[17] N. Soveiko and M. Nakhla, "Efficient capacitance extraction computations in wavelet domain," *IEEE Trans. Circ. Syst. I*, **47**(5), 684–701, May 2000.

[18] M. Toupikov, G. Pan, and B. Gilbert, "On nonlinear modeling of microwave devices using interpolating wavelets," *IEEE Trans. Microw. Theory Tech.*, **48**, 500–509, Apr. 2000.

[19] W. He and M. Lai, "Examples of bivariate nonseparable compactly supported orthogonal continuous wavelets," *IEEE Trans. Image Processing*, **9**, 949–953, 2000.

[20] I. Daubechies, I. Guskov, P. Schröder, and W. Sweldens, "Wavelets on irregular point sets," *Phil. Trans. R. Soc. Lond. A*, **357**(1760), 2397–2413, 1999.

[21] P. Schröder and W. Sweldens, "Spherical wavelets: efficiently representing functions on the sphere," *Computer Graphics Proceedings*, 161–172, 1995.

[22] I. Daubechies, *Ten Lectures on Wavelets*, SIAM, Philadelphia, 1992.

CHAPTER THREE

Basic Orthogonal Wavelet Theory

In Chapter 2 we saw how multiresolution analysis (MRA) works for the Haar system. A signal was decomposed into many components on different resolution levels. These components are mutually orthogonal. Despite their attractiveness, the Haar scalets and wavelets are not continuous functions. The discontinuities can create problems when applied to physical modeling. In this chapter we will construct many other orthogonal wavelets that are continuous and may even be smooth functions. Yet they preserve the same MRA and orthogonality as the Haar wavelets do.

The wavelet basis consists of scalets

$$\varphi_{m,n}(t) = 2^{m/2}\varphi(2^m t - n), \qquad m, n \in Z,$$

and wavelets

$$\psi_{m,n}(t) = 2^{m/2}\psi(2^m t - n), \qquad m, n \in Z.$$

3.1 MULTIRESOLUTION ANALYSIS

The study of orthogonal wavelets begins with the MRA. In this section we will show how an orthonormal basis of wavelets can be constructed starting from a such multiresolution analysis. Assume that a scalet φ is r times differentiable with rapid decay

$$\varphi^{(k)}(t) \leq C_{pk}(1 + |\, t\,|)^{-p}, \qquad k = 0, 1, 2, \ldots, r, \tag{3.1.1}$$
$$p \in Z, t \in R, C_{pk} - \text{constants}.$$

Thus we have defined a set, S_r, which will be used in the text;

$$\varphi \in S_r = \{\varphi : \varphi^{(k)}(t) \;\; \text{exist with rapid decay as in (3.1.1)}\}.$$

A multiresolution analysis of $L^2(R)$ is defined as a nested sequence of closed subspaces $\{V_j\}_{j \in Z}$ of $L^2(R)$, with the following properties [1]:

(1) $\cdots \subset V_{-1} \subset V_0 \subset \cdots \subset L^2(R)$.
(2) $f(\cdot) \in V_m \leftrightarrow f(2\cdot) \in V_{m+1}$.
(3) $f(t) \in V_0 \Rightarrow f(t+n) \in V_0$ for all $n \in Z$.
(4) $\bigcap_m V_m = 0$, closure$(\bigcup_m V_m) = L^2(R)$.
(5) There exists $\varphi(t) \in V_0$ such that set $\{\varphi(t-n)\}$ forms a Riesz basis of V_0.

A Riesz basis of a separable Hilbert space H is a basis $\{f_n\}$ that is close to being orthogonal. That is, there exists a bounded invertible operator which maps $\{f_n\}$ onto an orthonormal basis.

Let us explain these mathematical properties intuitively:

- In property (1) we form a nested sequence of closed subspaces. This sequence represents a causality relationship such that information at a given level is sufficient to compute the contents of the next coarser level.
- Property (2) implies that V_j is a dilation invariant subspace. As will be seen in later sections, this property allows us to build multigrid basis functions according to the nature of the solution. In the rapidly varying regions the resolution will be very fine, while in the slowly fluctuating regions the bases will be coarse.
- Property (3) suggests that V_j is invariant under translation (i.e., shifting).
- Property (4) relates residues or errors to the uniform Lipschitz regularity of the function, f, to be approximated by expansion in the wavelet bases.
- In property (5) the Riesz basis condition will be used to derive and prove convergence. The last two properties are more suitable for mathematicians; interested readers are referred to [1–3].

Clearly, $\sqrt{2}\varphi(2t-n)$ is an orthonormal basis for V_1, since the map $f \vdash \sqrt{2}f(2\cdot)$ is isometric from V_0 onto V_1. Since $\varphi \in V_1$, we have

$$\varphi(t) = \sum_k h_k \sqrt{2}\varphi(2t-k), \qquad \{h_k\} \in l^2,\ t \in R. \tag{3.1.2}$$

Equation (3.1.2) is called the dilation equation, and is one of the most useful equations in the field of wavelets. The MRA allows us to expand a function $f(t)$ in terms of basis functions, consisting of the scalets and wavelets. Any function $f \in L^2(R)$ can be projected onto V_m by means of a projection operator P_{V_m}, defined as $P_{V_m} f = f^m := \sum_n f_{m,n}\varphi_{m,n}$, where $f_{m,n}$ is the coefficient of expansion of f on the basis $\varphi_{m,n}$. From the previously listed MRA properties, it can be proved that $\lim_{m\to\infty} || f - f^m || = 0$, that is to say, that by increasing the resolution in MRA, a function can be approximated with any precision.

3.2 CONSTRUCTION OF SCALETS $\varphi(\tau)$

Haar wavelets are the simplest wavelet system, but their discontinuities hinder their effectiveness. Naturally people have found it useful to switch from a piecewise constant "box" to a piecewise linear "triangle." Unfortunately, the triangles are no longer orthogonal. Thus an orthogonalization procedure must be conducted, which leads to the Franklin wavelets.

3.2.1 Franklin Scalet

Consider a triangle function depicted in Fig. 3.1.

$$\theta(t) = (1 - |t - 1|)\chi_{[0,2]}(t).$$

This function is the convolution of two pulse functions of $\chi_{[0,1]}(t)$, where $\chi_{[0,1]}(t)$ is the characteristic function that is 1 in [0, 1] and 0 outside this interval. The Fourier transform of the pulse function can be obtained using the following relationships:

$$\{1(t) - 1(t-1)\} \leftrightarrow \frac{1}{s}(1 - e^{-s}) = \frac{1}{i\omega}(1 - e^{-i\omega}),$$

where $1(t)$ is the Heaviside step function. By the convolution theorem, the triangle has as its Fourier transform

$$\begin{aligned}\left(\frac{1 - e^{-i\omega}}{i\omega}\right)^2 &= e^{-i\omega}\left(\frac{e^{i\omega/2} - e^{-i\omega/2}}{i\omega}\right)^2 \\ &= e^{-i\omega}\left(\frac{\sin\omega/2}{\omega/2}\right)^2 \\ &= \hat{\theta}(\omega).\end{aligned}$$

Notice that $\theta(t)$ is centered at $t = 1$. Let us define $\theta_c(t) := \theta(t + 1)$, a triangle centered at $t = 0$ with a real spectrum of

$$\hat{\theta}_c(\omega) = \left(\frac{\sin\omega/2}{\omega/2}\right)^2.$$

Occasionally we will use $T(t) := \theta_c(t)$ to denote the triangle centered at the origin.

To find the orthogonal function $\varphi(t)$, we employ the isometric property of the Fourier transform. First, we may show that

$$\int_{-\infty}^{\infty} \varphi(t - n)\varphi(t)\,dt = \frac{1}{2\pi}\int_{-\infty}^{\infty} d\omega\hat{\varphi}(\omega)\overline{\hat{\varphi}(\omega)}e^{i\omega n}, \tag{3.2.1}$$

where the overbar denotes the complex conjugate.

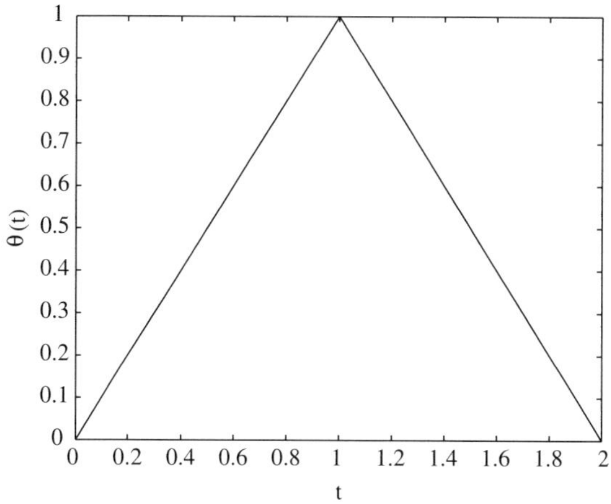

FIGURE 3.1 The triangle function $\theta(t)$.

Show.

$$
\begin{aligned}
\text{LHS} &= \int_{-\infty}^{\infty} \left[\frac{1}{2\pi} \int e^{i\Omega t} e^{-in\Omega} \hat{\varphi}(\Omega) d\Omega \right] \left[\frac{1}{2\pi} \int e^{i\omega t} \hat{\varphi}(\omega)\, d\omega \right] dt \\
&= \frac{1}{2\pi} \iiint d\omega d\Omega\, dt \hat{\varphi}(\omega) e^{-in\Omega} \hat{\varphi}(\Omega) \frac{1}{2\pi} e^{i(\omega+\Omega)t} \\
&= \frac{1}{2\pi} \int d\omega \hat{\varphi}(\omega) \int d\Omega e^{-in\Omega} \hat{\varphi}(\Omega) \frac{1}{2\pi} \int dt e^{i(\omega+\Omega)t} \\
&= \frac{1}{2\pi} \int d\omega \hat{\varphi}(\omega) \int d\Omega e^{-in\Omega} \hat{\varphi}(\Omega) \delta(\omega+\Omega) \\
&= \frac{1}{2\pi} \int d\omega \hat{\varphi}(\omega) \hat{\varphi}(-\omega) e^{in\omega},
\end{aligned}
$$

where $\delta(\cdot)$ is the Dirac delta. Since $\varphi(t)$ is real,

$$\hat{\varphi}(\omega) = \overline{\hat{\varphi}(-\omega)},$$

that is,

$$\hat{\varphi}(-\omega) = \overline{\hat{\varphi}(\omega)}.$$

Hence

$$\text{LHS} = \frac{1}{2\pi} \int d\omega \hat{\varphi}(\omega) \overline{\hat{\varphi}(\omega)} e^{in\omega}. \qquad \square$$

The orthogonality may be derived from the time domain by considering two basis functions. If $\varphi(t)$ and $\varphi(t-n)$ make an orthonormal system, then

$$\delta_{0,n} = \int_{-\infty}^{\infty} \varphi(t-n) \varphi(t)\, dt,$$

where $\delta_{0,n}$ is the Kronecker delta. By employing (3.2.1), we arrive at

$$\begin{aligned}\delta_{0,n} &= \frac{1}{2\pi}\int_{-\infty}^{\infty} d\omega e^{i\omega n}\hat{\varphi}(\omega)\overline{\hat{\varphi}(\omega)} \\ &= \frac{1}{2\pi}\sum_{k=-\infty}^{\infty}\int_0^{2\pi} |\,\hat{\varphi}(\omega+2k\pi)\,|^2 e^{i\omega n}\,d\omega \\ &= \frac{1}{2\pi}\int_0^{2\pi}\sum_k |\,\hat{\varphi}(\omega+2k\pi)\,|^2 e^{i\omega n}\,d\omega. \qquad (3.2.2)\end{aligned}$$

We define a periodic function

$$f(\omega) = |\,\hat{\varphi}^\dagger(\omega)\,|^2 := \sum_k |\,\hat{\varphi}(\omega+2k\pi)\,|^2.$$

The Fourier series of a periodic function with period of 2π is

$$f(\omega) = c_0 + \sum_{n=1}^{\pm\infty} c_n e^{i\omega n}. \qquad (3.2.3)$$

Comparing (3.2.2) with (3.2.3), we conclude that $c_0 = 1$, and $c_n = 0$ for $n \neq 1$. This conclusion can also be drawn from the uniqueness of the Fourier transform as follows. We know that

$$\frac{1}{2\pi}\int_0^{2\pi} e^{i\omega n}\,d\omega = \begin{cases}1 & \text{if } n = 0\\ 0 & \text{if } n \neq 0;\end{cases}$$

or equivalently

$$\frac{1}{2\pi}\int_0^{2\pi} e^{i\omega n}\,d\omega = \delta_{0,n}.$$

On the other hand, (3.2.2) suggests that $[\int_0^{2\pi}\sum_k |\,\hat{\varphi}(\omega+2k\pi)\,|^2 e^{i\omega n}\,d\omega]/2\pi = \delta_{0,n}$. From the uniqueness of the Fourier transform, we conclude that

$$|\,\hat{\varphi}^\dagger(\omega)\,|^2 := \sum_k |\,\hat{\varphi}(\omega+2k\pi)\,|^2 = 1. \qquad (3.2.4)$$

In the following paragraphs we will construct the scalet $\varphi(t)$ using translated triangles $\theta(t+1-n)$ as building blocks.

Since $\varphi \in V_0$, we have $\varphi(t) = \sum_n a_n\theta(t+1-n)$ for a sequence $\{a_n\} \in l^2$, meaning that $\sum_n |\,a_n\,|^2 < +\infty$. Taking the Fourier transform, we immediately have

$$\begin{aligned}\hat{\varphi}(\omega) &= \sum_n a_n e^{i\omega(1-n)}\hat{\theta}(\omega) \\ &= \sum_n a_n e^{-i\omega n}\hat{\theta}_c(\omega) \qquad (3.2.5)\\ &= \alpha(\omega)\hat{\theta}_c(\omega),\end{aligned}$$

where

$$\alpha(\omega) = \sum_n a_n e^{-i\omega n}.$$

Hence

$$|\hat{\varphi}^{\dagger}(\omega)|^2 = |\alpha(\omega)|^2 |\hat{\theta}_c^{\dagger}(\omega)|^2 = 1, \tag{3.2.6}$$

where

$$|\hat{\theta}_c^{\dagger}(\omega)|^2 = \sum_k |\hat{\theta}_c(\omega + 2k\pi)|^2.$$

Equation (3.2.6) can be used to find $\alpha(\omega)$. By definition, we have

$$\begin{aligned} |\hat{\varphi}^{\dagger}(\omega)|^2 &= \sum_k |\hat{\varphi}(\omega + 2k\pi)|^2 \\ &= \sum_k |\alpha(\omega + 2k\pi)\hat{\theta}_c(\omega + 2k\pi)|^2. \end{aligned}$$

Since

$$\alpha(\omega) = \sum_n a_n e^{-in\omega},$$

we have

$$\begin{aligned} \alpha(\omega + 2k\pi) &= \sum_n a_n e^{in(\omega + 2k\pi)} \\ &= \sum_n a_n e^{in\omega} \\ &= \alpha(\omega). \end{aligned}$$

Thus

$$|\hat{\varphi}^{\dagger}(\omega)|^2 = \left|\alpha(\omega)\right|^2 \sum_k \left|\hat{\theta}_c(\omega + 2k\pi)\right|^2 = |\alpha(\omega)|^2 |\hat{\theta}_c{}^{\dagger}(\omega)|^2.$$

Later in this section we show that $|\hat{\theta}_c{}^{\dagger}(\omega)|^2$ can be found in a closed form

$$|\hat{\theta}_c{}^{\dagger}(\omega)|^2 = \sum_k \left| \frac{\sin(\omega + 2k\pi)/2}{(\omega + 2k\pi)/2} \right|^4.$$

Therefore

$$|\alpha(\omega)|^2 = \frac{1}{\sum_k \left| \dfrac{\sin(\omega + 2k\pi)/2}{(\omega + 2k\pi)/2} \right|^4}.$$

It will be seen in the next paragraph that

$$\sum_k \left| \frac{\sin(\omega + 2k\pi)/2}{(\omega + 2k\pi)/2} \right|^4 = 1 - \frac{2}{3}\sin^2\frac{\omega}{2}. \tag{3.2.7}$$

Hence

$$\begin{aligned} \hat{\varphi}(\omega) &= \alpha(\omega)\hat{\theta}_c(\omega) \\ &= \alpha(\omega)\left(\frac{\sin\omega/2}{\omega/2}\right)^2 \\ &= \frac{1}{\sqrt{1 - \frac{2}{3}\sin^2\omega/2}}\left(\frac{\sin\omega/2}{\omega/2}\right)^2. \end{aligned} \tag{3.2.8}$$

Let us derive (3.2.7). The inverse Fourier transform of $e^{i\omega k}$ is

$$\begin{aligned} &\frac{1}{2\pi}\int_{-\pi}^{\pi} e^{i\omega x} e^{i\omega k} d\omega \\ &\quad = \frac{1}{2\pi}\int_{-\pi}^{\pi} d\omega e^{i\omega(x+k)} \\ &\quad = \frac{1}{2\pi}\frac{e^{i\omega(x+k)}}{i(x+k)}\Bigg|_{\omega=-\pi}^{\pi} \\ &\quad = \frac{e^{i(x+k)\pi} - e^{-i(x+k)\pi}}{2i}\frac{1}{\pi(x+k)} \\ &\quad = \frac{\sin\pi(x+k)}{\pi(x+k)}. \end{aligned}$$

Parseval's law relates the energy of a signal in the spatial domain and spectral domain as

$$\begin{aligned} \frac{1}{2\pi}\int_{-\pi}^{\pi} | e^{i\omega k} |^2 d\omega &= \sum_k \left| \frac{\sin\pi(x+k)}{\pi(x+k)} \right|^2 \\ &= | \sin\pi x |^2 \sum_k \frac{1}{| \pi(x+k) |^2}. \end{aligned}$$

Notice that the left-hand side of the previous equation is 1. So we have

$$\frac{1}{\sin^2\pi x} = \sum_k \frac{1}{[\pi(x+k)]^2}. \tag{3.2.9}$$

Taking the second derivative of the previous equation with respect to x, we obtain

$$\left(\frac{1}{\sin^2\pi x}\right)'' = 6\pi^2\frac{1 - \frac{2}{3}\sin^2\pi x}{\sin^4\pi x}$$

and

$$\left(\sum_k \frac{1}{[\pi(x+k)]^2}\right)'' = \frac{6}{\pi^2}\sum_k \frac{1}{(x+k)^4}.$$

Therefore

$$\sum_k \frac{1}{(\pi x + k\pi)^4} = \frac{1 - \frac{2}{3}\sin^2 \pi x}{\sin^4 \pi x}. \tag{3.2.10}$$

Letting $\pi x = \omega/2$, we obtain from this equation that

$$\sum_k \frac{1}{(k\pi + (\omega/2))^4} = \frac{1 - \frac{2}{3}\sin^2(\omega/2)}{\sin^4(\omega/2)} \tag{3.2.11}$$

which is equation (3.2.7).

The coefficients a_n in (3.2.5) can be evaluated numerically. As given in (3.2.5),

$$\begin{aligned}\hat{\varphi}(\omega) &= \sum_k a_k e^{-ik\omega}\hat{\theta}_c(\omega)\\ &= \left(\frac{\sin^2(\omega/2)}{(\omega/2)}\right)^2 \sum_k a_k e^{-i\omega k}.\end{aligned}$$

Using the time shift property of the Fourier transform, we obtain

$$\begin{aligned}\varphi(t) &= \sum_k a_k\theta(t+1-k),\\ a_k &= O(e^{-a|k|}).\end{aligned} \tag{3.2.12}$$

Notice again that $\theta(t+1) := \theta_c(t)$ is a triangle centered at $t = 0$, and its Fourier transform

$$\theta_c(\omega) = \left(\frac{\sin(\omega/2)}{(\omega/2)}\right)^2.$$

Coefficients a_k will be evaluated as follows: From the expression

$$\alpha(\omega) = \frac{1}{\sqrt{1 - \frac{2}{3}\sin^2(\omega/2)}} \tag{3.2.13}$$

$\alpha(\omega)$ is a periodic function of period 2π, which has the Fourier series

$$\sum_n a_n e^{-i\omega n} = \frac{1}{\sqrt{1 - \frac{2}{3}\sin^2(\omega/2)}}.$$

TABLE 3.1. First Ten Coefficients of $a_n = a_{-n}$ for the Franklin Scalet

a_0	1.29167548213672
a_1	−0.17466322755518
a_2	0.03521011276878
a_3	−0.00787442432698
a_4	0.00184794571482
a_5	−4.45921398374e-04
a_6	1.09576772871e-04
a_7	−2.72730550551e-05
a_8	6.85286905090e-06
a_9	−1.73457608425e-06

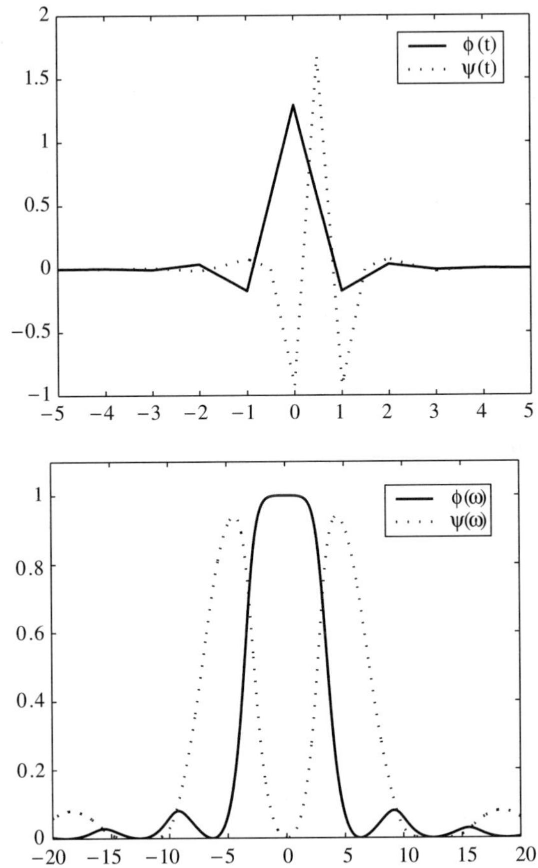

FIGURE 3.2 Franklin scalet φ and wavelet ψ.

By multiplying both sides by $e^{i\omega k}/2\pi$ and integrating, bearing in mind that

$$\frac{1}{2\pi}\int_{-\pi}^{\pi} e^{i\omega(k-n)}\,d\omega = \delta_{k,n},$$

we obtain

$$a_k = \frac{1}{2\pi}\int_{-\pi}^{\pi} \frac{e^{i\omega k}}{\sqrt{1-\frac{2}{3}\sin^2(\omega/2)}}\,d\omega = \frac{1}{\pi}\int_{0}^{\pi} \frac{\cos k\omega}{\sqrt{1-\frac{2}{3}\sin^2(\omega/2)}}\,d\omega.$$

This equation provides a numerical expression for the evaluation of a_n, which can be accomplished by imposing Gaussian–Legendre quadrature. The values of a_n are displayed in Table 3.1. Using these values of a_n and the translated triangle functions $\theta(t+1-n)$, the Franklin scalet is constructed according to (3.2.12). From the integral expression of a_k, we observe that $a_{-k} = a_k$. Also $\theta_c(t)$ is symmetric. Therefore the Franklin scalet is an even function. The Franklin wavelet is symmetric about $t = \frac{1}{2}$, and will be studied in the next section. The Franklin scalet and wavelets are depicted in Fig. 3.2.

3.2.2 Battle–Lemarie Scalets

The Franklin wavelets employ the triangle functions as building blocks in the construction of an orthogonal system. These triangles are continuous functions but not smooth; their derivatives are discontinuous at certain points. If we convolve the triangle with the box one more time, the resulting function will be smooth. The translations of this smooth function can then be used as building blocks to build smooth orthogonal wavelet systems. The greater the number of convolutions conducted, the smoother the building block functions become. This smoothness is achieved at the expense of larger support widths of the resulting scalets. In general, the B-spline of degree N is obtained by convolving the "box" N times. Hence

$$\hat{\theta}_N(\omega) = e^{-i\kappa(\omega/2)}\left(\frac{\sin(\omega/2)}{\omega/2}\right)^{N+1},$$

where

$$\kappa = \begin{cases} 0 & \text{if } N = \text{odd} \\ 1 & \text{if } N = \text{even} \end{cases},$$

and as such any shift by an integer can be ignored. We use integer translations of the basis functions, therefore only the half-integer shifts matter. The corresponding $\alpha_1(\omega)$ for $N = 1$ is the Franklin in (3.2.13). For $N = 2$, $\alpha_2(\omega) = \{\frac{1}{15}[2\cos^4(\omega/2) + 11\cos^2(\omega/2) + 2]\}^{-1/2}$. The resulting Battle–Lemarie wavelets are illustrated in Fig. 3.3.

Detailed construction of higher-order Battle–Lemarie wavelets is left to readers as an exercise problem in this chapter.

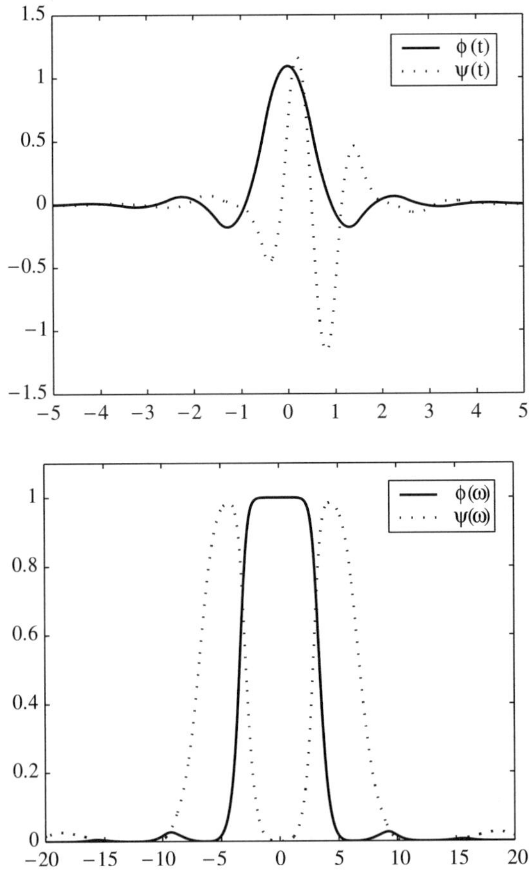

FIGURE 3.3 Battle–Lemarie scalet φ and wavelet ψ.

3.2.3 Preliminary Properties of Scalets

In the previous discussions we used the triangle functions as building blocks to generate the Franklin wavelets, according to $\varphi(t) = \sum_k a_k \theta_c(t-k)$. If the triangles are replaced by smoother building blocks, higher-order Battle–Lemarie wavelets may be obtained in the same manner. Unfortunately, the number of nonzero coefficients a_k are infinite, although a_k decays very rapidly, meaning that $a_k = O(e^{-a|k|})$. A challenging question arises: Is it possible to have a finite number of nonzero coefficients that generate orthogonal wavelets? This query leads us to the Daubechies wavelet.

We seek h_k in the dilation equation

$$\varphi(t) = \sum_k h_k \sqrt{2}\varphi(2t-k) \tag{3.2.14}$$

such that the orthogonality condition is satisfied. The derivation is performed in the Fourier domain. Taking the Fourier transform of $\varphi(t)$, we have

$$\begin{aligned}\hat{\varphi}(\omega) &= \frac{1}{\sqrt{2}} \sum h_k e^{-ik(\omega/2)} \hat{\varphi}\left(\frac{\omega}{2}\right) \\ &= \hat{h}\left(\frac{\omega}{2}\right) \hat{\varphi}\left(\frac{\omega}{2}\right),\end{aligned} \tag{3.2.15}$$

where $\hat{h}(\omega) := \sum h_k e^{-ik\omega}/\sqrt{2}$. Equation (3.2.15) is similar to (3.2.5) in nature, except that $\hat{\theta}_c(\omega) = ((\sin \omega/2)/(\omega/2))^2$ is given in the latter, while $\hat{\varphi}\,(\omega/2)$ remains unknown here. Equation (3.2.15) translates the orthonormal condition

$$|\,\hat{\varphi}^{\dagger}(\omega)\,|^2 = 1$$

into

$$\left|\,\hat{h}\left(\frac{\omega}{2}\right)\right|^2 + \left|\,\hat{h}\left(\frac{\omega}{2} + \pi\right)\right|^2 = 1. \tag{3.2.16}$$

Show. In fact we have

$$\begin{aligned}|\,\hat{\varphi}^{\dagger}(\omega)\,|^2 &= \sum_n |\,\hat{\varphi}(\omega + 2\pi n)\,|^2 \\ &= \sum_n \left|\,\hat{h}\left(\frac{\omega}{2} + n\pi\right)\hat{\varphi}\left(\frac{\omega}{2} + n\pi\right)\right|^2 \\ &= \sum_n \left|\,\hat{h}\left(\frac{\omega}{2} + n\pi\right)\right|^2 \left|\,\hat{\varphi}\left(\frac{\omega}{2} + n\pi\right)\right|^2.\end{aligned}$$

By definition,

$$\begin{aligned}\hat{h}\left(\frac{\omega}{2} + n\pi\right) &= \frac{1}{\sqrt{2}} \sum_k h_k e^{-ik(\omega + 2n\pi)/2} \\ &= \frac{1}{\sqrt{2}} \sum_k e^{-i(nk)\pi} h_k e^{-ik(\omega/2)} \\ &= \frac{1}{\sqrt{2}} \sum_k (-1)^{nk} h_k e^{-ik(\omega/2)} \\ &= \begin{cases} \hat{h}(\frac{\omega}{2}) & \text{if } n = 2m \\ \hat{h}(\frac{\omega}{2} + \pi) & \text{if } n = 2m + 1. \end{cases}\end{aligned}$$

As a matter of fact, for $n =$ odd, we have

$$\begin{aligned}\frac{1}{\sqrt{2}} \sum_k h_k e^{-ik[\omega + 2(2m+1)\pi]/2} &= \frac{1}{\sqrt{2}} \sum h_k e^{-ik[(\omega/2) + \pi]} \\ &= \hat{h}\left(\frac{\omega}{2} + \pi\right).\end{aligned}$$

By noticing that

$$\hat{\varphi}\left(\frac{\omega}{2} + n\pi\right) = \begin{cases} \hat{\varphi}(\frac{\omega}{2} + 2m\pi) & \text{if } n = 2m \\ \hat{\varphi}(\frac{\omega}{2} + (2m+1)\pi) & \text{if } n = 2m + 1, \end{cases}$$

we can refer to the orthogonal condition (3.2.6) and obtain

$$\begin{aligned}
1 &= \sum_n \left| \hat{h}\left(\frac{\omega}{2} + n\pi\right) \hat{\varphi}\left(\frac{\omega}{2} + n\pi\right) \right|^2 \\
&= \left| \hat{h}\left(\frac{\omega}{2}\right) \right|^2 \sum_m \left| \hat{\varphi}\left(\frac{\omega}{2} + 2m\pi\right) \right|^2 \\
&\quad + \left| \hat{h}\left(\frac{\omega}{2} + \pi\right) \right|^2 \sum_m \left| \hat{\varphi}\left(\frac{\omega}{2} + (2m+1)\pi\right) \right|^2 \\
&= \left| \hat{h}\left(\frac{\omega}{2}\right) \right|^2 \left| \hat{\varphi}^\dagger\left(\frac{\omega}{2}\right) \right|^2 + \left| \hat{h}\left(\frac{\omega}{2} + \pi\right) \right|^2 \left| \hat{\varphi}^\dagger\left(\frac{\omega}{2} + \pi\right) \right|^2 \\
&= \left| \hat{h}\left(\frac{\omega}{2}\right) \right|^2 + \left| \hat{h}\left(\frac{\omega}{2} + \pi\right) \right|^2,
\end{aligned}$$

where we have used $|\hat{\varphi}^\dagger(\cdot)|^2 = 1$. □

Note that $\hat{h}$ is a periodic function with period 2π. By setting $\omega = 0$ in (3.2.15), we arrive at

$$\hat{\varphi}(0) = \hat{h}(0)\hat{\varphi}(0).$$

Therefore

$$\hat{h}(0) = 1.$$

Setting $\omega = 0$ in (3.2.16) and using $\hat{h}(0) = 1$, we obtain

$$1 + |\hat{h}(\pi)|^2 = 1.$$

Therefore

$$\hat{h}(\pi) = 0.$$

In general,

$$\hat{h}((2m+1)\pi) = 0$$

and

$$\hat{h}(2m\pi) = 1.$$

Furthermore $\hat{h}(\omega/2)$ is a periodic function with period 4π.

3.3 WAVELET $\psi(\tau)$

After the scalets are obtained, we can create the corresponding wavelets $\psi(t)$. In this process we may take advantage of the MRA structure by choosing $\{\psi(t-n)\}$ as an orthonormal basis of W_0, which is the orthogonal complement of V_0 in V_1, namely

$$
\begin{aligned}
V_1 &= V_0 \bigoplus W_0 \\
V_2 &= V_1 \bigoplus W_1 \\
&\vdots
\end{aligned}
\tag{3.3.1}
$$

According to (3.3.1), $\psi(t)$ satisfies the orthogonality relations

$$\int \psi(t)\psi(t-n)\,dt = \delta_{0,n}, \tag{3.3.2}$$

$$\int \psi(t)\varphi(t-n)\,dt = 0. \tag{3.3.3}$$

If such a set $\{\psi(t)\}$ can be found, then

$$\psi_{m,n}(t) := 2^{m/2}\psi(2^m t - n)$$

is an orthogonal basis of W_m.

From the multiresolution analysis property (4) of Section 3.1, we have

$$\bigoplus_{m\in Z} W_m = L^2(R).$$

Hence $\{\psi_{mn}\}_{n,m\in Z}$ forms an orthogonal basis of $L^2(R)$. In the Fourier transform domain the two orthogonal equations (3.3.2) and (3.3.3) become, respectively,

$$\sum_{k=-\infty}^{\infty} |\,\hat{\psi}(\omega + 2k\pi)\,|^2 = 1, \tag{3.3.4}$$

$$\sum_{k=-\infty}^{\infty} \hat{\psi}(\omega + 2k\pi)\overline{\hat{\varphi}(\omega + 2k\pi)} = 0. \tag{3.3.5}$$

These two expressions can be arrived at in the same manner as $|\,\hat{\varphi}^{\dagger}\,|^2$ in the derivation of (3.2.4). Since $\psi(t) \in W_0$ and

$$W_0 \bigoplus V_0 = V_1,$$

it follows that $\psi(t) \in V_1$. Since $\psi(t) \in V_1$, $\psi(t)$ can be represented in terms of basis functions $\varphi(2t-k)$ in V_1, yielding the dilation equation

$$\psi(t) = \sum_k g_k \sqrt{2}\varphi(2t-k), \qquad g_k \in l^2, \tag{3.3.6}$$

where g_k is called the bandpass filter bank while h_k in (3.2.14) is referred to as the lowpass filter bank.

Next we will examine the relationship between g_k and h_k. By taking the Fourier transform of (3.3.6), we obtain

$$\begin{aligned}\hat{\psi}(\omega) &= \frac{1}{\sqrt{2}}\sum_k g_k e^{-ik(\omega/2)}\hat{\varphi}\left(\frac{\omega}{2}\right) \\ &= \hat{g}\left(\frac{\omega}{2}\right)\hat{\varphi}\left(\frac{\omega}{2}\right),\end{aligned} \tag{3.3.7}$$

where $\hat{g}(\omega/2) = \frac{1}{\sqrt{2}}\sum_k g_k e^{-ik(\omega/2)}$. The properties of $\hat{g}(\omega/2)$ are similar to those of $\hat{h}(\omega/2)$, namely

$$\left|\hat{g}\left(\frac{\omega}{2}\right)\right|^2 + \left|\hat{g}\left(\frac{\omega}{2}+\pi\right)\right|^2 = 1,$$

$$\hat{g}\left(\frac{\omega}{2}\right)\overline{\hat{h}\left(\frac{\omega}{2}\right)} + \hat{g}\left(\frac{\omega}{2}+\pi\right)\overline{\hat{h}\left(\frac{\omega}{2}+\pi\right)} = 0.$$

Show. Owing to this analogy, we will only show the second equation.

Following the steps in the derivation from (3.2.1) through (3.2.4), we have

$$\begin{aligned}0 &= \int_{-\infty}^{\infty} \psi(t)\varphi(t-n)\,dt \\ &= \frac{1}{2\pi}\int_{-\infty}^{\infty} e^{i\omega n}\hat{\psi}(\omega)\overline{\hat{\varphi}(\omega)}\,d\omega \\ &= \frac{1}{2\pi}\int_{\omega=0}^{2\pi}\sum_k \hat{\psi}(\omega+2k\pi)\overline{\hat{\varphi}(\omega+2k\pi)}e^{i\omega n}\,d\omega.\end{aligned}$$

From the uniqueness of the Fourier transform, we conclude that

$$\sum_k \hat{\psi}(\omega+2k\pi)\overline{\hat{\varphi}(\omega+2k\pi)} = 0.$$

Further simplifying the summation and using the tricks in (3.2.16), we have

$$\begin{aligned}0 &= \sum_k \hat{\psi}(\omega+2k\pi)\overline{\hat{\varphi}(\omega+2k\pi)} \\ &= \sum_k \left[\hat{g}\left(\frac{\omega}{2}+k\pi\right)\hat{\varphi}\left(\frac{\omega}{2}+k\pi\right)\right]\overline{\left[\hat{h}\left(\frac{\omega}{2}+k\pi\right)\hat{\varphi}\left(\frac{\omega}{2}+k\pi\right)\right]} \\ &= \hat{g}\left(\frac{\omega}{2}\right)\overline{\hat{h}\left(\frac{\omega}{2}\right)}\sum_m \hat{\varphi}\left(\frac{\omega}{2}+2m\pi\right)\overline{\hat{\varphi}\left(\frac{\omega}{2}+2m\pi\right)} \\ &\quad + \hat{g}\left(\frac{\omega}{2}+\pi\right)\overline{\hat{h}\left(\frac{\omega}{2}+\pi\right)}\sum_m \hat{\varphi}\left(\frac{\omega}{2}+(2m+1)\pi\right)\overline{\hat{\varphi}\left(\frac{\omega}{2}+(2m+1)\pi\right)} \\ &= \hat{g}\left(\frac{\omega}{2}\right)\overline{\hat{h}\left(\frac{\omega}{2}\right)}\left|\hat{\varphi}^{\dagger}\left(\frac{\omega}{2}\right)\right|^2 + \hat{g}\left(\frac{\omega}{2}+\pi\right)\overline{\hat{h}\left(\frac{\omega}{2}+\pi\right)}\left|\hat{\varphi}^{\dagger}\left(\frac{\omega}{2}+\pi\right)\right|^2 \\ &= \hat{g}\left(\frac{\omega}{2}\right)\overline{\hat{h}\left(\frac{\omega}{2}\right)} + \hat{g}\left(\frac{\omega}{2}+\pi\right)\overline{\hat{h}\left(\frac{\omega}{2}+\pi\right)},\end{aligned}$$

where $|\hat{\varphi}^{\dagger}(\cdot)|^2 = 1$ has been used. □

From the previous equation, we arrive at

$$\hat{g}\left(\frac{\omega}{2}\right)\overline{\hat{h}\left(\frac{\omega}{2}\right)} = -\hat{g}\left(\frac{\omega}{2}+\pi\right)\overline{\hat{h}\left(\frac{\omega}{2}+\pi\right)},$$

and consequently

$$\left|\hat{g}\left(\frac{\omega}{2}\right)\overline{\hat{h}\left(\frac{\omega}{2}\right)}\right|^2 = \left|\hat{g}\left(\frac{\omega}{2}+\pi\right)\overline{\hat{h}\left(\frac{\omega}{2}+\pi\right)}\right|^2. \tag{3.3.8}$$

Thus we obtain

$$\left|\hat{g}\left(\frac{\omega}{2}\right)\right| = \left|\overline{\hat{h}\left(\frac{\omega}{2}+\pi\right)}\right|.$$

By the equality above, the two sides of (3.3.8) become the

$$\text{LHS} = \left|\overline{\hat{h}\left(\frac{\omega}{2}+\pi\right)\hat{h}\left(\frac{\omega}{2}\right)}\right|^2,$$

$$\text{RHS} = \left|\overline{\hat{h}\left(\frac{\omega}{2}\right)\hat{h}\left(\frac{\omega}{2}+\pi\right)}\right|^2.$$

We could have chosen

$$\hat{g}\left(\frac{\omega}{2}\right) = \overline{\hat{h}\left(\frac{\omega}{2}+\pi\right)},$$

but it is not worth doing it this way; instead, we choose

$$\hat{g}\left(\frac{\omega}{2}\right) = \pm e^{-i(\omega/2)}\overline{\hat{h}\left(\frac{\omega}{2}+\pi\right)}. \tag{3.3.9}$$

On the other hand, if we define

$$\psi(t) := \sqrt{2}\sum_k h_{1-k}(-1)^k\varphi(2t-k),$$

we can find its Fourier transform immediately. Note that

$$\begin{aligned}\int_{-\infty}^{\infty}\varphi(2t-k)e^{-it\omega}\,dt &= \frac{1}{2}\int e^{-ik(\omega/2)}\varphi(2t-k)e^{-i(\omega/2)(2t-k)}d(2t-k)\\ &= \frac{1}{2}\hat{\varphi}\left(\frac{\omega}{2}\right)\cdot e^{-ik(\omega/2)}.\end{aligned}$$

As a result

$$\hat{\psi}(\omega) = \frac{1}{\sqrt{2}}\sum_k h_{1-k}e^{-ik\pi}e^{-ik(\omega/2)}\hat{\varphi}\left(\frac{\omega}{2}\right)$$

$$= \frac{1}{\sqrt{2}} \sum_k h_{1-k} e^{-i(\omega/2)} e^{i[(\omega/2)+\pi](1-k)} e^{-i\pi} \hat{\varphi}\left(\frac{\omega}{2}\right)$$

$$= e^{-i(\omega/2)} \left[\frac{1}{\sqrt{2}} \sum_\ell h_\ell e^{i[(\omega/2)+\pi]\ell} \right] \hat{\varphi}\left(\frac{\omega}{2}\right)(-1)$$

$$= e^{-i(\omega/2)} \overline{\hat{h}\left(\frac{\omega}{2}+\pi\right)} \hat{\varphi}\left(\frac{\omega}{2}\right)(-1)$$

$$= \pm\hat{g}\left(\frac{\omega}{2}\right) \hat{\varphi}\left(\frac{\omega}{2}\right). \tag{3.3.10}$$

This is consistent with (3.3.7). Hence the bandpass and lowpass filter are related by

$$g_k = (-1)^k h_{1-k}. \tag{3.3.11}$$

Sometimes we use $(-1)^{k-1}$ in (3.3.11), which makes the wavelet upside-down. However, this reversed wavelet possesses all required properties of a wavelet. Note that $\varphi(t)$ and $\psi(t)$ constructed in this way may have noncompact supports.

We conclude this section by quoting several theorems [3] and [4]. The proofs are quite abstract and are printed in a smaller font. Readers who are not interested in mathematical rigor may always skip material printed in smaller fonts without jeopardizing their understanding of the course.

Theorem 1. Assume that $\psi(t) \in S_r$, meaning that $\psi(t)$ has rth continuous derivatives and they are fast decaying according to (3.1.1); $\psi_{m,n}(t) := 2^{m/2}\psi(2^m t - n)$ form an orthonormal system in $L^2(R)$. Then $\psi(t)$ has rth zero moments, namely $\int_{-\infty}^{\infty} t^k \psi(t)\, dt = 0, k = 0, 1, \ldots, r$.

The significance of this theorem is its generality. For instance, the Battle–Lemarie wavelets of $N = 2$ are built from convolving the box function consequently twice. No zero moment requirement was forced explicitly. However, from the theorem, it is guaranteed that $\int \psi(t)t^\ell\, dt = 0,\ \ell = 0, 1$.

Proof. We prove the theorem by induction on k.

(1) $k = 0$, we wish to show that $\int_{-\infty}^{\infty} \psi(t)\, dt = 0$. $\exists N = 2^{j_0}k_0$ that $\psi(N) \neq 0$. Let $j \in Z$, $2^j N$ is an integer (all $j \geq j_0$). By orthogonality, we may write

$$0 = 2^{j/2} \int \psi(x)\overline{\psi(2^j x - 2^j N)}\, dx.$$

Using $y = 2^j x - 2^j N$, we have

$$x = 2^{-j}(y + 2^j N) = 2^{-j}y + N.$$

The previous inner product becomes

$$\int \psi(2^{-j}y + N)\overline{\psi(y)}\, dy = 0.$$

As $j \to \infty$,

$$\lim_{j\to\infty} \int \psi(2^{-j}y + N)\overline{\psi(y)}\,dy = \int \lim_{j\to\infty} \psi(2^{-j}y + N)\overline{\psi(y)}\,dy.$$

The change of the limit with integral is permitted because $|\,\psi(2^{-j}y + N)\overline{\psi(y)}\,| \leq c|\,\psi(y)\,|$, and Lebesgue dominated convergence allows the commutation. Thus

$$\begin{aligned} 0 &= \int \psi(N)\overline{\psi(y)}\,dy \\ &= \psi(N) \int \overline{\psi(y)}\,dy. \end{aligned}$$

Hence

$$\int_{-\infty}^{\infty} \overline{\psi(y)}\,dy = 0.$$

(2) Assume the identity is held for $k = 0, 1, \ldots, (n-1) < r$. Show that this is true for n:

$$\exists \psi^{(n)}(N) \neq 0$$

that

$$\psi(x) = \sum_{k=0}^{n} \psi^{(k)}(N) \frac{(x-N)^k}{k!} + r_n(x) \frac{(x-N)^n}{n!},$$

where the remainder $r_n(x) \to 0$ as $x \to N$. Using the substitution $y = 2^j(x - N)$, we have

$$\begin{aligned} 0 &= 2^j \int_{-\infty}^{\infty} \psi(x)\overline{\psi(2^j x - 2^j N)}\,dx \\ &= \int_{-\infty}^{\infty} dy \overline{\psi(y)} \left[\sum_{k=0}^{n} \psi^{(k)}(N) \frac{(2^{-j}y)^k}{k!} + r_n(2^{-j}y + N) \frac{(2^{-j}y)^n}{n!} \right]. \end{aligned}$$

By the assumption $\int t^k \psi(t)\,dt = 0$, we have

$$0 = \int_{-\infty}^{\infty} \left[\psi^{(n)}(N) \frac{(2^{-j}y)^n}{n!} + r_n(2^{-j}y + N) \frac{(2^{-j}y)^n}{n!} \right] \overline{\psi(y)}\,dy.$$

As $j \to \infty$, $r_n(2^{-j}y + N) \to r_n(N) \to 0$.
Therefore

$$\frac{(2^{-jn})}{n!} \cdot \psi^{(n)}(N) \int_{-\infty}^{\infty} y^n \overline{\psi(y)}\,dy = 0,$$

that is,

$$\int_{-\infty}^{\infty} y^n \overline{\psi(y)}\,dy = 0. \qquad \square$$

This leads to a more general theorem.

Theorem 2 (Vanishing Moments). Assume that φ and ψ form an orthogonal basis, and

$$|\varphi(t)| = O\left(\frac{1}{(1+t^2)^{1+(p/2)}}\right),$$

$$|\psi(t)| = O\left(\frac{1}{(1+t^2)^{1+(p/2)}}\right).$$

The following four statements are equivalent:

- The wavelet ψ has p vanishing moments.
- $\hat{\psi}(\omega)$ and its first $p-1$ derivatives are zero at $\omega = 0$.
- $\hat{h}(\omega)$ and its first $p-1$ derivatives are zero at $\omega = \pi$.
- For any $0 \leq n < p$,

$$q_n(t) = \sum_{k=-\infty}^{\infty} k^n \varphi(t-k)$$

  is a polynomial of degree n.

The proof of this theorem is referred to [3] and [4].

3.4 FRANKLIN WAVELET

In the previous section we derived the relationship between the bandpass filter g_k of the wavelets and the lowpass filter h_k of the scalets. Now we may construct the Franklin wavelet from Franklin scalets by applying the results from the previous section. We begin with

$$\varphi(t) = \sum_n a_n \theta(t+1-n) = \sum_n a_n T(t-n), \tag{3.4.1}$$

where $T(t) = \theta_c(t)$ is the triangle centered at the origin. Using the dilation equation and orthogonality, we have

$$\varphi_{0,0} = \varphi(t) = \sqrt{2}\sum_k h_k \varphi(2t-k) = \sum_k h_k \varphi_{1,k}$$

$$\langle \varphi_{0,0}, \varphi_{1,n} \rangle = \sum_k h_k \langle \varphi_{1,k}, \varphi_{1,n} \rangle = h_n. \tag{3.4.2}$$

Here we have used

$$\int \varphi(t-k)\varphi(t-n)\,dt = \delta_{k,n}$$

and

$$\varphi(\cdot) \in V_o \quad \leftrightarrow \quad \varphi(2\cdot) \in V_1.$$

Now from

$$\varphi(t) = \sum_k a_k \theta_c(t-k),$$

we obtain

$$\varphi_{1,n} = \sqrt{2}\varphi(2t-n) = \sqrt{2}\sum_\ell a_\ell \theta_c(2t-n-\ell). \tag{3.4.3}$$

Employing (3.4.2) in conjunction with (3.4.1) and (3.4.3), we arrive at

$$\begin{aligned} h_n &= \sqrt{2}\int_{-\infty}^{\infty} dt \sum_k a_k \theta_c(t-k) \sum_\ell a_\ell \theta_c(2t-n-\ell) \\ &= \sqrt{2}\sum_k a_k \sum_\ell a_\ell \int_{-\infty}^{\infty} \theta_c(t-k)\theta_c(2t-n-\ell)\, dt. \end{aligned}$$

Let $x = t - k$, then $t = x + k$ and $2t = 2(x+k)$. It follows that

$$\begin{aligned} h_n &= \sqrt{2}\sum_k a_k \sum_\ell a_\ell \int_{-\infty}^{\infty} \theta_c(x)\theta_c(2x+2k-n-\ell)\, dx \\ &= \sqrt{2}\sum_k a_k \sum_\ell a_\ell \int_{-\infty}^{\infty} \theta_c(x)\theta_c(2x-m)\, dx \\ &= \sqrt{2}\sum_k a_k Z(k,n), \end{aligned} \tag{3.4.4}$$

where $m = n + \ell - 2k$, and

$$\begin{aligned} Z(k,n) = \tfrac{1}{24}a_{|2k-n-2|} + \tfrac{1}{4}a_{|2k-n-1|} + \tfrac{5}{12}a_{|2k-n|} \\ + \tfrac{1}{4}a_{|2k-n+1|} + \tfrac{1}{24}a_{|2k-n+2|}. \end{aligned} \tag{3.4.5}$$

Equations (3.4.4) and (3.4.5) may be derived by considering the integral of the two-triangle product

$$\int_{-\infty}^{\infty} \theta_c(x)\theta_c(2x-m)\, dx = \tfrac{1}{24}\delta_{m,-2} + \tfrac{1}{4}\delta_{m,-1} + \tfrac{5}{12}\delta_{m,0} + \tfrac{1}{4}\delta_{m,1} + \tfrac{1}{24}\delta_{m,2}.$$

As a result

$$h_n = \sqrt{2}\sum_{k=-\infty}^{\infty} a_k Z(k,n)$$

$$= \sqrt{2} \sum_k a_k \sum_\ell a_\ell \left[\tfrac{1}{24}\delta_{\ell+n-2k,-2} + \tfrac{1}{4}\delta_{\ell+n-2k,-1} + \tfrac{5}{12}\delta_{\ell+n-2k,0} \right.$$

$$\left. + \tfrac{1}{4}\delta_{\ell+n-2k,1} + \tfrac{1}{24}\delta_{\ell+n-2k,2} \right]$$

$$\approx \sqrt{2} \sum_{k=-20}^{20} a_k Z(k, n).$$

Since $a_{-k} = a_k$ for Franklin, we use absolute signs in the subscripts of (3.4.5).

In general, the wavelet is given by the dilation equation

$$\psi(t) = \sqrt{2} \sum g_k \varphi(2t - k), \tag{3.4.6}$$

where

$$g_k = (-1)^k h_{1-k}$$

or

$$g_k = (-1)^{k\pm 1} h_{1-k}.$$

Equation (3.4.6) suggests that the Franklin wavelet can be evaluated from the existing Franklin scalet. However, often the Franklin wavelet is written in a superposition of translations of the triangle $\theta_c(2t-1)$; that is to say, θ_c compressed by 2 and centered at $\frac{1}{2}$, as $\psi(t) = \sum_\ell b_\ell \theta_c(2t - 1 - \ell)$. By substituting $\varphi(t) = \sum_n a_n \theta_c(t - n)$ in (3.4.6), we obtain

$$\psi(t) = \sum_k g_k \sqrt{2} \sum_n a_n \theta_c(2t - k - n)$$

$$= \sum_k \sum_n \sqrt{2}\, g_k\, a_n \theta_c(2t - k - n).$$

Setting $k + n = \ell + 1$, we arrive at

$$\psi(t) = \sum_\ell \sqrt{2} \left(\sum_n a_n g_{\ell-n+1} \right) \theta_c(2t - \ell - 1).$$

Hence

$$b_\ell = \sqrt{2} \sum_n a_n g_{\ell-n+1}.$$

Noticing that

$$g_k = (-1)^{k-1} h_{1-k},$$

TABLE 3.2. Coefficients for the Franklin Scalet and Wavelet

n	h_n	b_n
0	0.8176521786	1.6819434057
1	0.3972972937	−0.9271597915
2	−0.0691016686	0.0049495286
3	−0.0519452675	0.0774411477
4	0.0169708517	0.0237802960
5	0.0099904577	−0.0196404228
6	−0.0038831035	−0.0046406534
7	−0.0022018736	0.0041267166
8	0.0009232981	0.0011996559
9	0.0005116014	−0.0009810930
10	−0.0002242677	−0.0002914430
11	−0.0001226728	0.0002350121
12	0.0000553460	0.0000728832
13	0.0000300063	−0.0000576869
14	−0.0000138152	−0.0000182940
15	−0.0000074427	0.0000143267

we finally arrive at

$$b_\ell = \sqrt{2} \sum_n (-1)^{\ell-n} h_{n-\ell} a_n.$$

To efficiently compute $\{b_\ell\}$, we must truncate the summation at proper locations. In fact the decay of $\{a_n\}$ is rapid. For $n = 15$, $|\, a_n \,| < 2 \cdot 10^{-9}$, so it may be truncated with 16 terms. The numerical data for h_n and b_n are tabulated in Table 3.2, while the resulting Franklin wavelets are depicted in Fig. 3.2. By the same token, the resulting Battle–Lemarie wavelets are constructed and plotted in Fig. 3.3.

3.5 PROPERTIES OF SCALETS $\hat{\varphi}(\omega)$

The scalets $\varphi(t - n)$ are orthonormal. Furthermore $\hat{\varphi}(\omega)$ is bounded, and $\hat{\varphi}(\omega)$ is continuous at $\omega = 0$. In general,

$$\hat{\varphi}(2k\pi) = \delta_{0,k},$$

that is,

$$\hat{\varphi}(0) = 1$$

and

$$\hat{\varphi}(2k\pi) = 0 \quad \text{if } k \neq 0.$$

This property can be seen from the Franklin scalet

$$\hat{\varphi}(\omega) = \left(\frac{\sin(\omega/2)}{\omega/2}\right)^2 \left\{1 - \frac{2}{3}\sin^2\left(\frac{\omega}{2}\right)\right\}^{-1/2}.$$

In this section we will prove the basic and most useful property of $\hat{\varphi}(\omega)$ for scalets in general, that is, $\hat{\varphi}(0) = 1$. The proof is printed in the following paragraph, and it lasts several pages.

Proof. Consider the characteristic function

$$\hat{f}(\omega) = \chi(\omega) = \begin{cases} 1 & \text{if } 0 \le \omega \le 1 \\ 0 & \text{elsewhere.} \end{cases} \tag{3.5.1}$$

Its inverse Fourier transform is

$$\begin{aligned} f(t) &= \frac{1}{2\pi}\frac{e^{it} - 1}{it} \\ &= \frac{1}{2\pi} e^{i(t/2)} \frac{\sin(t/2)}{t/2}. \end{aligned}$$

The projection of $f(t)$ onto V_m is

$$f_m(t) = \sum_n \langle f, \varphi_{m,n} \rangle \varphi_{m,n}(t),$$

where

$$\varphi_{m,n}(t) = 2^{(m/2)}\varphi(2^m t - n).$$

Using Parseval's theorem, we have

$$\| f_m \|^2 = \sum_n | \langle f, \varphi_{m,n} \rangle |^2. \tag{3.5.2}$$

First, let us evaluate the Fourier transform of $\varphi_{m,n}(t)$,

$$\begin{aligned} &\int_{-\infty}^{\infty} [2^{(m/2)}\varphi(2^m t - n)] e^{-i\omega t}\, dt \\ &= 2^{-m} \int_{-\infty}^{\infty} 2^{(m/2)}\varphi(2^m t - n) e^{-i((\omega/2^m))(2^m t - n)} e^{-i(\omega n/2^m)} d(2^m t - n) \\ &= \frac{1}{2^{m/2}} e^{-i\omega n/2^m} \int \varphi(u) e^{-i(\omega/2^m)u}\, du \\ &= \frac{1}{2^{m/2}} e^{-i\omega n/2^m} \hat{\varphi}\left(\frac{\omega}{2^m}\right). \end{aligned}$$

Thus

$$\begin{aligned} \| f_m \|^2 &= \sum_n | \langle f, \varphi_{m,n} \rangle |^2 \\ &= \sum_n \left| \frac{1}{2\pi} \int \hat{f}(\omega) 2^{-m/2} e^{-i(\omega n/2^m)} \hat{\varphi}\left(\frac{\omega}{2^m}\right) d\omega \right|^2 \end{aligned}$$

$$= \frac{1}{(2\pi)^2} \frac{1}{2^m} \sum_n \left| \int \hat{f}(\omega)\hat{\varphi}\left(\frac{\omega}{2^m}\right) e^{-i(\omega n/2^m)} \, d\omega \right|^2$$

$$= \frac{1}{2^m} \sum_n \left| \frac{1}{2\pi} \int_0^1 d\omega e^{-i(\omega n/2^m)} \hat{\varphi}\left(\frac{\omega}{2^m}\right) \right|^2, \tag{3.5.3}$$

where we have used (3.5.1) and

$$\langle f(t), p(t) \rangle = \frac{1}{2\pi} \langle F(\omega), P(\omega) \rangle.$$

Equation (3.5.3) can be considered in two different ways.

(1) A function defined on [0, 1] can always be extended into a periodic function with $[-2^m\pi, 2^m\pi]$ as one period, and be analyzed as a periodic function. We have a finite power signal (periodic) and a discrete spectrum. As a result the Fourier coefficient of a periodic function $q(\omega)$ is

$$c_n = \frac{1}{2\ell} \int_{-\ell}^{\ell} q(\omega) e^{-i(n\pi\omega/\ell)} \, d\omega.$$

Let $q(\omega) = \hat{f}(\omega)\hat{\varphi}(\omega/2^m)$ and $\ell = 2^m\pi$. It follows that

$$c_n = \frac{1}{2 \cdot 2^m \pi} \int_{-2^m\pi}^{2^m\pi} \hat{f}(\omega)\hat{\varphi}\left(\frac{\omega}{2^m}\right) e^{-i(n\pi\omega/\pi 2^m)} \, d\omega$$

$$= \frac{1}{2 \cdot 2^m \pi} \int_{-2^m\pi}^{2^m\pi} \hat{f}(\omega)\hat{\varphi}\left(\frac{\omega}{2^m}\right) e^{-i(n\omega/2^m)} \, d\omega. \tag{3.5.4}$$

For a periodic function $q(\omega)$ with period $[-\ell, \ell]$, we can always write

$$q(\omega) = \sum_n c_n e^{i(n\pi/\ell)\omega} = \sum_n c_n e^{i(n/2^m)\omega}.$$

Thus the inner product

$$\frac{1}{2\pi} \langle q(\omega), q(\omega) \rangle = \frac{1}{2\pi} \int_{-\ell}^{\ell} \overline{\sum_k c_k e^{i(k/2^m)\omega}} \sum_r c_r e^{i(r/2^m)\omega} \, d\omega$$

$$= \frac{1}{2\pi} \sum_k \sum_r \overline{c_k} c_r \int_{-2^m\pi}^{2^m\pi} e^{-i[(k-r)/2^m]\omega} \, d\omega$$

$$= \frac{1}{2\pi} 2\pi 2^m \sum_k | c_k |^2$$

$$= 2^m \sum_k | c_k |^2. \tag{3.5.5}$$

In fact, by letting $n = r - k$, $\beta = 1/2^m$ in the equation above, the integral becomes

$$\int_{-\pi/\beta}^{\pi/\beta} e^{in\beta\omega} \, d\omega = \frac{e^{in\pi} - e^{-in\pi}}{in\beta}$$

$$
\begin{aligned}
&= 2\frac{e^{in\pi} - e^{-in\pi}}{2i}\frac{1}{n\beta} \\
&= 2\pi\frac{\sin n\pi}{n\pi}2^m \\
&= \begin{cases} 0 & \text{if } n \neq 0 \\ 2^m 2\pi & \text{if } n = 0. \end{cases}
\end{aligned}
$$

On the other hand, we may evaluate the inner product directly. Namely

$$
\begin{aligned}
\frac{1}{2\pi}\langle q(\omega), q(\omega)\rangle &= \frac{1}{2\pi}\int_{-\ell}^{\ell} \overline{\hat{f}(\omega)\hat{\varphi}\left(\frac{\omega}{2^m}\right)}\hat{f}(\omega)\varphi\left(\frac{\omega}{2^m}\right)\,d\omega \\
&= \frac{1}{2\pi}\int_{-2^m\pi}^{2^m\pi} \mid \hat{f}(\omega)\hat{\varphi}\left(\frac{\omega}{2^m}\right) \mid^2 d\omega. \qquad (3.5.6)
\end{aligned}
$$

Furthermore, from (3.5.5) and (3.5.4), this inner product is equal to

$$
\begin{aligned}
2^m \sum_k \mid c_k \mid^2 &= 2^m \sum_k \left| \frac{1}{2\pi 2^m}\int_{-2^m\pi}^{2^m\pi} \hat{f}(\omega)\hat{\varphi}\left(\frac{\omega}{2^m}\right) e^{-i(k/2^m)\omega}\,d\omega \right|^2 \\
&= \frac{1}{(2\pi)^2}\frac{1}{2^m}\sum_k \left| \int_{-2^m\pi}^{2^m\pi} \hat{f}(\omega)\hat{\varphi}\left(\frac{\omega}{2^m}\right) e^{-i(k/2^m)\omega}\,d\omega \right|^2. \qquad (3.5.7)
\end{aligned}
$$

A comparison of (3.5.5), (3.5.7) versus (3.5.6) leads to

$$
\begin{aligned}
(2\pi)2^m \sum_k \mid c_k \mid^2 &= \frac{1}{2\pi}\frac{1}{2^m}\sum_k \left| \int_{-2^m\pi}^{2^m\pi} \hat{f}(\omega)\hat{\varphi}\left(\frac{\omega}{2^m}\right) e^{-i(k/2^m)\omega}\,d\omega \right|^2 \\
&= \langle q(\omega), q(\omega)\rangle \\
&= \int_{-2^m\pi}^{2^m\pi} \left| \hat{f}(\omega)\hat{\varphi}\left(\frac{\omega}{2^m}\right) d\omega \right|^2.
\end{aligned}
$$

Returning to (3.5.3), we obtain

$$
\begin{aligned}
\| f_m \|^2 &= \frac{1}{(2\pi)^2}\frac{1}{2^m}\sum_n \left| \int \hat{f}(\omega)\hat{\varphi}\left(\frac{\omega}{2^m}\right) e^{-i(n/2^m)\omega}\,d\omega \right|^2 \\
&= \frac{1}{2\pi}\int_{-2^m\pi}^{2^m\pi} \left| \hat{f}(\omega)\hat{\varphi}\left(\frac{\omega}{2^m}\right) \right|^2 d\omega \\
&= \frac{1}{2\pi}\int_0^1 \left| \hat{\varphi}\left(\frac{\omega}{2^m}\right) \right|^2 d\omega,
\end{aligned}
$$

where we have used the characteristic function $f(\omega)$ to reduce the integral limits. As $m \to \infty$,

$$
\| f \|^2 = \lim_{m\to\infty} \| f_m \|^2 = \frac{1}{2\pi}\int_0^1 \mid \hat{\varphi}(0) \mid^2 d\omega.
$$

(2) The norm may be evaluated in the time domain

$$f(t) = \frac{1}{2\pi}\frac{e^{it}-1}{it} = \frac{1}{2\pi}e^{i(t/2)}2\frac{e^{i(t/2)}-e^{-i(t/2)}}{2it} = \frac{e^{i(t/2)}}{2\pi}\frac{\sin(t/2)}{t/2},$$

$$\|f\| = \int_{-\infty}^{\infty} |\, f(t)\,|^2\,dt = \left(\frac{1}{2\pi}\right)^2 \int_{-\infty}^{\infty}\left(\frac{\sin(t/2)}{t/2}\right)^2 dt$$

$$= \left(\frac{1}{2\pi}\right)^2 \cdot 2\pi = \frac{1}{2\pi}.$$

Hence

$$\int_0^1 |\,\hat{\varphi}(0)\,|^2\, d\omega = 1.$$

That is, the integrand $\hat{\varphi}(0) = 1$, namely

$$1 = \hat{\varphi}(0) = \int_{-\infty}^{\infty} \varphi(t)e^{-i\omega t}\,dt|_{\omega=0} = \int_{-\infty}^{\infty} \varphi(t)\,dt.$$

The previous equation indicates that the scalet has a d.c. component of unity. From the frequency dilation equation $\hat{\varphi}(\omega) = \hat{h}(\omega/2)\hat{\varphi}(\omega/2)$ we obtain

$$\hat{\varphi}(2\pi) = \hat{\varphi}(\pi)\hat{h}(\pi) = 0 \qquad \text{because } \hat{h}(\pi) = 0,$$

$$\hat{\varphi}(4\pi) = \hat{\varphi}(2\pi)\hat{h}(2\pi) = 0 \qquad \text{because } \hat{\varphi}(2\pi) = 0,$$

$$\cdots \quad \cdots$$

By induction and with the help of (3.2.3) we may show that

$$\hat{\varphi}(2k\pi) = \hat{\varphi}(k\pi)\hat{h}(k\pi) = 0, \qquad k \neq 0.$$

Hence

$$\hat{\varphi}(2k\pi) = \delta_{0,k}.$$

□

In summary, the scalet $\varphi(t)$ has the following properties:

(1) $\sum_k |\,\hat{\varphi}(\omega + 2\pi k)\,|^2 = 1$, orthogonality.
(2) $\int_{-\infty}^{\infty} \varphi(t)\,dt = 1 = \hat{\varphi}(0)$.
(3) $\hat{\varphi}(2\pi k) = \delta_{0,k}$.
(4) $\hat{\varphi}(\omega) = \hat{h}(\omega/2)\hat{\varphi}(\omega/2)$ with $\hat{h}(\omega/2)$ of period 4π,

$$\left|\hat{h}\left(\frac{\omega}{2}\right)\right|^2 + \left|\hat{h}\left(\frac{\omega}{2}+\pi\right)\right|^2 = 1$$

$$\hat{h}(0) = 1$$

$$\hat{h}(\pi) = 0.$$

(5) $\sum_{n=-\infty}^{\infty} \varphi(t-n) = 1.$
Note that $\sum_{n=-\infty}^{\infty} \varphi(t-n)$ is a periodic function of period 1. Therefore it can be expanded into the Fourier series

$$\sum_{n=-\infty}^{\infty} \varphi(t-n) = \sum_{m} c_m e^{i2\pi mt}.$$

Multiplying both sides by $e^{-i2\pi kt}$ and integrating over $\int_0^1 dt$, we may take advantage of orthogonality to get

$$\begin{aligned} c_k &= \int_0^1 \sum_{n=-\infty}^{\infty} \varphi(t-n)e^{-i2\pi kt}\, dt \\ &= \sum_{n=-\infty}^{\infty} \int_0^1 \varphi(t-n)e^{-i2\pi kt}\, dt \\ &= \int_{-\infty}^{\infty} \varphi(t)e^{-i2\pi kt}\, dt \\ &= \hat{\varphi}(2\pi k) = \delta_{0,k}. \end{aligned} \tag{3.5.8}$$

As a result

$$\sum_{-\infty}^{\infty} \varphi(t-n) = \sum_{k} c_k e^{i2\pi k} = c_0 = 1.$$

(6) $\hat{\psi}(\omega) = e^{-i\omega/2}\overline{\hat{h}\,[(\omega/2)+\pi]}\hat{\varphi}\,(\omega/2).$

Of course, the last property is for wavelets, but it is stated here for ease of reference. These properties will be used in later sections of this chapter to construct the Daubechies and Coifman wavelets.

3.6 DAUBECHIES WAVELETS

In contrast with the infinitely supported Franklin or Battle–Lemarie, the Daubechies wavelets are compactly supported orthogonal systems. In the construction of Daubechies wavelets, we seek a finite set of nonzero coefficients h_k in the dilation equation

$$\varphi(t) = \sum_{k} h_k \sqrt{2}\varphi(2t-k).$$

Recalling that

$$\hat{h}\left(\frac{\omega}{2}\right) = \sum_{k} \frac{h_k}{\sqrt{2}} e^{-i(\omega/2)k}$$

and the properties of $\hat{\varphi}(\omega)$,

$$\hat{h}(0) = 1,$$
$$\hat{h}(\pi) = 0,$$

we obtain

(i) $1 = \sum_k \frac{h_k}{\sqrt{2}}$.

(ii) $0 = \sum_k \frac{(-1)^k h_k}{\sqrt{2}}$.

Furthermore

$$\begin{aligned}
\delta_{n,0} &= \int \varphi(t)\varphi(t-n)\,dt \\
&= \int \sum_k \sum_l h_k h_l \sqrt{2}\varphi(2t-k)\sqrt{2}\varphi(2t-2n-l)\,dt \\
&= \sum_k \sum_l h_k h_l \int \varphi_{1,k}\varphi_{1,2n+l}\,dt \\
&= \sum_k \sum_l h_k h_l \delta_{k,2n+l} \\
&= \sum_k h_k h_{k-2n}.
\end{aligned}$$

Namely

(iii) $\sum_k h_k h_{k-2n} = \delta_{0,n}$.

Let us use a trial-and-error method to find a set of nonzero lowpass filter coefficients h_k under conditions of (i) through (iii):

(1) Set $h_k = 0$ for $k \neq 0, 1$, and find h_0, h_1. From properties (i) through (iii) we have

$$\begin{cases} h_0 + h_1 = \sqrt{2} \\ h_0 - h_1 = 0 \\ h_0^2 + h_1^2 = 1. \end{cases}$$

The solution is $h_0 = h_1 = 1/\sqrt{2}$, which leads to the Haar system

$$\varphi(t) = (\varphi(2t) + \varphi(2t-1)).$$

(2) Set $h_k = 0$ for $k \neq 0, 1, 2$; find h_0, h_1, h_2:

$$\begin{cases} h_0 + h_1 + h_2 = \sqrt{2} \\ h_0 - h_1 + h_2 = 0 \\ h_0^2 + h_1^2 + h_2^2 = 1, & n = 0 \text{ in (iii)} \\ \qquad h_0 h_2 = 0, & n = -1 \text{ in (iii)}. \end{cases}$$

The last equation suggests $h_2 = 0$, or $h_0 = 0$. Again, the Haar system is the solution.

(3) Set $h_k = 0$ for $k \neq 0, 1, 2, 3$, and solve for h_0, h_1, h_2, h_3,

$$\begin{cases} h_0 + h_1 + h_2 + h_3 = \sqrt{2} \\ h_0 - h_1 + h_2 - h_3 = 0 \\ h_0^2 + h_1^2 + h_2^2 + h_3^2 = 1, & n = 0 \text{ in (iii)} \\ \qquad h_0 h_2 + h_1 h_3 = 0, & n = -1 \text{ in (iii)}. \end{cases} \tag{3.6.1}$$

From the first two equations we obtain

$$h_0 + h_2 = \frac{1}{\sqrt{2}} \Rightarrow h_2 = \frac{1}{\sqrt{2}} - h_0,$$

$$h_1 + h_3 = \frac{1}{\sqrt{2}} \Rightarrow h_3 = \frac{1}{\sqrt{2}} - h_1.$$

From the third equation, we have

$$h_0^2 + h_1^2 + \left(\frac{1}{\sqrt{2}} - h_0\right)^2 + \left(\frac{1}{\sqrt{2}} - h_1\right)^2 = 1,$$

that is,

$$\left(h_0 - \frac{\sqrt{2}}{4}\right)^2 + \left(h_1 - \frac{\sqrt{2}}{4}\right)^2 = \left(\frac{1}{2}\right)^2.$$

This equation represents a circle in the $h_0 - h_1$ plane that is centered at $(\sqrt{2}/4, \sqrt{2}/4)$ with radius $1/2$, and passes through the origin. The last equation in (3.6.1) gives

$$h_0\left(\frac{1}{\sqrt{2}} - h_0\right) + h_1\left(\frac{1}{\sqrt{2}} - h_1\right) = 0,$$

that is,

$$\left(h_0 - \frac{1}{2\sqrt{2}}\right)^2 + \left(h_1 - \frac{1}{2\sqrt{2}}\right)^2 = \left(\frac{1}{2}\right)^2,$$

which represents the same circle as the third equation does. This means that the four equations are not independent. Daubechies introduced ν, such that

$$\begin{aligned} h_0 &= \frac{\nu(\nu-1)}{D}, \\ h_1 &= \frac{1-\nu}{D}, \\ h_2 &= \frac{1+\nu}{D}, \\ h_3 &= \frac{\nu(1+\nu)}{D}, \end{aligned} \tag{3.6.2}$$

where

$$D = \sqrt{2}(1+\nu^2), \qquad \nu \in R. \tag{3.6.3}$$

We can verify that for any ν, the four equations in (3.6.1) are all satisfied. We need one more equation to specify ν, which will be obtained as follows. The wavelets have the frequency domain expression

$$\hat{\psi}(\omega) = \hat{g}\left(\frac{\omega}{2}\right)\hat{\varphi}\left(\frac{\omega}{2}\right).$$

We have also had the zero moment

$$\int_{-\infty}^{\infty} t\psi(t)\,dt = 0.$$

From

$$\hat{\psi}(\omega) = \int_{-\infty}^{\infty} \psi(t)e^{-i\omega t}\,dt,$$

we have the derivative

$$\hat{\psi}'(\omega) = \int -it\psi(t)e^{-i\omega t}\,dt.$$

Hence

$$\begin{aligned} \hat{\psi}'(0) &= -i\int t\psi(t)\,dt \\ &= 0. \end{aligned}$$

But

$$\hat{\psi}'(0) = \hat{\psi}'(\omega)|_{\omega=0}$$

and

$$\hat{\psi}(\omega) = \hat{g}\left(\frac{\omega}{2}\right)\hat{\varphi}\left(\frac{\omega}{2}\right).$$

Thus

$$0 = \hat{\psi}'(0) = \tfrac{1}{2}[\hat{g}'(0)\hat{\varphi}(0) + \hat{g}(0)\hat{\varphi}'(0)].$$

Since

$$\hat{g}(0) = \hat{h}(\pi) = 0$$

and

$$\hat{\varphi}(0) = 1,$$

we arrive at

$$\hat{g}'(0) = 0.$$

Using (3.3.9) in the previous equation, we arrive at

$$\begin{aligned}
\hat{g}\left(\frac{\omega}{2}\right) &= e^{-i(\omega/2)}\overline{\hat{h}(\frac{\omega}{2}+\pi)} = \frac{1}{\sqrt{2}}\sum_k h_k e^{i(\omega/2)(k-1)}e^{ik\pi},\\
\hat{g}'\left(\frac{\omega}{2}\right) &= \frac{1}{\sqrt{2}}\sum_k h_k \frac{i}{2}(k-1)e^{i(\omega/2)(k-1)}(-1)^k,\\
\hat{g}'(0) &= \frac{i}{2\sqrt{2}}\sum_k h_k(k-1)(-1)^k,
\end{aligned}$$

that is,

$$\sum_k (-1)^k k h_k - \sum (-1)^k h_k = 0.$$

Therefore the additional equation to specify h_k is

(4)

$$\sum_k (-1)^k k h_k = 0. \tag{3.6.4}$$

Equation (3.6.4) and equations (3.6.2) through (3.6.3) lead to

$$\begin{aligned}
&-h_1 + 2h_2 - 3h_3 = 0,\\
&-\frac{1-\nu}{D} + 2\frac{1+\nu}{D} - 3\frac{\nu(1+\nu)}{D} = 0,
\end{aligned}$$

$$3\nu^2 = 1,$$
$$\nu = \pm\frac{1}{\sqrt{3}}.$$

As a consequence

$$h_0 = \frac{\nu(\nu-1)}{D} = \frac{-1/\sqrt{3}[(-1/\sqrt{3})-1]}{4\sqrt{2}/3} = \frac{3[(1/3)+(1/\sqrt{3})]}{4\sqrt{2}} = \frac{1+\sqrt{3}}{4\sqrt{2}},$$
$$h_1 = \frac{1+(1/\sqrt{3})}{4\sqrt{2}/3} = \frac{3+\sqrt{3}}{4\sqrt{2}},$$
$$h_2 = \frac{3-\sqrt{3}}{4\sqrt{2}},$$
$$h_3 = \frac{1-\sqrt{3}}{4\sqrt{2}}.$$

Therefore we obtain

$$\begin{aligned}\varphi(t) &= \sum_{n=0}^{3} h_n\sqrt{2}\varphi(2t-n)\\ &= \sum_{n=0}^{3} c_n\varphi(2t-n),\end{aligned} \tag{3.6.5}$$

where

$$\begin{aligned}c_0 &= \frac{1+\sqrt{3}}{4},\\ c_1 &= \frac{3+\sqrt{3}}{4},\\ c_2 &= \frac{3-\sqrt{3}}{4},\\ c_3 &= \frac{1-\sqrt{3}}{4}.\end{aligned} \tag{3.6.6}$$

The corresponding scalet and wavelet are illustrated in Fig. 3.4. It is expected that higher order wavelets are smoother, but their supports are wider. Higher-order Daubechies wavelets can be derived in the same way presented here. The coefficients are tabulated in Table 3.3 and will be employed to construct the Daubechies scalets and wavelets of different orders. In general, for Daubechies scalets $h_n = 0$ for $n < 0$ and $n > 2N + 1$; the support $\varphi = [0, 2N - 1]$, and the support $\psi = [1 - N, N]$, where N is the order, or the number of vanishing moments, meaning that, $\int_{-\infty}^{\infty} t^k\psi(t)\,dt = 0, k = 0, 1, \ldots, N - 1$. A detailed discussion may be found in [1].

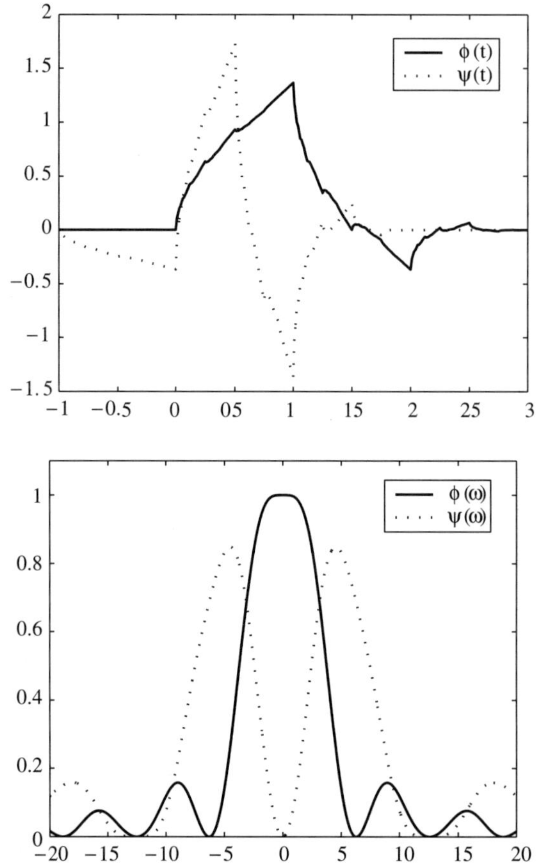

FIGURE 3.4 Daubechies scalet φ and wavelet ψ $(N = 2)$.

TABLE 3.3. Coefficients for Compactly Supported Daubechies Wavelets

	n	$h_n/\sqrt{2}$		n	$h_n/\sqrt{2}$
$N = 2$	0	3.415063509461097E-001	$N = 4$	0	1.629017140256491E-001
	1	5.915063509461096E-001		1	5.054728575459143E-001
	2	1.584936490538904E-001		2	4.461000691304508E-001
	3	−9.150635094610961E-002		3	−1.978751311782236E-002
				4	−1.322535836845199E-001
$N = 3$	0	2.352336038920818E-001		5	2.180815023708858E-002
	1	5.705584579157218E-001		6	2.325180053549088E-002
	2	3.251825002631161E-001		7	−7.493494665180714E-003
	3	−9.546720778416371E-002			
	4	−6.041610415519814E-002			
	5	2.490874986844184E-002			

TABLE 3.3. (*Continued*)

	n	$h_n/\sqrt{2}$
$N=5$	0	1.132094912917792E-001
	1	4.269717713525141E-001
	2	5.121634721295983E-001
	3	9.788348067390437E-002
	4	−1.713283576914677E-001
	5	−2.280056594177345E-002
	6	5.485132932106696E-002
	7	−4.413400054179146E-003
	8	−8.895935050977097E-003
	9	2.358713969533956E-003
$N=6$	0	7.887121600145074E-002
	1	3.497519070376180E-001
	2	5.311318799408694E-001
	3	2.229156614650182E-001
	4	−1.599932994460615E-001
	5	−9.175903203014797E-002
	6	6.894404648737192E-002
	7	1.946160485416436E-002
	8	−2.233187416509466E-002
	9	3.916255761485099E-004
	10	3.378031181463931E-003
	11	−7.617669028012678E-004
$N=7$	0	5.504971537280798E-002
	1	2.803956418127441E-001
	2	5.155742458180708E-001
	3	3.321862411055367E-001
	4	−1.017569112313153E-001
	5	−1.584175056403090E-001
	6	5.042323250469193E-002
	7	5.700172257986757E-002
	8	−2.689122629484261E-002
	9	−1.171997078210231E-002
	10	8.874896189679887E-003
	11	3.037574977010931E-004
	12	−1.273952359093552E-003
	13	2.501134265612311E-004
$N=8$	0	3.847781105407849E-002
	1	2.212336235761367E-001
	2	4.777430752138948E-001
	3	4.139082662112025E-001
	4	−1.119286766690355E-002
	5	−2.008293163905141E-001
	6	3.340970462194946E-004
	7	9.103817842366542E-002
	8	−1.228195052284944E-002
	9	−3.117510332514109E-002
	10	9.886079648352046E-003
	11	6.184422409816434E-003
	12	−3.443859628442111E-003
	13	−2.770022744794268E-004
	14	4.776148556496479E-004
	15	−8.306863068663424E-005
$N=9$	0	2.692517479466241E-002
	1	1.724171519069747E-001
	2	4.276745321796965E-001
	3	4.647728571831284E-001
	4	9.418477475317015E-002
	5	−2.073758809009324E-001
	6	−6.847677451236382E-002
	7	1.050341711395198E-001
	8	2.172633773061817E-002
	9	−4.782363206009557E-002
	10	1.774464066183207E-004
	11	1.581208292625636E-002
	12	−3.339810113138862E-003
	13	−3.027480287545116E-003
	14	1.306483640247298E-003
	15	1.629073356760951E-004
	16	−1.781648795107705E-004
	17	2.782275701717495E-005
$N=10$	0	1.885857879611486E-002
	1	1.330610913968807E-001
	2	3.727875357431275E-001
	3	4.868140553667018E-001
	4	1.988188708845045E-001
	5	−1.766681008969463E-001
	6	−1.385549393604189E-001
	7	9.025464309758936E-002
	8	6.580149355052253E-002
	9	−5.048328559835650E-002
	10	−2.082962404377542E-002
	11	2.348490704871285E-002
	12	2.550218483929039E-003
	13	−7.589501167907126E-003
	14	9.866626824928449E-004
	15	1.408843295102602E-003
	16	−4.849739199258639E-004
	17	−8.235450304483359E-005
	18	6.617718342564559E-005
	19	−9.379207813735183E-006

3.7 COIFMAN WAVELETS (COIFLETS)

An orthonormal wavelet system with compact support is called the Coifman wavelet system of order L if $\varphi(t)$ and $\psi(t)$ have $L-1$ and L vanishing moments, respectively,

$$\int dtt^l \psi(t) = 0, \qquad l = 0, 1, \ldots, L-1, \tag{3.7.1}$$

$$\int dtt^l \varphi(t) = 0, \qquad l = 1, 2, \ldots, L-1. \tag{3.7.2}$$

As usual, the scalet still has a normalized d.c. component

$$\int dt\varphi(t) = 1. \tag{3.7.3}$$

The nonzero support of the Coiflets of order $L = 2K$ is $3L - 1$.

Consider the case $L = 2$. Notice that (3.7.1) states the vanishing moments for the wavelets, and (3.7.3) is the normalization of the scalet, with respect to the d.c. component. Both of these two equations are shared by other wavelets. The unique property of the Coiflets is contained in (3.7.2), namely the vanishing moments of the scalets. This property can be shown to yield

$$\sum kh_k = 0.$$

Show. From basic wavelet theory

$$\hat{\varphi}(\omega) = \int e^{-i\omega t}\varphi(t)\,dt.$$

By taking a derivative with respect to ω,

$$\begin{aligned}
\hat{\varphi}'(\omega) &= \int (-it)e^{-i\omega t}\varphi(t)\,dt \\
\hat{\varphi}'(0) &= -i\int t\varphi(t)\,dt \\
&= 0
\end{aligned} \tag{3.7.4}$$

due to the zero moment of (3.7.1). On the other hand, we have

$$\begin{aligned}
\hat{\varphi}(\omega) &= \hat{h}\left(\frac{\omega}{2}\right)\hat{\varphi}\left(\frac{\omega}{2}\right) \\
\hat{\varphi}'(\omega) &= \frac{1}{2}\left[\hat{h}'\left(\frac{\omega}{2}\right)\hat{\varphi}\left(\frac{\omega}{2}\right) + \hat{h}\left(\frac{\omega}{2}\right)\hat{\varphi}'\left(\frac{\omega}{2}\right)\right] \\
\hat{\varphi}'(0) &= \frac{1}{2}\left[\hat{h}'(0)\hat{\varphi}(0) + \hat{h}(0)\hat{\varphi}'(0)\right].
\end{aligned} \tag{3.7.5}$$

In (3.7.5) the left-hand side is zero due to (3.7.4). On the right-hand side $\hat{\varphi}(0) = 1$ and $\hat{h}(0) = 1$. Hence

$$\hat{h}'(0) = 0. \tag{3.7.6}$$

Noticing that

$$\hat{h}(\omega) = \frac{1}{\sqrt{2}} \sum_k h_k e^{-i(\omega/2)k},$$

and taking a derivative and then setting the argument to zero, we have

$$\hat{h}'(\omega) = \frac{1}{\sqrt{2}} \sum_k h_k \left(-i\frac{k}{2}\right) e^{-i(\omega/2)k},$$

$$\hat{h}'(0) = \frac{-i}{2\sqrt{2}} \sum_k k h_k.$$

By virtue of (3.7.6) we obtain

$$\sum_k k h_k = 0. \tag{3.7.7}$$

□

Recall that in Section 3.6 we had three equations for the filter bank coefficients:

(i) $\sum_k h_k/\sqrt{2} = 1$.

(ii) $\sum_k \frac{(-1)^k}{\sqrt{2}} h_k = 0$.

(iii) $\sum_k h_k h_{k-2n} = \delta_{0,n}$.

The two compactly support wavelets, Daubechies and Coifman, have similarities and distinctions in the governing equations. Table 3.4 summarizes and compares the nature of these wavelets. Equations (i), (ii), (iii), (3.6.4), and (3.7.7) are sufficient to solve the h_k for the Coiflets of order 2. They form a set of nonlinear equations that are shown explicitly below:

$$\begin{aligned}
h_{-2} + h_{-1} + h_0 + h_1 + h_2 + h_3 &= \sqrt{2} && \text{(i)},\\
h_{-2} - h_{-1} + h_0 - h_1 + h_2 - h_3 &= 0 && \text{(ii)},\\
h_{-2}^2 + h_{-1}^2 + h_0^2 + h_1^2 + h_2^2 + h_3^2 &= 1 && \text{(iii) with } n = 0,\\
h_{-2}h_0 + h_{-1}h_1 + h_0h_2 + h_1h_3 &= 0 && \text{(iii) with } n = 1,\\
h_{-2}h_2 + h_{-1}h_3 &= 0 && \text{(iii) with } n = 2,\\
-2h_{-2} + h_{-1} - h_1 + 2h_2 - 3h_3 &= 0 && \text{by (3.6.4)},\\
-2h_{-2} - h_{-1} + h_1 + 2h_2 + 3h_3 &= 0 && \text{by (3.7.7)}.
\end{aligned}$$

The six-variable nonlinear equations for Coiflets of $L = 2$ were solved numerically using the software package Maple, yielding two sets of solutions:

TABLE 3.4. Comparison between the Daubechies and Coifman Wavelets

Daubechies	Basic Equations	Coifman
$h_0 + h_1 + h_2 + h_3 = \sqrt{2}$	$\sum \frac{h_k}{\sqrt{2}} = 1$	$h_{-2} - h_{-1} + h_0 + h_1 + h_2 + h_3 = \sqrt{2}$
$h_0 - h_1 + h_2 - h_3 = 0$	$\sum \frac{(-1)^k}{\sqrt{2}} h_k = 0$	$h_{-2} - h_1 + h_0 - h_1 + h_2 - h_3 = 0$
	$\hat{h}(\pi) = 0$	
$h_0^2 + h_1^2 + h_2^2 + h_3^2 = 1$		$h_{-2}^2 + h_{-1}^2 + h_0^2 + h_1^2 + h_2^2 + h_3^2 = 1$
$h_0 h_2 + h_1 h_3 = 0$	$\sum_k h_k h_{k-2n} = \delta_{0,n}$	$h_{-2} h_0 + h_{-1} h_1 + h_0 h_2 + h_1 h_3 = 0$
	$\int \varphi_{0,0} \varphi_{0,n} = \delta_{0,n}$	$h_{-2} h_2 + h_{-1} h_3 = 0$
$-h_1 + 2h_2 - 3h_3 = 0$	$\sum (-1)^k k h_k = 0$	$-2h_{-2} + h_{-1} - h_1 + 2h_2 - 3h_3 = 0$
	$\hat{g}(0) = 0$	
	$\int t\psi(t) = 0$	
Not applicable	$\sum k h_k = 0$	$-2h_{-2} - h_{-1} + h_1 + 2h_2 + 3h_3 = 0$
	$\hat{h}(0) = 0$	
	$\int t\varphi(t) = 0$	$h_2 = h_{-2}$
$[0, 2N - 1]$	supp φ	$[-L, 2L - 1]$
$[1 - N, N]$	supp ψ	$[1 - 1.5L, 1.5L]$

$$
\begin{aligned}
\frac{h_{-2}}{\sqrt{2}} &= 0.1139297284707685 \\
\frac{h_{-1}}{\sqrt{2}} &= 0.07357027152923154 \\
\frac{h_0}{\sqrt{2}} &= 0.272140543058463 \\
\frac{h_1}{\sqrt{2}} &= 0.602859456941536 \\
\frac{h_2}{\sqrt{2}} &= 0.113929728470768 \\
\frac{h_3}{\sqrt{2}} &= -0.176429728470768
\end{aligned}
\tag{3.7.8}
$$

and

$$
\begin{aligned}
\frac{h_{-2}}{\sqrt{2}} &= -0.05142972847076846 \\
\frac{h_{-1}}{\sqrt{2}} &= 0.238929728470768 \\
\frac{h_0}{\sqrt{2}} &= 0.6028594569415369 \\
\frac{h_1}{\sqrt{2}} &= 0.2721405430584631 \\
\frac{h_2}{\sqrt{2}} &= -0.05142972847076846 \\
\frac{h_3}{\sqrt{2}} &= -0.01107027152923154.
\end{aligned}
\tag{3.7.9}
$$

TABLE 3.5. Filter Bank of Coiflets

	n	$h_n/\sqrt{2}$
$L = 2$	−2	−5.142972847076846E-2
	−1	2.389297284707680E-1
	0	6.028594569415369E-1
	1	2.721405430584631E-1
	2	−5.142972847076846E-2
	3	−1.107027152923154E-2
$L = 4$	−4	1.158759673871687E-2
	−3	-2.932013798346857E-2
	−2	−4.763959031008130E-2
	−1	2.730210465347667E-1
	0	5.746823938568640E-1
	1	2.948671936956192E-1
	2	−5.408560709171150E-2
	3	−4.202648046077160E-2
	4	1.674441016327951E-2
	5	3.967883612962012E-3
	6	−1.289203356140660E-3
	7	−5.095053991076440E-4
$L = 6$	−6	−2.682418670922068E-3
	−5	5.503126707831385E-3
	−4	1.658356047917034E-2
	−3	−4.650776447872699E-2
	−2	-4.322076356021191E-2
	−1	2.865033352736475E-1
	0	5.612852568703300E-1
	1	3.029835717728243E-1
	2	−5.077014075488885E-2
	3	−5.819625076158550E-2
	4	2.443409432116696E-2
	5	1.122924096203786E-2
	6	−6.369601011048820E-3
	7	−1.820458915566242E-3
	8	7.902051009575940E-4
	9	3.296651737931827E-4
	10	−5.019277455327665E-5
	11	−2.446573425530812E-5
$L = 8$	−8	6.309612114309465E-4
	−7	−1.152225143769972E-3
	−6	−5.194525163470320E-3
	−5	1.136246148326480E-2
	−4	1.886723856956300E-2
	−3	−5.746424190192710E-2
	−2	−3.965265296244911E-2
	−1	2.936674050161005E-1
	0	5.531264550395490E-1
$L = 8$	1	3.071573096678857E-1
	2	−4.711273752389571E-2
	3	−6.803811467802060E-2
	4	2.781363695846951E-2
	5	1.773583142270309E-2
	6	−1.075631615503724E-2
	7	4.001010844950537E-3
	8	2.652664913530499E-3
	9	8.955939276952845E-4
	10	−4.165001950941707E-4
	11	−1.838296167136254E-4
	12	4.408022661597205E-5
	13	2.208284691230834E-5
	14	−2.304919162676503E-6
	15	−1.262179179994623E-6
$L = 10$	−10	−1.499645228345962E-4
	−9	2.535527523580332E-4
	−8	1.540286725995231E-3
	−7	−2.941078164035692E-3
	−6	−7.164112349410080E-3
	−5	1.655218330649289E-2
	−4	1.991901719798436E-2
	−3	−6.499837825472325E-2
	−2	−3.680255347446876E-2
	−1	2.980959014587191E-1
	0	5.475082713540365E-1
	1	3.097002590784202E-1
	2	−4.386731482823612E-2
	3	−7.464442013283965E-2
	4	2.919469277528069E-2
	5	2.310457227706682E-2
	6	1.397129268638197E-2
	7	−6.476749751505850E-3
	8	4.781116799130641E-3
	9	1.719383484385501E-3
	10	−1.174947934413533E-3
	11	−4.508222400696236E-4
	12	2.134457875036282E-4
	13	9.924691139878545E-5
	14	−2.914684388622115E-5
	15	−1.504031798197687E-5
	16	2.616809660013098E-6
	17	1.457029123551625E-6
	18	−1.148199649902970E-7
	19	−6.791060627322360E-8

The second set of the solutions can be derived in irrational numbers [5] as

$$\begin{aligned} \frac{h_{-2}}{\sqrt{2}} &= \frac{1-\sqrt{7}}{32} \\ \frac{h_{-1}}{\sqrt{2}} &= \frac{5+\sqrt{7}}{32} \\ \frac{h_0}{\sqrt{2}} &= \frac{7+\sqrt{7}}{16} \\ \frac{h_1}{\sqrt{2}} &= \frac{7-\sqrt{7}}{16} \\ \frac{h_2}{\sqrt{2}} &= \frac{1-\sqrt{7}}{32} \\ \frac{h_3}{\sqrt{2}} &= \frac{-3+\sqrt{7}}{32}. \end{aligned} \tag{3.7.10}$$

The coefficients of higher-order Coiflets can be generated in a similar way and they are tabulated in Table 3.5. The coefficients in (3.7.9) through (3.7.10) and Table 3.5

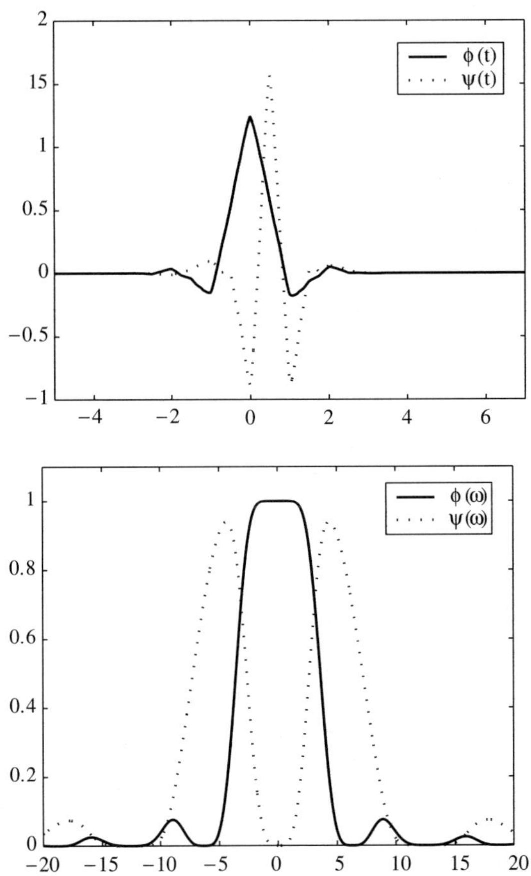

FIGURE 3.5 Coifman scalet φ and wavelet $\psi (L = 4)$.

will be used to construct the Coifman scalets and wavelets according to the dilation equations

$$\varphi(x) = \sqrt{2}\sum_k h_k \varphi(2x - k),$$

$$\psi(x) = \sqrt{2}\sum_k g_k \varphi(2x - k).$$

The resultant Coifman scalet and wavelet of order $L = 4$ are plotted along with their Fourier transform (in magnitude) in Fig. 3.5.

3.8 CONSTRUCTING WAVELETS BY RECURSION AND ITERATION

3.8.1 Construction of Scalets

The Cascade Algorithm. Suppose that φ is known at integer points $2t = k$. The dilation equation (3.6.5) then provides φ at the half-integer points. Repeating the procedure, we obtain φ at the quarter-integers, and so forth. Ultimately we obtain $\varphi(t)$ at all dyadic points of $t = k/2^j$.

***Example 1 Box Function on* $[0, 1]$.** We have two nonzero coefficients h_k

$$h_0 = h_1 = \frac{1}{\sqrt{2}}, \quad \varphi(t) = \sum_{n=0}^{1} h_n \sqrt{2}\varphi(2t - n).$$

The iterative procedure produces the invariant values, which are the Haar scalet.

***Example 2 Daubechies Wavelets of N* $= 2$.** Equation (3.6.5) provides nonzero values of $\varphi(\cdot)$ on $[0, 3]$. At integer points, we have from (3.6.5),

$$\varphi(1) = \frac{3+\sqrt{3}}{4}\varphi(1) + \frac{1+\sqrt{3}}{4}\varphi(2),$$

$$\varphi(2) = \frac{1-\sqrt{3}}{4}\varphi(1) + \frac{3-\sqrt{3}}{4}\varphi(2),$$

namely

$$\begin{pmatrix} \varphi(1) \\ \varphi(2) \end{pmatrix} = \begin{pmatrix} L_{11} & L_{12} \\ L_{21} & L_{22} \end{pmatrix} \begin{pmatrix} \varphi(1) \\ \varphi(2) \end{pmatrix}, \tag{3.8.1}$$

where

$$L_{ij} = c_{2i-j}$$

with nonzero c_0, c_1, c_2, and c_3 given by equation (3.6.6). We can consider (3.8.1) as an eigenvalue problem

$$\lambda | X\rangle = L\,|X\rangle.$$

Its eigenvalues $\lambda = 1, \frac{1}{2}$ can be found from

$$\begin{vmatrix} \dfrac{3+\sqrt{3}}{4} - \lambda & \dfrac{1+\sqrt{3}}{4} \\ \dfrac{1-\sqrt{3}}{4} & \dfrac{3-\sqrt{3}}{4} - \lambda \end{vmatrix} = 0,$$

that is,

$$\lambda^2 - \tfrac{3}{2}\lambda + \tfrac{1}{2} = 0.$$

However, $\lambda = \frac{1}{2}$ does not correspond to (3.8.1). Therefore $\lambda = 1$ is the only choice. For $\lambda = 1$, the corresponding eigenvector can be found from

$$\begin{bmatrix} \dfrac{\sqrt{3}-1}{4} & \dfrac{\sqrt{3}+1}{4} \\ -\dfrac{\sqrt{3}-1}{4} & -\dfrac{\sqrt{3}+1}{4} \end{bmatrix} \begin{bmatrix} \phi(1) \\ \phi(2) \end{bmatrix} = 0,$$

that is,

$$\begin{pmatrix} \phi(1) \\ \phi(2) \end{pmatrix} = \begin{pmatrix} 1 \\ \sqrt{3}-2 \end{pmatrix}.$$

From the normalization condition of $\sum_{n=-\infty}^{\infty} \varphi(t-n) = 1$ and the support $\varphi(\cdot) = [0, 3]$, we obtain

$$\phi(1) + \phi(2) = 1.$$

Simple algebraic operations give the normalized eigenvector

$$\begin{pmatrix} \phi(1) \\ \phi(2) \end{pmatrix} = \begin{pmatrix} \dfrac{1+\sqrt{3}}{2} \\ \dfrac{1-\sqrt{3}}{2} \end{pmatrix}. \tag{3.8.2}$$

The exact values of $\phi(1)$ and $\phi(2)$ are employed to construct the positive sampling functions that are the basis functions of the sampling-biorthogonal time domain (SBTD) method in Chapter 5.

The previous equation provides the accurate values of φ at integer points. These values may be employed to evaluate values of $\varphi(x)$ at half-integer points by using equation (3.6.5), namely

$$\begin{aligned} \varphi(t) &= \sum \sqrt{2} h_n \varphi(2t-n) \\ &= c_0\varphi(2t) + c_1\varphi(2t-1) + c_2\varphi(2t-2) + c_3\varphi(2t-3). \end{aligned}$$

As a result the half-integer values are

$$\begin{aligned}
\varphi\left(\tfrac{1}{2}\right) &= c_0\varphi(1), \\
\varphi\left(1\tfrac{1}{2}\right) &= c_1\varphi(2) + c_2\varphi(1), \\
\varphi\left(2\tfrac{1}{2}\right) &= c_3\varphi(2).
\end{aligned}$$

The quarter-integer values may be obtained from the half-integer values as

$$\begin{aligned}
\varphi\left(\tfrac{1}{4}\right) &= c_0\varphi\left(\tfrac{1}{2}\right), \\
\varphi\left(1\tfrac{1}{4}\right) &= c_0\varphi\left(2\tfrac{1}{2}\right) + c_1\varphi\left(1\tfrac{1}{2}\right) + c_2\varphi\left(\tfrac{1}{2}\right), \\
\varphi\left(1\tfrac{3}{4}\right) &= c_1\varphi\left(2\tfrac{1}{2}\right) + c_2\varphi\left(1\tfrac{1}{2}\right) + c_3\varphi\left(\tfrac{1}{2}\right), \\
\varphi\left(2\tfrac{1}{4}\right) &= c_2\varphi\left(2\tfrac{1}{2}\right) + c_3\varphi\left(1\tfrac{1}{2}\right), \\
\varphi\left(2\tfrac{3}{4}\right) &= c_3\varphi\left(2\tfrac{1}{2}\right).
\end{aligned}$$

Repeating this process, the values at any dyadic fraction points may be found.

Example 3 For Daubechies wavelet of order $N = 3$, there are six nonzero filter coefficients, namely $h_n \neq 0$ only for $n = 0, 1, 2, \ldots, 5$. Find the scalets $\varphi(t)$ at $t = 1, 2, 3, 4$.

Solution Using the dilation equation, we have

$$\begin{aligned}
\varphi(t) &= \sum_{n=0}^{5} c_n\varphi(2t - n) \\
&= c_0\varphi(2t) + c_1\varphi(2t-1) + c_2\varphi(2t-2) \\
&\quad + c_3\varphi(2t-3) + c_4\varphi(2t-4) + c_5\varphi(2t-5),
\end{aligned}$$

where $c_i = \sqrt{2}h_i, i = 0, 1, \ldots, 5$ and h_i are tabulated in Table 3.5. The support $\varphi = [0, 2N-1] = [0, 5]$. Hence

$$\begin{aligned}
\varphi(1) &= c_0\varphi(2) + c_1\varphi(1), \\
\varphi(2) &= c_0\varphi(4) + c_1\varphi(3) + c_2\varphi(2) + c_3\varphi(1), \\
\varphi(3) &= c_2\varphi(4) + c_3\varphi(3) + c_4\varphi(2) + c_5\varphi(1), \\
\varphi(4) &= c_4\varphi(4) + c_5\varphi(3),
\end{aligned}$$

that is,

$$\begin{bmatrix} \varphi(1) \\ \varphi(2) \\ \varphi(3) \\ \varphi(4) \end{bmatrix} = \begin{bmatrix} c_1 & c_0 & 0 & 0 \\ c_3 & c_2 & c_1 & c_0 \\ c_5 & c_4 & c_3 & c_2 \\ 0 & 0 & c_5 & c_4 \end{bmatrix} \begin{bmatrix} \varphi(1) \\ \varphi(2) \\ \varphi(3) \\ \varphi(4) \end{bmatrix}.$$

From the equation above, $\lambda = 1$ is an eigenvalue, namely

$$\begin{aligned} A \mid \varphi\rangle &= \lambda \mid \varphi\rangle \\ &= 1 \mid \varphi\rangle. \end{aligned}$$

The corresponding eigenvector with unit norm may be found using Matlab or other software, and is given by

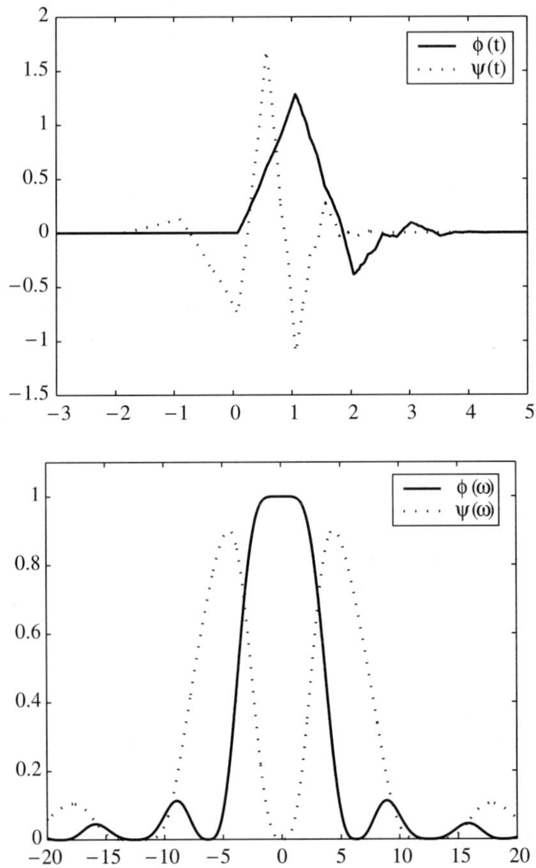

FIGURE 3.6 Daubechies scalet φ and wavelet ψ ($N = 3$).

$$| \varphi \rangle = \begin{bmatrix} 0.9554 \\ -0.2866 \\ 0.0708 \\ 0.0031 \end{bmatrix}.$$

After normalization under the condition

$$\varphi(1) + \varphi(2) + \varphi(3) + \varphi(4) = 1,$$

we obtain

$$\begin{bmatrix} \varphi(1) \\ \varphi(2) \\ \varphi(3) \\ \varphi(4) \end{bmatrix} = \begin{bmatrix} 1.2864 \\ -0.3859 \\ 0.0953 \\ 0.0042 \end{bmatrix}.$$

One may compare these values with the corresponding Daubechies φ of order 3 in Fig. 3.6.

The Iteration Method. Both scalet and wavelet can be constructed using the filter bank of h_k in Table 3.3 for the Daubechies wavelets and in Table 3.5 for the Coifman wavelets. The iterative method is very simple and easy to implement, as shown by the attached FORTRAN program. The major steps are outlined below.

- Initiate values within the nonzero support, which is $[0, 2N - 1]$ for the Daubechies scalets of order N.
- Begin iterative procedures, using the dilation equation.
- Set up a precision and the program stops when the error is bounded within the precision.
- A wavelet is constructed based on the obtained scalets. No iteration is necessary.

```
c-----Example of Daubechies' Wavelet and Scalet (N=2)
      Parameter (N = 128, NIT = 60)
      Real*3 f1(-6*N:4*N), F2(-6*N:4*N),H,X,P1,P2,P3,P4
      P1 = (1.D00 + DSQRT (3.D00))/4.D00
      P2 = (3.D00 + DSQRT (3.D00))/4.D00
      P3 = (3.D00 - DSQRT (3.D00))/4.D00
      P4 = (1.D00 - DSQRT (3.D00))/4.D00
      H = 2.D00/N
      K = 1.D00/H + 0.5D0
c-----Initiate the scalets
     DO I = 0, N/2
       F1 (I) = DFLOAT (I)*H
     ENDDO
     DO I = 1, N/2
       F1 (I + N/2) = 1.300 - DFLOAT (I) *H
     ENDDO
```

```
c-----Iterative Procedure for Scalet
      DO IT = 1, NIT
       DO I = 0, 2*N
         F2 (I) = P1*F1 (2*I) + P2*F1 (2*I-K) + P3*F1 (2*I-2*K)
                                              + P4*F1 (2*I-3*K)
      ENDDO
      DO I = -6*N, 4*N
       F1 (I) = F2 (I)
       ENDDO
      ENDDO
c-----Wavelets
     OPEN (1, FILE = 'scalf.dat', STATUS = 'UNKNOWN')
     OPEN (2, FILE = 'wavel.dat', STATUS = 'UNKNOWN')
     DO I = -N, 2*N
       X = I * H
       F2 (I) = P4*F1 (2*I + 2*K) - P3*F1 (2*I + K) + P2*F1 (2*I) -
                P1*F1 (2*I-K)
       WRITE (1,*) X,      F1 (I)
       WRITE (2,*) X,      F2 (I)
     ENDDO
     CLOSE (1, STATUS='KEEP')
     CLOSE (2, STATUS='KEEP')
STOP
END
```

3.8.2 Construction of Wavelets

Using available $\varphi(t)$ values, we construct the corresponding wavelet function $\psi(t)$ by the dilation equation. For the Daubechies $N = 2$, we have

$$\begin{aligned}\psi(t) &= \sum_k \sqrt{2} g_k \varphi(2t-k) \\ &= \sum (-1)^k h_{1-k} \sqrt{2} \varphi(2t-k) \\ &= c_3 \varphi(2t+2) - c_2 \varphi(2t+1) + c_1 \varphi(2t) - c_0 \varphi(2t-1).\end{aligned}$$

The corresponding Daubechies wavelets of $N = 2$ has a support $[-1, 2]$. Higher-order Daubechies wavelets may be constructed using more nonzero coefficients, which are listed in Table 3.3. It is expected that the higher the wavelet order is, the smoother its shape and the wider its support. In general, for order N, the Daubechies scalets and wavelets have support of $[0, 2N-1]$ and $[1-N, N]$, respectively. Notice that the Daubechies scalets and wavelets do not have explicit expressions. As a result no one knows the exact value of

$$\varphi(\sqrt{2}) = ?$$

3.9 MEYER WAVELETS

In the previous sections, the Battle–Lemarie, Daubechies, and Coifman wavelets were derived and expressed in the time (spatial) domain. Logically one may ask why we do not construct a wavelet system in the transform domain. In communication theory we often encounter with band-limited signals, including the famous Shannon sampling function of sinc $(t) = \sin \pi t/\pi t$, which has a bandwidth of $\omega \in [-\pi, \pi)$. In this section we will see that the sinc function is a scalet in the Meyer wavelet family.

3.9.1 Basic Properties of Meyer Wavelets

To obtain Meyer wavelets, we begin by constructing the scalets directly in the Fourier transform domain. Recall that in the previous sections we had arrived at the following properties:

(i) $\sum_k |\hat{\varphi}(\omega + 2k\pi)|^2 = 1$, or equivalently

$$\int \varphi(t)\varphi(t-n)\,dt = \delta_{0,n}. \tag{3.9.1}$$

(ii) $\hat{\varphi}(\omega) = \hat{h}(\omega/2)\hat{\varphi}(\omega/2) \leftrightarrow V_0 \subseteq V_1$, and

$$\hat{h}\left(\frac{\omega}{2} + 2\pi\right) = \hat{h}\left(\frac{\omega}{2}\right).$$

(iii) $\hat{\varphi}(\omega)$ is continuous at $\omega = 0$, that is,

$$\lim_{\omega \to 0} \hat{\varphi}(\omega) = \hat{\varphi}(0) = 1.$$

The raised cosine functions are widely employed in baseband pulse shaping of digital communication systems [6]. In contrast to the sharp edges of the sinc functions, these band-limited signals have smooth frequency windows up to $[-2\pi, 2\pi)$. Do they form either orthogonal systems or wavelets? These questions triggered genius work of Meyer. For Meyer wavelets we seek $\hat{\varphi}(\omega)$ such that the support region of the transform domain scalet is within $[-2\pi, 2\pi]$, namely

$$\operatorname{supp} \hat{\varphi}(\omega) \subseteq [-2\pi, 2\pi].$$

From (i) we have

$$|\hat{\varphi}(\omega)|^2 + |\hat{\varphi}(\omega - 2\pi)|^2 = 1, \qquad 0 \le \omega \le 2\pi.$$

This situation is due to the confined interval of ω.

Lemma. Scalet $\hat{\varphi}(\omega)$ has support of $[-a, a]$, $\pi \le a \le 4\pi/3$ if $\hat{\varphi}$ satisfies (i) and (ii).

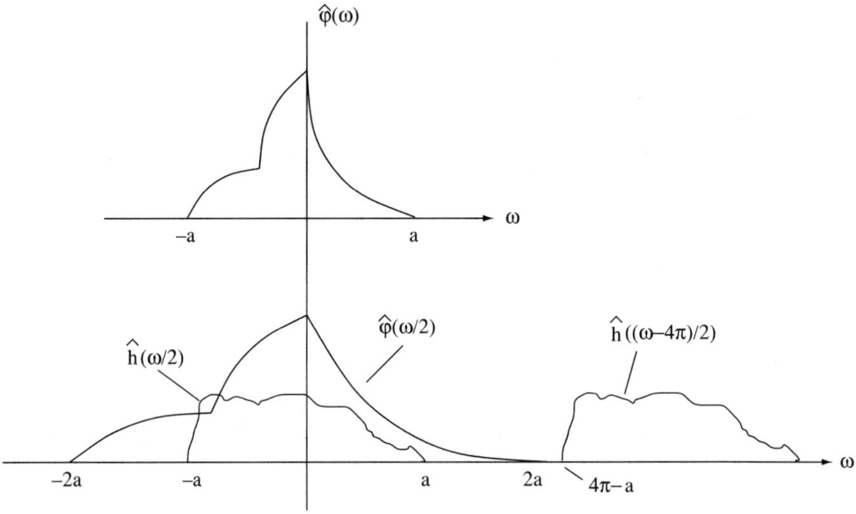

FIGURE 3.7 Additional bandwidth of Meyer scalets.

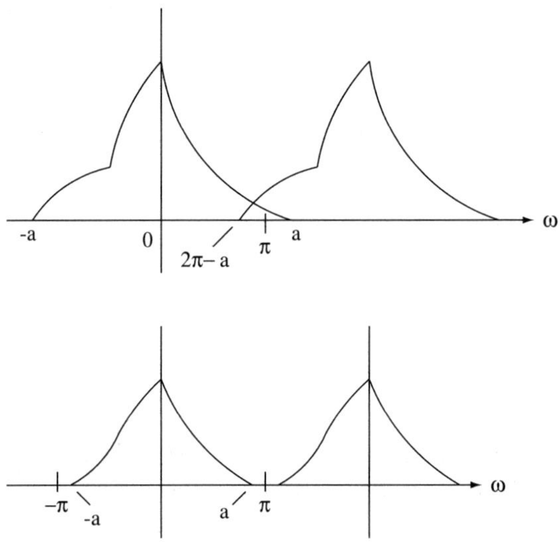

FIGURE 3.8 Minimum bandwidth of Meyer scalets.

Proof. By the assumption that $\hat{\varphi}(\omega)$ has a support $[-a, a]$, we immediately see that:

(a) $\hat{\varphi}(\omega/2)$ has support $[-2a, 2a]$.
(b) $\hat{h}(\omega/2) = 0$ in $(a, 2a)$.

Claim (a) is straightforward. Claim (b) may be justified with the help of Fig. 3.7. Note that from (ii) $\hat{\varphi}(\omega) = \hat{h}(\omega/2)\hat{\varphi}(\omega/2)$ where $\hat{\varphi}(\omega/2)$ is a stretched version of $\hat{\varphi}(\omega)$. The requirement of $\hat{\varphi}(\omega) = 0$ for $\omega > a$ forces $\hat{h}(\omega/2) = 0$ in $(a, 2a)$. Because $\hat{h}(\omega/2)$ is of period 4π, namely $\hat{h}[(\omega + 4\pi)/2] = \hat{h}(\omega/2)$, we have

$$4\pi - a \geq 2a;$$

otherwise, $\hat{h}(\cdot)$ would be nonzero between $(a, 2a)$. This argument is explained in Fig. 3.7. Therefore

$$a \leq \tfrac{4}{3}\pi.$$

On the other hand, if $a < \pi$, then the condition

$$|\hat{\varphi}(\omega)|^2 + |\hat{\varphi}(\omega - 2\pi)|^2 = 1$$

cannot be satisfied, and some gaps are inevitable. In the gap region, $|\hat{\varphi}(\omega)|^2 + |\hat{\varphi}(\omega - 2\pi)|^2 = 0$ as illustrated in Fig. 3.8. In conclusion, we have proved that $\hat{\varphi}(\omega)$ has support $[-a, a]$ and $\pi \leq a \leq \frac{4}{3}\pi$. □

Theorem 3. Let $p(\zeta)$ be a distribution, or a nonnegative function on real axis, R, such that:

(i) $p(\zeta) \geq 0$.
(ii) $\operatorname{supp} p(\zeta) \subseteq [-\varepsilon, \varepsilon]$, $0 \leq \varepsilon \leq \pi/3$; the inverse Fourier transform $\mathcal{F}^{-1}p(\omega) := \hat{p}(t)$ is unity at the origin, namely $\hat{p}(t)|_{t=0} = 1$.
(iii) $\int_{-\pi/3}^{\pi/3} p(\zeta)\, d\zeta = 1$.

Define a function

$$\hat{\varphi}(\omega) := \left[\int_{\omega-\pi}^{\omega+\pi} p(\zeta) d\zeta\right]^{1/2}. \tag{3.9.2}$$

Then $\hat{\varphi}(\omega)$ possesses the following properties:

(i) $\operatorname{supp} |\hat{\varphi}(\omega)|^2 \subseteq [-\pi - \varepsilon, \pi + \varepsilon]$.
(ii) $|\hat{\varphi}(\omega)| = 1$ for

$$|\omega| \leq \pi - \varepsilon. \tag{3.9.3}$$

(iii) $\sum_k |\hat{\varphi}(\omega + 2k\pi)|^2 = 1$,

and the corresponding function $\varphi(t) = \mathcal{F}^{-1}(\hat{\varphi}(\omega))$ is a scalet. Rigorous proof of the previous proposition is rather mathematical. Interested readers are referred to [7].

Instead of proving the theorem, we will provide some intuitive explanations here for $\varepsilon = \pi/3$.

Let us show that the orthogonality (i) of (3.9.1) holds, meaning that

$$\sum_k \left| \hat{\varphi}(\omega + 2k\pi) \right|^2 = 1.$$

Show. If $\omega \le -4\pi/3$,

$$\hat{\varphi}(\omega) = \left[\int_{\omega-\pi}^{\omega+\pi} p(\zeta)\, d\zeta \right]^{1/2} = \left[\int_{\omega-\pi}^{-\pi/3} p(\zeta)\, d\zeta \right]^{1/2} = 0$$

because the integral is off the support. If $\omega \ge 4\pi/3$,

$$\hat{\varphi}(\omega) = \left[\int_{\omega-\pi}^{\omega+\pi} p(\zeta)\, d\zeta \right]^{1/2} = \left[\int_{\pi/3}^{\omega+\pi} p(\zeta)\, d\zeta \right]^{1/2} = 0$$

by the same token. Now we can verify that condition (i) holds:

$$\begin{aligned}
| \hat{\varphi}(\omega - 2\pi) |^2 + | \hat{\varphi}(\omega) |^2 &= \int_{\omega-3\pi}^{\omega-\pi} p(\zeta)\, d\zeta + \int_{\omega-\pi}^{\omega+\pi} p(\zeta)\, d\zeta \\
&= \int_{\omega-3\pi}^{\omega+\pi} p(\zeta)\, d\zeta \\
&= \int_{\omega-3\pi}^{-\pi/3} p(\zeta)\, d\zeta + \int_{-\pi/3}^{\pi/3} p(\zeta)\, d\zeta + \int_{\pi/3}^{\omega+\pi} p(\zeta)\, d\zeta \\
&= 1.
\end{aligned}$$

□

It can be verified that the function defined by (3.9.2) possesses other properties of the scalets. For instance, we may show the validity of (ii) of (3.9.1), namely the dilation equation in the Fourier domain. That is,

$$\hat{\varphi}(\omega) = \hat{h}\left(\frac{\omega}{2}\right) \hat{\varphi}\left(\frac{\omega}{2}\right). \tag{3.9.4}$$

Show. Let us choose

$$\overline{\hat{h}\left(\frac{\omega}{2}\right)} = \sum_k \hat{\varphi}(\omega + 4k\pi). \tag{3.9.5}$$

Then

$$\hat{h}\left(\frac{\omega}{2}\right) = 0$$

if

$$\frac{4\pi}{3} < |\omega| < \frac{8\pi}{3}.$$

This is verified by the following steps:

(a) $4\pi/3 < \omega < 8\pi/3$. From (3.9.5) one has

$$\overline{\hat{h}\left(\frac{\omega}{2}\right)} = \hat{\varphi}(\omega) + \hat{\varphi}(\omega - 4\pi) = 0 + \hat{\varphi}(\Omega)|_{\Omega \le -4\pi/3} = 0.$$

Hence the RHS of (3.9.4) is zero. The LHS of (3.9.4) also equals zero because $\hat{\varphi}(\omega)$ is beyond the support of $|\omega| < 4\pi/3$. Therefore (3.9.4) holds.

(b) $0 < \omega < 4\pi/3$. The RHS of (3.9.4),

$$\begin{aligned}\hat{h}\left(\frac{\omega}{2}\right)\hat{\varphi}\left(\frac{\omega}{2}\right) &= \sum_k \hat{\varphi}(\omega + 4k\pi)\hat{\varphi}\left(\frac{\omega}{2}\right)\\ &= \hat{\varphi}(\omega) \qquad \text{for } |\omega| < \frac{2}{3}\pi\end{aligned}$$

because of (ii) of (3.9.3), $\hat{\varphi}(\omega/2) = 1$ for $|\omega| < 2\pi/3$, and the only nonzero term in the summation is

$$\hat{\varphi}(\omega + 4k\pi)\big|_{k=0} = \hat{\varphi}(\omega). \qquad \square$$

Theorem 4. The wavelet corresponding to the scalet of (3.9.2) is

$$\begin{aligned}\hat{\psi}(\omega) &= e^{-i(\omega/2)}\left[\int_{(\omega/2)-\pi}^{\omega-\pi} p(\zeta)d\zeta\right]^{1/2} && \text{for } \omega \in [\pi - \varepsilon, 2\pi + 2\varepsilon],\\ \hat{\psi}(\omega) &= e^{-i(\omega/2)}\left[\int_{\omega+\pi}^{(\omega/2)+\pi} p(\zeta)d\zeta\right]^{1/2} && \text{for } \omega \in [-2\pi - 2\varepsilon, -\pi + \varepsilon].\end{aligned} \tag{3.9.6}$$

A rigorous proof can be found in [8]. In the next few pages we will verify (3.9.6). Readers may skip the verification without a loss of comprehension.

Verification. The corresponding wavelet, from (3.3.10), is

$$\begin{aligned}\hat{\psi}(\omega) &= e^{-i\omega/2}\overline{\hat{h}\left(\frac{\omega}{2}+\pi\right)}\hat{\varphi}\left(\frac{\omega}{2}\right)\\ &= e^{-i\omega/2}\sum_k \hat{\varphi}(\omega + (2k+1)2\pi)\hat{\varphi}\left(\frac{\omega}{2}\right)\\ &= e^{-i\omega/2}[\hat{\varphi}(\omega + 2\pi) + \hat{\varphi}(\omega - 2\pi)]\hat{\varphi}\left(\frac{\omega}{2}\right)\\ &= e^{-i\omega/2}\left\{\left[\int_{\omega+\pi}^{\omega+3\pi} p(\zeta)\,d\zeta\right]^{1/2}\right.\\ &\quad \left. + \left[\int_{\omega-3\pi}^{\omega-\pi} p(\zeta)\,d\zeta\right]^{1/2}\right\}\left(\int_{(\omega/2)-\pi}^{(\omega/2)+\pi} p(\zeta)\,d\zeta\right)^{1/2}.\end{aligned} \tag{3.9.7}$$

In the equation above we used the fact that

$$\overline{\hat{h}\left(\frac{\omega}{2}\right)} = \sum_k \hat{\varphi}(\omega + 4k\pi);$$

thus

$$\overline{\hat{h}\left(\frac{\omega}{2} + \pi\right)} = \overline{\hat{h}\left(\frac{\omega + 2\pi}{2}\right)} = \sum_k \hat{\varphi}(\omega + 2\pi + 4k\pi).$$

Let us derive $\hat{\psi}(\omega)$ of (3.9.6) for positive and negative ω, respectively.

Case $\boldsymbol{\omega \geq 0}$**.** For positive frequency the first term in the braces of (3.9.7),

$$\int_{\omega+\pi}^{\omega+3\pi} p(\zeta)\, d\zeta = 0.$$

Hence

$$\begin{aligned} \hat{\psi}(\omega) &= e^{-i\omega/2} \left\{ \int_{\omega-3\pi}^{\omega-\pi} p(\zeta)\, d\zeta \int_{(\omega/2)-\pi}^{(\omega/2)+\pi} p(\zeta)\, d\zeta \right\}^{1/2} \\ &= e^{-i\omega/2} \left\{ \int_{(\omega/2)-\pi}^{\omega-\pi} p(\zeta)\, d\zeta \right\}^{1/2}. \end{aligned} \tag{3.9.8}$$

This second equality in the previous equation can be shown by dividing ω into four intervals:

$0 \leq \omega \leq \pi - \varepsilon$

$\pi - \varepsilon \leq \omega < \pi + \varepsilon$

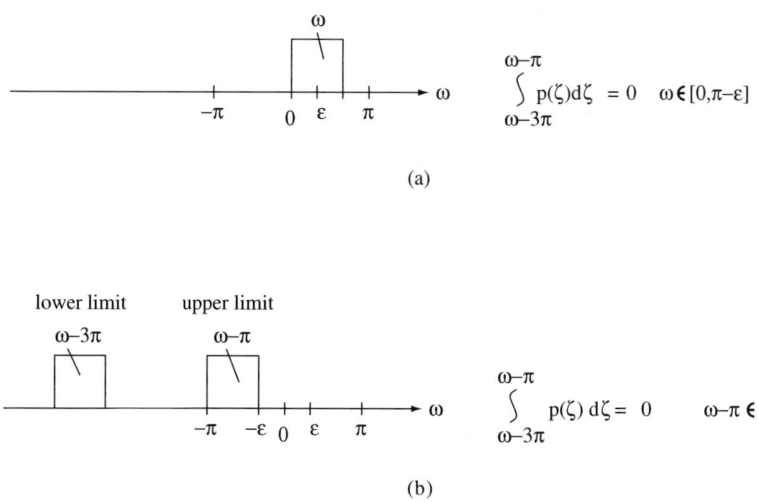

FIGURE 3.9 Meyer wavelets, interval (i).

$$\pi + \varepsilon < \omega < 2\pi + 2\varepsilon$$
$$2\pi + 2\varepsilon < \omega.$$

In the following discussions, we will always use the fact that supp $p(\zeta) = [-\varepsilon, \varepsilon]$. For ease of geometric explanation, we refer to Figs. 3.9 to 3.12.

(i) In the first interval, the possible ω varies within $[0, \pi - \varepsilon]$ as marked in Fig. 3.9*a*, and the upper limit of the integral, $\omega - \pi$, is marked in Fig. 3.9*b* along with that of lower limit $\omega - 3\pi$. Clearly,

$$\int_{\omega-3\pi}^{\omega-\pi} p(\zeta)\, d\zeta = 0.$$

This is because the upper limit of the integral, $(\omega - \pi) \in [-\pi, -\varepsilon]$, which is beyond the lower limit of supp $p(\zeta) = [-\varepsilon, \varepsilon]$. This vanishing integral leads to a zero product in (3.9.8), so $\hat{\psi}(\omega) = 0$.

(ii) In the second interval, we refer to Fig. 3.10:

$$\int_{\omega-3\pi}^{\omega-\pi} p(\zeta)\, d\zeta \int_{(\omega/2)-\pi}^{(\omega/2)+\pi} p(\zeta)\, d\zeta = \int_{(\omega/2)-\pi}^{\omega-\pi} p(\zeta)\, d\zeta.$$

As illustrated in Fig. 3.10*b*, the integration of $\int_{(\omega/2)-\pi}^{(\omega/2)+\pi}$ covers $[-\varepsilon, \varepsilon]$ for all possible ω in the lower and upper limits. Thus we have

$$\int_{(\omega/2)-\pi}^{(\omega/2)+\pi} p(\zeta)\, d\zeta = 1.$$

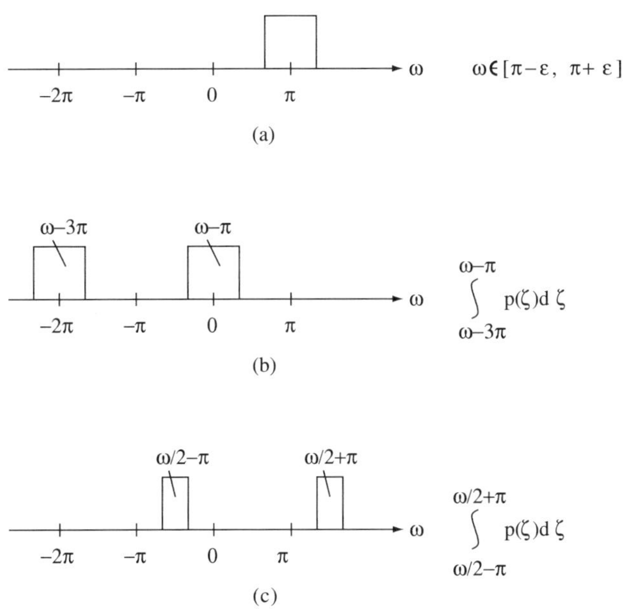

FIGURE 3.10 Meyer wavelets, interval (ii).

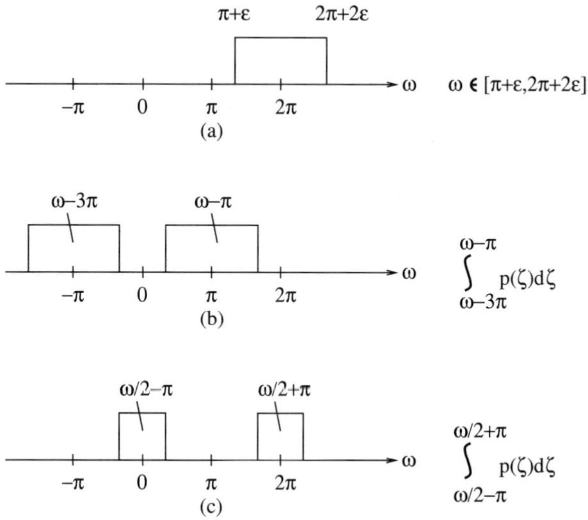

FIGURE 3.11 Meyer wavelets, interval (iii).

The common nonzero contributions are only

$$\int_{(\omega/2)-\pi}^{\omega-\pi} p(\zeta)\, d\zeta.$$

(iii) In the third interval, we refer to Fig. 3.11.

$$\int_{\omega-3\pi}^{\omega-\pi} p(\zeta)\, d\zeta \int_{(\omega/2)-\pi}^{(\omega/2)+\pi} p(\zeta)\, d\zeta = \int_{(\omega/2)-\pi}^{\omega-\pi} p(\zeta)\, d\zeta.$$

This is because the integral of $\int_{\omega-3\pi}^{\omega-\pi}$ covers $[-\varepsilon, \varepsilon]$ for all possible lower and upper limits as depicted in Fig. 3.11*b*. As a result

$$\int_{\omega-3\pi}^{\omega-\pi} p(\zeta)\, d\zeta = 1.$$

Hence the common nonzero contributions are only

$$\int_{(\omega/2)-\pi}^{\omega-\pi} p(\zeta)\, d\zeta.$$

(iv) The fourth interval, as the first, gives no contribution. This is due to the fact that the lower limit of the integral, $(\omega/2) - \pi = \varepsilon$ as in Fig. 3.12.

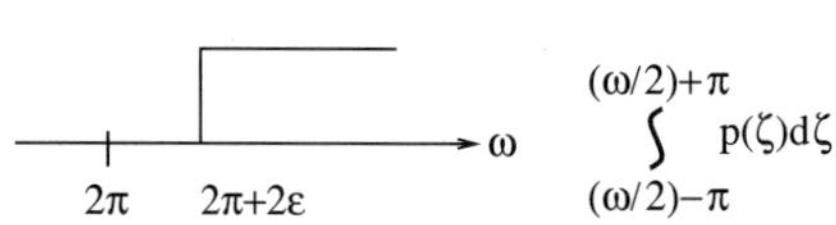

FIGURE 3.12 Meyer wavelets, interval (iv).

Combining (i) through (iv), we obtain

$$\hat{\psi}(\omega) = e^{-i\omega/2}\left\{\int_{(\omega/2)-\pi}^{\omega-\pi} p(\zeta)\,d\zeta\right\}^{1/2}.$$

Case $\omega < 0$. For negative frequency, one may repeat the procedures that have been carried out for $\omega \geq 0$, and arrive at

$$\hat{\psi}(\omega) = e^{-i\omega/2}\left\{\int_{\omega+\pi}^{(\omega/2)+\pi} p(\zeta)\,d\zeta\right\}^{1/2}. \tag{3.9.9}$$

We conclude that the Meyer scalet $\hat{\varphi}(\omega)$ is given by (3.9.2) with a support of $[-\pi-\varepsilon, \pi+\varepsilon]$, and the Meyer wavelet $\hat{\psi}(\omega)$ is given by (3.9.6) with a support of $[-2\pi-2\varepsilon, -\pi+\varepsilon] \cup [\pi-\varepsilon, 2\pi+2\varepsilon]$. The following examples will help us to digest the previous derivations.

3.9.2 Meyer Wavelet Family

In this subsection we present the Shannon wavelets and raised cosine wavelets, which are the most popular in the Meyer family.

Shannon Wavelet

CASE 1. SHANNON SCALET. Choose $p = \delta(\omega)$, the Dirac delta function. We may easily verify that all of the three requirements are met:

(i) $\delta(\omega) \geq 0,\ \mathcal{F}^{-1}\{\delta(\omega)\} := \delta(t)|_{t=0} = 1.$
(ii) $\operatorname{supp}\delta(\omega) \subseteq [-\varepsilon, \varepsilon],\ 0 \leq \varepsilon \leq \pi/3.$
(iii) $\int_{-\pi/3}^{\pi/3} \delta(\omega)\,d\omega = 1.$

Now from Theorem 3,

$$|\,\hat{\varphi}(\omega)\,|^2 = \int_{\omega-\pi}^{\omega+\pi} \delta(\zeta)d\zeta = \begin{cases} 1 & \omega-\pi < 0 < \omega+\pi \\ 0 & \text{otherwise.} \end{cases}$$

The simplest choice of $\hat{\varphi}(\omega)$ according to the equation above is

$$\hat{\varphi}(\omega) = 1 \qquad \text{for } -\pi < \omega < \pi.$$

The time domain scalet is

$$\varphi(t) = \mathcal{F}^{-1}\{\hat{\varphi}(\omega)\} = \frac{\sin \pi t}{\pi t}.$$

This is the Shannon sampling function.

CASE 2. SHANNON WAVELET. Notice that

$$\psi(t) \in W_0 \perp V_0$$

and

$$W_0 \oplus V_0 = V_1,$$

which means that

$$\psi(t) \in V_1.$$

Thus

$$\psi(t) = \sum_n g_n \sqrt{2}\varphi(2t - n),$$

where

$$g_n = (-1)^{n-1} h_{1-n}.$$

The bandpass g_n is related to lowpass h_{1-n}, which can be obtained by

$$h_n = \langle \varphi_{0,0}, \varphi_{1,n} \rangle.$$

A quick alternative way of constructing $\psi(t)$ is

$$\psi(t) = 2\varphi(2t - 1) - \varphi\left(t - \tfrac{1}{2}\right). \tag{3.9.10}$$

Show. We can easily verify the following three equations

$$\begin{aligned} \langle \varphi(t), \psi(t) \rangle &= 0, \\ \langle \psi(t), \psi(t) \rangle &= 1, \\ \langle \psi(t - m)\psi(t - n) \rangle &= 0, \qquad m \neq n. \end{aligned} \tag{3.9.11}$$

Proof 1. A direct approach is to represent $\psi(t)$ in terms of $\varphi(t)$ using (3.9.10). One may verify that the requirements (3.9.11) are all satisfied.

Proof 2.

$$\hat{\psi}(\omega) = e^{-i(\omega/2)} \left[\hat{\varphi}\left(\frac{\omega}{2}\right) - \hat{\varphi}(\omega) \right].$$

Notice that $\hat{\varphi}(\omega)$ has support $[-\pi, \pi)$, and $\hat{\varphi}(\omega/2)$ has support $[-2\pi, 2\pi)$. Therefore $\hat{\psi}(\omega)$ has support $[-2\pi, -\pi) \cup [\pi, 2\pi)$, which is disjoint with $[-\pi, \pi)$ of $\hat{\varphi}(\omega)$. As a result $\langle \hat{\varphi}(\omega), \hat{\psi}(\omega) \rangle = 0$. From Parseval's law

$$\langle \varphi(t), \psi(t) \rangle = \frac{1}{2\pi} \langle \hat{\varphi}(\omega), \hat{\psi}(\omega) \rangle = 0. \qquad \square$$

Figure 3.13 illustrates the Shannon scalet (the sinc function) and wavelet.

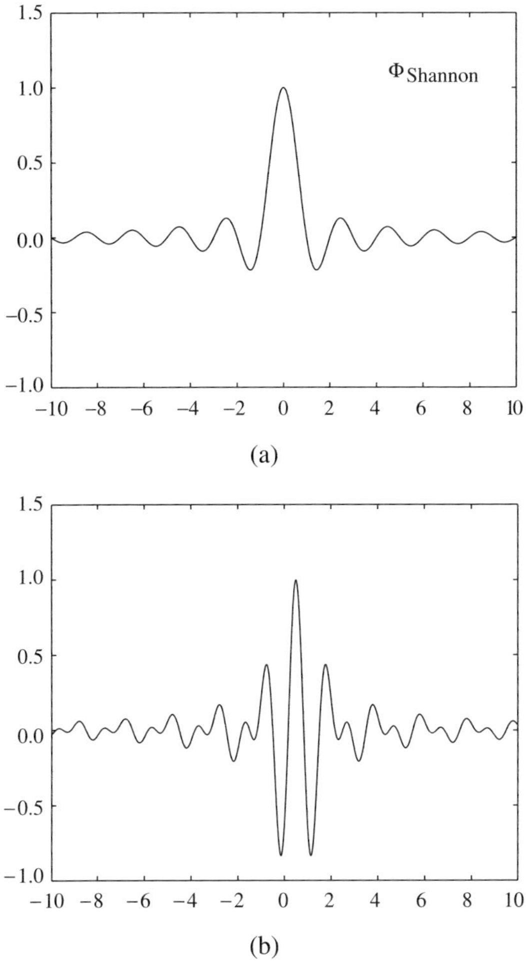

FIGURE 3.13 (*a*) Shannon scalet, namely the sinc function; (*b*) Shannon wavelet.

The next example, the raised cosine wavelet, is perhaps the most practical example in the family of Meyer wavelets. The raised cosine pulses have been used for several decades in digital communication to eliminate the intersymbol interference by means of their orthogonality.

Raised Cosine Wavelet. The raised cosine wavelet was proposed by Meyer. Given here is the scalet

$$\hat{\varphi}(\omega) = \begin{cases} 1, & |\omega| \leq \frac{2}{3}\pi \\ \cos[\frac{\pi}{2}(\frac{3}{2\pi}|\omega| - 1)], & \frac{2\pi}{3} < |\omega| \leq \frac{4\pi}{3} \\ 0 & \text{otherwise.} \end{cases} \tag{3.9.12}$$

Find the corresponding distribution $p(\omega)$, and verify the wavelet $\hat{\psi}(\omega)$

$$e^{i(\omega/2)}\,\hat{\psi}(\omega) = \begin{cases} -\sin\frac{3}{8}\omega, & -\frac{8}{3}\pi < \omega \le -\frac{4}{3}\pi \\ -\cos\frac{3}{4}\omega, & -\frac{4}{3}\pi < \omega \le -\frac{2}{3}\pi \\ -\cos\frac{3}{4}\omega, & \frac{2}{3}\pi < \omega \le \frac{4}{3}\pi \\ \sin\frac{3}{8}\omega, & \frac{4}{3}\pi < \omega \le \frac{8}{3}\pi \\ 0 & \text{otherwise.} \end{cases} \tag{3.9.13}$$

Solution This is a special case of $\varepsilon = \pi/3$. From

$$\hat{\varphi}(\omega) = \left\{ \int_{\omega-\pi}^{\omega\,|\,\pi} p(\zeta)\,d\zeta \right\}^{1/2}$$

we have

$$\hat{\varphi}^2(\omega) = \int_{\omega-\pi}^{\omega+\pi} p(\zeta)\,d\zeta.$$

Differentiating by the proper procedure for differentiation of an integral, we have

$$2\hat{\varphi}\hat{\varphi}'(\omega) = p(\omega+\pi) - p(\omega-\pi). \tag{3.9.14}$$

(i) Case $\omega > 0$. For $\omega > 0$ we have in (3.9.12) $|\,\omega\,| = \omega$. Therefore

$$-2\cos\left[\frac{\pi}{2}\left(\frac{3}{2\pi}\omega - 1\right)\right]\sin\left[\frac{\pi}{2}\left(\frac{3}{2\pi}\omega - 1\right)\right]\frac{3}{4} = p(\omega+\pi) - p(\omega-\pi).$$

Note that the left-hand side is readily simplified by the double angle formula obtained from trigonometry. On the other hand, $p(x) \neq 0$ only for $|\,x\,| \le \pi/3$. In the case of $2\pi/3 < \omega < 4\pi/3$ in (3.9.12), p is outside its support, so $p(\omega+\pi) = 0$. As a result we must drop the first term of the right-hand side, yielding

$$p(\omega-\pi) = \frac{3}{4}\sin\left[\pi\left(\frac{3}{2\pi}\omega - 1\right)\right].$$

Changing the argument from $\omega - \pi$ to ω, we have

$$p(\omega) = \begin{cases} \frac{3}{4}\sin\left(\pi\left(\frac{3\omega}{2\pi} + \frac{1}{2}\right)\right), & 0 \le \omega \le \frac{\pi}{3} \\ 0, & \omega > \frac{\pi}{3}. \end{cases} \tag{3.9.15}$$

(ii) Case $\omega < 0$. By replacing $-\omega$ for $|\,\omega\,|$ in (3.9.12), we obtain from (3.9.14),

$$-2\cos\left[\frac{\pi}{2}\left(\frac{3}{2\pi}\omega + 1\right)\right]\sin\left[\frac{\pi}{2}\left(\frac{3}{2\pi}\omega + 1\right)\right]\frac{3}{4} = p(\omega+\pi) - p(\omega-\pi).$$

Since $\omega < 0$, we ought to drop the second term on the right-hand side in view of the range of ω in (3.9.12). This results in

$$p(\omega + \pi) = -\frac{3}{4}\sin\left[\pi\left(\frac{3}{2\pi}\omega + 1\right)\right]$$

or

$$p(\omega) = \frac{3}{4}\sin\left[\pi\left(-\frac{3}{2\pi}\omega + \frac{1}{2}\right)\right], \qquad -\frac{\pi}{3} < \omega \le 0. \tag{3.9.16}$$

In combining (3.9.15) and (3.9.16), we finally have

$$p(\omega) = \begin{cases} \frac{3}{4}\sin\left[\pi\left(\frac{3}{2\pi}|\omega| + \frac{1}{2}\right)\right], & |\omega| < \frac{\pi}{3} \\ 0 & \text{otherwise.} \end{cases}$$

Next we evaluate the wavelet in the transform domain. Recall earlier in this section that the wavelet can be expressed in terms of p as follows:

For $\omega > 0$,

$$\hat{\psi}(\omega) = e^{-i\omega/2}\left\{\int_{(\omega/2)-\pi}^{\omega-\pi} p(\zeta)d\zeta\right\}^{1/2}, \qquad \omega \in [\pi - \varepsilon, 2\pi + \varepsilon].$$

Consider the integral

$$I := \int_{(\omega/2)-\pi}^{\omega-\pi} p(\Omega)d\Omega = \int_{(\omega/2)-\pi}^{\omega-\pi} \frac{3}{4}\sin\left(\frac{\pi}{2} + \frac{3\Omega}{2}\right) d\Omega.$$

Note that $p(\Omega) \neq 0$ only for $-\pi/3 < \Omega \le \pi/3$. Let

$$\alpha = \frac{3\Omega}{2} + \frac{\pi}{2}, \quad \text{then} \quad d\alpha = \frac{3}{2}d\Omega.$$

The nonzero interval

$$\Omega : -\frac{\pi}{3} \to \frac{\pi}{3} \quad \Rightarrow \quad \alpha : 0 \to \pi.$$

The corresponding lower and upper limits are

$$\Omega = \frac{\omega}{2} - \pi \;\Rightarrow\; \alpha = \frac{3}{2}\left(\frac{\omega}{2} - \pi\right) + \frac{\pi}{2} = \frac{3\omega}{4} - \pi,$$

$$\Omega = \omega - \pi \;\Rightarrow\; \alpha = \frac{3}{2}(\omega - \pi) + \frac{\pi}{2} = \frac{3\omega}{2} - \pi.$$

Formally, the integral

$$I = \frac{2}{3}\frac{3}{4}\int_{\frac{3}{4}\omega-\pi}^{\frac{3}{2}\omega-\pi} \sin\alpha d\alpha = -\frac{1}{2}\cos\alpha\Big|_{\frac{3}{4}\omega-\pi}^{\frac{3}{2}\omega-\pi}. \tag{3.9.17}$$

The previous equation is only a formal expression of the integral I. The integral limits must be carefully assigned because of the confined nonzero support of $p(\Omega)$. To this end, we divide the integral I into four subintegrals as depicted in Fig. 3.14–3.16, and write

$$I = I_1 + I_2 + I_3 + I_4.$$

(i) $0 < \omega \leq 2\pi/3$. The upper limit $\alpha : -\pi \to 0$, as indicated in Fig. 3.14. Hence $I_1 = 0$.

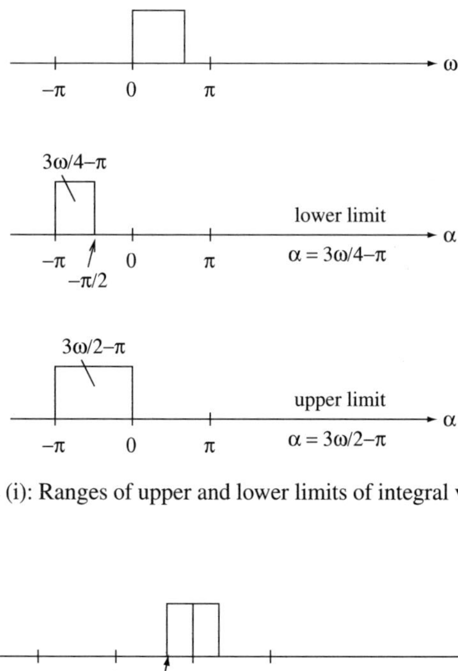

FIGURE 3.14 Case (i): Ranges of upper and lower limits of integral variable α for $0 < \omega \leq 2\pi/3$.

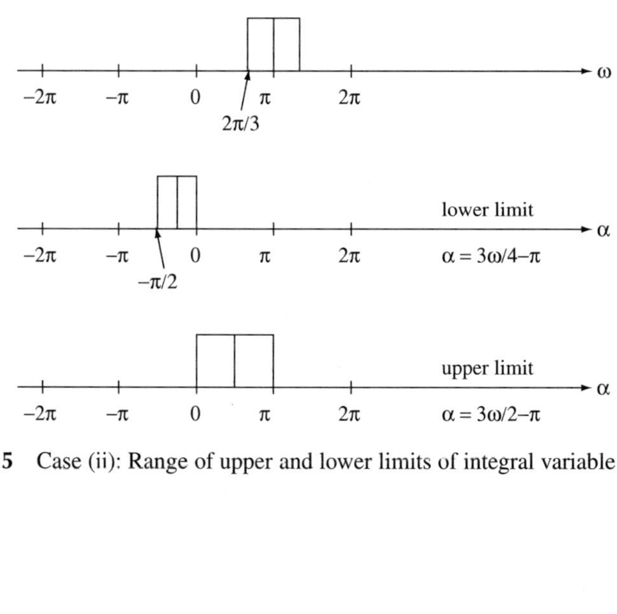

FIGURE 3.15 Case (ii): Range of upper and lower limits of integral variable α for $2\pi/3 < \omega \leq 4\pi/3$.

(ii) $2\pi/3 < \omega \leq 4\pi/3$. As indicated in Fig. 3.15, the lower limit α ranges $-\pi/2 \to 0$ and the upper limit α ranges $0 \to \pi$. Taking into account the nonzero support of $\alpha : 0 \to \pi$, we obtain

$$\begin{aligned} I_2 &= -\tfrac{1}{2}\cos\alpha\Big|_0^{(3\omega/2)-\pi} = \tfrac{1}{2}\big[1 - \cos\big(\tfrac{3}{2}\omega - \pi\big)\big] \\ &= \tfrac{1}{2}\big[\cos\big(\tfrac{3}{2}\omega\big) + 1\big] = \cos^2\big(\tfrac{3}{4}\omega\big). \end{aligned}$$

(iii) $4\pi/3 < \omega \leq 8\pi/3$. As indicated in Fig. 3.16, the lower limit α ranges $0 \to \pi$. That is, it ranges fully in the nonzero support. Therefore we take the complete expression as the lower limit. The upper limit α ranges: $\pi \to 2\pi$. Taking into account the nonzero support of $\alpha : 0 \to \pi$, we adopt the upper limit $\alpha = \pi$. Therefore

$$\begin{aligned} I_3 &= -\tfrac{1}{2}\cos\alpha\Big|_{(3\omega/4)-\pi}^{\pi} \\ &= -\tfrac{1}{2}\big[-1 - \cos\big(\tfrac{3}{4}\omega - \pi\big)\big] \\ &= \tfrac{1}{2}\big(1 + \cos\big(\tfrac{3}{4}\omega - \pi\big)\big) = \tfrac{1}{2}\big[1 - \cos\tfrac{3}{4}\omega\big] \\ &= \sin^2\big(\tfrac{3}{8}\omega\big). \end{aligned}$$

(iv) $\omega > 8\pi/3$. It can be verified easily that the lower limit $\alpha \geq \pi$. Therefore

$$I_4 = 0.$$

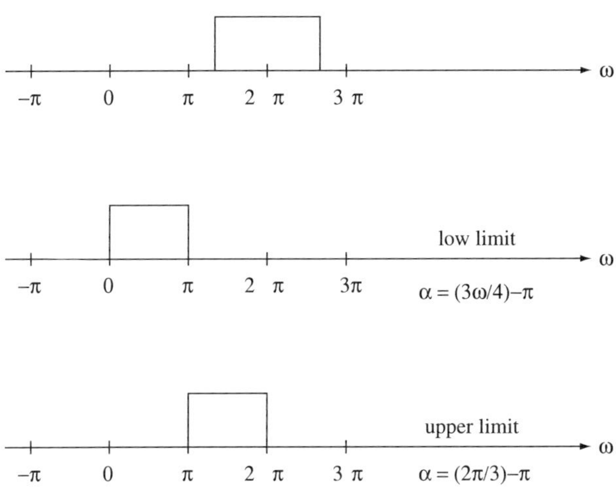

FIGURE 3.16 Case (iii): Ranges of lower and upper limits of integral variable α for $4\pi/3 < \omega \leq 8\pi/3$.

In conclusion, we may write

$$\hat{\psi}(\omega) = e^{-i(\omega/2)} \begin{cases} -\cos\frac{3}{4}\omega, & \frac{2\pi}{3} < \omega \le \frac{4}{3}\pi \\ \sin\frac{3}{8}\omega, & \frac{4\pi}{3} < \omega \le \frac{8}{3}\pi \\ 0 & \text{otherwise.} \end{cases}$$

Notice that in the square root, $\{\int p(\Omega)d\Omega\}^{1/2}$, a negative sign in $\sqrt{\cos^2(3\omega/4)}$ and a positive sign in $\sqrt{\sin^2(3\omega/4)}$ were chosen to make $\hat{\psi}(\omega)$ continuous at $\omega = 4\pi/3$.

Now for $\omega < 0$, we have from (3.9.16), (3.9.6) that

$$\begin{aligned} I &= \int_{\omega+\pi}^{(\omega/2)+\pi} \frac{3}{4}\sin\left[\pi\left(-\frac{3}{2\pi}\Omega + \frac{1}{2}\right)\right] d\Omega \\ &= \frac{3}{4}\int_{\omega+\pi}^{(\omega/2)+\pi} \sin\left(\frac{\pi}{2} - \frac{3}{2}\Omega\right) d\Omega. \end{aligned}$$

Again, we recall that $p(\Omega) \neq 0$ only for $-\pi/3 \le \Omega \le \pi/3$. Let $\beta = \pi/2 - 3\Omega/2$. $-\pi/3 < \Omega \le \pi/3 \Rightarrow \pi \ge \beta \ge 0$. The corresponding lower and upper limits for variable Ω and β are

$$\Omega = \omega + \pi, \qquad \beta = \frac{\pi}{2} - \frac{3}{2}(\omega + \pi) = -\pi - \frac{3}{2}\omega,$$

$$\Omega = \frac{\omega}{2} + \pi, \qquad \beta = \frac{\pi}{2} - \frac{3}{2}\left(\frac{\omega}{2} + \pi\right) = -\pi - \frac{3}{4}\omega.$$

Formally, we have

$$I = \frac{1}{2}\int_{-\pi-(3\omega/4)}^{-\pi-(3\omega/2)} \sin\beta d\beta = -\frac{1}{2}\cos\beta\Big|_{-(3\omega/4)-\pi}^{-(3\omega/2)-\pi}. \tag{3.9.18}$$

It can be seen immediately that if we replace $-\omega$ with ω in (3.9.18) we obtain (3.9.17). Therefore similar expressions in integrals I_2 and I_3 can be derived for negative ω. The detailed work is left as an exercise for the reader. When taking square roots in (3.9.9), we need to select the right sign, which leads to

$$\hat{\psi}(\omega) = e^{-i\omega/2} \begin{cases} -\cos\frac{3}{4}\omega, & -\frac{4}{3}\pi < \omega \le -\frac{2}{3}\pi \\ -\sin\frac{3}{8}\omega, & -\frac{8}{3}\pi < \omega \le -\frac{4}{3}\pi \\ 0 & \text{otherwise.} \end{cases}$$

If the positive sign were selected, we would not be able to meet the orthogonality condition between the scalets and wavelets, namely

$$\sum_{k=-\infty}^{\infty} \hat{\psi}(\omega + 2k\pi)\overline{\hat{\varphi}(\omega + 2k\pi)} = 0.$$

In summary,

$$e^{i(\omega/2)}\,\hat{\psi}(\omega)=\begin{cases}-\sin\frac{3}{8}\omega, & -\frac{8}{3}\pi<\omega\le-\frac{4}{3}\pi\\ -\cos\frac{3}{4}\omega, & -\frac{4}{3}\pi<\omega\le-\frac{2}{3}\pi\\ -\cos\frac{3}{4}\omega, & \frac{2}{3}\pi<\omega\le\frac{4}{3}\pi\\ \sin\frac{3}{8}\omega, & \frac{4}{3}\pi<\omega\le\frac{8}{3}\pi\\ 0 & \text{otherwise.}\end{cases}$$

The nonzero support for $\hat{\psi}(\omega)$ of the raised cosine wavelet is a special case of Theorem 4 with $\varepsilon=\pi/3$. A plot of the Meyer raised cosine scalet $\hat{\varphi}(\omega)$ and wavelet $\hat{\psi}(\omega)$ is illustrated in Fig. 3.17. Readers can verify that in addition to (3.3.5), the

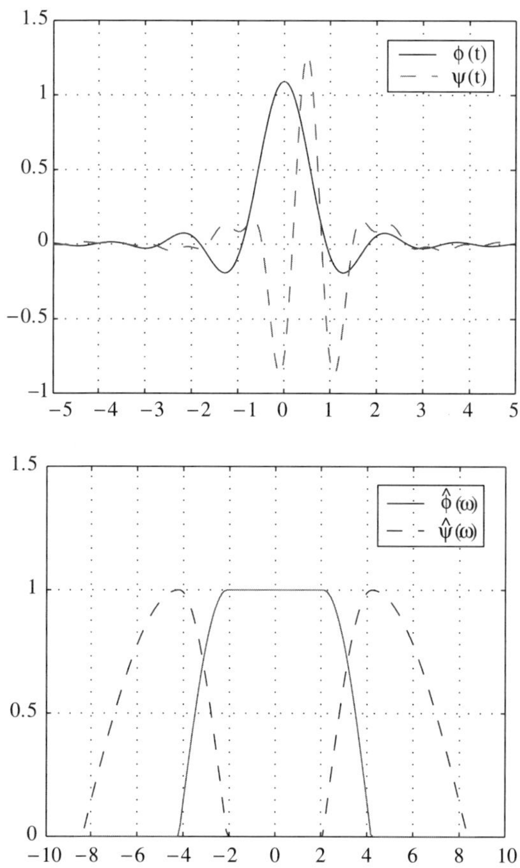

FIGURE 3.17 Raised cosine scalet and wavelet.

wavelet also satisfies

$$\sum_k | \hat{\psi}(\omega + 2k\pi) |^2 = 1.$$

The corresponding scalet $\varphi(t)$ and wavelet $\psi(t)$ are also plotted in Fig. 3.17.

3.9.3 Other Examples of Meyer Wavelets

The following examples are the individual probability density functions (pdf) that can be employed in (3.9.2) to produce the corresponding scalets in the Meyer family [9]. It seems that they possess more mathematical elegance than they do usefulness in engineering applications.

Example 1

$$p(\omega) = \begin{cases} \frac{1}{2}\varepsilon, & |\omega| < \varepsilon, 0 < \varepsilon \le \frac{\pi}{3} \\ 0 & \text{otherwise.} \end{cases}$$

Example 2

$$p(\omega) = \begin{cases} \frac{|\varepsilon - \omega|}{\varepsilon}, & |\omega| < \varepsilon \\ 0 & \text{otherwise.} \end{cases}$$

Example 3

$$p(\omega) = \begin{cases} C_\varepsilon e^{-\varepsilon^2/(\varepsilon^2 - \omega^2)}, & |\omega| \le \varepsilon \le \pi/3 \\ 0 & \text{otherwise.} \end{cases}$$

The time domain scalet belongs to C_∞. It can be shown that the kth derivative is bounded by

$$\varphi^{(k)}(t) \le \frac{C_{pk}}{(1 + |t|)^p}.$$

3.10 MALLAT'S DECOMPOSITION AND RECONSTRUCTION

Mallat's decomposition and reconstruction algorithm relates the expansion coefficients of a function at different levels [10]. Using this algorithm, one can develop a fast wavelet transform (FWT).

3.10.1 Reconstruction

For space V_1 we have two distinct orthonormal bases

1. $\{\sqrt{2}\varphi(2t - \ell)\}_{\ell=-\infty}^{\infty}$, V_1.
2. $\{\varphi(t - n), \psi(t - m)\}_{n,m=-\infty}^{\infty}$, $V_0 \oplus W_0$.

Hence for $\forall f \in V_1$ we may write

$$f(t) = \sum_{\ell} a_\ell^1 \sqrt{2}\varphi(2t - \ell) \tag{3.10.1}$$

$$= \sum_n a_n^0 \varphi(t - n) + \sum_n b_n^0 \psi(t - n). \tag{3.10.2}$$

By the standard dilation equations

$$\varphi(t - n) = \sqrt{2} \sum_k h_k \varphi(2t - 2n - k),$$

$$\psi(t - n) = \sqrt{2} \sum_k (-1)^k h_{1-k} \varphi(2t - 2n - k).$$

We substitute these two equations into (3.10.2) to get

$$\begin{aligned} f(t) &= \sum_n a_n^0 \sqrt{2} \sum_k h_k \varphi(2t - 2n - k) \\ &\quad + \sum_n b_n^0 \sqrt{2} \sum_k (-1)^k h_{1-k} \varphi(2t - 2n - k) \\ &= \sum_n a_n^0 \sum_\ell h_{\ell-2n} \sqrt{2}\varphi(2t - \ell) + \sum_n b_n^0 \sum_\ell (-1)^\ell h_{1-\ell+2n} \sqrt{2}\varphi(2t - \ell) \\ &= \sum_\ell \left\{ \sum_n h_{\ell-2n} a_n^0 + \sum_n (-1)^\ell h_{1-\ell+2n} b_n^0 \right\} \sqrt{2}\varphi(2t - \ell), \end{aligned}$$

where $\ell = 2n + k$. Comparing with (3.10.1), we obtain

$$a_\ell^1 = \sum_n h_{\ell-2n} a_n^0 + \sum_n (-1)^\ell h_{1-\ell+2n} b_n^0. \tag{3.10.3}$$

Equation (3.10.3) suggests an algorithm that computes a coefficient from both the scalets and wavelets on a lower level (e.g., level 0).

3.10.2 Decomposition

In contrast to reconstruction, decomposition works from a high level down to a low level. We start with an expansion at level zero

$$f(t) = \sum_n [a_n \varphi(t - n) + b_n \psi(t - n)]. \tag{3.10.4}$$

Multiplying (3.10.4) by $\varphi(t - m)$ and integrating both sides, we have

$$\int f(t)\varphi(t - m)\, dt = a_m,$$

where we have employed the orthogonality conditions

$$\langle \varphi_{0,n} \varphi_{0,m} \rangle = \delta_{m,n},$$
$$\langle \psi_{0,n} \varphi_{0,m} \rangle = 0.$$

Thus

$$\begin{aligned}
a_n^0 &= \int_{-\infty}^{\infty} f(t)\varphi(t-n)\,dt \\
&= \int_{-\infty}^{\infty} f(t) \sum_k h_k \sqrt{2}\varphi(2t-2n-k)\,dt \\
&= \sum_k h_k \int_{-\infty}^{\infty} f(t)\sqrt{2}\varphi(2t-(2n+k))\,dt \\
&= \sum_k h_k \langle f, \varphi_{1,2n+k} \rangle \\
&= \sum_k h_k a_{2n+k}^1 \\
&= \sum_l a_l^1 h_{l-2n},
\end{aligned}$$

or equivalently

$$a_n^0 = \sum_k a_k^1 h_{k-2n}.$$

Similarly

$$b_n^0 = \sum a_k^1 (-1)^k h_{1-k+2n}.$$

The previous two equations can be generalized to relate levels J and $J-1$, namely

$$a_n^{J-1} = \sum_k a_k^J h_{k-2n}, \tag{3.10.5}$$

$$b_n^{J-1} = \sum a_k^J (-1)^k h_{1-k+2n}. \tag{3.10.6}$$

Repeating these processes, we have the following diagrams respectively:

Decomposition

$$\begin{array}{ccccccccc}
 & \nearrow b^{N-1} & \nearrow b^{N-2} & \nearrow \cdots & \nearrow b_n^0 \\
a_n^N \longrightarrow & a^{N-1} \longrightarrow & a^{N-2} \rightarrow & \cdots \rightarrow & a_n^0.
\end{array}$$

Reconstruction

$$\begin{array}{ccccccccc} b_n^0 & \searrow & b_n^1 & \searrow & \cdots & & b_n^{M-1} & & \searrow \\ a_n^0 & \longrightarrow & a_n^1 & \longrightarrow & a_n^2 & \cdots & a_n^{M-1} & \longrightarrow & a_n^M. \end{array}$$

Thus we need to evaluate only once the coefficients from $f(t)$ at the finest scale, or highest level, of interest.

3.11 PROBLEMS

3.11.1 Exercise 1

1. For $\varphi(t) = \sum_n a_n \theta(t + 1 - n)$ of the scalet of the Franklin wavelets,

  (a) Evaluate a_n for $n = 0, 1, \ldots, 9$.

  (b) Plot the scalet $\varphi(t)$.

2. The Franklin wavelets are also referred to as the Battle–Lemarie wavelets of the first order ($N = 1$), which are derived from the B-splines of the first order. In general

$$\hat{\theta_N}(\omega) = e^{i[(N+1)\omega/2]} \left(\frac{\sin \omega/2}{\omega/2}\right)^{N+1}.$$

We have derived the associated summation for $N = 1$, namely

$$|\,\hat{\theta}^{\dagger}(\omega)\,|^2 = 1 - \frac{2}{3}\sin^2\left(\frac{\omega}{2}\right) = \frac{\left(1 + 2\cos^2(\omega/2)\right)}{3}.$$

Derive the corresponding expression in terms of $\cos^4(\omega/2), \cos^2(\omega/2)$ for $N = 2$.

3.11.2 Exercise 2

1. Use the Fourier transform to verify that

$$\int_{-\infty}^{\infty} dt \left(\frac{\sin t}{t}\right)^3 = \frac{3\pi}{4}$$

and

$$\int_{-\infty}^{\infty} dt \left(\frac{\sin t}{t}\right)^4 = \frac{2\pi}{3}.$$

2. Given a time domain function $f(t) = \exp\{-(a - ib)t^2\}$, find its Fourier transform.

3. Show that

$$\sum_k \hat{\psi}(\omega + 2k\pi)\overline{\hat{\varphi}(\omega + 2k\pi)} = 0.$$

4. Show that

$$\langle f, p \rangle = \frac{1}{2\pi} \langle F(\omega), P(\omega) \rangle,$$

where

$$\langle f, p \rangle = \int dt \overline{f(t)} p(t)$$

and

$$\langle F(\omega), P(\omega) \rangle = \int d\omega \overline{F}(\omega) P(\omega).$$

The Fourier transform pair is defined as

$$\begin{cases} P(\omega) = \int dt p(t) e^{-j\omega t} \\ p(t) = \frac{1}{2\pi} \int d\omega P(\omega) e^{j\omega t}. \end{cases}$$

5. The scalet can be expressed as the "filter banks" of

$$\varphi(x) = \sum_{n=-\infty}^{\infty} h_n \varphi_{1,n},$$

where

$$\varphi_{1,n} = \sqrt{2}\varphi(2x - n).$$

Evaluate the first 16 filters of the Franklin wavelets, that is,

$$h_n, \qquad n = 0, 1, 2, \ldots, 15.$$

6. The Franklin wavelet can be represented as

$$\psi(x) = \sum_k b_k \theta_c(2x - k - 1),$$

where

$$\theta_c(x) = \begin{cases} 1 - |x| & \text{for } |x| \leq 1 \\ 0 & \text{otherwise.} \end{cases}$$

(a) Evaluate $b_k, k = 0, 1, 2, \ldots, 15$.

(b) Plot $\psi(x)$.

3.11.3 Exercise 3

1. Construct the Battle–Lemarie wavelets of $N = 2$, given

$$|\hat{\theta}_2^\dagger|^2 = \frac{1}{15}\left[2\cos^4\left(\frac{\omega}{2}\right) + 11\cos^2\left(\frac{\omega}{2}\right) + 2\right].$$

   (a) Evaluate a_n.
   (b) Construct the scalet $\varphi(t)$, and plot $\varphi(t)$.
   (c) Construct the wavelet $\psi(t)$.

   You may verify your answers with the figures in the text.

2. For the Franklin wavelets that you have constructed, verify numerically
   (a) $\int \varphi(t)dt = 1$.
   (b) $\int \psi(t)dt = 0$.
   (c) $\int \varphi(t)\varphi(t-1)dt = 0$.
   (d) $\int \psi(t)\varphi(t)dt = 0$.

3. Using the iterative algorithm, compute and plot the Daubechies wavelets for $N = 2$. You may compare your figures with those in the text.

4. Plot in the frequency domain:
   (a) The Franklin scalets and wavelets.
   (b) The Battle–Lemarie scalets and wavelets of order $N = 2$.
   (c) The Daubechies scalets and wavelets of order $N = 2$.

3.11.4 Exercise 4

1. Calculate the exact values of $\varphi(\frac{1}{2}), \varphi(\frac{1}{4}), \psi(\frac{1}{2})$, and $\psi(\frac{1}{8})$ for Daubechies wavelets, $N = 2$.

2. Construct and plot Daubechies' wavelets φ and ψ of $N = 3$. Verify your $\psi(t)$ and $\varphi(t)$ numerically by calculating

$$\int \varphi(t)\, dt = ?$$

$$\int \psi(t)\, dt = ?$$

   Note: You may need to use Table 3.3.

3. Construct and plot the Coifman wavelets φ and ψ of $L = 4$. Use the φ and ψ that you have obtained to evaluate

$$\int \cos t\varphi(t)\, dt = ?$$

$$\int t^3\psi(t)\, dt = ?$$

   Note: You may find Table 3.5 useful.

4. The Shannon scalets and wavelets are respectively

$$\varphi(t) = \frac{\sin \pi t}{\pi t},$$

$$\psi(t) = \frac{\sin 2\pi(t - 1/2) - \sin \pi(t - 1/2)}{\pi(t - 1/2)}.$$

Plot both $\varphi(t)$ and $\psi(t)$.

5. Prove analytically that $\int \varphi(t)\psi(t)\, dt = 0$.

6. The dilation equations for the scalets and wavelets are, respectively,

$$\varphi(x) = \sum_k h_k \sqrt{2}\varphi(2t - k),$$

$$\psi(x) = \sum_k g_k \sqrt{2}\varphi(2t - k).$$

For Daubechies' wavelets of order $N = 2$

$$h_0 = \tfrac{1+\sqrt{3}}{4\sqrt{2}} = 0.4829629,$$

$$h_1 = \tfrac{3+\sqrt{3}}{4\sqrt{2}} = 0.8365163,$$

$$h_2 = \tfrac{3-\sqrt{3}}{4\sqrt{2}} = 0.2241439,$$

$$h_3 = \tfrac{1-\sqrt{3}}{4\sqrt{2}} = -0.1294095.$$

Find the nonzero bandpass filter coefficients g_k.

7. Compute the first moment of Daubechies scalet of $N = 2$, $M_1 := \int_0^3 x\varphi(x)\, dx = ?$ Then evaluate $\varphi(M_1) = ?$, $\varphi(M_1 + 1) = ?$, $\varphi(M_1 + 2) = ?$

8. Derive and plot the raised cosine scalets and wavelets.

(a) Derive $\hat{\psi}(\omega)$ using $\hat{\varphi}(\omega)$ of (3.9.12).

(b) Analytically derive $\varphi(t)$ and $\psi(t)$ from $\hat{\varphi}(\omega)$ and $\hat{\psi}(\omega)$.

(c) Plot $\varphi(t)$ and $\psi(t)$.

BIBLIOGRAPHY

[1] I. Daubechies, *Ten Lectures on Wavelets*, SIAM, Philadelphia, 1992.

[2] Y. Meyer, *Wavelets: Algorithms and Applications*, R. D. Ryan, Transl., SIAM, Philadelphia, 1993.

[3] S. Mallat, *A Wavelet Tour of Signal Processing*, Academic Press, New York, 1998.

[4] G. Walter and X. Shen, *Wavelets and Other Orthogonal Systems*, 2nd ed., CRC Press, New York, 2001.

[5] J. Tian and R. Wells Jr., "Vanishing moments and biorthogonal wavelet systems," *Mathematics in Signal Processing* IV, J. McWhirter and I. Proudler, Eds., Oxford University Press, Oxford, 1998, pp. 301–314.

[6] K. Shanmugam, *Digital and Analog Communication Systems,* John Wiley, New York, 1979.

[7] G. Walter, "Wavelet subspaces with an oversampling property," *Indag. Mathem.*, N.S., **4** (4), 499–507, 1993.

[8] G. Walter, "Translation and dilation invariance in orthogonal wavelets," *Appl. Comput. Harmon. Anal.*, **1**, 344–349, 1994.

[9] P. Lemarie and Y. Meyer, "Ondelettes et bases Hilbertiennes," *Rev. Math. Iberoamericana*, **2**, 1–18, 1986.

[10] S. Mallat, "A theory for multiresolution signal decomposition: The wavelet representation," *IEEE Trans. Pattern Anal. Mach. Intell.*, **11** (7), 674–693, July 1989.

CHAPTER FOUR

Wavelets in Boundary Integral Equations

Numerical treatment of integral equations can be found in classic books [1, 2]. In this chapter the integral equations obtained from field analysis of electromagnetic wave scattering, radiating, and guiding problems are solved by the wavelet expansion method [3–7]. The integral equations are converted into a system of linear algebraic equations. The subsectional bases, namely the pulses or piecewise sinusoidal (PWS) modes, are replaced by a set of orthogonal wavelets. In the numerical example we demonstrate that while the PWS basis yields a full matrix, the wavelet expansion results in a nearly diagonal or nearly block-diagonal matrix; both approaches result in very close answers. However, as the geometry of the problem becomes more complicated, and consequently the resulting matrix size increases greatly, the advantages of having a nearly diagonal matrix over a full matrix will become more profound.

4.1 WAVELETS IN ELECTROMAGNETICS

Galerkin's method is a zero residual method if the basis functions are orthogonal and complete, and thus Galerkin's method with orthogonal basis functions is generally more accurate and rapidly convergent. Two types of orthogonal basis functions are frequently utilized for electromagnetic field computation. Mode expansion method (or mode-matching method) has often been applied to solve problems due to various discontinuities in waveguides, finlines, and microstrip lines. Generally, this technique is useful when the geometry of the structure can be identified as consisting of two or more regions, which each belongings to a separable coordinate system. The basic idea in the mode expansion procedure is to expand the unknown fields in the individual regions in terms of their respective normal modes. In fact the mode expansion method is identical to Galerkin's method which uses the normal mode functions as the basis functions. Quite often the normal modes are made of the classical orthogonal series systems such as trigonometric, Legendre, Bessel, Hermite, and Chebyshev. Owing to the orthogonality of the normal modes, a sparse system of linear algebraic equations is expected to be generated by the mode expansion method. For general cases of arbitrary geometries and material distributions, however, the mode functions are often too difficult to be constructed.

The second class of orthogonal basis functions consists of a group of subsectional bases, each of which is defined only in a given subsection of the solution domain. An advantage of the subsectional bases is the localization property, that is, each of the expansion coefficients affects the approximation of the unknown function only over a subdomain of the region of interest. Thus, often not only does this class of computations simplify the computation, but it also leads easily to convergent solutions. In the subsectional basis systems, generally, only partial orthogonality can be attained; only the pair of bases whose supporting regions do not overlap are orthogonal. Moreover the higher the continuity order of the constructed bases is rendered, the larger the required supporting region. Hence there exists a trade-off between the orthogonality and continuity for the subsectional basis systems.

Even if complete orthogonal bases with higher-order continuity are hard to build, the subsectional bases with certain continuity order can be constructed widely (e.g. by using polynomial interpolation functions). The finite element method, which has been universally applied in engineering, is a subsectional basis method. So is the boundary element method. Because of the kind of orthogonality, or, say, localization that exists in subsectional basis systems, the differential operator equations may yield sparse systems of linear algebraic equations by using subsectional bases. However, it is also noted that the subsectional basis systems do not necessarily convert the integral operator equations into sparse systems of linear algebraic equations.

Orthogonal wavelets have several properties that are fascinating for electromagnetic field computations. First, wavelets are sets of orthonormal bases of $L^2(R)$. They are problem-independent orthogonal bases and thus are suitable for numerical computations for general cases. Second, the trade-off between orthogonality and continuity is well balanced in orthogonal wavelet systems because now the orthogonality always holds, whether the supporting regions are overlapping or not. One can build an orthogonal wavelet system with any order of regularity, expecting larger supporting regions as higher orders of regularity are selected. Third, in addition to the advantages of the traditional orthogonal basis systems, orthogonal wavelets have a cancellation property such that they are much more certain to yield sparse systems of linear algebraic equations.

Furthermore orthogonal wavelets have localization properties in both the spatial and spectral domains. Therefore the decorrelation of the expansion coefficients occurs both in the space and Fourier domains. Nevertheless, according to the theory of multigrid processing, one can improve convergence by operating on both fine and coarse grids to reduce both the "high-frequency" and "low-frequency" component errors between the approximate and exact solutions. In contrast, the traditional way of operating only on fine grids reduces only the "high-frequency" component. The expansion with subsectional bases actually is equivalent to the expansion on the finest scale only (in fact, the pulse function is equivalent to the scalet of Haar's bases). On the contrary, the multiresolution analysis implemented by wavelet expansion provides a multigrid method. Finally, the pyramid scheme employed in the wavelet analysis provides fast algorithms.

4.2 LINEAR OPERATORS

Functional spaces and linear operators were presented systematically and rigorously in Chapter 1. In this section we will only quote the minimum prerequisite knowledge for the method of moment applications.

INNER PRODUCT $\langle f, g\rangle$. An inner product $\langle f, g\rangle$ on a complex linear space is a complex-valued scalar satisfying

$$\langle f, g\rangle = \overline{\langle g, f\rangle}$$

$$\langle \alpha f + \beta g, h\rangle = \overline{\alpha}\langle f, h\rangle + \overline{\beta}\langle g, h\rangle$$

$$\langle f, f\rangle = \|f\|^2 \begin{cases} > 0 & \text{if } f \neq 0 \\ = 0 & \text{if } f = 0, \end{cases}$$

where the overbar denotes the complex conjugate.

OPERATOR L. The linear operator L and its corresponding equation are given as

$$Lf = g.$$

For instance, the Poisson equation is

$$-\epsilon \nabla^2 \phi = \rho,$$

where the linear operator

$$L = -\epsilon \nabla^2 .$$

The adjoint L^a is defined by

$$\langle Lf, g\rangle = \langle f, L^a g\rangle.$$

An adjoint operator is self-adjoint if $L^a = L$.

The inverse operator of L is denoted as L^{-1}. For instance, the formal solution to (4.3.1) is

$$f = L^{-1} g.$$

In numerical computations we use a matrix to represent a linear operator.

4.3 METHOD OF MOMENTS (MoM)

Consider an operator equation

$$Lf = g, \tag{4.3.1}$$

where L is a linear operator, f is the unknown function, and g is a given excitation. We first expand the unknown function $f(x)$ in terms of the basis functions $f_n(x)$ with unknown coefficients α_n, namely

$$f = \sum_n \alpha_n f_n.$$

Thus

$$L \sum_n \alpha_n f_n = g.$$

Multiplying both sides of (4.3.1) by the weighting (testing) function w_m and taking the inner product $\langle \cdot, \cdot \rangle$, we obtain

$$\sum_n \alpha_n \langle w_m, Lf_n \rangle = \langle w_m, g \rangle.$$

In matrix form, it appears as

$$[l_{mn}] |\, \alpha \rangle = |g\rangle, \tag{4.3.2}$$

where

$$|\, \alpha \rangle = \begin{pmatrix} \alpha_1 \\ \alpha_2 \\ \vdots \\ \alpha_N \end{pmatrix}_{N\times 1} \qquad \longleftarrow \text{unknown},$$

$$[l_{mn}] = \begin{pmatrix} \langle w_1, Lf_1, \rangle & \langle w_1, Lf_2 \rangle & \cdots \\ \langle w_2, Lf_1, \rangle & \langle w_2, Lf_2 \rangle & \cdots \\ & \cdots & \end{pmatrix}_{N\times N} \qquad \longleftarrow \text{evaluated},$$

and

$$|\, g \rangle = \begin{pmatrix} \langle w_1, g \rangle \\ \langle w_2, g \rangle \\ \vdots \end{pmatrix}_{N\times 1}.$$

Formally, equation (4.3.2) is solved to yield

$$|\, \alpha \rangle = [l_{mn}]^{-1} \,|g\rangle.$$

There are two kinds of popular schemes in the method of moments:

(i) *Pulse-delta scheme.* Basis functions f_n = pulse functions, and testing function
$w_m = \delta(x_m - x)$, Dirac delta function (point matching).

(ii) *Galerkin scheme* $w_m = f_m$.

The pulse-delta scheme is equivalent to the rectangular rule in the numerical integration, and the Galerkin scheme is a zero residual method.

Example Charged Conducting Plate (zero thickness). A charged plate is depicted in Fig. 4.1. Find the charge distribution.

Solution The electrostatic potential at any point (x, y, z) in space is given by

$$V(x, y, z) = \int_{-a}^{a} dx' \int_{-a}^{a} dy' \frac{\sigma(x', y')}{4\pi\epsilon |\mathbf{r} - \mathbf{r}'|}$$

with the unknown charge density $\sigma(x', y')$. An integral equation of the first kind is then formulated as

$$V(\mathbf{r}) = \int_{v} G(\mathbf{r}, \mathbf{r}')\sigma(\mathbf{r}')d^3\mathbf{r}',$$

where the Green's function is $G(\mathbf{r}, \mathbf{r}') = 1/4\pi\epsilon |\mathbf{r} - \mathbf{r}'|$, and the potential on the plate surface is a constant V.

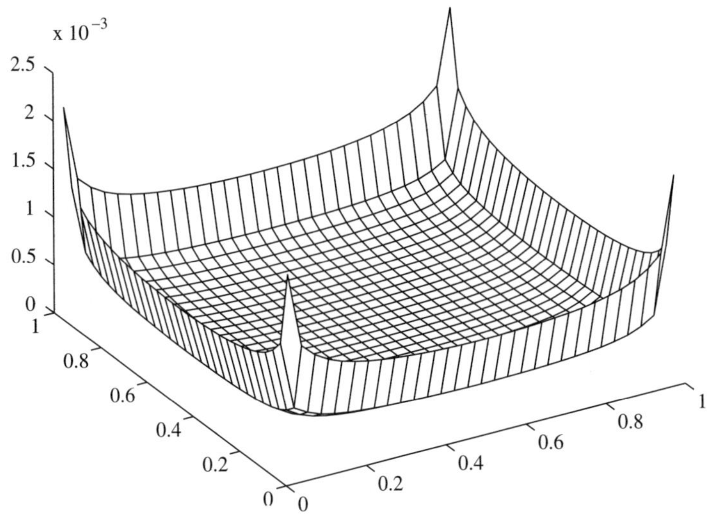

FIGURE 4.1 Charge distribution q/ϵ on a square plate.

The capacitance of the plate can be found by

$$C = \frac{q}{V} = \frac{1}{V}\int_{-a}^{a} dx \int_{-a}^{a} dy \sigma(x, y).$$

Therefore the main problem is to solve $\sigma(x', y')$ of the integral equation by the MoM, namely

$$\int_{-a}^{a} G(\mathbf{r}, \mathbf{r}')\sigma(\mathbf{r}')\, ds' = V,$$

where $G(\mathbf{r}, \mathbf{r}')$ is also called the integral kernel. We present this example because of its simple Green's function, and clear physical meaning. The numerical procedures may be outlined in the following steps:

(i) Define the basis functions to be the pulse functions

$$f_n(x', y') = \begin{cases} 1, & (x', y') \text{on } \Delta S_n \\ 0, & \text{on all other } \Delta S_m, m \neq n, \end{cases}$$

that is, α_n applies only on ΔS_n.

(ii) Approximate the charge by

$$\sigma(x, y) \approx \sum_{n=1}^{N} \alpha_n f_n.$$

(iii) Convert the operator equation into a matrix equation. The operator equation is

$$L\sigma = V,$$

or in the explicit form

$$\int G(\mathbf{r}, \mathbf{r}') \left(\sum_n \alpha_n f_n \right) ds' = V.$$

Applying the linearity of the operator, we obtain

$$\sum_n \alpha_n \int_{\Delta S_n} f_n \frac{dx'\, dy'}{4\pi\epsilon\sqrt{(x - x')^2 + (y - y')^2 + (z - z')^2}} = V(x, y).$$

Take the inner product of the equation above with the testing function

$$w_m(x, y) = \delta(x_m - x)\, \delta(y_m - y),$$

where (x_m, y_m) is the midpoint of the patch ΔS_m. The corresponding system of equations is formed with

$$\begin{aligned} \text{RHS} = \langle w_m, V \rangle &= \int V(x, y)\, \delta(x_m - x)\, \delta(y_m - y)\, dx\, dy \\ &= V(x_m, y_m) \end{aligned}$$

and

$$\begin{aligned} \text{LHS} &= \sum_n \alpha_n \int_{\Delta S_n} dx'\, dy' \int dx\, dy \frac{\delta(x_m - x)\, \delta(y_m - y)}{4\pi\epsilon\sqrt{(x - x')^2 + (y - y')^2}} \\ &= \sum_n \alpha_n \int_{\Delta S_n} dx'\, dy' \frac{1}{4\pi\epsilon\sqrt{(x_m - x')^2 + (y_m - y')^2}}. \end{aligned}$$

Thus a matrix equation converted from the operator equation is

$$\begin{pmatrix} l_{11} & l_{12} & \cdots & l_{1N} \\ l_{21} & l_{22} & \cdots & l_{2N} \\ & \vdots & & \\ l_{N1} & l_{N2} & \cdots & l_{NN} \end{pmatrix} \begin{pmatrix} \alpha_1 \\ \alpha_2 \\ \vdots \\ \alpha_N \end{pmatrix} = \begin{pmatrix} V(x_1, y_1) \\ V(x_2, y_2) \\ \vdots \\ V(x_N, y_N) \end{pmatrix}.$$

In choosing pulse basis functions, the charge is assumed to be a constant over a subarea (patch), namely

$$l_{mn} = \int_{\Delta S_n} dx'\, dy' \frac{1}{4\pi\epsilon\sqrt{(x_m - x')^2 + (y_m - y')^2}}.$$

(iv) In handling integral equations, singularity occurs when the field point (x, y), in this case (x_m, y_m), lies with in the domain of integration, ΔS_n. For the diagonal element, $l_{mn} (m = n)$, the integrand experiences a singularity which must be treated carefully. Analytical removal, pole extraction by an asymptote, and folding technique are among the popular methods for handling singularities [8, 9, 10]. In the present case the method of analytic removal is applicable:

$$\begin{aligned} l_{11} &= \int_{x_1 - b}^{x_1 + b} dx \int_{y_1 - b}^{y_1 + b} dy \frac{1}{4\pi\epsilon\sqrt{(x - x_1)^2 + (y - y_1)^2}} \\ &= \int_{-b}^{b} dx \int_{-b}^{b} dy \frac{1}{4\pi\epsilon\sqrt{x^2 + y^2}} \\ &= \frac{b}{4\pi\epsilon} \int_{-1}^{1} du \int_{-1}^{1} dv \frac{1}{\sqrt{u^2 + v^2}}, \end{aligned}$$

where

$$\int_{-1}^{1} du \int dv \frac{1}{\sqrt{u^2 + v^2}} = 8 \int_{\theta=0}^{\pi/4} \int_{\rho=0}^{1/\cos\theta} \frac{\rho\, d\rho\, d\theta}{\rho}$$

$$= 8 \int_0^{\pi/4} \frac{d \sin\theta}{\cos^2\theta} = 8 \int_0^{1/\sqrt{2}} \frac{d\tau}{1-\tau^2}$$
$$= 8 \cdot \frac{1}{2} \int_0^{1/\sqrt{2}} \left(\frac{1}{1-\tau} + \frac{1}{1+\tau} \right) d\tau$$
$$= 8 \ln \sqrt{3 + 2\sqrt{2}} = 8 \ln(1 + \sqrt{2}).$$

The numerical solution of the charge distribution on the plate is depicted in Fig. 4.1.

4.4 FUNCTIONAL EXPANSION OF A GIVEN FUNCTION

In the previous section the MoM was briefly discussed. The MoM is a powerful numerical algorithm, and it has been employed to solve electromagnetic problems for a half-century. Unfortunately, the MoM matrix is full. With the help of wavelets, one can obtain sparse impedance matrices.

We begin with the expansion of a given function in the wavelet bases. It is easier to expand a given function in a wavelet basis than to expand an unknown function in wavelets while solving the corresponding integral equation by the method of moments. The experience we gain here will be applied in the wavelet-based MoM.

From the multiresolution analysis (MRA), the nested subspaces can be decomposed as

$$\begin{aligned} V_{m+1} &= W_m \oplus V_m \\ &= W_m \oplus W_{m-1} + V_{m-1} \\ &= W_{m-1} \oplus W_{m-2} \oplus W_{m-3} \oplus \cdots \end{aligned}$$

and

$$\bigoplus_{m \in Z} W_m = L^2(R).$$

Therefore $\{\psi_{m,n}\}_{m,n \in Z}$ is an orthonormal (o.n.) basis of $L^2(R)$. For all $f(x) \in L^2(R)$, we have

$$f(x) = \sum_{m,n} \langle f(x), \psi_{m,n}(x) \rangle \psi_{m,n}(x).$$

In practice, we can only approximate a given physical phenomenon with finite precision. Mathematically the approximation is to project a function from the L^2 onto a subspace $V_{m+1} = V_m \oplus W_m$, namely

$$f(x) \simeq A_{m+1} f(x) := \sum_n s_n^{m+1} \varphi_{m+1,n}(x),$$

where $s_n^{m+1} = \langle f(x), \varphi_{m+1,n} \rangle$, $A_{m+1} f(x)$ is the approximation of $f(x)$ at resolution level 2^{m+1} and A_{m+1} is the projection operator. As $m \to \infty$,

$$A_m f(x) = f(x).$$

Since

$$V_{m+1} = V_m \oplus W_m,$$

it follows that

$$f(x) \simeq A_{m+1} f(x) = A_m f(x) + B_m f(x),$$

where

$$B_m f(x) = \sum_n d_n^m \psi_{m,n}(x)$$

and

$$d_n^m = \langle f(x), \psi_{m,n}(x) \rangle.$$

Continuing the process, we obtain

$$A_{m+1} f(x) = A_{m_1} f(x) + \sum_{m'=m_1}^{m} B_{m'} f(x), \tag{4.4.1}$$

where m_1 is a prescribed number, representing the lowest resolution level.

Example Expand $f(x)$ in terms of Daubechies wavelets $N = 2$, where $f(x) = 1 - |\,x\,|$ for $|\,x\,| \leq 1$.

Solution The function $f(x)$ is defined on $[-1, 1]$, but the Daubechies are with $\operatorname{supp}\{\varphi\} = [0, 3]$ and $\operatorname{supp}\{\psi\} = [-1, 2]$. Therefore we cannot use $\varphi_{0,0}$ nor $\psi_{0,0}$ because they are too wide. Let us choose

$$f(x) \sim f^4 \in V_4.$$

By the MRA

$$V_4 = V_2 \oplus W_2 \oplus W_3,$$

where the lowest resolution level $m_1 = 2$. Thus

$$\begin{aligned} f(x) \sim & \sum_n \langle f(x), \varphi_{2,n}(x) \rangle \varphi_{2,n}(x) \\ & + \sum_p \langle f(x), \psi_{2,p}(x) \rangle \psi_{2,p}(x) + \sum_k \langle f(x), \psi_{3,k}(x) \rangle \psi_{3,k}(x), \end{aligned}$$

where

$$\varphi_{j,n}(x) = 2^{j/2} \varphi(2^j x - n),$$

$$\psi_{j,k}(x) = 2^{j/2} \psi(2^j x - k).$$

From the supports of supp $\{\varphi\}$, supp $\{\psi\}$ and the scale, we have

$$\text{supp}\,\{\varphi_{2,0}(x)\} = \left[0, \tfrac{3}{4}\right]$$

$$\text{supp}\,\{\psi_{2,0}(x)\} = \left[-\tfrac{1}{4}, \tfrac{1}{2}\right]$$

$$\text{supp}\,\{\psi_{3,0}(x)\} = \left[-\tfrac{1}{8}, \tfrac{1}{4}\right].$$

It can be easily verified:

(1) For $j = 2$, $\varphi_{2,-6}(x)$ is the leftmost scalet that intercepts -1 and

$$\text{supp}\,\{\varphi_{2,-6}(x)\} = \left[-\tfrac{6}{4}, -\tfrac{3}{4}\right].$$

Note that $\varphi_{2,-5}(x)$ also intersects -1, with supp $\{\varphi_{2,-5}(x)\} = [-5/4, -2/4]$. However, it is only next to the leftmost scalet. We find $n = -6$ by substituting n into

$$2^2 x - (-n) = 0 \qquad \text{(left edge)},$$

$$2^2 x - (-n) = 3 \qquad \text{(right edge)}.$$

The integer n is selected such that x solved from the right edge equation is just on the right of -1. When $n = -7$ is used in the left and right edge equations, the resultant interval does not intersect -1. In the same way, the rightmost scalet that intercepts 1 is found as $\varphi_{2,3}(x)$ and supp $\{\varphi_{2,3}(x)\} = [3/4, 6/4]$.

(2) For $j = 3$, the leftmost $\psi_{3,n}(x)$ that intercepts -1 is $n = -9$. In fact supp $\{\psi_{3,-9}(x)\} = [-10/8, -7/8]$, as solved from

$$2^3 x - (-9) = -1 \Rightarrow x = -\tfrac{10}{8},$$

$$2^3 x - (-9) = 2 \Rightarrow x = -\tfrac{7}{8}.$$

In the same manner, the rightmost $\psi_{3,n}$ that intercepts 1 is $n = 8$.

(3) For $j = 2$, $\psi_{2,-5}(x)$ is the leftmost basis that intercepts -1 and

$$\text{supp}\,\{\psi_{2,-5}(x)\} = \left[-\tfrac{6}{4}, -\tfrac{3}{4}\right].$$

The rightmost basis that intercepts 1 is $\psi_{2,4}$ and supp $\{\psi_{2,4}(x)\} = [3/4, 6/4]$.

In conclusion,

$$f(x) \approx \sum_{n=-6}^{3} \langle f, \varphi_{2,n}\rangle \varphi_{2,n}(x) + \sum_{m=-5}^{4} \langle f, \psi_{2,m}\rangle \psi_{2,m}(x) + \sum_{k=-9}^{8} \langle f, \psi_{3,k}\rangle \psi_{3,k}(x).$$

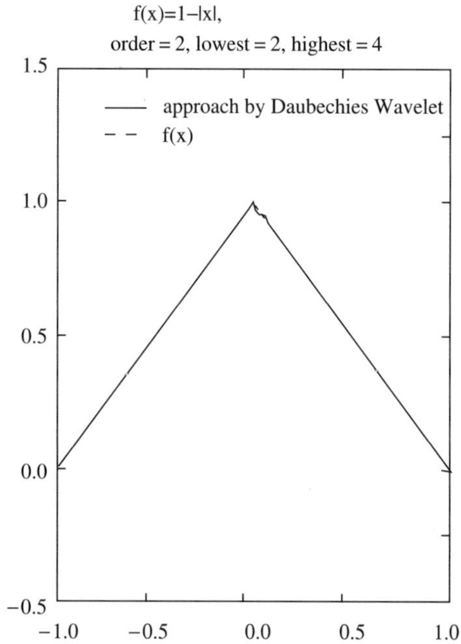

FIGURE 4.2 Expansion of $f(x)$ in Daubechies wavelets of $N = 2$.

Figure 4.2 depicts the function $f(x)$ and its wavelet expansion in V_4. It can be seen that the two agree well.

4.5 OPERATOR EXPANSION: NONSTANDARD FORM

We solve integral equations and expand a 1D function $f(x')$ in terms of wavelet basis functions by (4.4.1). We substitute this expansion into the integral equation, and then test using Galerkin's procedure for the unprimed variable x. The corresponding matrix represents an approximation of the operator. In most cases only linear operators are discussed. An integral operator T is given as

$$(Tf)(x) = \int K(x, y) f(y)\, dy, \tag{4.5.1}$$

where $K(x, y)$ is the integral kernel. For instance, the integral equation

$$\int G(x, x')\sigma(x')\, dx' = V(x)$$

is that for a 1D problem $\sigma(x')$ with a 2D kernel $K(x, y) = G(x, x')$.

4.5.1 Operator Expansion in Haar Wavelets

Expanding the kernel into a two-dimensional Haar series, we have

$$K(x, y) = \sum_{I,I'} \alpha_{II'} \psi_I(x) \psi_{I'}(y) + \sum_{I,I'} \beta_{II'} \psi_I(x) \varphi_{I'}(y) + \sum_{I,I'} \gamma_{II'} \varphi_I(x) \psi_{I'}(y), \tag{4.5.2}$$

where

$$\begin{aligned} \alpha_{II'} &= \iint K(x, y) \psi_I(x) \psi_{I'}(y)\, dx\, dy, \\ \beta_{II'} &= \iint K(x, y) \psi_I(x) \varphi_{I'}(y)\, dx\, dy, \\ \gamma_{II'} &= \iint K(x, y) \varphi_I(x) \psi_{I'}(y)\, dx\, dy. \end{aligned} \tag{4.5.3}$$

The previous expansion may be classified as two categories: standard form and non-standard form.

CASE 1. STANDARD FORM

$$I = I_{-j,k}, \quad I_{-j,k} = [2^j k, 2^j (k+1)],$$
$$I' = I_{-j',k'}, \quad I_{-j',k'} = [2^{j'} k, 2^{j'} (k'+1)].$$

Note that the combination of II' in (4.5.2) and (4.5.3) experiences all possible levels, with $j = j'$ and $j \neq j'$.

CASE 2. NONSTANDARD FORM

$$I = I_{-j,k}, \qquad I_{-j,k} = [2^j k, 2^j (k+1)],$$
$$I' = I_{-j,k'} \qquad \text{(instead of } I_{-j',k'}\text{)}.$$

In contrast to the standard form, only II' with equal levels appear in (4.5.2) and (4.5.3).

In this section only the nonstandard form is studied. To simplify the notation, we use

$$\alpha^j_{k,k'} = \alpha_{I_{j,k}, I_{j,k'}},$$
$$\beta^j_{k,k'} = \beta_{I_{j,k}, I_{j,k'}},$$
$$\gamma^j_{k,k'} = \gamma_{I_{j,k}, I_{j,k'}}.$$

Equation (4.5.2) is referred to as the nonstandard form. Substituting (4.5.2) into (4.5.1), we obtain

$$\begin{aligned} T(f)(x) &= \sum_I \psi_I(x) \sum_{I'} \alpha_{II'} \int \overbrace{\psi_{I'}(y) f(y)\, dy}^{d_{I'}} \\ &\quad + \sum_I \psi_I(x) \sum_{I'} \beta_{II'} \int \overbrace{\varphi_{I'}(y) f(y)\, dy}^{s_{I'}} \\ &\quad + \sum_I \varphi_I(x) \sum_{I'} \gamma_{II'} \int \overbrace{\psi_{I'}(y) f(y)\, dy}^{d_{I'}}, \end{aligned}$$

where I and I' always have the same length of $[2^j k, 2^j(k+1)]$, and $I = I_{-j,k}$, $I' = I_{-j,k'}$ are understood.

Define a projection operator

$$P_{-j} f = \sum_k \langle f, \varphi_{I_{-j,k}} \rangle \varphi_{I_{-j,k}}, \qquad j = 0, 1, \ldots, n.$$

The integral operator T is then approximated by T_0, according to the projection P_0 of prespecified precision

$$Tf \sim T_0 f = P_0(T(P_0 f)) \quad \text{or} \quad T \sim T_0 = P_0 T P_0,$$

where the first P_0 represents testing and the second P_0 is for expansion. The nonstandard decomposition yields

$$\begin{aligned} P_0 T P_0 &= \{(P_0 - P_{-1}) T \overbrace{(P_0 - P_{-1})}^{Q_{-1}} + (P_0 - P_{-1}) T P_{-1} \\ &\quad + P_{-1} T (P_0 - P_{-1})\} + P_{-1} T P_{-1} \\ &= \{Q_{-1} T Q_{-1} + Q_{-1} T P_{-1} + P_{-1} T Q_{-1}\} + P_{-1} T P_{-1}, \end{aligned}$$

where the last term on the right-hand side is similar to the left-hand side, but is one level down. The first equal mark in the previous operator equation can be verified by

$$a^2 = (a-b)^2 + (a-b)b + b(a-b) + b^2$$

with $a \leftrightarrow P_0, b \leftrightarrow P_{-1}$, and no commutation is allowed.

Repeating this process, on the $-1, -2, \ldots,$ levels, we have

$$\begin{aligned} P_0 T P_0 &= \sum_{j=1}^{n} (P_{-j+1} T P_{-j+1} - P_{-j} T P_{-j}) + P_{-n} T P_{-n} \\ &= \sum_{j=1}^{n} \{P_{-j+1} - P_{-j}) T \overbrace{(P_{-j+1} - P_{-j})}^{Q_{-j}} + (P_{-j+1} - P_{-j}) T P_{-j} \end{aligned}$$

$$+ P_{-j}T(P_{-j+1} - P_{-j})\} + P_{-n}TP_{-n} \cdots$$

$$= \sum_{j=1}^{n} [Q_{-j}TQ_{-j} + Q_{-j}TP_{-j} + P_{-j}TQ_{-j}] + P_{-n}TP_{-n}. \quad (4.5.4)$$

Equation (4.5.4) is for decomposing T_0 into a summation of contributions from different levels of wavelets and scales referred to as the telescopic series. The formulas, derived in this subsection based on Haar, apply to general wavelet systems.

4.5.2 Operator Expansion in General Wavelet Systems

We now readily to expand operators in general wavelet systems, either compactly or infinitely supported. Let T be a linear operator

$$T : L^2(R) \to L^2(R).$$

The projection operator

$$P_j : L^2(R) \to V_j$$

provides that

$$(P_j f)(x) = \sum_k \langle f, \varphi_{j,k} \rangle \varphi_{j,k}(x).$$

Expanding T in telescopic series, we obtain

$$T = \sum_{j \in Z} Q_j T Q_j + Q_j T P_j + P_j T Q_j,$$

where

$$Q_j = P_{j+1} - P_j.$$

Let T be an integral operator

$$(Tf)(x) = \int K(x, x') f(x')\, dx'$$

$$T \sim T_j = P_j T P_j,$$

then the kernel $K(x, x')$ can be expanded in a nonstandard form

$$K(x, x') = \sum_{m=m_l}^{m_{h-1}} \sum_{n,k'} \{\alpha^m_{n,k'} \psi_{m,n}(x) \psi_{m,k'}(x') + \beta^m_{n,k'} \psi_{m,n}(x) \varphi_{m,k'}(x')$$

$$+ \gamma^m_{n,k'} \varphi_{m,n}(x) \psi_{m,k'}(x')\} + \sum_{n,k'} s^{m_l}_{n,k'} \varphi_{m_l,n}(x) \varphi_{m_l,k'}(x'), \quad (4.5.5)$$

where

$$\alpha_{n,k'}^{m} = \langle K'(x, x'), \psi_{m,n}(x)\psi_{m,k'}(x')\rangle,$$
$$\beta_{n,k'}^{m} = \langle K'(x, x'), \psi_{m,n}(x)\varphi_{m,k'}(x')\rangle,$$
$$\gamma_{n,k'}^{m} = \langle K'(x, x'), \varphi_{m,n}(x)\psi_{m,k'}(x')\rangle,$$
$$s_{n,k'}^{m_l} = \langle K'(x, x'), \varphi_{m_l,n}(x)\varphi_{m_l,k'}(x')\rangle. \tag{4.5.6}$$

This leads to a fast wavelet transform algorithm, which will be discussed later in Section 4.8. For further in depth information, readers are referred to [3].

4.5.3 Numerical Example

A plane wave is impinging on a conducting screen with two slots as shown in Fig. 4.3, where the two slots are at $[0.1\lambda, 1.1\lambda]$ and $[-0.1\lambda, -1.1\lambda]$. Find the magnetic current $\mathbf{M} = \mathbf{E} \times \hat{n}$ on the slots [5].

Formulation

(1) Using the equivalence principle, we can close the slots with magnetic current $\mathbf{M} = -\hat{n} \times \mathbf{E}$.

(2) Applying image theory, we can put the image magnetic current, and then remove the conducting plane

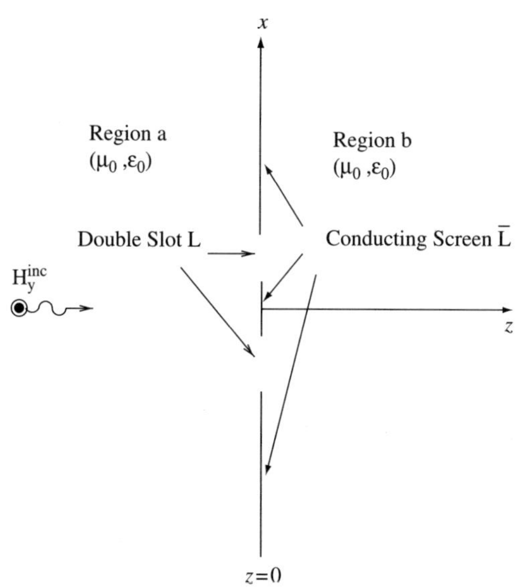

FIGURE 4.3 Diffraction of two apertures.

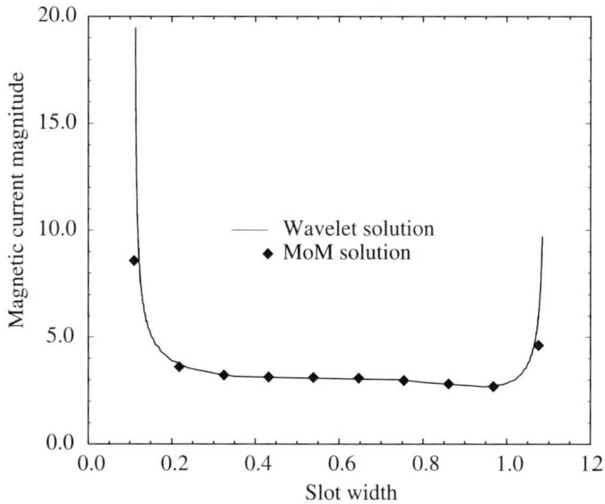

FIGURE 4.4 Magnitude of magnetic current.

$$\begin{aligned}\mathbf{M} &= \mathbf{M}^+ + \mathbf{M}^- \\ &= 2M_y\hat{y}.\end{aligned}$$

Equivalently, there is a source consisting of two magnetic current sheets in free space.

(3) In the wave equations

$$\begin{aligned}\nabla \times \nabla \times \mathbf{E} - k^2\mathbf{E} &= -j\omega\mu\mathbf{J} - \nabla \times \mathbf{M}, \\ \nabla \times \nabla \times \mathbf{H} - k^2\mathbf{H} &= -j\omega\epsilon\mathbf{M} + \nabla \times \mathbf{J},\end{aligned}$$

duality theorem has been applied. We choose the second equation above for the formulation, namely

$$\nabla \times \nabla \times \mathbf{H} - k^2\mathbf{H} = -j\omega\epsilon\mathbf{M}.$$

Using a vector identity, we have

$$\nabla(\nabla \cdot \mathbf{H}) - \nabla^2\mathbf{H} - k^2\mathbf{H} = -j\omega\epsilon\mathbf{M}.$$

Because of the zero divergence of the magnetic field, we obtain

$$\begin{aligned}(\nabla^2 + k^2)\mathbf{H}^s &= j\omega\epsilon\mathbf{M} \\ &= j\frac{\omega^2\mu\epsilon}{\omega\mu}\mathbf{M} \\ &= j\frac{k}{\eta}\mathbf{M}(\mathbf{r}'),\end{aligned}$$

where $\eta = \sqrt{\frac{\mu}{\epsilon}}$ is the intrinsic impedance. Thus the formal solution of the scattered field

$$\mathbf{H}^s = \int \overline{\overline{G}} \cdot j\frac{k}{\eta}\mathbf{M}(\mathbf{r}')dr',$$

where the dyadic Green's function, $\overline{\overline{G}}$, satisfies

$$\nabla \times \nabla \times \overline{\overline{G}} + \mu\epsilon\frac{\partial^2}{\partial t^2}\overline{\overline{G}} = -\overline{\overline{I}}\,\delta(\mathbf{r} - \mathbf{r}').$$

In the frequency domain

$$\overline{\overline{G}} = \left(\overline{\overline{I}} + \frac{\nabla\nabla}{k^2}\right) G_0$$

with

$$G_0 = -\frac{1}{4\pi|\,\mathbf{r} - \mathbf{r}'\,|}e^{jk|\,\mathbf{r}-\mathbf{r}'\,|}.$$

For the scalar case we have

$$H_y^s = \frac{jk}{\eta}\int_V (2M_y)G_0\,dv'.$$

For 2D structures the 2D free-space Green's function is

$$G_0 = -\frac{1}{4j}H_0^{(2)}(k|\,\boldsymbol{\rho} - \boldsymbol{\rho}'\,|).$$

It follows that

$$\begin{aligned} H_y^s(x) &= \frac{-k}{2\eta}\int H_0^{(2)}\left(-k\sqrt{(x-x')^2 + (z-z')^2}\right) M_y(x')\,dx' \\ &= \frac{-k}{2\eta}\int H_0^{(2)}(k|\,x - x'\,|)M_y(x')\,dx'. \end{aligned}$$

Applying the boundary condition

$$H_y^{in}(x) + H_y^s(x)|_{z=0} = 0,$$

we end with

$$\int H_0^{(2)}(k|\,x - x'\,|)M_y(x')\,dx' = \frac{\eta}{2k}H^{in}(x).$$

Using $x = \lambda u$, normalized by wavelength, we arrive at

$$\int H_0^{(2)}(k\lambda |u - u'|) M_y(u')\lambda \, du' = \frac{\eta}{2k} H^{in}(u)$$

or

$$\int H_0^{(2)}(2\pi |u - u'|) M_y(u') \, du' = \frac{\eta}{\pi} H^{in}(u).$$

For convenience, we use x instead of u

$$\int_L H_0^{(2)}(2\pi |x - x'|) M_y(x') \, dx' = \frac{\eta}{\pi} H_y^{in}(x), \tag{4.5.7}$$

where x has been normalized by wavelength λ, and

$$L = [-1.1, -0.1] \bigcup [0.1, 1.1].$$

Edge Treatment. In boundary value problems edges must be properly handled, or else the solution can have a nonphysical meaning. If the edges are not treated, the solution is oscillatory in nature, and this behavior is inaccurate.

Consider the LHS of (4.5.7)

$$\begin{aligned} \text{LHS} &= \left(\int_a^b + \int_c^d \right) (H_0^{(2)} M_y(x') \, dx') \\ &= \int_c^d [H_0^{(2)}(2\pi |x - x'|) + H_0^{(2)}(2\pi |x + x'|)] M_y(x') \, dx', \end{aligned}$$

where $a = -1.1$, $b = -0.1$, $c = 0.1$, and $d = 1.1$.

First, let us take a close look at the integral $\int_c^d$. Near the right edge of the slot, some of the basis functions, say $\varphi_{j,n}$, may not be completely supported in the interval (c, d). There seem to be two choices for us.

(1) Chop off $\varphi_{j,n}(x)$ for the portion $x > d$, (denoted as $\varphi_{j,n}^c(x)$). However, this basis is incomplete. Therefore we have destroyed the orthogonality, and

$$\int \varphi_{j,n}^c(x) \varphi_{i,m}(x) \, dx \neq 0.$$

(2) Remove the incomplete basis functions from the expansion. However, the interval will no longer be covered completely by the basis functions.

It turns out that neither of the previous ideas work. To solve the integral equations on bounded intervals using wavelets as basis functions, the treatment of edges must be carried out with caution. There are several techniques, and we list the most commonly used ones below:

- Coordinate transformation.
- Periodic wavelets.
- Intervallic wavelets.
- Weighted wavelets.

Here we apply coordinate transformation to

$$I = \int_a^b f(x')G(x, x')\,dx'. \tag{4.5.8}$$

Using the transform

$$x = \frac{b-a}{\pi}\tan^{-1}t + \frac{b+a}{2},$$

we will map $x : [a, b]$ to $t : (-\infty, +\infty)$.

The Jacobian is

$$\frac{dx}{dt} = \frac{b-a}{\pi}\frac{1}{1+t^2}.$$

As a result

$$I = \frac{b-a}{\pi}\int_{-\infty}^{\infty} f(x(t'))G(t, t')\frac{1}{1+t'^2}\,dt'.$$

Since the wavelets are defined on the real line R as

$$\bigoplus_{m\in Z} W_m = L^2(R),$$

no edge exists in the transformed domain. Thus, in the transform domain t, we can allocate the scalets and wavelets as much as we like since the interval is not bounded. Physically, those basis functions of large t in magnitude are compressed in the original physical coordinates. The rapidly varying M_y near the two edges in physical space has been stretched horizontally in the transform domain. Hence the expansion approximates the function better. Figure 4.5 shows the basis functions in the original coordinate system. It can be seen clearly that more basis functions are placed near the two edges of the slot so that the singular behavior of the fields there is modeled more precisely.

In the case where the incident wave is not normal to the screen, there will be two integrals in the integral equation. The same transform can be applied to both of them. How far should we put the wavelet bases in the t-axis? One could set up a stop criterion in terms of the relative error of the consequent solutions with a different number of wavelet bases.

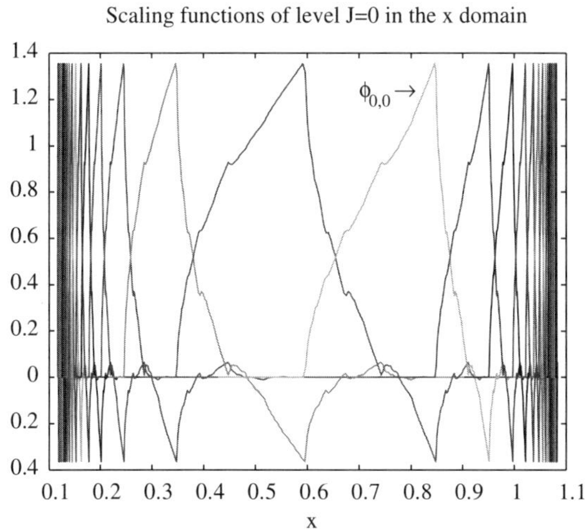

FIGURE 4.5 Compressed Daubechies scalets as basis functions.

Matrix Equations. We will discuss the problem in the physical space, although it applies to the transform domain as well. Let us assume that the unknown function $f(x) \in L^2(R)$ is projected to the highest resolution subspace as $f^{m_h} \in V_{m_h}$:

$$\begin{aligned} f(x) \stackrel{\triangle}{=} M_y(x) &= \sum_{m=m_l}^{m_{h-1}} \sum_n M_{m,n}^{\psi} \psi_{m,n}(x) + \sum_n M_{m_l,n}^{\varphi} \varphi_{m_l,n}(x) \\ &= \sum_n (M_{0,n}^{\psi} \psi_{0,n}(x) + M_{0,n}^{\varphi} \varphi_{0,n}(x)) + \sum_n M_{1,n}^{\psi} \psi_{1,n}(x) \\ &\quad + \sum_n M_{2,n}^{\psi} \psi_{2,n}(x) + \sum_n M_{3,n}^{\psi} \psi_{3,n}(x), \end{aligned} \tag{4.5.9}$$

where $M_{m,n}^{\psi}$, $M_{m_l,n}^{\varphi}$ are unknown, and we use $m_l = 0, m_h = 4$ to be specific.

The Green's function, according to Eq. (4.5.5), is

$$\begin{aligned} G(x, x') = &\sum_{m=m_l}^{m_h-1} \sum_{n,k'} \{\alpha_{n,k'}^{m} \psi_{m,n}(x) \psi_{m,k'}(x') \\ &+ \beta_{n,k'}^{m} \psi_{m,n}(x) \varphi_{m,k'}(x') + \gamma_{n,k'}^{m} \varphi_{m,n}(x) \psi_{m,k'}(x')\} \\ &+ \sum_{n,k'} s_{n,k'}^{m_l} \varphi_{m_l,n}(x) \varphi_{m_l,k'}(x'). \end{aligned} \tag{4.5.10}$$

Substituting (4.5.9) and (4.5.10) into (4.5.8) and making the inner product with the testing functions $\varphi_{m_l,p}(x)$, $\psi_{m_l,p}(x)$ and $\psi_{m,p}(x)$ according to the Galerkin procedure, we obtain a set of system equations. To be more specific yet not too tedious, we assume that $m_l = 0$ and $m_h = 2$. For the case of $\int dx\varphi_{0,p}(x)$ testing, we arrive at

$$\begin{aligned}
&\int dx\varphi_{m_l,p}(x)\int G(x,x')f(x')\,dx' \\
&\quad= \int dx\,dx'\varphi_{0,p}(x)\sum_{n,k'}\{[\alpha^0_{n,k'}\psi_{0,n}(x)\psi_{0,k'}(x') \\
&\qquad+ \beta^0_{n,k'}\psi_{0,n}(x)\varphi_{0,k'}(x') + \gamma^0_{n,k'}\varphi_{0,n}(x)\psi_{0,k'}(x') + s^0_{n,k'}\varphi_{0,n}(x)\varphi_{0,k'}(x') \\
&\qquad+ \alpha^1_{n,k'}\psi_{1,n}(x)\psi_{1,k'}(x') + \beta^1_{n,k'}\psi_{1,n}(x)\varphi_{1,k'}(x') + \gamma^1_{n,k'}\varphi_{1,n}(x)\psi_{1,k'}(x')]\} \\
&\sum_{q'}[M^{\varphi}_{0,n}\varphi_{0,q'}(x') + M^{\psi}_{0,n}\psi_{0,q'}(x') + M^{\psi}_{1,n}\psi_{1,q'}(x')] \\
&\quad= \sum_{n,k',q'}\alpha^0_{n,k'}[M^{\varphi}_{0,n}\int dx'\psi_{0,k'}(x')\varphi_{0,q'}(x')\int dx\psi_{0,n}(x)\varphi_{0,p}(x) \\
&\qquad+ M^{\psi}_{0,n}\int dx'\psi_{0,k'}(x')\psi_{0,q'}(x')\int dx\psi_{0,n}(x)\varphi_{0,p}(x) \\
&\qquad+ M^{\psi}_{1,n}\int dx'\psi_{0,k'}(x')\psi_{1,q'}(x')\int dx\psi_{0,n}(x)\varphi_{0,p}(x)] + \beta^0_{n,k'}\{\cdots,
\end{aligned}$$

where α, β, γ, s are pre-evaluated according to (4.5.6). The inner product of the right-hand side with testing function $\varphi_{0,p}(x)$ will result in a complex number in general. Thus we arrive at set of algebraic equations.

4.6 PERIODIC WAVELETS

4.6.1 Construction of Periodic Wavelets

Consider a periodic function with period 1, namely

$$f(x+1) = f(x) \Leftrightarrow f(x-1) = f(x).$$

Then, the wavelet coefficients on a given scale j

$$\langle f, \psi_{j,k}\rangle = \langle f, \psi_{j,k+2^j}\rangle.$$

Show.

$$\begin{aligned}
\text{RHS} &= \int f(x)2^{j/2}\psi(2^j x - k - 2^j)\,dx \\
&= \int f(x)\psi[2^j(x-1) - k]2^{j/2}\,dx \\
&= \int f(u+1)\psi(2^j u - k)2^{j/2}\,du
\end{aligned}$$

$$= \int f(u)\psi(2^j u - k)2^{j/2}\,du$$
$$= \langle f, \psi_{j,k}\rangle$$
$$= \text{LHS}. \qquad \square$$

Thus a periodic MRA on [0, 1] can be constructed by periodizing the basis functions as

$$\varphi^p_{j,k} = \sum_{\ell \in Z} \varphi_{j,k}(x+\ell) \quad \text{for } 0 \le k < 2^j, \quad j > 0,$$
$$\psi^p_{j,k} = \sum_{\ell \in Z} \psi_{j,k}(x+\ell) \quad \text{for } 0 \le k < 2^j, \quad j > 0,$$

where superscript p stands for periodic.

The subspace V^p_j has a dimension of 2^j, and

$$V^p_j = \text{span}\{\varphi^p_{j,k}, k \in Z\}.$$

We do not consider the cases $j < 0$, meaning the stretched wavelets. For $j \le 0$, it can be shown that

$$\varphi^p_{j,k}(x) = 2^{j/2} \sum_{\ell} \varphi(2^j x - k + 2^j \ell) = 2^{-j/2}, \qquad \text{constant}, \tag{4.6.1}$$
$$\psi^p_{j,k}(x) = 2^{j/2} \sum_{\ell} \psi(2^j x - k + 2^j \ell) = 0.$$

We prove (4.6.1) in two steps, using mathematical induction although other proofs are also possible.

Proof. First, we show that (4.6.1) is held for $j = 0$, that is, $\sum_{\ell=-\infty}^{\infty} \varphi(x+\ell) = 1$.

$$\begin{aligned}
\text{LHS} &= \sum_{\ell=\infty}^{-\infty} \varphi(x-\ell) \\
&= \sum_{\ell} \sum_{n} h_n \sqrt{2}\varphi(2x - 2\ell - n) \\
&= \sum_{\ell} \sum_{m} \sqrt{2} h_{m-2\ell} \varphi(2x - m) \\
&= \sum_{m} \left(\sum_{\ell} \sqrt{2} h_{m-2\ell} \right) \varphi(2x - m),
\end{aligned} \tag{4.6.2}$$

where we have used $m = 2\ell + n \Rightarrow n = m - 2\ell$.

Since we had in Chapter 3,

$$\hat{h}(\pi) = 0,$$

it follows that

$$0 = \sum \frac{h_n}{\sqrt{2}} e^{-i\frac{\omega}{2}n}|_{\omega=2\pi} = \frac{1}{\sqrt{2}} \sum (-1)^n h_n.$$

Therefore

$$\sum_n h_{2n} = \sum_n h_{2n+1}.$$

From $\hat{h}(0) = 1$, we obtain $\sum_n h_n = \sqrt{2}$. As a result

$$\sum_n h_{2n} = \sum_n h_{2n+1} = \frac{1}{\sqrt{2}}.$$

Hence we have from (4.6.2),

$$\sum_l \varphi(x - l) = \sum_m \varphi(2x - m) = \sum_n \varphi(4x - n) = \cdots = \text{const.}$$

Since $\int_{-\infty}^{\infty} \varphi(x)\,dx = 1$, the constant is necessarily equal to 1.

Second, we assume that

$$2^{j/2} \sum_\ell \varphi(2^j x - k + 2^j \ell) = 2^{-j/2}, \qquad j < 0,$$

and we wish to show that

$$2^{(j-1)/2} \sum_\ell \varphi(2^{j-1} x - k + 2^{j-1} \ell) = 2^{-(j-1)/2}. \tag{4.6.3}$$

Applying the dilation equation to (4.6.3), we have

$$\begin{aligned}
\text{LHS} &= 2^{(j-1)/2} \sum_\ell \sum_n \sqrt{2} h_n \varphi(2^j x - 2k + 2^j \ell - n) \\
&= 2^{j/2} \sum_n \sum_\ell h_n \varphi(2^j x + 2^j \ell - 2k - n).
\end{aligned}$$

Let $2k + n = p$, then $n = p - 2k$. Thus

$$\begin{aligned}
\text{LHS} &= \sum_p h_{p-2k} \sum_\ell 2^{j/2} \varphi(2^j x - p + 2^j \ell) \\
&= \sum_p h_{p-2k} 2^{-j/2}.
\end{aligned}$$

The equality is achieved by the induction assumption. Hence

$$\begin{aligned}
\text{LHS} &= 2^{-j/2} \sum_p h_{p-2k} \\
&= 2^{-j/2} \sqrt{2} \\
&= 2^{-(j-1)/2}.
\end{aligned}$$

□

In summary

(1) For large j, the wavelets are greatly compressed within [0, 1]. Hence

$$\varphi^p_{j,k}(x) = \varphi_{j,k}(x).$$

(2) In contrast, for small enough j, $\varphi_{j,k}(x)$ is chopped into pieces of length 1, which are shifted onto [0, 1] and added up, yielding the periodic wavelets.

The constructed periodic wavelets of Coifman, Daubechies, and Franklin are depicted in Figs. 4.6, 4.7, and 4.8.

4.6.2 Properties of Periodic Wavelets

It can be verified that $\psi^p_{j,k}, \varphi^p_{j,k}$ form an orthonormal basis system possessing the same MRA properties as the regular wavelets do, for example,

$$\begin{aligned}
\langle \psi^p_{j,k}, \varphi^p_{j,k} \rangle &= \int_0^1 \psi^p_{j,k}(x)\varphi^p_{j,k}(x)\,dx = 0 \\
\langle \psi^p_{j,k}, \psi^p_{j,k'} \rangle &= \delta_{k,k'} \\
V^p_0 &\subset V^p_1 \subset V^p_2 \subset \cdots \\
W^p_j \bigoplus V^p_j &= V^p_{j+1}.
\end{aligned}$$

V^p_j has 2^j basis functions $\{\varphi^p_{j,k}; k = 0, 1, \ldots, 2^j - 1\}$, and $L^2([0, 1])$ contains the following basis functions

(1) $\varphi^p_{0,0}(x) = 1$
(2) $\psi_{0,0}(x)$
(3) $\psi^p_{1,0}(x)$
(4) $\psi^p_{1,1}(x)$
(5) $\psi^p_{2,0}(x)$
(6) $\psi^p_{2,1}(x)$
(7) $\psi^p_{2,2}(x)$
(8) $\psi^p_{2,3}(x)$

Let us show that $\langle \psi^p_{j,k}, \psi^p_{j,k'} \rangle = \delta_{k,k'}$.

Proof.

$$\begin{aligned}
\langle \psi^p_{j,k}, \psi^p_{j,k'} \rangle &= \int_0^1 dx \psi^p_{j,k}(x)\ \psi^p_{j,k'}(x) \\
&= \int_0^1 dx \sum_\ell 2^{j/2}\psi(2^j x + 2^j \ell - k) \sum_{\ell'} 2^{j/2}\psi(2^j x + 2^j \ell' - k')
\end{aligned}$$

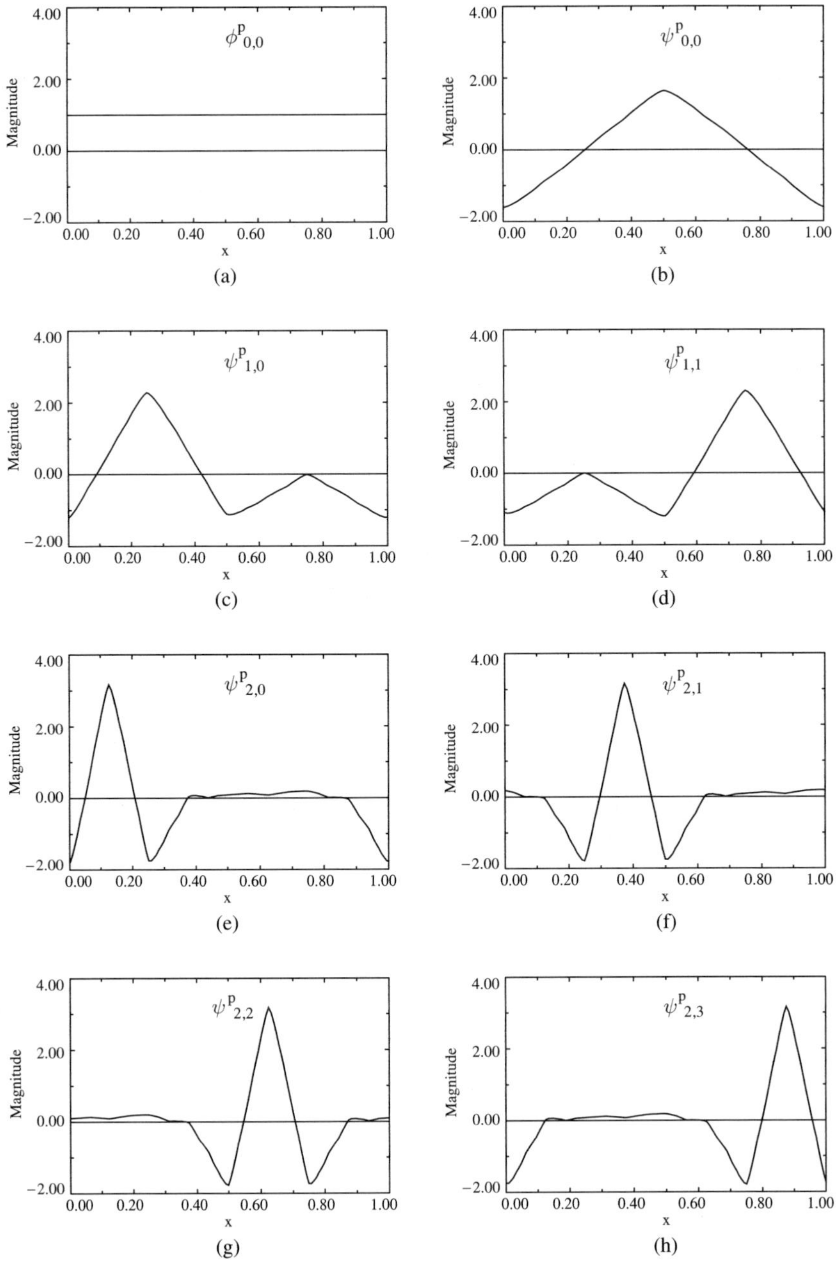

FIGURE 4.6 Periodic Coifman wavelets.

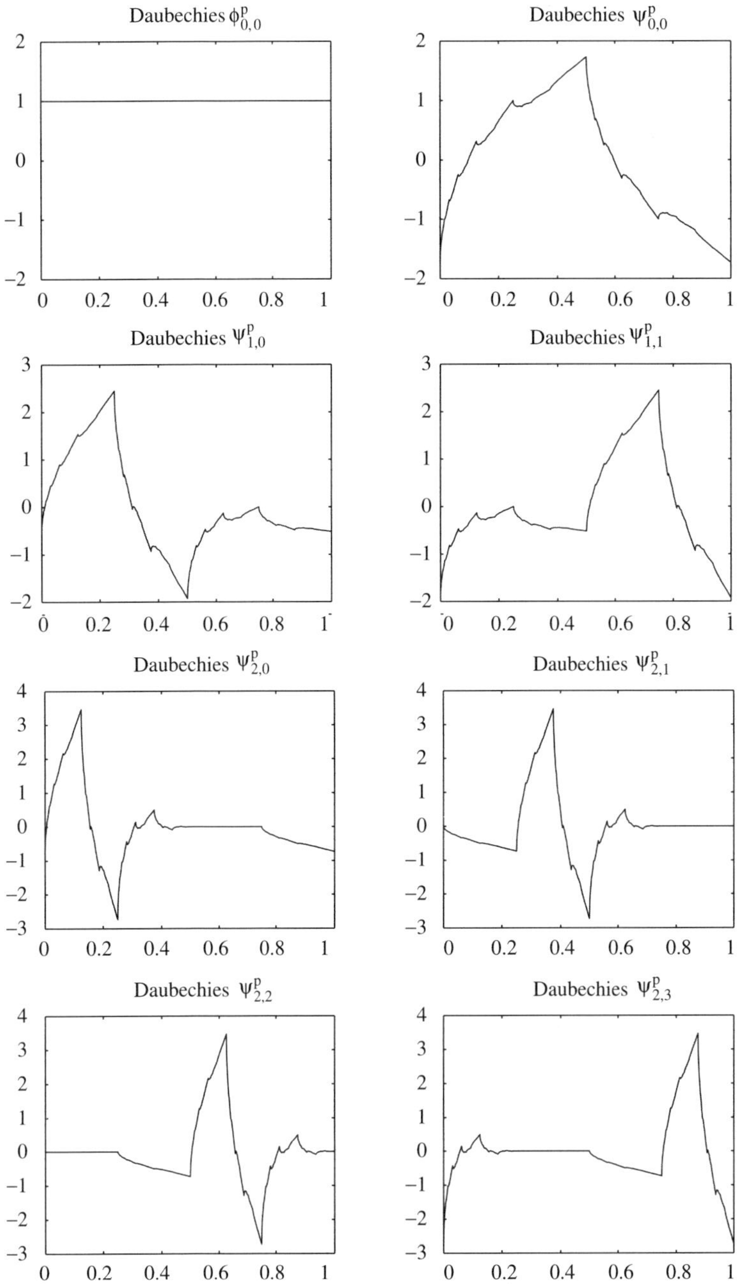

FIGURE 4.7 Periodic Daubechies wavelets.

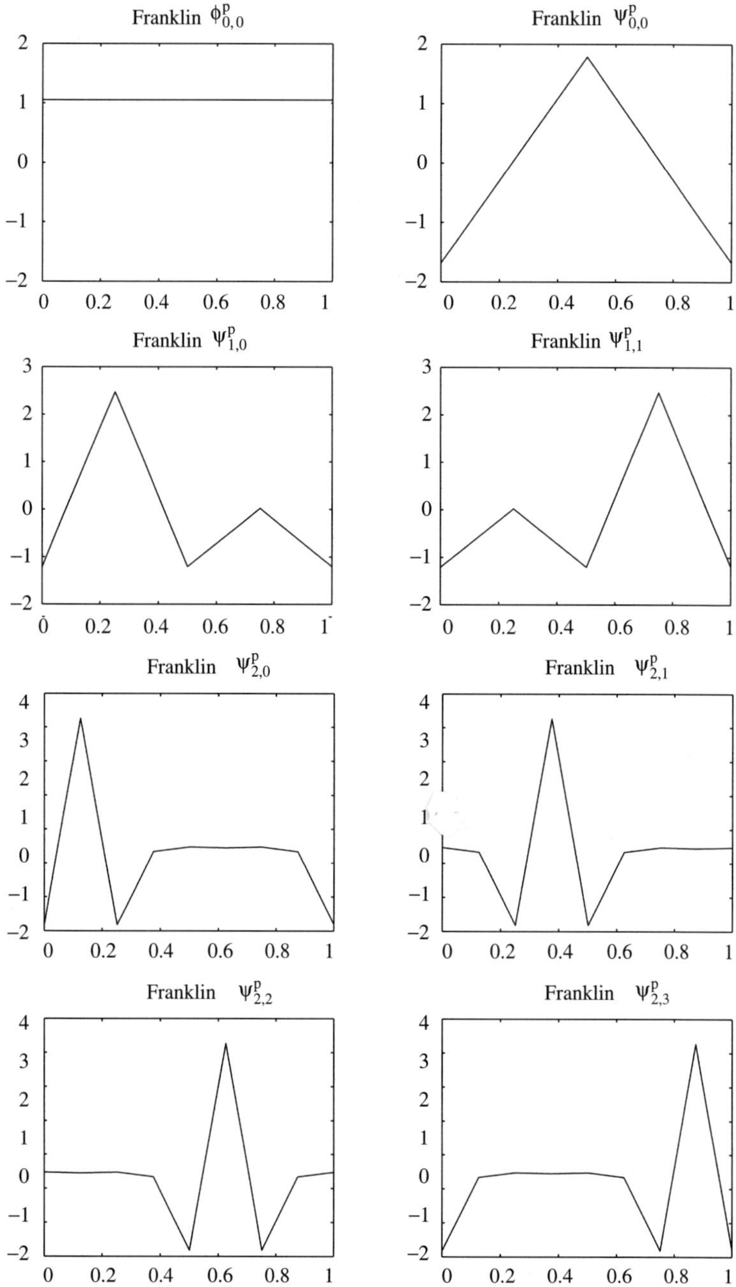

FIGURE 4.8 Periodic Franklin wavelets.

The unknown is expanded in terms of basis functions as

$$J_z = \sum_n \alpha_n f_n.$$

The weighting functions are $\delta(\boldsymbol{\rho} - \boldsymbol{\rho}_m)$, testing at the midpoint (x_m, y_m) of each ΔC_m.

Thus the integral equation is converted into a matrix equation

$$[\ell_{mn}]|\,\alpha_n\rangle = |\,g_m\rangle,$$

where

$$\begin{aligned} g_m &= E_z^i(x_m, y_m) \\ \ell_{mn} &= \frac{k\eta}{4}\int_{\Delta C_n} H_0^{(2)}\left[k\sqrt{(x - x_m)^2 + (y - y_m)^2}\right] d\ell \\ &\approx \frac{\eta}{4}(k\,\Delta C_n) H_0^{(2)}\left[k\sqrt{(x - x_m)^2 + (y - y_m)^2}\right], \qquad m \neq n, \end{aligned}$$

and the diagonal elements are

$$\ell_{nn} \approx \frac{\eta}{4}(k\,\Delta C_n)\left[1 - j\frac{2}{\pi}\ln\left(\frac{\gamma k\,\Delta C_n}{4e}\right)\right], \tag{4.7.2}$$

where the small argument approximation of $H_0^{(2)}(\cdot)$ has been applied, and $\gamma = e^{0.577215660}$ is Euler's constant.

CASE 2. WAVELET APPROACH The integral equation to be solved is

$$E^i(\boldsymbol{\rho}) = \frac{k\eta}{4}\int J(\boldsymbol{\rho}')H_0^{(2)}(k|\,\boldsymbol{\rho} - \boldsymbol{\rho}'\,|)\,d\ell'$$

where the subscript z has been dropped to simplify the notation. Note that the wavelets are defined on the straight line while the equation is formed on an ellipse. Using the parametric form

$$\begin{cases} x = a\cos\theta = \frac{\lambda}{4}\cos\theta \\ y = b\sin\theta = \lambda\sin\theta, \end{cases}$$

we obtain

$$\begin{aligned} |\,\boldsymbol{\rho} - \boldsymbol{\rho}'\,| &= \sqrt{a^2(\cos\theta - \cos\theta')^2 + b^2(\sin\theta - \sin\theta')^2} \\ d\ell' &= \sqrt{(a\sin\theta')^2 + (b\cos\theta')^2}\,d\theta' \\ &= \frac{\lambda}{4}\sqrt{1 + 15\cos^2\theta'}\,d\theta' \\ J(\boldsymbol{\rho}') &= J(\boldsymbol{\rho}(\theta')). \end{aligned}$$

Thus, with normal incidence of $\varphi_i = 0$, Eq. (4.7.1) becomes

$$\frac{\eta}{4}\left(\frac{k\lambda}{4}\right)\int_0^{2\pi} J(\boldsymbol{\rho}(\theta'))H_0^{(2)}\left(\frac{k\lambda}{4}\sqrt{(\cos\theta-\cos\theta')^2+16(\sin\theta-\sin\theta')^2}\right)$$
$$\cdot\sqrt{1+15\cos^2\theta'}\,d\theta' = e^{jk(\lambda/4)\cos\theta},$$

or

$$\frac{\eta}{4}\cdot\frac{\pi}{2}\int_0^{2\pi} J(\boldsymbol{\rho}(\theta'))H_0^{(2)}\left(\frac{\pi}{2}\sqrt{(\cos\theta-\cos\theta')^2+16(\sin\theta-\sin\theta')^2}\right)$$
$$\cdot\sqrt{1+15\cos^2\theta'}\,d\theta' = e^{j(\pi/2)\cos\theta}.$$

The integrand is periodic with period 2π. To obtain periodic functions of period 1, we use

$$\theta' = 2\pi\xi', \quad d\theta' = 2\pi\,d\xi'.$$

It follows that

$$\frac{\eta\pi^2}{4}\int_0^1 J(\xi')H_0^{(2)}\left(\frac{\pi}{2}\sqrt{(\cos\theta-\cos\theta')^2+16(\sin\theta-\sin\theta')^2}\right)$$
$$\cdot\sqrt{1+15\cos^2\theta(\xi')}\,d\xi' = e^{j(\pi/2)\cos\theta(\xi)}.$$

We prefer dimensionless expressions of the equation above. Expand the unknown in terms of periodic wavelets

$$J(\xi') = \sum_{n=0}^{N} a_n g_n(\xi),$$

where

$$\begin{aligned}
g_0(\xi) &= 1 \\
&= \varphi_{0,0}^P(\xi) \\
g_1(\xi) &= \psi_{0,0}^P(\xi) \\
g_2(\xi) &= \psi_{1,0}^P(\xi) \\
&= \psi^P(2\xi) \\
g_3(\xi) &= \psi_{1,1}^P(\xi) \\
&= \psi^P(2\xi-1)
\end{aligned}$$

$$
\begin{aligned}
&= \psi^p\left[2\left(\xi - \tfrac{1}{2}\right)\right] \\
&= g_2\left(\xi - \tfrac{1}{2}\right) \\
g_4(\xi) &= \psi_{2,0}(\xi) \\
g_5(\xi) &= \psi_{2,1}(\xi) \\
g_6(\xi) &= \psi_{2,2}(\xi) \\
g_7(\xi) &= \psi_{2,3}(\xi) \\
\cdots &= \psi_{3,0}(\xi) \\
\cdots &= \psi_{3,1}(\xi) \\
&= \vdots \\
g_{2^4}(\xi) &= \psi_{3,7}(\xi).
\end{aligned}
$$

We have a total of $2^{(3+1)} = 16$ basic functions. The first eight bases were depicted in Fig. 4.6, 4.7, or 4.8, depending on the selected wavelet.

Using Galerkin's procedure, we obtain the matrix equation

$$[\ell_{mn}]|\, a_n\rangle = |\, g_m\rangle,$$

where

$$
\begin{aligned}
\ell_{mn} = \frac{\eta\pi^2}{4} \int_0^1 \int_0^1 & g_m(\xi) g_n(\xi') H_0^{(2)}(\xi, \xi') \\
& \times \sqrt{1 + 15\cos^2(2\pi\xi')}\, d\xi'\, d\xi, \qquad m \neq n.
\end{aligned}
$$

The diagonal elements ℓ_{nn} have singularities, and care must be exercised. The following four methods are commonly employed [10, 12]:

(1) Singularity removal by analytical means.
(2) Extraction of the singularity (numerical and analytical).
(3) Folding technique (numerical).
(4) Generalized and hybrid Gaussian quadrature (numerical).

There are also some simple ways to avoid the singularity of $\xi' = \xi$. One way is to apply several tricks:

(1) Use different quadrature points for the unprimed and primed integrals.
(2) For the magnetic field integral equations (MFIE), simply drop the contribution of $\xi' = \xi$.

(3) Break one integral into two pieces, with the singular point $\xi = s$ as the breaking point and Gaussian quadrature never takes values at the end points. Thus, for the case of $m = n$, we could use

$$\int_0^1 \left[\left(\int_0^s + \int_s^1 \right) d\xi' \right] d\xi.$$

The resulting impedance matrix and the radar cross section of the elliptic cylinder appear as plotted in Figs. 4.10 and 4.11.

FIGURE 4.10 Magnitude of coefficient matrix using wavelet expansion.

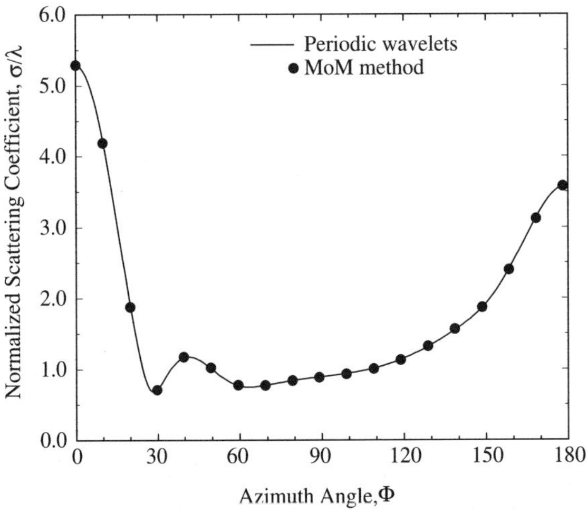

FIGURE 4.11 Scattering coefficient for conducting elliptic cylinder.

It should be mentioned that this problem can be worked out using standard wavelets. As long as the boundary curve has a closed contour, there is no need to employ the intervallic wavelets, nor the periodic wavelets. The standard wavelets are sufficient. In this case at the left edge, portions of the wavelets that are beyond the interval are circularly shifted to the right edge. This procedure is similar to the circular convolution in the discrete Fourier transform.

4.8 FAST WAVELET TRANSFORM (FWT)

4.8.1 Discretization of Operation Equations

Using Galerkin's procedure and wavelet basis functions, an operator equation

$$(\mathcal{T} f)(x) = g(x)$$

can be discretized into an algebraic equation at level j as

$$[\mathcal{T}^j] | c^j\rangle = | g^j\rangle$$

with a sparse coefficient matrix. Define a matrix

$$[\mathcal{T}^j] := P_j \mathcal{T} P_j, \tag{4.8.1}$$

where the matrix element

$$\mathcal{T}^j_{kk'} = \langle \varphi_{j,k}, \mathcal{T}(\varphi_{j,k'})\rangle \tag{4.8.2}$$

and the vector component

$$g^j_k = \langle \varphi_{j,k}, g\rangle. \tag{4.8.3}$$

The unknown $f(x)$ has been approximated by

$$f^j(x) = P_j f(x) = \sum c^j_k \varphi_{j,k}(x),$$

where $c^j_k = \langle f(x), \varphi_{j,k}\rangle$. The advantages of wavelets are the MRA, zero moments, orthogonality, localization, and the sparse coefficient matrix $[\mathcal{T}^j]$. However, the evaluation of matrix elements in wavelet formulation is much more involved than in the MoM, mainly because of poor regularity, highly oscillatory behavior, and a lack of closed form expressions of the wavelets.

In this section we will discuss the FWT using the Franklin (or more generally the Battle–Lemarie) wavelets, although the technique is applicable to other wavelets. The Franklin wavelets have computational simplicity, symmetry, and approximately closed form. As a result the computational cost of a matrix filled by the Franklin wavelets is almost the same as that of MoM by the triangle basis functions. The

Battle–Lemarie wavelets can be expressed in terms of the B-spline

$$\begin{cases} \varphi(x) = \sum_k a_k \theta_N(x-k) \\ \psi(x) = \sum_k b_k \theta_N(2x-k-1), \end{cases}$$

where a_k and b_k were listed in Table 3.1 and Table 3.2 with

$$b_k = \sqrt{2} \sum g_n a_{k-n}.$$

A special but important case of $N = 1$ is the Franklin wavelets

$$\begin{cases} \varphi(x) = \sum_k a_k \theta_c(x-k) \\ \psi(x) = \sum b_k \theta_c(2x-k-1), \end{cases}$$

where θ_c is the triangle

$$\theta_c(x) = \begin{cases} 1 - |x|, & |x| < 1 \\ 0 & \text{otherwise}. \end{cases}$$

For many problems the computational domain is confined. Thus we may need to construct on $L^2([0, 1])$ the periodic wavelets

$$\varphi^p_{j,k} = \sum_{l \in Z} \varphi_{j,k}(x+l),$$

$$\psi^p_{j,k} = \sum_{l \in Z} \psi_{j,k}(x+l).$$

4.8.2 Fast Algorithm

The fast algorithm is based on the assumption that the impedance matrix has been obtained at the finest resolution level:

$$\begin{aligned} \mathcal{T}^j_{k,k'} &= \langle \varphi_{j,k}, \mathcal{T}(\varphi_{j,k'}) \rangle \\ &= \left\langle \sum_i a_i \theta_c(2^j x - k - i) \mathcal{T}, \left[\sum_{i'} a_{i'} \theta_c(2^j x - k' - i') \right] \right\rangle \\ &= 2^j \sum_i \sum_{i'} a_i a_{i'} \langle \theta_c(2^j x - k - i), \mathcal{T}[\theta_c(2^j x - k' - i')] \rangle \\ &= \sum_i \sum_{i'} a_i a_{i'} Z^j_{k+i,k'+i'}, \end{aligned}$$

where

$$Z^j_{i,k} = 2^j \langle \theta_c(2^j x - i), \mathcal{T}[\theta_c(2^j x - k)] \rangle.$$

Because the B-splines are continuous (and smooth if $N > 1$) with relatively small support and with closed forms, the evaluation of $Z^j_{k,i}$ is much easier than direct evaluation of

$$\mathcal{T}^j_{kk'} = \langle \varphi_{j,k}, \mathcal{T}(\varphi_{j,k'}) \rangle.$$

If the domain is bounded, we need to use periodic wavelets. Correspondingly

$$\mathcal{T}^{(p)j}_{kk'} = \sum_l \sum_{l'} \mathcal{T}^j_{k+2^j l, k'+2^j l'}.$$

4.8.3 Matrix Sparsification Using FWT

The continuous operator and discrete operator are related by

$$\mathcal{T} = \lim_{j\to\infty} (P_j \mathcal{T} P_j) = \lim_{j\to\infty} (\mathcal{T}^j),$$

where $\mathcal{T}^j$ is the approximation of $\mathcal{T}$ projected on V_j and tested in V_j. There are two different methods toward the sparsification of an existing impedance matrix.

Nonstandard Form. Since

$$V_{j+1} = V_j \bigoplus W_j,$$

$$\mathcal{T}^{j+1} = \begin{bmatrix} A^{jj} & B^{jj} \\ C^{jj} & \mathcal{T}^j \end{bmatrix},$$

where

$$A^{jj}_{kk'} = \langle \psi_{j,k}, \mathcal{T}(\psi_{j,k'}) \rangle,$$

$$B^{jj}_{kk'} = \langle \psi_{j,k}, \mathcal{T}(\varphi_{j,k'}) \rangle,$$

$$C^{jj}_{kk'} = \langle \varphi_{j,k}, \mathcal{T}(\psi_{j,k'}) \rangle, \tag{4.8.4}$$

$\mathcal{T}^j_{kk'}$ has been previously defined in (4.8.2). Matrix $A^{jj}_{kk'}$ is very sparse because both of the expansion and testing functions are wavelets. Matrices $B^{jj}_{kk'}$ and $C^{jj}_{kk'}$ are composed of a mix of scalet and wavelet. Matrix $\mathcal{T}^j_{kk'}$ is dense because both of the expansion and testing functions are scalets. These submatrices represent the interaction between the sources and fields in different subspaces. Figure 4.12 illustrates the following FWT procedures in a schematic overview.

Submatrix A^{jj} can be evaluated as follows.

$$A^{jj}_{kk'} = \langle \psi_{j,k}, \mathcal{T}(\psi_{j,k'}) \rangle$$

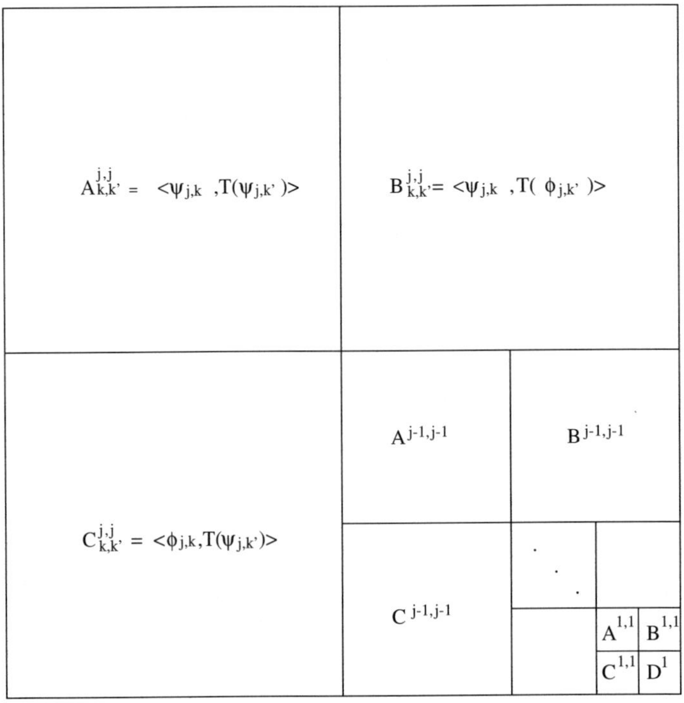

FIGURE 4.12 Nonstandard form representation of a decomposed matrix.

$$
\begin{aligned}
&= \left\langle \sum_n g_{n-2k}\varphi_{j+1,n}, \mathcal{T}\sum_m g_{m-2k'}\varphi_{j+1,m} \right\rangle \\
&= \sum_m \sum_n g_{n-2k} g_{m-2k'} \langle \varphi_{j+1,n}, \mathcal{T}(\varphi_{j+1,m}) \rangle .
\end{aligned}
$$

The development of detailed steps relies on the two equations

$$
\varphi_{j,k} = \sum_n h_{n-2k}\varphi_{j+1,n}, \tag{4.8.5}
$$

$$
\psi_{j,k} = \sum_n g_{n-2k}\varphi_{j+1,n}, \tag{4.8.6}
$$

where h and g are the lowpass and bandpass filter coefficients, respectively.

Show.

$$
\begin{aligned}
\varphi_{j,k} &= 2^{j/2}\varphi(2^j x - k) \\
&= 2^{j/2}\varphi(u) \\
&= 2^{j/2}\sum_m h_m \sqrt{2}\varphi(2u - m)
\end{aligned}
$$

$$= 2^{(j+1)/2} \sum_m h_m \varphi[2(2^j x - k) - m]$$

$$= 2^{(j+1)/2} \sum_m h_m \varphi[2^{j+1} x - (2k + m)].$$

Letting $n = 2k + m$, it follows that $m = n - 2k$. Thus

$$\varphi_{j,k} = \sum_n h_{n-2k} 2^{(j+1)/2} \varphi(2^{j+1} x - n)$$

$$= \sum_n h_{n-2k} \varphi_{j+1,n}. \qquad \square$$

Equation (4.8.6) can be shown in the same manner.

Now the submatrix

$$A_{k,k'}^{j,j} = \langle \psi_{j,k}, \mathcal{T}(\psi_{j,k'}) \rangle$$

$$= \left\langle \left(\sum_n g_{n-2k} \varphi_{j+1,n} \right), \mathcal{T} \left(\sum_m g_{m-2k'} \varphi_{j+1,m} \right) \right\rangle$$

$$= \sum_n \sum_m g_{n-2k} g_{m-2k'} \langle \varphi_{j+1,n}, \mathcal{T}(\varphi_{j+1,m}) \rangle.$$

From the definition of $\mathcal{T}_{kk'}^{j+1}$ we arrive at

$$A_{k,k'}^{jj} = \sum_n \sum_m g_{n-2k} g_{m-2k'} \mathcal{T}_{n,m}^{j+1}. \tag{4.8.7}$$

Similarly

$$B_{kk'}^{jj} = \langle \psi_{j,k}, \mathcal{T}(\varphi_{j,k'}) \rangle = \left\langle \sum_n g_{n-2k} \varphi_{j+1,n}, \mathcal{T} \left(\sum_m h_{m-2k'} \varphi_{j+1,m} \right) \right\rangle$$

$$= \sum_n \sum_m g_{n-2k} h_{m-2k'} \mathcal{T}_{n,m}^{j+1}, \tag{4.8.8}$$

$$C_{k,k'}^{jj} = \langle \varphi_{j,k}, \mathcal{T}(\psi_{j,k'}) \rangle$$

$$= \sum_n \sum_m h_{n-2k} g_{m-2k'} \mathcal{T}_{n,m}^{j+1},$$

$$\mathcal{T}_{k,k'}^{j} = \sum_n \sum_m h_{n-2k} h_{m-2k'} \mathcal{T}_{n,m}^{j+1}. \tag{4.8.9}$$

Utilizing (4.8.7) through (4.8.9), we obtain the updated matrix, which is sparser than the previous matrix, particularly in the upper-left quarter.

Repeating the previous procedures to the submatrix $\mathcal{T}^j$, we obtain

$$A^{j-1,j-1}, B^{j-1,j-1}, C^{j-1,j-1}, \mathcal{T}^{j-1}.$$

Next we decompose $\mathcal{T}^{j-1}$ into

$$A^{j-2,j-2}, B^{j-2,j-2}, C^{j-2,j-2}, \mathcal{T}^{j-2}.$$

Such a procedure decomposes a matrix $\mathcal{T}^{j+1}$ into a telescopic structure known as the nonstandard form, as depicted in Fig. 4.12.

Standard Form. We may further improve the matrix sparsity in the nonstandard form as follows. Let us consider the lower-left quarter, C^{jj}. Notice the fact that

$$\begin{aligned} A^{\ell-1,j}_{k,k'} &= \langle \psi_{\ell-1,k}, \mathcal{T}(\psi_{j,k'})\rangle \\ &= \langle \sum_n g_{n-2k}\varphi_{\ell,n}, \mathcal{T}(\psi_{j,k'})\rangle \\ &= \sum_n g_{n-2k}\langle \varphi_{\ell,n}, \mathcal{T}(\psi_{j,k'})\rangle \\ &= \sum_n g_{n-2k} C^{\ell,j}_{n,k'} \qquad \ell = j, j-1, \ldots, 1. \end{aligned}$$

For instance, we can fill out the 512×1024 matrix of $A^{j-1,j}$, using the entries in $C^{j,j}$ of 1024×1024. Hence

$$C^{\ell,j} = \begin{pmatrix} A^{\ell-1,j} \\ C^{\ell-1,j} \end{pmatrix}, \qquad \ell = j, j-1, \ldots, 1.$$

As shown in the chart of Fig. 4.13,

$$\begin{aligned} C^{\ell-1,j}_{k,k'} &= \langle \varphi_{\ell-1,k}, \mathcal{T}(\psi_{j,k'})\rangle \\ &= \left\langle \sum_n h_{n-2k}\varphi_{\ell,n}, \mathcal{T}(\psi_{j,k'}) \right\rangle \\ &= \sum_n h_{n-2k}\langle \varphi_{\ell,n}, \mathcal{T}(\psi_{j,k'})\rangle = \sum_n h_{n-2k} C^{\ell,j}_{n,k'}, \qquad \ell = j, j-1, \ldots, 1. \end{aligned}$$

Note that the lower level implies a wider support of the basis and a fewer number of elements in the expansion of the unknown. This process can be repeated, as

$$\begin{aligned} A^{j-2,j}_{k,k'} &= \sum_n g_{n-2k} C^{j-1,j}_{n,k'}, \\ C^{j-2,j}_{k,k'} &= \sum_n h_{n-2k} C^{j-1,j}_{n,k'}, \end{aligned}$$

and can be ended with

$$\begin{aligned} A^{0,j}_{k,k'} &= \sum_n g_{n-2k} C^{1,j}_{n,k'}, \\ C^{0,j}_{k,k'} &= \sum_n h_{n-2k} C^{1,j}_{n,k'}. \end{aligned}$$

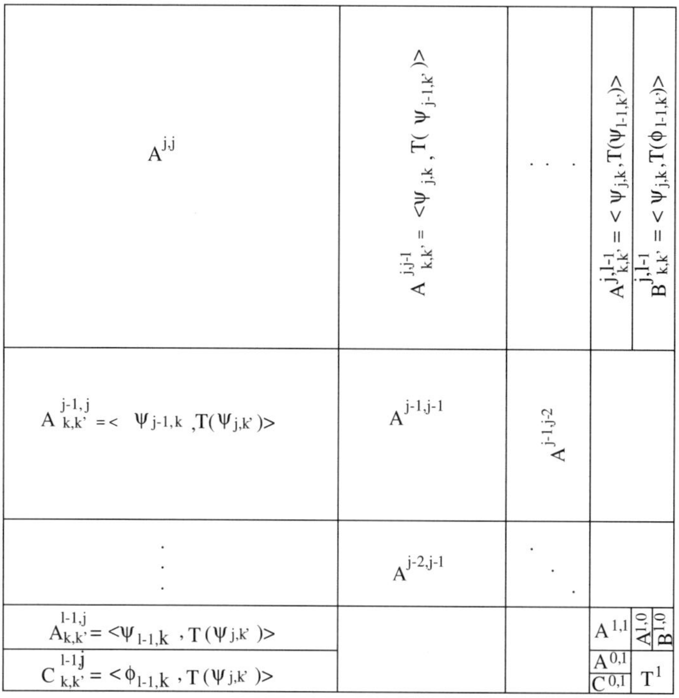

FIGURE 4.13 Standard form representation of a decomposed matrix.

In the same manner, the upper-right quarter

$$B^{j,\ell} = [A^{j,\ell-1} B^{j,\ell-1}], \qquad \ell = j, j-1, \ldots, 1.$$

The submatrices in the preceding equations are

$$\begin{aligned} A^{j,l-1}_{k,k'} &= \langle \psi_{j,k}, \mathcal{T}(\psi_{l-1,k'}) \rangle \\ &= \left\langle \psi_{j,k}, \mathcal{T}\left(\sum_n g_{n-2k'} \varphi_{l,n} \right) \right\rangle \\ &= \sum_n g_{n-2k'} \langle \psi_{j,k}, \mathcal{T}(\varphi_{l,n}) \rangle \\ &= \sum_n g_{n-2k'} B^{j,l}_{k,n} \end{aligned}$$

and

$$B^{j,l-1}_{k,k'} = \langle \psi_{j,k}, \mathcal{T}(\varphi_{l-1,k'}) \rangle$$

$$= \sum_n h_{n-2k'} \langle \psi_{j,k}, \mathcal{T}(\varphi_{l,n}) \rangle$$

$$= \sum_n h_{n-2k'} B_{k,n}^{j,l}.$$

Finally, we are ready to attack the lower-right block $\mathcal{T}^j$.

The decomposition procedure for $\mathcal{T}^j$ is the same as that for $\mathcal{T}^{j+1}$, except that the matrix dimensions are now twice as small in each direction. The resultant matrix is shown in Fig. 4.13.

4.9 APPLICATIONS OF THE FWT

We apply the FWT to sparsify the impedance matrix of an integral equation formulation referring to this approach as the quasi-dynamic method (QDM). This goes beyond the quasi-static method but is still not a full-wave approach. This algorithm extracts the frequency-dependent circuit parameters, $L(f)$ and $R(f)$ [13]. We will discuss the formulation briefly and present some relevant results. Detailed implementation of the QDM will be illustrated in Chapter 9.

4.9.1 Formulation

The QDM neglects the displacement current inside the conductors, and it also neglects the transverse current within the transmission lines. As a consequence inside the conductors

$$(\nabla^2 + k^2) J_z = 0,$$

where

$$k^2 = \omega^2 \mu \left(\epsilon + i \frac{\sigma}{\omega} \right) \approx i \omega \mu \sigma.$$

Hence we have the diffusion equation

$$(\nabla^2 + i \omega \mu \sigma) J_z = 0. \tag{4.9.1}$$

In the exterior region the transverse electromagnetic (TEM) assumption is used, namely

$$\nabla^2 A_z = 0. \tag{4.9.2}$$

There are two boundary conditions for tangential magnetic fields and normal magnetic fields

$$\begin{cases} H_t^{(1)} = H_t^{(2)} \\ B_n^{(1)} = B_n^{(2)}, \end{cases}$$

where superscripts (1) and (2) denote interior and exterior, respectively. We use J_z inside the conductor and A_z outside. They must be related at the boundary. It can be shown (see Exercise 8) that the previous boundary conditions lead to

$$\begin{cases} \dfrac{\partial J_z}{\partial n} = i\omega\sigma \dfrac{\partial A_z}{\partial n} \\ \dfrac{\partial J_z}{\partial l} = i\omega\sigma \dfrac{\partial A_z}{\partial l}, \end{cases}$$

where $\partial/\partial n$ and $\partial/\partial l$ are the normal and tangential derivatives, respectively. It follows that

$$J_z = i\omega\sigma A_z - \frac{I_q^{dc}}{S_q},$$

where I_q^{dc} and S_q are the d.c. (direct current) density and cross section of line q.

Employing Green's identity, we can convert a surface integral into a boundary line integral by

$$\iint ds(\phi\nabla^2\psi + \nabla\phi \times \nabla\psi) = \oint dl\phi\frac{\partial\psi}{\partial n}.$$

Letting $\psi = J_z, \phi = 1$ and utilizing (4.9.1), we obtain the total current on a wire:

$$\begin{aligned} I &= \iint_s ds J_z \\ &= \frac{1}{-i\omega\mu\sigma}\iint \nabla^2 J_z\, ds \\ &= \frac{i}{\omega\mu\sigma}\oint dl \frac{\partial J_n}{\partial n}. \end{aligned}$$

4.9.2 Circuit Parameters

The distributed circuit parameters, namely resistance R and inductance L, can be extracted from the field solutions as

$$\begin{aligned} R &= \frac{2P_d}{|\, I \,|^2} \\ &= \frac{\iint E_z J_z^* \, ds}{|\iint J_z \, ds\,|^2} \\ &= \frac{1}{\sigma}\frac{\iint |\, J_z \,|^2 \, ds}{|\iint J_z \, ds\,|^2}. \end{aligned} \tag{4.9.3}$$

The numerator

$$\iint ds J_z J_z^* = \iint ds J_z \left(\frac{i}{\omega\mu\sigma} \nabla^2 J_z \right)^*$$
$$= \frac{-i}{\omega\mu\sigma} \iint J_z \nabla^2 J_z^*.$$

Similarly

$$\iint ds J_z J_z^* = \frac{i}{\omega\mu\sigma} \iint J_z^* \nabla^2 J_z$$
$$\iint ds J_z J_z^* = \frac{1}{2} \frac{i}{\omega\mu\sigma} \iint ds (J_z^* \nabla^2 J_z - \nabla^2 J_z^* J_z)$$
$$= \frac{1}{2} \frac{i}{\omega\mu\sigma} \oint dl \left[J_z^* \frac{\partial J_z}{\partial n} - J_z \frac{\partial J_z^*}{\partial n} \right]$$
$$= \frac{1}{w\mu\sigma} \oint dl \, \mathrm{Im} \left\{ J_z \frac{\partial J_z^*}{\partial n} \right\}.$$

The denominator of (4.9.3),

$$| I |^2 = \left| \int ds J_z \right|^2 = \left| \frac{1}{w\mu\sigma} \oint dl \frac{\partial J_z}{\partial n} \right|^2.$$

Hence

$$R = \frac{1}{\sigma} \frac{(1/\omega\mu\sigma) \oint_{\mathrm{all}} dl \, \mathrm{Im}\{J_z (\partial J_z^* / \partial n)\}}{(1/\omega^2\mu^2\sigma^2) | \oint_{\mathrm{line}} dl (\partial J_z / \partial n) |^2}$$
$$= \omega\mu \frac{\oint dl \, \mathrm{Im}\{J_z (\partial J_z^* / \partial n)\}}{| \oint dl (\partial J_z / \partial n) |^2}. \tag{4.9.4}$$

In a similar way we can derive the inductance per unit length

$$L = \frac{4W_m}{| I |^2} = \mu \frac{\iint ds (\mathbf{H} \cdot \mathbf{H}^*)}{| I |^2}$$

from the stored magnetic energy

$$W_m = \tfrac{1}{4} L I^2.$$

It can be shown (Exercise 8) that

$$L = -\mu \frac{\oint_{\text{all wires}} dl \, \mathrm{Re} \left\{ (I_q^{dc} / S_q) (\partial J_z^\star / \partial n) \right\}}{| \oint_{\text{signal wire}} dl (\partial J_z / \partial n) |^2}.$$

The mutual resistance and inductance

$$R_{ij} = \frac{1}{2}\left(R_{ii} + R_{jj} - 2\frac{P_d}{I_q^2}\right),$$

$$L_{ij} = \frac{1}{2}\left(l_{ii} + L_{jj} - 4\frac{W_m}{I_q^2}\right),$$

where I_q = current on line i, while $-I_q$ is that on line j.

Show. We derive the two-conductor case, and the extension to N-conductor case is straightforward. The power dissipation due to resistance is

$$\begin{aligned} P_d &= \tfrac{1}{2}\langle I|\,R\,|I\rangle \\ &= \tfrac{1}{2}(I_1 I_2)\begin{pmatrix} R_{11} & R_{12} \\ R_{21} & R_{22} \end{pmatrix}\begin{pmatrix} I_1 \\ I_2 \end{pmatrix}, \end{aligned}$$

where $R_{12} = R_{21}$. Thus

$$P_d = \tfrac{1}{2}(R_{11}I_1^2 + 2R_{12}I_1I_2 + R_{22}I_2^2).$$

Assuming that $I_1 = I_q$ and $I_2 = -I_q$, we obtain

$$\begin{aligned} P_d &= \tfrac{1}{2}(R_{11}I_q^2 + R_{22}I_q^2 - 2R_{12}I_q^2), \\ I_q^2 R_{12} &= \tfrac{1}{2}(R_{11}I_q^2 + R_{22}I_q - 2P_d). \end{aligned} \qquad \square$$

Here we must obtain the self-resistances from (4.9.4) before we can evaluate the mutual resistance. Mutual inductance can be derived in the same fashion. Even though field quantities are obtained, digital engineers prefer to use equivalent circuit parameters, which can be easily incorporated into the SPICE models. In general, one needs to convert the solutions (either quasi-static, or full-wave) into the S-parameters, Z-parameters, or frequency-dependent circuit parameters $L(f)$, $C(f)$, $R(f)$, and $G(f)$.

4.9.3 Integral Equations and Wavelet Expansion

$$\oint_{\text{all}} dl' G_0(l,l')\frac{\partial J_z(l')}{\partial n'} = \oint_{\text{all}} dl' \left[J_z(l') + \frac{I_q}{S_q}\right]' \left[\frac{\partial G_0(l,l')}{\partial n'} - \frac{1}{2}\,\delta(l-l')\right],$$

$$\oint_{\text{line } q} dl' G_i(l,l')\frac{\partial J_z(l')}{\partial n'} = \oint_{\text{line } q} dl' J_z(l')' \left[\frac{\partial G_i(l,l')}{\partial n'} + \frac{1}{2}\,\delta(l-l')\right],$$

where G_0 and G_i are Green's functions derived from (4.9.1) and (4.9.2), respectively.

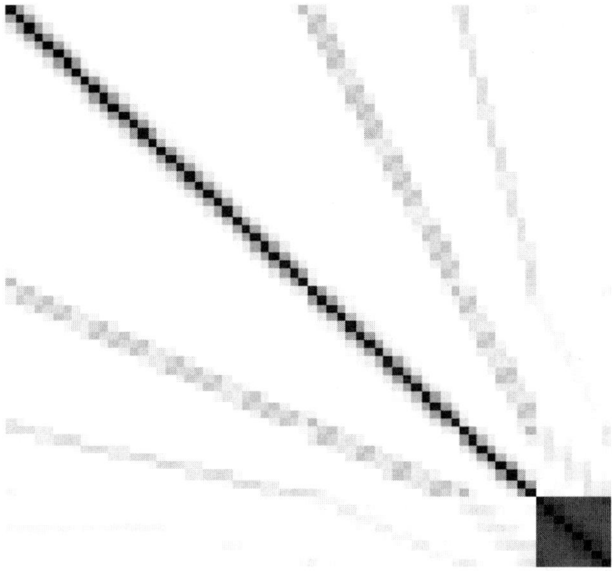

FIGURE 4.14 Impedance matrix of standard form from fast wavelet transform. (Courtesy: X. Zhu, G. Lei, and G. Pan, *J. Comput. Phys.*, **132**, 299–311, 1997.)

We expand the unknown in terms of basis functions with unknown coefficients

$$J_z = \sum_m J_m B_m(l),$$

$$\frac{\partial J_z}{\partial n} = \sum_m K_m B_m(l),$$

where $B_m(l)$ are wavelet basis functions.

4.9.4 Numerical Results

The results are obtained using the Coifman wavelets. It was recognized that the Coifman scalets provided fast evaluations of the matrix entries because of the one-point quadrature formula associated with the zero-moment property of the Coifman scalets. Detailed discussions can be found in Chapter 9. Here, in Fig. 4.14, we demonstrate the sparse FWT matrix.

4.10 INTERVALLIC COIFMAN WAVELETS

In order to expand an unknown function, which is defined on [0, 1] in terms of periodic wavelets, the function itself must have equal value at the two end points, 0 and 1. This condition is rather restrictive, and it has limited the application of the periodic wavelets. An alternate approach, that is less restrictive, is the use of intervallic wavelets. Let us sketch the construction of orthogonal intervallic scalets on [0, 1], which is a modification of the approach in [14] and [15]. This approach converts a regular (unbounded) wavelet into its corresponding intervallic (bounded) wavelet within the domain [0, 1]. The discussion focuses on Coiflets, but it applies to other wavelets as well.

4.10.1 Intervallic Scalets

Consider an orthonormal basis of the Coifman scalet $\varphi(x)$. The zero-moment property that was discussed in Chapter 3 is

$$\int x^l \varphi(x)\, dx = 0, \qquad l = 1, 2, \ldots, 2K - 1,$$

where $2K$ is the order of the wavelets. The nonzero support is $6K - 1$, namely

$$\operatorname{clos}\{x\colon \varphi(x) \neq 0\} = [-2K, 4K - 1].$$

For $\varphi(2^j x - k)$, we need to have $(2^j x - k) \in [-2K, 4K - 1]$. Therefore

$$x \in [2^{-j}(-2K + k), 2^{-j}(4K + k - 1)].$$

Let us denote the corresponding interval as

$$B_{j,k} := \operatorname{clos}\{x\colon \varphi_{j,k}(x) \neq 0\} = [2^{-j}(-2K + k), 2^{-j}(4K + k - 1)]. \tag{4.10.1}$$

We divide the regular (unbounded) wavelets into three groups:

(1) The left group, S_j^L, intercepting the left boundary point 0.
(2) The right group, S_j^R, intercepting the right boundary point 1.
(3) Completely situated within the support. No treatment is necessary.

The two groups, S_j^L and S_j^R are treated in a similar manner.

The Left Group. Define a set consisting of integers $S_j^L = \{k : 0 \in B_{j,k}\}$. To find $k \in S_j^L$, we solve

$$\begin{cases} 2^{-j}(-2K + k) \leq 0 \\ 2^{-j}(4K + k - 1) \geq 0. \end{cases} \tag{4.10.2}$$

It follows that

$$\begin{cases} k \leq 2K \\ k \geq -4K + 1. \end{cases}$$

Factor 2^{-j} does not play any role in finding k since all of through by this factor. Equation (4.10.2) can be divided. Therefore

$$-4K + 1 \leq k \leq 2K.$$

The two equal signs represent the two wavelets that touch but do not intercept the point 0. The first summation of the expansion in terms of scalets is

$$\sum_{-4K+2}^{2K-1} \cdot$$

The Right Group. In a similar fashion, for the end point 1,

$$\begin{cases} 2^{-j}(-2K+k) \le 1 \\ 2^{-j}(4K+k-1) \ge 1 \end{cases} \Rightarrow \begin{cases} k \le 2^j + 2K \\ k \ge 2^j - 4K + 1, \end{cases}$$

or equivalently

$$2^j - 4K + 1 \le k \le 2^j + 2K.$$

Again, we need to rule out the two wavelets that touch point 1. We will use the summation

$$\sum_{k=2^j-4K+2}^{2^j+2K-1} \cdot$$

We wish to build the left basis functions from the wavelets in the left group. Ideally these wavelets are orthogonal to the wavelets in the central group, and orthogonal to the right basis functions. Finally they are orthonormal among themselves within the group. The construction begins with the expansion of the monomials

$$x^r|_{[0,1]} = \left(\sum_{k=-4K+2}^{2K-1} + \sum_{2K}^{2^j-4K+1} + \sum_{2^j-4K+2}^{2^j+2K-1} \right) \langle x^r, \varphi_{j,k} \rangle \varphi_{j,k}(x)|_{[0,1]},$$
$$r = 0, 1, \ldots, 2K-1. \tag{4.10.3}$$

Obviously we must have a sufficiently high level j in order to construct the intervallic wavelets. To separate the left group from the the right group, the central group must not be empty, namely

$$2K \le 2^j - 4K + 1,$$

that is,

$$6K \le 2^j + 1.$$

For instance, if

$$K = 1, \qquad j \ge 3$$
$$K = 2, \qquad j \ge 4$$

$$K = 3, \qquad j \geq 5$$
$$K = 4, \qquad j \geq 5$$
$$\dots$$

As derived in (4.10.1) that the support

$$B_{j,k} = \text{clos}\{x: \varphi_{j,k}(x) \neq 0\} = [2^{-j}(-2K + k), 2^{-j}(4K + k - 1)].$$

The equation above reveals that if j is large enough, a particular $\varphi_{j,k}$ can intersect at most one of the two endpoints. We may define the set of indexes

$$S_j = \{k: B_{j,k} \bigcap (0, 1) \neq 0\}.$$

The set S_j can be classified into three subsets:

$S_j^L = \{k: 0 \in B_{j,k}\}$, these wavelets intersect 0.

$S_j^R = \{k: 1 \in B_{j,k}\}$, these wavelets intersect 1.

$S_j^I = \{k: B_{j,k} \in (0, 1)\}$, these wavelets are completely contained in (0, 1).

For sufficiently large j, S_j^L and S_j^R are disjoint, and

$$S_j = S_j^L \bigcup S_j^I \bigcup S_j^R.$$

Explicitly

$$\begin{aligned} S_j^L &= \{k: 2^j - 4K \leq k \leq 2K - 1\}, \\ S_j^I &= \{k: 2K \leq k \leq 2^j - 4K + 1\}, \\ S_j^R &= \{k: 2^j - 4K + 2 \leq k \leq 2^j + 2K - 1\}. \end{aligned} \tag{4.10.4}$$

Wavelets with the translation indexes $k \in S_j^I$ reside completely within (0, 1) and do not need any special treatment. In contrast, those in S_j^L and S_j^R are incomplete and must be reconstructed, forming the left- and right-edge basis functions, respectively. One approach to making the edge bases is to employ monomials: $x^0, x, x^2, \dots, x^{r-1}$. Figure 4.15 is a plot of these curves with $r = 4$. For any monomial x^r, $r \leq 2K - 1, r \in Z$, we have

$$x^r = \sum_k \langle x^r, \varphi_{j,k} \rangle \varphi_{j,k}(x), 0 \leq r \leq 2K - 1, \tag{4.10.5}$$

where $\varphi_{j,k}(x)$ is unrestricted, namely

$$x \in R.$$

In the special case that $r = 0$, equation (4.10.5) becomes

$$1 = \sum_k \varphi_{j,k}(x).$$

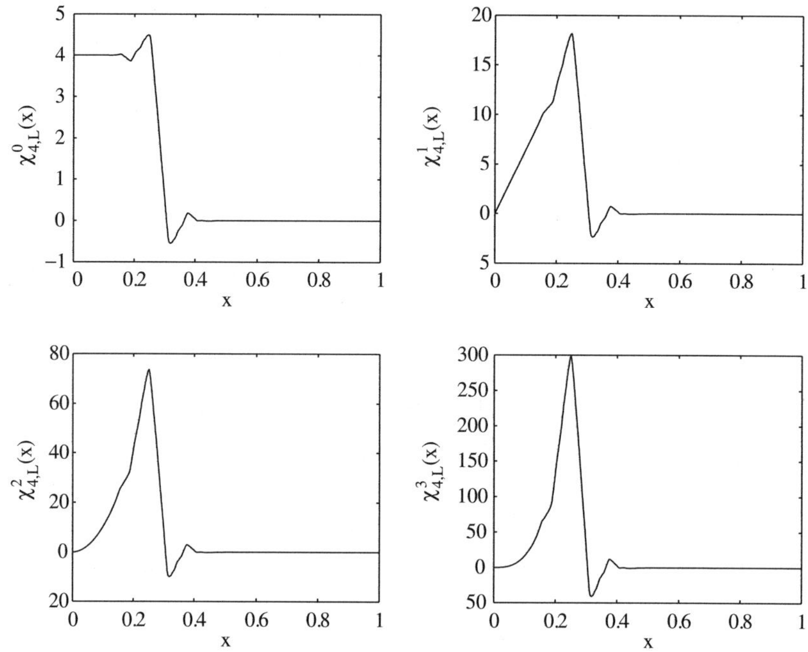

FIGURE 4.15 Left-edge basis before orthonormalization.

Next, if $x \in [0, 1]$,

$$x^r|_{[0,1]} = \sum_k \langle x^r, \varphi_{j,k} \rangle \varphi_{j,k}(x)|_{[0,1]}. \tag{4.10.6}$$

To provide a geometric explanation of the equation above near the left endpoint 0, we sketch a diagram in Fig. 4.16, in which x^2 is considered. In this example the summation in (4.10.6) only consists of two terms.

$\langle x^2, \varphi_{j,-2} \rangle\, \varphi_{j,-2}(x)|_{[0,1]} = 0$, dropped from (4.10.6) because $\varphi_{j,-2}$ is beyond $[0, 1]$.

$\langle x^2, \varphi_{j,-1} \rangle\, \varphi_{j,-1}(x)|_{[0,1]}$ is a nonzero term in (4.10.6) and $\varphi_{j,-1}$ is a partial segment.

$\langle x^2, \varphi_{j,0} \rangle\, \varphi_{j,0}(x)|_{[0,1]}$ is a nonzero term in (4.10.6) and $\varphi_{j,0}$ is a partial segment.

$\langle x^2, \varphi_{j,1} \rangle\, \varphi_{j,1}(x)|_{[0,1]}$ is excluded from (4.10.6) because $\varphi_{j,1}$ is a complete wavelet.

The new basis functions, $x^r_{j,L}$ $(r = 0, 1, \ldots, 2K - 1)$ are defined as

$$x^r_{j,L} = \sum_{k \in S^l_j} \langle x^r, \varphi_{j,k} \rangle \varphi_{j,k}(x)|_{[0,1]}.$$

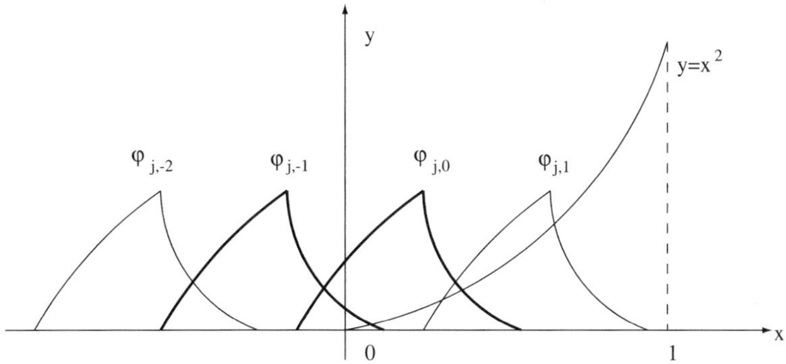

FIGURE 4.16 Schematic sketch in construction of a left-edge basis.

For the example of Fig. 4.16,

$$
\begin{array}{ll}
r=0 & x^0_{j,L} = \langle x^0, \varphi_{j,-1}\rangle \varphi_{j,-1}(x)|_{[0,1]} + \langle x^0, \varphi_{j,0}\rangle \varphi_{j,0}(x)|_{[0,1]} \\
r=1 & x^1_{j,L} = \langle x, \varphi_{j,-1}\rangle \varphi_{j,-1}(x)|_{[0,1]} + \langle x, \varphi_{j,0}\rangle \varphi_{j,0}(x)|_{[0,1]} \\
r=2 & x^2_{j,L} = \langle x^2, \varphi_{j,-1}\rangle \varphi_{j,-1}(x)|_{[0,1]} + \langle x^2, \varphi_{j,0}\rangle \varphi_{j,0}(x)|_{[0,1]} \\
r=3 & \cdots
\end{array}
$$

The new left basis is orthogonal to any of the central basis functions. This is due to the fact that an incomplete φ is orthogonal to any complete (interior) φ of the same level. (Verify this statement for yourself, and note that any two incomplete φ are no longer orthogonal to each other.) The left basis is a summation with each term being an incomplete φ multiplied by a coefficient. These left basis functions, prior to orthonormalization, are plotted in Fig. 4.15.

In a similar fashion the right group of basis functions $x^r_{j,R}$ are constructed and they are illustrated in the Fig. 4.17. The intervallic scalets and regular scalets satisfy the following orthogonal relations:

$$
\begin{aligned}
&\langle x^r_{j,L}(x), \varphi_{j,k}(x)\rangle = 0, \\
&\langle x^r_{j,R}(x), \varphi_{j,k}(x)\rangle = 0, \\
&\langle x^r_{j,R}(x), x^r_{j,L}(x)\rangle = 0, \qquad r = 0, 1, \ldots, 2k-1.
\end{aligned}
$$

These equations indicate that the left basis and interior basis (regular) at the same level j are orthogonal. By the same token, the right basis and interior basis functions at the same level j are orthogonal. The left basis and right basis are orthogonal because they are disjoint. Notice that orthogonality does not hold among the left basis functions. Nonetheless, they can be orthonormalized by linear transformation, such as by the Schmidt–Cramer procedure. In conclusion, wavelets of the left, central, and right group are mutually orthogonal. Every basis is orthonormal within the central

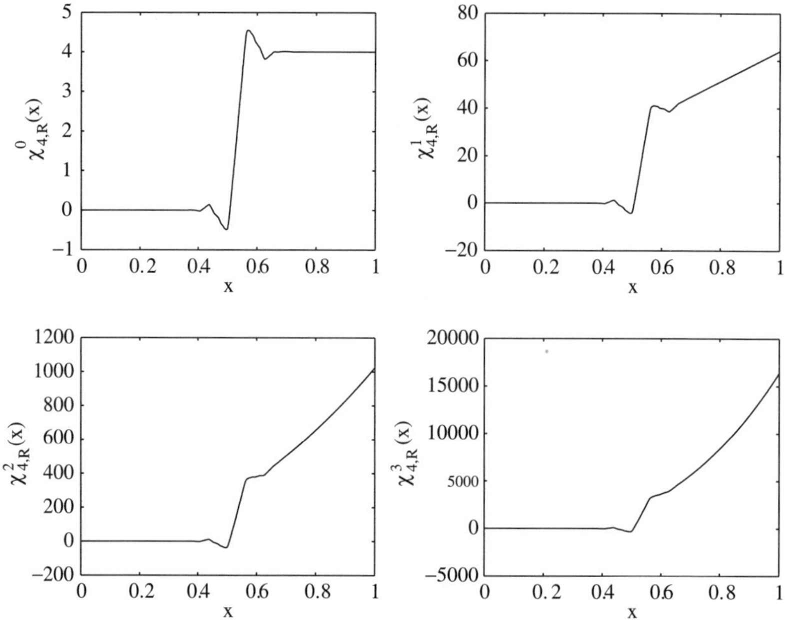

FIGURE 4.17 Right-edge basis before orthonormalization.

group. However, wavelets within the left group are not yet orthonormal to each other, nor are wavelets within the right group.

Return to (4.10.3),

$$x^r|_{[0,1]} = \sum_k \langle x^r, \varphi_{j,k}\rangle \varphi_{j,k}(x)|_{[0,1]}$$

$$= \left(\sum_{k=-4K+2}^{2K-1} + \sum_{2K}^{2^j-4K+1} + \sum_{2^j-4K+2}^{2^j+2K-1} \right) \langle x^r, \varphi_{j,k}\rangle \varphi_{j,k}(x)|_{[0,1]}.$$

Define new basis functions with normalized coefficients

$$X^r_{j,L} = 2^{j[r+(1/2)]} x^r_{j,L}$$
$$= 2^{j[r+(1/2)]} \sum_{k=-4K+2}^{2K-1} \langle x^r, \varphi_{j,k}\rangle \varphi_{j,k}(x)|_{[0,1]} \qquad (4.10.7)$$

and

$$X^r_{j,R} = 2^{j(r+\frac{1}{2})} \sum_{k=2^j-4K+2}^{2^j+2K-1} \langle x^r, \varphi_{j,k}\rangle \varphi_{j,k}(x)|_{[0,1]}, \qquad r = 0, 1, \ldots, 2K-1.$$

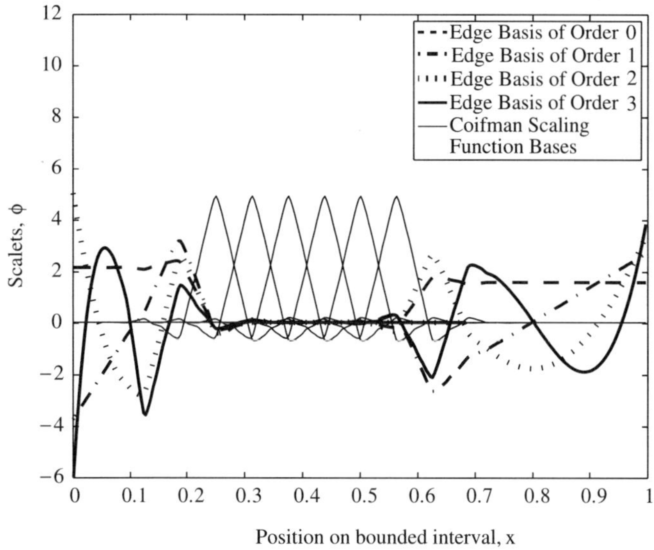

FIGURE 4.18 Coifman intervallic scalets at level 4.

The left group and right group must be disjoint, namely

$$-2K + (2^j - 4K + 2) \geq (2K - 1) + 4K - 1,$$

from (4.10.1). It follows that

$$2^j \geq 12K - 4.$$

For

$$K = 2,$$
$$j \geq 5.$$

Figure 4.18 depicts the Coifman intervallic scalets at level 4, for use in the solution of integral equations. In this figure, these can be seen clearly the left group, right group, and central group of the basis functions. Note that the left and right groups slightly overlap.

Hence

$$2^{j/2}(2^j x)^r = X^r_{j,L} + 2^{j[r+(1/2)]} \sum_{k=2K}^{2^j-4K+1} \langle x^r, \varphi_{j,k}\rangle \varphi_{j,k}(x)|_{[0,1]} + X^r_{j,R}.$$

Without risk of confusion, we abuse the notation by denoting $X^r_{j,L}$ and $X^r_{j,R}$ as $x^r_{j,L}$ and $x^r_{j,R}$ in the rest of this section. The new subspaces $\tilde{V}_j$

$$\tilde{V}_j = \overline{\{x^r_{j,L}\}}_{r\leq 2K-1} \bigcup \overline{\{\varphi_{j,k}(x)|_{[0,1]}\}}_{k=2K}^{2^j-4K+1} \bigcup \overline{\{x^r_{j,R}\}}_{r\leq 2K-1}.$$

That is, these subspaces are from the linear span of the basis functions. The collections

$$\{x^r_{j,L}\}_{r\le 2K-1}$$

$$\{x^r_{j,R}\}_{r\le 2K-1}$$

$$\{\varphi_{j,k}(x)|_{[0,1]}\}_{k=2K}^{2^j-4K+1}$$

are mutually orthogonal. It can be proved that

$$\tilde{V}_j \subset \tilde{V}_{j+1} \subset \tilde{V}_{j+2} \subset \cdots$$

and $\tilde{V}_j$ form an MRA of $L^2([0,1])$. Therefore all functions in the collections are linearly independent and can be used as bases. In order to form an orthonormal basis, we need to orthogonalize $x^r_{j,L}$ (and $x^r_{j,R}$), $r = 0, 1, \ldots, 2K - 1$. The remaining tasks are to orthogonalize the basis functions within the left and right group. The traditional Schmidt–Cramer orthogonalization can be applied, or equivalent matrix manipulation may be conducted.

Case 1. Schmidt–Cramer Orthogonalization From a set of linearly independent bases $\{a_1, \ldots, a_n\}$, we can generate a set of orthonormal bases $\{b_1, \ldots, b_n\}$ as follows:

$$\begin{aligned} b_1 &= a_1, \\ b_2 &= a_2 - \frac{\langle a_2, b_1\rangle}{\langle b_1, b_1\rangle} b_1, \\ &\vdots \\ b_{k+1} &= a_{k+1} - \frac{\langle a_{k+1}, b_1\rangle}{\langle b_1, b_1\rangle} b_1 - \cdots - \frac{\langle a_{k+1}, b_k\rangle}{\langle b_k, b_k\rangle} b_k. \end{aligned} \tag{4.10.8}$$

Case 2. Matrix Approach Define the new orthonormal (o.n.) basis

$$\varphi^r_{j,L} = \sum_{p=0}^{2K-1} a^L_{r,p} x^p_{j,L}, \tag{4.10.9}$$

where the matrix $A = \{a_{r,p}\}_{2K\times 2K}$ is yet to be found. Consider a matrix $Q = \{q_{m,n}\}$, where

$$q_{m,n} = \langle x^m_{j,L}, x^n_{j,L}\rangle.$$

The orthonormality condition of $\langle \varphi^r_{j,L}, \varphi^s_{j,L}\rangle = \delta_{rs}$ is then $AQA^t = I$. It can be shown that matrix Q is symmetrical and positively definite. Hence we can apply Cholesky decomposition, namely

$$Q = CC^\dagger,$$

where $C^{\dagger}$ is the complex conjugate and transpose of C. The coefficient matrix in (4.10.9) is

$$A = C^{-1}.$$

The previous matrix approach is outlined for the left edge basis, but it can be employed to orthonormalize the right edge basis functions as well

$$\varphi^r_{j,R} = \sum_{p=0}^{2K-1} a^R_{r,p} x^p_{j,R}.$$

The orthonormalized left basis functions and right basis functions are presented in Figs. 4.19 and 4.20, respectively.

The dimension of $\tilde{V}_j$ is

$$\dim \tilde{V}_j = 2^j - 2K + 2.$$

In fact we have the left-edge basis, which consists of $2K$ basis functions

$$\varphi^r_{j,L}, \qquad r = 0, 1, 2, \ldots, 2K - 1.$$

They may be rearranged as $\varphi^{2K-k}_{j,L}$, $k = 1, 2, \ldots, 2K$. There are $2K$ basis functions in the right edge group, $\varphi^r_{j,R}$, and a quantity of $(2^j - 4K + 1) - (2K) + 1 = 2^j - 6K + 2$

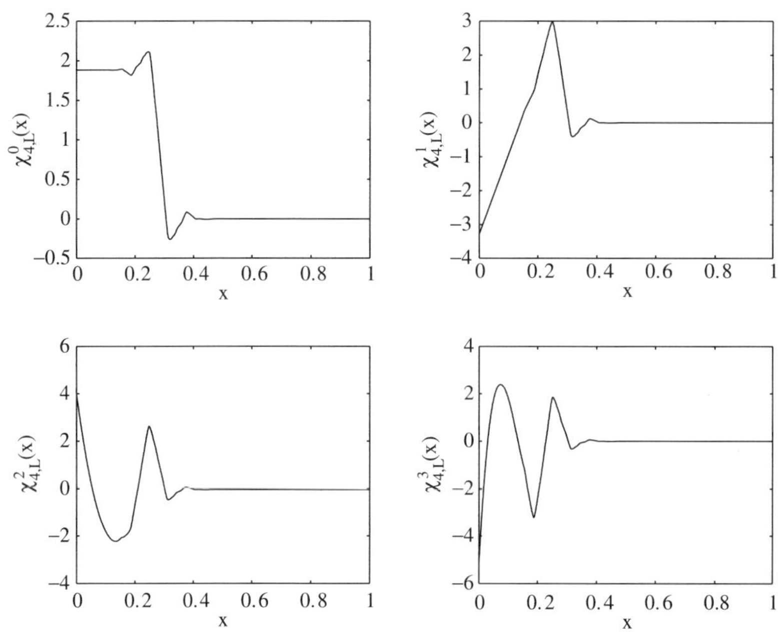

FIGURE 4.19 Left-edge basis after orthonormalization.

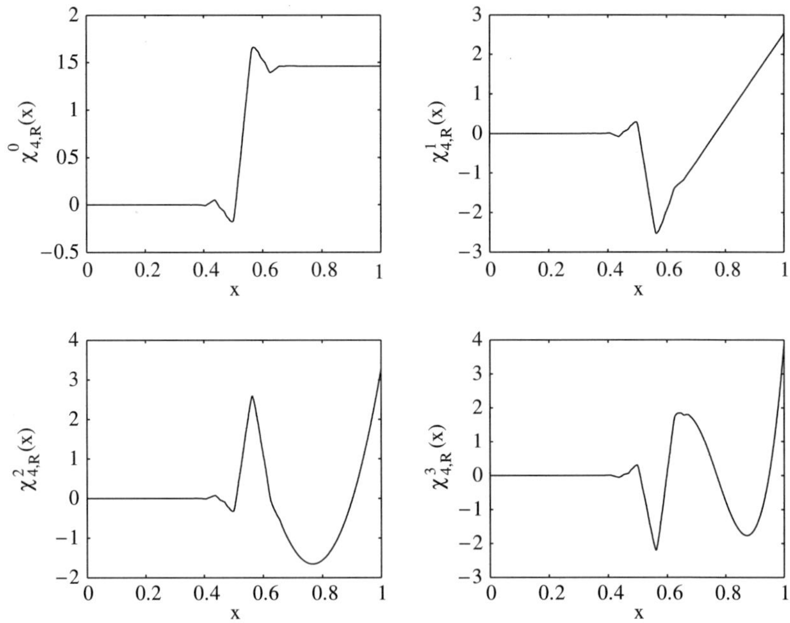

FIGURE 4.20 Right-edge basis after orthonormalization.

central basis functions from

$$\sum_{2K}^{2^j-4K+1}.$$

Thus

$$\phi_{j,k} = \begin{cases} \varphi^k_{j,L}, & k = 0, 1, 2, \ldots, 2K-1 \\ \varphi_{j,k}, & k = 2K, \ldots, 2^j - 4K + 1 \\ \varphi^{k-(2^j-4K+2)}_{j,R}, & k = 2^j - 4K + 2, \ldots, 2^j - 2K + 1. \end{cases} \tag{4.10.10}$$

The number of total bases in $\tilde{V}_j$ is $2K + (2^j - 6K + 2) + 2K = 2^j - 2K + 2$, that is to say, $\dim \tilde{V}_j = 2^j - 2K + 2$.

4.10.2 Intervallic Wavelets on [0, 1]

The derivation of intervallic wavelets is much more complicated than that of intervallic scalets. In the text, only the major results are outlined. Detailed discussions can be found in the Appendix to this chapter.

Since

$$\tilde{V}_{j+1} = \tilde{V}_j \oplus \tilde{W}_j,$$

immediately we obtain

$$\dim \tilde{W}_j = \dim \tilde{V}_{j+1} - \dim \tilde{V}_j = 2^j.$$

There are certain $\psi_{j,k}$ that are both completely supported within [0, 1] and belong to $\tilde{V}_{j+1}$. For instance, these conditions are satisfied when k belongs to the set

$$\{k: 3K - 1 \le k \le 2^j - 3K\}.$$

Note that

$$\begin{aligned}
\operatorname{clos}\{x: \varphi(x) \neq 0\} &= [-2K, 4K - 1],\\
\operatorname{clos}\{x: \psi(x) \neq 0\} &= [1 - 3K, 3K].
\end{aligned}$$

Thus these functions $\psi_{j,k}$ are in $\tilde{W}_j$. Comparing this equation against the dimension of $\tilde{W}_j = 2^j$, we may need an additional $6K - 2$ functions. Approximately one-half of them are located near the left endpoint; the other half are near the right endpoint.

To find the remaining functions, we must identify functions in $\tilde{V}_{j+1}$ that cannot be written as combinations of either functions in $\tilde{V}_j$ or in $\psi_{j,k}$, which we have identified. From the MRA, in $L^2(R)$, any basis in $\tilde{V}_{j+1}$

$$\varphi_{j+1,k}(x) = \sum_m h_{k-2m}\varphi_{j,m}(x) + \sum_m g_{k-2m}\psi_{j,m}(x). \tag{4.10.11}$$

Show. Let

$$\varphi_{j+1,m}(x) = \sum_k (a_k\varphi_{j,k} + b_k\psi_{j,k}).$$

Multiplying both sides by $\int dx\varphi_{j,p}$, we obtain

$$\begin{aligned}
\text{LHS} &= \int dx\varphi_{j+1,m}\varphi_{j,p}\\
&= \int dx\varphi_{j+1,m}\sum_n \varphi_{j+1,n}h_{n-2p}\\
&= \sum_n h_{n-2p}\int dx\varphi_{j+1,m}\varphi_{j+1,n}\\
&= h_{m-2p}\\
\text{RHS} &= \sum_k (a_k\langle\varphi_{j,k},\varphi_{j,p}\rangle + b_k\langle\psi_{j,k},\varphi_{j,p}\rangle)\\
&= a_p.
\end{aligned}$$

In the derivations above, we have used (4.8.5) and (4.8.6). The second summation in (4.10.11) can be derived in the same manner. □

The left-edge bases of $\tilde{V}_{j+1}$ can be obtained from (4.10.11) using $k = 8K - 1, 8K - 2, \ldots, 2K + 1$. In this sequence of functions, every second one is linearly dependent on the previous ones (modulo functions in $\tilde{V}_j$). The additional functions in $\tilde{V}_j$ at the left point can now be written as

$$\psi_{j,L}^{r} = \varphi_{j+1,8K-2r+1} - \sum_{l} \langle \varphi_{j+1,8K-2r+1}, \phi_{j,l} \rangle \phi_{j,l};$$

similarly

$$\psi_{j,R}^{r} = \varphi_{j+1,2^{j+1}-10K+2r} - \sum_{l} \langle \varphi_{j+1,2^{j+1}-10K+2r}, \phi_{j,l} \rangle \phi_{j,l},$$

where $\phi_{j,l}$ are the intervallic scalets and were defined in (4.10.10). Finally

$$\psi_{j,k} = \begin{cases} \psi_{j,L}^{k}, & k = 1, 2, \ldots, 3K - 1 \\ \psi_{j,k}, & k = 3K + 1, 3K + 2, \ldots, 2^j - 3K + 1 \\ \psi_{j,R}^{k-2^j+3K}, & k = 2^j - 3K + 2, \ldots, 2^j. \end{cases}$$

For more information, readers are referred to the Appendix to this chapter.

4.11 LIFTING SCHEME AND LAZY WAVELETS

Before beginning, let us determine the notation. We will always assume the interval to be [0, 1] and a set of points $\{x_{j,k} \mid j \in J, k \in K(j)\}$. Here j denotes the level of scalets. One can think of $K(j)$ as a general index set. The index k ranges from 0 to 2^j. We assume that $K(j) \subset K(j + 1)$. In the refinement relations, $0 \leq k < 2^j + 1$, while $0 \leq l < 2^{j+1}+1$; N and $\tilde{N}$ denote numbers of vanishing moments for ordinary wavelets and for dual wavelets.

4.11.1 Lazy Wavelets

Given a set of points on an interval, one can formally associate scalets ϕ functions and dual scalets $\tilde{\phi}$ with the "lazy wavelet" [16]

$$\begin{aligned} \phi_{j,k} &= \delta_{k,k'}, \\ \tilde{\phi}_{j,k} &= \delta(x - x_k), \\ \psi_{j,k} &= \phi_{j+1,2k+1}, \\ \tilde{\psi}_{j,k} &= \tilde{\phi}_{j+1,2k+1}. \end{aligned}$$

Formally these wavelets are biorthogonal, but $\tilde{\phi}_{j,k}$ does not belong to space L^2, while $\phi_{j,k}$ is zero. It is definite that $N = \tilde{N} = 0$. The lazy wavelet transform is an orthogonal transform that does nothing.

For every scalet $\phi_{j,k}$, coefficients $\{h_{j,k,l}\}$ exist, such that the scalet satisfies a generalized refinement relationship

$$\phi_{j,k} = \sum_l h_{j,k,l}\phi_{j+1,l}.$$

Each scalet satisfies different refinement relations. For the lazy wavelets $h_{j,k,l} = \delta_{2k,l}$. The dual scalets satisfy refinement relations with coefficients $\{\tilde{h}_{j,k,l}\} = \delta_{2k,l}$.

For the wavelets we have the refinement relationship

$$\psi_{j,k} = \sum_l g_{j,k,l}\phi_{j+1,l}.$$

This is also true for the dual wavelets. The dual wavelets $\tilde{\psi}_{j,k}$ are biorthogonal to the wavelets, namely

$$\langle \psi_{j,k}, \tilde{\psi}_{j',k'} \rangle_w = \delta_{j,j'}\,\delta_{k,k'}.$$

4.11.2 Lifting Scheme Algorithm

The basic idea that inspired the name is to start from a simple or trivial multiresolution analysis and build a new, more preferable one. In doing so, we leave the dual scalet untouched. A new dual wavelet $\tilde{\psi}_{j,k}$ is built by taking the old wavelet $\tilde{\psi}^0_{j,k}$ and adding up linear combinations of dual scalets on the same level. This results in

$$\tilde{\psi}_{j,m} = \tilde{\psi}^0_{j,m} - \sum_l \tilde{s}_{j,k,l}\tilde{\phi}_{j+1,l}.$$

Here we have chosen the constants $\tilde{s}_{j,k,l}$ such that

$$\int_0^1 w(x)\tilde{\psi}_{j,m}(x)\,dx = 0,$$

$$\int_0^1 w(x)x\tilde{\psi}_{j,m}(x)\,dx = 0,$$

$$\int_0^1 w(x)x^2\tilde{\psi}_{j,m}(x)\,dx = 0,$$

$$\int_0^1 w(x)x^3\tilde{\psi}_{j,m}(x)\,dx = 0,$$

to preserve $N = 4$ moments.

Results of "Dual Lifting" of Lazy Wavelets

(1) The dual scalet remains the same after lifting.

(2) The dual wavelet changed. Now it has $N = 4$ vanishing moments, but is still a combination of δ-functions. New coefficients of the refinement relation are

$$\tilde{g}_{j,m,l} = \delta_{2m+1,l} + \sum_{k=0}^{2^j} \tilde{s}_{j,k,m}\, \delta_{l,2k}.$$

(3) The primary scalet changed too. New coefficients of the refinement relation are

$$h_{j,k,l} = \delta_{2k,l} + \sum_{m=0}^{2^j} \tilde{s}_{j,k,m}\, \delta_{l,2m+1}.$$

(4) The primary wavelet changed. It still obeys its old refinement relations, but now with respect to new scalets.

The remarkable thing happened with the primary scalet. Before the lifting, it was a simple pulse function. After the lifting, it becomes the interpolating scalet, and locally it is a polynomial of $\tilde{N} - 1$ order; however, the dual scalet is still a δ-function, and the primary wavelet does not preserve any moment. Next we can consider the lifting of the primary wavelet. As for the case of the dual wavelet, a new primary wavelet $\psi_{j,m}$ is built by taking the old wavelet $\psi^0_{j,m}$ and adding on linear combinations of primary scalets on the same level. This results in

$$\psi_{j,m} = \psi^0_{j,m} - \sum_l s_{j,k,l} \phi_{j+1,l}.$$

Here we choose the constants $s_{j,k,l}$ such that

$$\int_0^1 w(x)\psi_{j,m}(x)\, dx = 0,$$

$$\int_0^1 w(x)x\psi_{j,m}(x)\, dx = 0,$$

$$\int_0^1 w(x)x^2\psi_{j,m}(x)\, dx = 0,$$

$$\int_0^1 w(x)x^3\psi_{j,m}(x)\, dx = 0,$$

in order to preserve $\tilde{N} = 4$ moments.

Results of Lifting of Interpolating Wavelets

(1) The primary scalet remains the same after lifting.
(2) The primary wavelet has changed. Now it has $\tilde{N} = 4$ vanishing moments, and it is a smooth function.
(3) The dual scalet has also changed. New coefficients of the refinement relationship are

$$\tilde{h}_{j,k,l} = \delta_{2k,l} + s_{j,k,l}.$$

(4) The dual wavelet has changed, but it still obeys its old refinement relations from the previous step, with respect to new scalets.

4.11.3 Cascade Algorithm

The manner in which we will construct the primary scalets and wavelets is clear. Now the question is how we will construct the dual scalets after the lifting of interpolating wavelets. To do so, we use the cascade algorithm.

Cascade Algorithm. Suppose that we want to build $\tilde{\phi}_{j_0,k_0}$. First, we define a Kronecker sequence $\{\lambda_{j_0,k} = \delta_{j_0,k_0} \mid k \in K(j_0)\}$. Then we generate sequences $\{\lambda_{j,k} \mid k \in K(j)\}$ for $j > j_0$ by recursively applying the formula

$$\lambda_{j+1,l} = \sum_{k=0}^{2^j} \tilde{h}_{j,k,l} \lambda_{j,k}.$$

The limit functions satisfy

$$\lim_{j \to \infty} \lambda_{j,k} = \tilde{\phi}_{j_0,k_0}.$$

After dual scalets are built, dual wavelets can be constructed by using refinement relationships for dual wavelets, where matrices $\tilde{g}_j$ are determined by dual lifting.

Matrices $\tilde{h}_j$ and $\tilde{g}_j$ depend on index j, and they are different for different values of j. To construct these matrices, we have to perform the primary lifting and dual lifting of the lazy wavelets along each step of the cascade algorithm. The application of the lazy wavelets and lifting scheme can be found in [17].

4.12 GREEN'S SCALETS AND SAMPLING SERIES

In this section we will construct the bases of $L^2(R)$ from Green's functions [18]. These bases are either orthogonal scalets or sampling functions. In the former case we obtain the wavelet bases, while in the latter we build up the multiscale sampling expansions.

The traditional wavelets are difficult to use in the differential equations. This is because a differential operator $\mathcal{D}$ usually takes a function $f \in V_0$ completely outside of any space in the multiresolution decomposition $\{V_m\}$. In regard such as to scalets $\varphi(t)$, $\varphi'(t)$ cannot in general be represented by a series of the basis functions, namely

$$\varphi'(t) \neq \sum_n \alpha_{m,n}\varphi(2^m t - n)$$

for any m. One exception is the Shannon wavelet, where $\varphi'(t)$ is expanded in terms of $\varphi(2^m t - n)$. Nonetheless, such a series expansion converges very slowly, in contrast to the case of Fourier series. For a partial sum

$$f_p(t) = \sum_{|k|\le p} c_k e^{ikt},$$

$\mathcal{D}(f_p)$ maps the partial sum onto a trigonometric polynomial of the same degree

$$f_p'(t) = i \sum_{|k|\le p} k c_k e^{ikt}.$$

However, for certain differential operators there seems to be a natural way to define scalets and sampling functions from Green's functions. Because these orthogonal scalets or sampling functions are the superposition of Green's functions, they satisfy the Poisson or Helmholtz equations with prespecified boundary conditions. They can be very powerful in representing the unknown functions of an electromagnetics problem.

4.12.1 Ordinary Differential Equations (ODEs)

First, let us consider an ordinary differential operator $\mathcal{D}$ and the corresponding ODE:

$$\mathcal{D}y(t) = f(t). \tag{4.12.1}$$

The corresponding Green's function is

$$g(t) = \tfrac{1}{2}\,\mathrm{sgn}\,(t),$$

satisfying

$$\mathcal{D}g = \delta(t).$$

The solution to (4.12.1) in the convolutional form is

$$y = g * f.$$

Note that

$$g \notin L^1(R),$$

so the convolution may not exist.

If we plan to find a solution by means of the sampling theorem with sampling function $\varphi(t)$, we express the RHS of (4.12.1) as

$$f(t) = \sum_n f(n)\varphi(t-n).$$

Thus

$$y(t) = \sum_n u_n \varphi(t-n)$$

and

$$\begin{aligned} y'(t) &= \sum_n u_n \varphi'(t-n) \\ &= \sum_n u_n \sum_k \gamma_{k-n} \varphi(t-k) \end{aligned}$$

provided that $\varphi' \in V_0$. However, this condition does not usually hold. Even in the Shannon theorem, in which it does hold, the final answer involves the solution of the discrete convolution

$$\sum_k \gamma_{k-n} u_n = f(k).$$

In the rest of the section we will use the Green's function to construct a scalet or a sampling function. From Chapter 3 we have learned that the orthogonality of $\varphi(t-n)$ corresponds to Eq. (3.2.4), namely

$$\sum_k |\hat{\varphi}(\omega + 2\pi k)|^2 = 1.$$

In a similar manner the sampling property requires that

$$\sum_k \hat{\varphi}(\omega + 2\pi k) = 1.$$

The previous properties will be employed to convert a Green's function into a scalet or a sampling function.

Example 1 Derive the sampling function $\varphi_1(t)$ from the 1D Green's function, satisfying the condition

$$\frac{d}{dt} g_1(t) = -\delta(t).$$

Solution Taking the Fourier transform on both sides, we obtain

$$i\omega \hat{g}_1(\omega) = -1,$$

which is

$$\hat{g}_1(\omega) = \frac{-1}{i\omega}.$$

Let us define a scalet in the frequency domain

$$\hat{\varphi}_1(\omega) := \frac{\hat{g}_1(\omega)}{[\sum_k \mid \hat{g}_1(\omega + 2\pi k) \mid^2]^{1/2}}. \tag{4.12.2}$$

Clearly, as will be seen in Example 2,

$$\sum_k \mid \hat{\varphi}_1(\omega + 2k\pi) \mid^2 = 1;$$

namely $\{\varphi_1(t-n)\}$ forms an orthogonal system. The denominator in (4.12.2) can be simplified by the trick, Eq. (3.2.9), we performed in Chapter 3 while constructing the Franklin wavelet:

$$\begin{aligned}
\sum_k \frac{1}{\mid i(\omega + 2k\pi) \mid^2} &= \sum_k \frac{1}{\mid \omega + 2k\pi \mid^2} \\
&= \frac{1}{4} \sum_k \frac{1}{[\pi(f+k)]^2} \\
&= \frac{1}{4\sin^2(\pi f)} \\
&= \frac{1}{4\sin^2(\omega/2)}.
\end{aligned} \tag{4.12.3}$$

Hence

$$\begin{aligned}
\hat{\varphi}_1(\omega) &= \hat{g}_1(\omega) 2 \mid \sin\frac{\omega}{2} \mid \\
&= -\frac{\mid \sin(\omega/2) \mid}{i(\omega/2)}.
\end{aligned} \tag{4.12.4}$$

Interestingly we recognize that $g_1 \notin L^2(R)$ but that

$$\varphi_1 \in L^2(R).$$

Next we may derive the time-domain expression of the sampling function $\varphi_1(t)$. Noting that the full-wave rectifier $\mid \sin(\omega/2) \mid$ is a periodic even function with period 2π, we can see that it has a Fourier series

$$\mid \sin\frac{\omega}{2} \mid = \sum_m a_m e^{im\omega}$$

$$a_m = -\frac{1}{\pi}\frac{1}{m^2 - (1/2)^2}.$$

It follows that

$$\hat{\varphi}_1(\omega) = \frac{2}{\pi} \sum_k \frac{1}{k^2 - (1/4)} \cdot \frac{e^{ik\omega}}{i\omega}.$$

The corresponding time-domain expression is given by

$$\varphi_1(t) = \frac{2}{\pi} \sum_k \frac{1}{k^2 - 1/4} g_1(t+k).$$

Example 2 Show that $\{\varphi_1(t-n)\}$ forms an orthogonal system.

Show. From (4.12.2),

$$\hat{\varphi}_1(\omega) = \frac{\hat{g}(\omega)}{\sqrt{\sum_k |\, \hat{g}(\omega + 2k\pi)\,|^2}}$$

$$\begin{aligned}\hat{\varphi}_1(\omega + 2n\pi) &= \frac{\hat{g}(\omega + 2n\pi)}{\sqrt{\sum_k |\, \hat{g}(\omega + 2(n+k)\pi)\,|^2}} \\ &= \frac{\hat{g}(\omega + 2n\pi)}{\sqrt{\sum_k |\, \hat{g}(\omega + 2k\pi)\,|^2}}.\end{aligned}$$

Hence

$$\sum_n |\, \hat{\varphi}_1(\omega + 2n\pi)\,|^2 = \frac{\sum_n |\, \hat{g}(\omega + 2n\pi)\,|^2}{\sum_k |\, \hat{g}(\omega + 2k\pi)\,|^2} = 1. \qquad \square$$

This implies that $\{\varphi_1(t-n)\}$ forms an orthogonal system. It can be verified that φ_1 satisfies the dilation equation

$$\hat{\varphi}_1(\omega) = \left|\cos\frac{\omega}{4}\right| \hat{\varphi}_1\left(\frac{\omega}{2}\right).$$

Show. Using (4.12.4), we have

$$\left|\cos\frac{\omega}{4}\right| \hat{\varphi}_1\left(\frac{\omega}{2}\right) = \left|\cos\frac{\omega}{4}\right| \frac{|\sin(\omega/4)\,|}{i(\omega/4)} = \frac{|\sin(\omega/2)\,|}{i(\omega/2)} = \hat{\varphi}_1(\omega).$$

As a result φ_1 can be used to construct the MRA or to decompose V_m in the usual manner. The corresponding wavelet is

$$\hat{\psi}_1(\omega) = e^{-i(\omega/2)} \left|\sin\frac{\omega}{4}\right| \hat{\varphi}_1\left(\frac{\omega}{2}\right). \qquad \square$$

Example 3 Given a differential operator

$$\mathcal{L} = a^2 + \mathcal{D}^2 \tag{4.12.5}$$

with the corresponding Green's function g_2 satisfying

$$\mathcal{L} g_2 = -\delta,$$

find the scalet $\hat{\psi}_2$.

Solution The Fourier transform of (4.12.5) is

$$(a^2 - \omega^2)\hat{g}_2(\omega) = -1.$$

Therefore

$$\hat{g}_2(\omega) = \frac{-1}{a^2 - \omega^2}.$$

Note that $\hat{g}_2 \notin L^2(R)$ because of the singularity of $\hat{g}_2(\omega)$. In the time domain

$$g_2(t) = \frac{-\sin at}{a} \operatorname{sgn}(t).$$

For this example, we will build the sampling function instead of the scalet

$$\begin{aligned}\hat{\varphi}_2(\omega) &:= \frac{\hat{g}_2(\omega)}{\sum_k \hat{g}_2(\omega + 2\pi k)} \\ &= \frac{1/a^2 - \omega^2}{\sum_k 1/(a^2 - (\omega + 2k\pi)^2)}.\end{aligned}$$

The summation in the previous equation has a closed form expression due to the fact that

$$\frac{1}{a^2 - \omega^2} = \frac{1}{2a}\left(\frac{1}{\omega + a} - \frac{1}{\omega - a}\right)$$

and

$$\sum_k \frac{1}{\omega + a + 2k\pi} = \frac{1}{2\tan(\omega + a)/2},$$

which is obtained by integrating (3.2.9). It follows that

$$\begin{aligned}\sum_k \frac{1}{a^2 - (\omega + 2k\pi)^2} &= \frac{1}{2a}\left\{\sum_k \frac{1}{\omega + a + 2k\pi} - \sum_k \frac{1}{\omega - a + 2k\pi}\right\} \\ &= \frac{1}{4a}\left[\frac{\cos\frac{\omega+a}{2}}{\sin\frac{\omega+a}{2}} - \frac{\cos\frac{\omega-a}{2}}{\sin\frac{\omega-a}{2}}\right] \\ &= -\frac{1}{4a}\frac{\sin(\frac{\omega+a}{2} - \frac{\omega-a}{2})}{\frac{1}{2}(\cos a - \cos\omega)} \\ &= -\frac{\sin a}{2a}\frac{1}{\cos a - \cos\omega}.\end{aligned}$$

Therefore

$$\hat{\varphi}_2(\omega) = -\frac{2a}{\sin a} \cdot \frac{\cos a - \cos\omega}{a^2 - \omega^2}.$$

The nice feature of $\hat{\varphi}_2(\omega)$ is that the singularities at $\omega = \pm a$ have been removed. Therefore

$$\hat{\varphi}_2(\omega) \in L^2(R).$$

Unfortunately, $\varphi_2(t)$ does not satisfy the dilation equation and as such cannot be used for an MRA decomposition. The reason underlying for this is that the operator $\mathcal{L}$ is not homogeneous, that is to say,

$$a^2 + \mathcal{D}^2 = P(i\mathcal{D}) \neq \lambda^{-2} P(i\lambda\mathcal{D})$$

where P is a polynomial. Nonetheless, its translation still forms a generating set for a translation-invariant subset of $L^2(R)$.

Example 4 Homogeneous second-order differential operator $\mathcal{D}^2$. The Green's function is

$$g_3(t) = -\left|\frac{t}{2}\right|,$$

and its Fourier transform is

$$\hat{g}_3(\omega) = \frac{1}{\omega^2}.$$

The sampling function is a special case of $a = 0$ in Example 3:

$$\hat{\varphi}_3(\omega) = \left(\frac{\sin(\omega/2)}{\omega/2}\right)^2,$$

which is the sampling function of the B-spline of order 1. The wavelet was discussed in Chapter 3. If we wish to obtain the scalet from the operator $\mathcal{D}^2$, we use

$$\begin{aligned}
\hat{\varphi}_4(\omega) &= \frac{\hat{g}(\omega)}{\left\{\sum_k |\,\hat{g}(\omega + 2k\pi)\,|^2\ \right\}^{1/2}} \\
&= \frac{1/\omega^2}{\left\{\sum_k 1/(\omega + 2k\pi)^4\right\}^{1/2}} \\
&= \frac{\sin^2(\omega/2)}{(\frac{\omega}{2})^2\sqrt{1 - \frac{2}{3}\sin^2(\omega/2)}},
\end{aligned}$$

where we have used (3.2.11) to obtain the summation. The reader may recognize that this is the Franklin scalet of (3.2.8). Indeed, we can use wavelets to solve the differential equation

$$\mathcal{D}^2 u = f.$$

Let us approximate the excitation f in V_m:

$$f_m(t) = \sum_n f_{mn} 2^{m/2} \varphi_4(2^m t - n).$$

The solution in terms of the Green's function will be

$$u_m = g * f_m$$

if $g_4(t) \in L^1$, or is at least bounded. Unfortunately, $g_4(t)$ is unbounded. To get around this, we expand $g_4(t)$ in terms of the scalet

$$g_4(t) = \sum_n g_n \varphi_4(t - n),$$

where the coefficients g_n (not the bandpass filter) are the Fourier coefficients of the function

$$a(\omega) = \frac{4 \sin^2(\omega/2)}{\sqrt{1 - \frac{2}{3}\sin^2(\omega/2)}}.$$

It can be verified that the same g_m works on other scales, namely

$$g_4(t) = \sum_n g_n 2^{-m} \varphi_4(2^m t - n)$$

The solution, after many manipulations, is

$$u_m(t) = \sum_n \sum_k f_{mk} g_{k-n} 2^{-[m+(1/2)]} \theta(2^m t - n),$$

where

$$\theta(t) = \varphi_4(t) * \varphi_4(t)$$

is the sampling function, due to the orthogonality and symmetry of φ_4.

4.12.2 Partial Differential Equations (PDEs)

The theory of linear ordinary differential equations (ODEs) is rather complete. There is no special need to develop numerical computations in terms of integral equations and Green's functions. On the other hand, there is wide interest in numerical solutions of PDEs. There are many homogeneous PDEs that can give us the dilation equations that allow changes in scale. Major difficulties arise in higher dimensions, because the constructed scalets will not automatically be $\in L^2(R^d)$; it is hard to obtain closed-form expressions for scalets and sampling functions.

We begin with quite a general case of a differential operator $P(i\mathcal{D})$ on R^d. We assume that the operator is homogeneous, that is to say, that terms in the polynomial P have the same degree. This class of equations include Poisson's equation

$$\nabla^2\varphi = -\frac{\rho}{\epsilon}$$

and the wave equation

$$(\nabla^2 + k^2)\varphi = 0.$$

However, the diffusion equation

$$(\nabla^2 - i\mu\sigma\omega)\varphi = 0$$

does not belong to the homogeneous class.

The governing equation of Green's function is

$$P(i\mathcal{D})g(\mathbf{r}, \mathbf{r}') = -\delta(\mathbf{r}).$$

By taking the Fourier transform, we obtain the spectral domain expression

$$\hat{g}(\mathbf{k}) = -\frac{1}{P(\mathbf{k})},$$

where $\mathbf{k}$ is the wave number vector.

Following the steps and procedures for ODEs, we obtain the sampling function

$$\begin{aligned}\hat{\varphi}(\mathbf{k}) &= \frac{\hat{g}(\mathbf{k})}{\sum_{\mathbf{n}\in Z^d} \hat{g}(\mathbf{k} + 2\mathbf{n}\pi)} \\ &= \frac{1/P(\mathbf{k})}{\sum_n 1/P(\mathbf{k} + 2\mathbf{n}\pi)} \\ &= \frac{1}{1 + \sum_{\mathbf{n}\neq 0} P(k)/P(\mathbf{k} + 2\mathbf{n}\pi)}, \end{aligned} \tag{4.12.6}$$

where $\mathbf{n}$ is a set of integers.

From (4.12.6) it is clear that $\hat{\varphi}$ is bounded when P has isolated zeros. It is not necessarily true that $\hat{\varphi} \in L^2$. For the wave equation, P may have nonisolated zeros, implying that $\hat{\varphi}$ is not even bounded. The dilation equation can be found by the same method as in the ODE cases. This is true because

$$P\left(\frac{\mathbf{k}}{2}\right) = \frac{P(\mathbf{k})}{2^d}.$$

The 2D Poisson Equation. The Poisson equation relates the electric potential $U(\boldsymbol{\rho})$ to the charge density $\sigma(\boldsymbol{\rho})$ by

$$\nabla^2 U(\boldsymbol{\rho}) = -\frac{\sigma(\boldsymbol{\rho})}{\epsilon},$$

where $\boldsymbol{\rho} \in R^2$. The corresponding Green's function satisfies

$$\nabla^2 G(\boldsymbol{\rho}) = -\delta(\boldsymbol{\rho}), \qquad \boldsymbol{\rho} \in R^2.$$

Taking the Fourier transform with respect to x and y consequently, we have

$$\left[(i\kappa_x)^2 + \frac{\partial^2}{\partial y^2}\right]\tilde{G}(u, y) = -\delta(y)$$

$$\Rightarrow$$

$$(u^2 + v^2)\tilde{G}(u, v) = 1$$

$$\tilde{G}(u, v) = \frac{1}{\kappa^2},$$

where

$$\begin{cases} \kappa^2 = u^2 + v^2 \\ u = \kappa \cos\alpha = \kappa_x \\ v = \kappa \sin\alpha = \kappa_y. \end{cases}$$

The Green's function is obtained by taking the inverse Fourier transform as

$$\begin{aligned} G(x, y) &= \frac{1}{(2\pi)^2} \iint \tilde{G}(u, v) e^{iux+ivy}\, du\, dv \\ &= \frac{1}{(2\pi)^2} \iint \tilde{G} e^{i\kappa\rho[\cos\alpha\cos\theta + \sin\alpha\sin\theta]} \kappa\, d\kappa\, d\theta \\ &= \frac{1}{(2\pi)^2} \int_0^\infty \int_{-\pi}^{\pi} \frac{e^{i\kappa\rho\cos(\alpha-\theta)}}{\kappa} d\alpha d\kappa \\ &= \frac{1}{2(2\pi)^2} \int_{-\infty}^\infty \int_{-\pi}^{\pi} \frac{e^{ik\rho\cos(\alpha-\theta)}}{|\kappa|} d\alpha d\kappa. \end{aligned}$$

Let us define the inverse Fourier transform of $1/\kappa$:

$$\begin{aligned} a(\tau) &:= \frac{1}{2\pi} \int_{-\infty}^\infty \frac{e^{i\kappa\tau}}{|\kappa|} d\kappa \\ &= -\frac{1}{\pi}(\gamma + \ln|\tau|), \end{aligned}$$

where τ stands for $\rho\cos(\alpha-\theta)$ as the outcome of the exponential integral. Hence the Green's function is

$$G(x, y) = \frac{1}{4\pi}\int_{-\pi}^{\pi} a(\rho\cos\alpha)d\alpha.$$

Here $a(\tau)$ is employed to construct an orthogonal scalet $\varphi(\tau)$ and wavelet $\psi(\tau)$. The 1D Fourier transform of $\varphi(\tau)$ is given by

$$\begin{aligned}\hat{\varphi}(\kappa) &= \frac{\hat{a}(\kappa)}{\left\{\sum_n |\,\hat{a}(|\,\kappa\,| + 2n\pi)\,|^2\right\}^{1/2}} \\ &= \frac{1/|\,\kappa\,|}{\sqrt{\sum_n 1/||\,\kappa\,| + 2n\pi\,|^2}} \\ &= \frac{1/|\,\kappa\,|}{(2|\,\sin\,|\,\kappa\,|/2|)^{-1}} \\ &= \frac{|\,\sin\kappa/2\,|}{|\,\kappa/2\,|},\end{aligned}$$

where the summation in the denominator is carried out according to Eq. (4.12.3). Following Example 1, we may derive the scalet and wavelet in the spatial domain.

Noticing that

$$\frac{\hat{\varphi}(\kappa)}{\hat{\varphi}(\kappa/2)} = \frac{|\,\sin\kappa/2\,|}{|\,\kappa/2\,|}\cdot\frac{|\,\kappa/4\,|}{|\,\sin\kappa/4\,|} = \left|\cos\frac{\kappa}{4}\right|,$$

we find

$$\hat{\varphi}(\kappa) = \left|\cos\frac{\kappa}{4}\right|\hat{\varphi}\left(\frac{\kappa}{2}\right), \tag{4.12.7}$$

which is the dilation equation in the spectral (Fourier transform) domain. Comparing (4.12.7) with the equation in Chapter 3

$$\hat{\varphi}(\omega) = \hat{h}\left(\frac{\omega}{2}\right)\hat{\varphi}\left(\frac{\omega}{2}\right),$$

we see that the lowpass filter here is

$$\hat{h}\left(\frac{\omega}{2}\right) = \left|\cos\frac{\omega}{4}\right|.$$

Consider $f(x) = |\,\cos x/4\,|$, an even periodic function of period $2\ell = 4\pi$. Its Fourier series

$$\begin{aligned}f(x) &= a_0 + \sum_{n=1}^{\infty}\left(a_n\cos\frac{n\pi}{\ell}x + b_n\sin\frac{n\pi}{\ell}x\right) \\ &= \sum_{n=-\infty}^{\infty} c_n e^{-i(n\pi/\ell)x}.\end{aligned} \tag{4.12.8}$$

Multiplying both sides of (4.12.8) by $e^{i(m\pi/\ell)x}$ and integrating, we obtain

$$\text{LHS} = \int_{-\ell}^{\ell} f(x) e^{i(m/\ell)\pi x}\, dx$$

and

$$\begin{aligned}\text{RHS} &= \sum_{n=-\infty}^{\infty} c_n \int_{-\ell}^{\ell} e^{i[(m-n)/\ell]}\, dx \\ &= 2\ell c_m.\end{aligned}$$

Hence

$$\begin{aligned}
c_m &= \frac{1}{2\ell}\int_{-\ell}^{\ell} f(x) e^{i(m/\ell)\pi x} \\
&= \frac{1}{2\ell}\int_{-\ell}^{\ell} \left|\cos\frac{x}{4}\right| e^{i(m/\ell)}\, dx \\
&= \frac{1}{4\pi}\int_{-2\pi}^{2\pi} \frac{e^{i(x/4)} + e^{-i(x/4)}}{2} \cdot e^{i(m/2)x}\, dx \\
&= \frac{1}{8\pi}\int_{-2\pi}^{2\pi} \left[e^{i[(2m+1)/4]x} + e^{i[(2m-1)/4]x}\right] dx \\
&= \frac{1}{2\pi i}\left[\frac{1}{2m+1} e^{i[(2m+1)/4]x} + \frac{1}{2m-1} e^{i[(2m-1)/4]x}\right]_{-2\pi}^{2\pi} \\
&= \frac{1}{\pi}\left\{\frac{1}{2m+1}\sin\left[\frac{\pi}{2}(2m+1)\right] + \frac{1}{2m-1}\sin\left[\frac{\pi}{2}(2m-1)\right]\right\} \\
&= \frac{1}{\pi}\left\{\frac{(-1)^m}{2m+1} - \frac{(-1)^m}{2m-1}\right\} \\
&= -\frac{2}{\pi}\frac{(-1)^m}{4m^2-1}.
\end{aligned}$$

Therefore

$$\cos\frac{\kappa}{4} = \frac{2}{\pi}\sum_{n=-\infty}^{\infty} \frac{(-1)^n}{1-4n^2} e^{-i(\kappa/2)n}.$$

Returning to the dilation equation (4.12.7), we have

$$\hat{\varphi}(\kappa) = \frac{2}{\pi}\sum_{n=-\infty}^{\infty} \frac{(-1)^n}{1-4n^2} e^{-i(\kappa/2)n}\hat{\varphi}\left(\frac{\kappa}{2}\right).$$

The inverse Fourier transform leads to

$$\varphi(\tau) = \frac{4}{\pi}\sum_{n=-\infty}^{\infty}\frac{(-1)^n}{1-4n^2}\varphi(2\tau - n).$$

The dilation relation (3.3.10) can be used to obtain the wavelet

$$\hat{\psi}(\omega) = e^{-i(\kappa/2)}\overline{\hat{h}\left(\frac{\omega}{2}+\pi\right)}\hat{\varphi}\left(\frac{\omega}{2}\right).$$

That is,

$$\begin{aligned}\hat{\psi}(\kappa) &= e^{-i(\kappa/2)}\left|\cos\left(\frac{\kappa}{4}+\frac{\pi}{2}\right)\right|\left|\frac{\sin(\kappa/4)}{\kappa/4}\right| \\ &= e^{-i(\kappa/2)}\frac{|\sin^2\kappa/4|}{|\kappa/4|} = e^{-i(\kappa/2)}\frac{(1-\cos\kappa/2)}{|\kappa/2|} \\ &= \frac{1}{|\kappa|}(-1+2e^{-i(\kappa/2)} - e^{-i\kappa}).\end{aligned}$$

Hence the wavelet

$$\begin{aligned}\psi(\tau) &= -a(\tau) + 2a\left(\tau-\tfrac{1}{2}\right) - a(\tau-1) \\ &= \frac{1}{\pi}\{\ln|\tau| - 2\ln|\tau-\tfrac{1}{2}| + \ln|\tau-1|\} \\ &= \frac{1}{\pi}\ln\left|\frac{\tau(\tau-1)}{(\tau-\frac{1}{2})^2}\right|.\end{aligned}$$

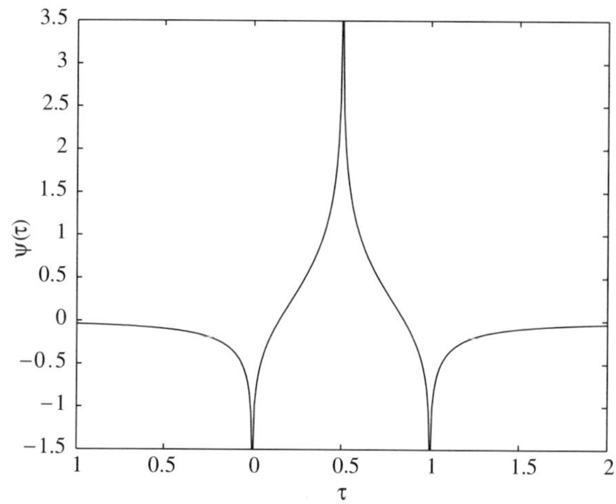

FIGURE 4.21 Wavelet constructed from Green's function of the Poisson equation.

Figure 4.21 shows the constructed wavelet based on the Green's function of the 2D Poisson equation. The immediate application of these Green's scalets and wavelets has yet to be found.

4.13 APPENDIX: DERIVATION OF INTERVALLIC WAVELETS ON [0, 1]

Consider the detail space $W_j[0,1]$, $j \geq j_0$, that is the orthogonal complement of $V_j[0,1]$ in $V_{j+1}[0,1]$, namely

$$V_{j+1}[0,1] = V_j[0,1] \oplus W_j[0,1].$$

First, let us evaluate the dimensions of the related spaces. As was given in the previous section,

$$\dim V_j[0,1] = 2^j - 2K + 2.$$

Hence

$$\dim V_{j+1}[0,1] = 2^{j+1} - 2K + 2$$

and

$$\begin{aligned} \dim W_j[0,1] &= \dim V_{j+1}[0,1] - \dim V_j[0,1] \\ &= 2^j. \end{aligned} \tag{4.13.1}$$

Now let us identify the wavelets that are completely supported within [0, 1] and also belong to $V_{j+1}[0,1]$. From multiresolution analysis (MRA)

$$\psi_{j,k}(x) = \sum_m g_{m-2k}\varphi_{j+1,m}(x).$$

The support of $\psi_{j,k}(x)$ can be determined as

$$1 - 3K \leq 2^j x - k \leq 3K,$$

where $2K$ is the order of the wavelets. It follows that

$$(1 - 3K + k)2^{-j} \leq x \leq (3K + k)2^{-j}.$$

For $x \in [0,1]$, we have, on one hand,

$$(1 - 3K + k)2^{-j} \geq 0,$$

or equivalently

$$k \geq 3K - 1.$$

On the other hand,

$$(3K + k)2^{-j} \leq 1,$$

that is,

$$k \leq 2^j - 3K.$$

As a result

$$3K - 1 \leq k \leq 2^j - 3K. \tag{4.13.2}$$

The wavelets $\psi_{j,k}$ satisfying the equation above will be supported completely within [0, 1].

For the scalets $\varphi_{j+1,m}$ that are completely within [0, 1]:

$$m \in S_{j+1} = \{m: 2K \leq m \leq 2^{j+1} - 4K + 1\}$$

In combination of

$$k \in \{k: 3K - 1 \leq k \leq 2^j - 3K\},$$

we will show that for these m and k,

$$g_{m-2k} \neq 0.$$

Show. For the Coifman wavelets it can be shown

$$g_l \neq 0, \quad 2 - 4K \leq l \leq 2K + 1. \tag{4.13.3}$$

In fact, from [19],

$$h_l \neq 0, \quad -2K \leq l \leq 4K - 1. \tag{4.13.4}$$

Using

$$g_m = (-1)^m h_{1-m},$$

we obtain (4.13.3) immediately. □

Now

$$2 - 4K \leq m - 2k \leq 2K + 1,$$

that is,

$$2 - 4K + 2k \leq m \leq 2K + 1 + 2k. \tag{4.13.5}$$

In the conjunction of (4.13.5) and

$$3K - 1 \leq k \leq 2^j - 3K,$$

We have from the left-hand side of (4.13.5),

$$\begin{aligned} 2 - 4K + 2k &\geq 2 - 4K + 2(3K - 1) \\ &= 2K \end{aligned}$$

and

$$\begin{aligned} 2 - 4K + 2k &\leq 2 - 4K + 2(2^j - 3K) \\ &= 2^{j+1} - 10K + 2. \end{aligned}$$

The right-hand side of (4.13.5) gives

$$\begin{aligned} 2K + 1 + 2k &\geq 2K + 1 + 2(3K - 1) \\ &= 8K - 1 \end{aligned}$$

and

$$\begin{aligned} 2K + 1 + 2k &\leq 2K + 1 + 2(2^j - 3K) \\ &= 2^{j+1} - 4K + 1. \end{aligned}$$

The previous derivation reveals that

$$\begin{aligned} &m \geq 2 - 4K + 2k \geq 2K \\ &m \leq 2K + 1 + 2k \leq 2^{j+1} - 4K + 1, \end{aligned}$$

namely

$$2K \leq m \leq 2^{j+1} - 4K + 1.$$

Therefore

$$m \in S_{j+1}.$$

As a consequence

$$\psi_{j,k}(x) \in V_{j+1}[0, 1]$$

provided that k satisfies

$$3K - 1 \leq k \leq 2^j - 3K.$$

These wavelets are in fact in $W_j[0, 1]$. The dimension of $W_j[0, 1]$ was derived in (4.13.1) as

$$\dim W_j[0, 1] = 2^j.$$

Hence we still need to find the remaining

$$2^j - [(2^j - 3K) - (3K - 1) + 1] = 6K - 2$$

basis functions. Roughly one-half of them is for the left endpoint, and the other half is for the right endpoint.

Let us find the remaining functions. First, we will identify basis functions in $V_{j+1}[0, 1]$ that cannot be represented in terms of the basis functions in $V_j[0, 1]$, nor in terms of $\psi_{j,k}$ that satisfy (4.13.2). We define coefficient

$$\alpha^r_{j+1,m} := 2^{(j+1)[r+(1/2)]}\langle x^r, \varphi_{j+1,m}\rangle, \qquad r = 0, 1, \ldots, 2k - 1.$$

From the (4.10.7) we denote the coefficient

$$\alpha^r_{j,m} := 2^{j[r+(1/2)]}\langle x^r, \varphi_{j,m}\rangle, \qquad r = 0, 1, \ldots, 2k - 1.$$

Then from (4.10.7) we have

$$x_{j+1,L}^r = 2^{(j+1)[r+(1/2)]} \sum_{m=4k+2}^{2k+1} \langle x^r, \varphi_{j+1,m} \rangle \varphi_{j+1,m} \mid_{[0,1]} \tag{4.13.6}$$

$$= \sum_m \alpha_{j+1,m} \varphi_{j+1,m} \mid_{[0,1]} \tag{4.13.7}$$

$$= \sum_m 2^{(j+1)[r+(1/2)]} \langle x^r, \varphi_{j+1,m} \rangle \varphi_{j+1,m}(x)|_{[0,1]}$$

$$= \sum_m \alpha_{j+1,m}^r \varphi_{j+1,m}(x)|_{[0,1]}$$

$$= \sum_m \alpha_{j+1,m}^r \left[\sum_k h_{m-2k} \varphi_{j,k}(x) + \sum_k g_{m-2k} \psi_{j,k}(x) \right]_{[0,1]}$$

$$= \sum_k \left(\sum_m \alpha_{j+1,m}^r h_{m-2k} \right) \varphi_{j,k}(x)|_{[0,1]}$$

$$+ \sum_k \left(\sum_m \alpha_{j+1,m}^r g_{m-2k} \right) \psi_{j,k}(x)|_{[0,1]}, \tag{4.13.8}$$

where we have applied (4.10.11), that

$$\varphi_{j+1,m}(x) = \sum_k h_{m-2k} \varphi_{j,k}(x) + g_{m-2k} \psi_{j,k}(x).$$

We examine the first term in the previous equation:

$$I = \sum_k \left(\sum_m \alpha_{j+1,m}^r h_{m-2k} \right) \varphi_{j,k}(x)|_{[0,1]}$$

$$= \sum_k \left(2^{r+(1/2)} \sum_m 2^{j[r+(1/2)]} \langle x^r, \varphi_{j+1,m} \rangle h_{m-2k} \right) \varphi_{j,k}(x)|_{[0,1]}$$

$$= 2^{r+(1/2)} \sum_k \left(2^{j[r+(1/2)]} \left\langle x^r, \sum_m h_{m-2k} \varphi_{j+1,m} \right\rangle \right) \varphi_{j,k}(x)|_{[0,1]}$$

$$= 2^{r+(1/2)} \sum_k (2^{j[r+(1/2)]} \langle x^r, \varphi_{j,k} \rangle) \varphi_{j,k}(x)|_{[0,1]}$$

$$= 2^{r+(1/2)} \sum_k \alpha_{j,k}^r \varphi_{j,k}(x)|_{[0,1]}$$

$$= 2^{r+(1/2)} x_{j,L}^r$$

where we have used (4.8.5), that

$$\sum_m h_{m-2k} \varphi_{j+1,m} = \varphi_{j,k}.$$

The second term in Eq. (4.13.8),

$$
\begin{aligned}
II &= \sum_k \left(\sum_m \alpha^r_{j+1,m} g_{m-2k} \right) \psi_{j,k}(x)|_{[0,1]} \\
&= \sum_k \left(2^{r+(1/2)} \sum_m 2^{j[r+(1/2)]} \langle x^r, \varphi_{j+1,m} \rangle g_{m-2k} \right) \psi_{j,k}(x)|_{[0,1]} \\
&= \sum_k 2^{r+(1/2)} \left\langle x^r, \sum_m g_{m-2k} \varphi_{j+1,m} \right\rangle 2^{j[r+(1/2)]} \psi_{j,k}(x)|_{[0,1]} \\
&= 2^{r+(1/2)} \sum_k \langle x^r, \psi_{j,k} \rangle \psi_{j,k}(x)|_{[0,1]} \\
&= 0.
\end{aligned}
$$

In the previous equation we employed (4.8.6) and

$$
\langle x^r, \psi_{j,k}(x) \rangle = 0, \; r = 0, 1, 2, \ldots, L-1.
$$

The normalization factor $2^{r+(1/2)}$ can be found as follows:

$$
\langle x^r, \varphi_{j,k}(x) \rangle = \int x^r 2^{j/2} \varphi(2^j x - k)\, dx
$$

Let $2^j x - k = t$; we have $x = 2^{-j}(t+k),\; dx = 2^{-j}\, dt$. Thus

$$
\begin{aligned}
\langle x^r, \varphi_{j,k}(x) \rangle &= \int 2^{-jr} (t+k)^r 2^{j/2} \varphi(t) 2^{-j}\, dt \\
&= 2^{-j[r+(1/2)]} \int dt (t^r + C^1_r t^{r-1} k + C^2_r t^{r-2} k^2 + \cdots + k^r) \varphi(t) \\
&= 2^{-j[r+(1/2)]} k^r.
\end{aligned}
\tag{4.13.9}
$$

First we note that

$$
x^r_{j+1,L} = 2^{r+(1/2)} x^r_{j,L} - Q,
$$

where Q involves only those terms that are in

$$
\operatorname{clos}\{\varphi_{j+1,m}\}_{m \in S^I_{j+1}}.
$$

The functions $x^r_{j+1,L}$ and $x^r_{j+1,R}$ are linear combinations of basis functions in $V_j[0,1]$ and basis functions from the collection $\operatorname{clos}\{\varphi_{j+1,k}\}_{k \in S^I_{j+1}}$. This implies the fact that certain functions in $\operatorname{clos}\{\varphi_{j+1,k}\}_{k \in S^I_{j+1}}$ cannot be represented by linear combinations of the basis functions $\varphi_{j,k} \in V_j[0,1]$ and $\psi_{j,k} \in W_j[0,1]$ identified in the previous paragraph.

Recall that

$$
S^I_{j+1} = \{k \colon 2K + 1 \le k \le 2^{j+1} - 4K\}.
$$

From the MRA, $\forall\varphi \in L^2(R)$, we have

$$\varphi_{j+1,k}(x) = \sum_m h_{k-2m}\varphi_{j,m}(x) + \sum_m g_{k-2m}\psi_{j,m}(x).$$

(1) Let $k = 2K$,

$$\varphi_{j+1,2K}(x) = \sum_m h_{k-2m}\varphi_{j,m}(x) + \sum_m g_{K-2m}\psi_{j,m}(x).$$

The nonzero lowpass filter h_i, $-2K \le i \le 4K - 1$ as given in (4.13.4), that is,

$$-2K \le 2K - 2m \le 4K - 1.$$

It follows that

$$1 - K \le m \le 2K.$$

The bandpass filter $g_i \neq 0$ when $2 - 4K \le i \le 2K + 1$, as provided in (4.13.3). Thus

$$2 - 4K \le 2K - 2m \le 2K + 1,$$

that is,

$$0 \le m \le 3K - 1.$$

Hence

$$\begin{aligned}
\varphi_{j+1,2K}(x) &= \sum_{m=1-K}^{2K} h_{2K-2m}\varphi_{j,m}(x) + \sum_{m=0}^{3K-1} g_{2K-2m}\psi_{j,m}(x) \\
&= h_{4K-2}\varphi_{j,1-K}(x) + h_{4K-4}\varphi_{j,2-K}(x) \\
&\quad + h_{4K-6}\varphi_{j,3-K}(x) + \cdots + h_{-2K}\varphi_{j,2K}(x) \\
&\quad + g_{2K}\psi_{j,0}(x) + g_{2K-2}\psi_{j,1}(x) \\
&\quad + g_{2K-4}\psi_{j,2}(x) + \cdots + g_{2-4K}\psi_{j,3K-1}(x).
\end{aligned}$$

(2) Let $k = 2K + 1$,

$$\varphi_{j+1,2K+1}(x) = \sum_m h_{2K+1-2m}\varphi_{j,m}(x) + \sum_m g_{2K+1-2m}\psi_{j,m}(x).$$

The nonzero lowpass filter h_i, $-2K \le i \le 4K - 1$ as given in (4.13.4), that is,

$$-2K \le 2K + 1 - 2m \le 4K - 1.$$

It follows that

$$1 - K \le m \le 2K.$$

The nonzero bandpass filter $g_i \neq 0$ when $2 - 4K \le i \le 2K + 1$, as provided in (4.13.3). Thus

$$2 - 4K \le 2K + 1 - 2m \le 2K + 1,$$

or equivalently

$$0 \le m \le 3K - 1.$$

Hence

$$\begin{aligned}
\varphi_{j+1,2K+1}(x) &= \sum_{m=1-K}^{2K} h_{2K+1-2m}\varphi_{j,m}(x) + \sum_{m=0}^{3K-1} g_{2K+1-2m}\psi_{j,m}(x) \\
&= h_{4K-1}\varphi_{j,1-K}(x) + h_{4K-3}\varphi_{j,2-K}(x) \\
&\quad + h_{4K-5}\varphi_{j,3-K}(x) + \cdots + h_{1-2K}\varphi_{j,2K}(x) \\
&\quad + g_{2K+1}\psi_{j,0}(x) + g_{2K-1}\psi_{j,1}(x) \\
&\quad + g_{2K-3}\psi_{j,2}(x) + \cdots + g_{3-4K}\psi_{j,3K-1}(x).
\end{aligned}$$

(3) The next basis is for $k = 2K + 2$,

$$\varphi_{j+1,2K+2}(x) = \sum_m h_{2K+2-2m}\varphi_{j,m} + \sum_m g_{2K+2-2m}\psi_{j,m}(x).$$

Similarly for the lowpass filter we now have

$$-2K \le 2K + 2 - 2m \le 4K - 1,$$

that is,

$$\begin{aligned}
3 - 2K &\le 2m \le 4K + 2, \\
2 - K &\le m \le 2K + 1.
\end{aligned}$$

The bandpass filters require that

$$\begin{aligned}
&2 - 4K \le 2K + 2 - 2m \le 2K + 1, \\
&1 - 2K \le K + 1 - m \le K + \tfrac{1}{2}, \\
&1 \le m \le 3K.
\end{aligned}$$

Thus

$$\begin{aligned}
\varphi_{j+1,2K+2}(x) &= \sum_{m=2-K}^{2K+1} h_{2K+2-2m}\varphi_{j,m}(x) + \sum_{m=1}^{3K} g_{2K+2-2m}\psi_{j,m}(x) \\
&= h_{4K-2}\varphi_{j,2-K}(x) + h_{4K-4}\varphi_{j,3-K}(x) \\
&\quad + \cdots + h_{-2K}\varphi_{j,2K+1}(x) \\
&\quad g_{2K}\varphi_{j,1}(x) + g_{2K-2}\psi_{j,2}(x) + \cdots + g_{2-4K}\psi_{j,3K}(x).
\end{aligned}$$

(4) For the third basis $\varphi_{j+1,2K+3}$,

(i) The lowpass filters satisfy

$$\begin{aligned}
-2K &\le 2K + 3 - 2m \le 4K - 1, \\
2 - K &\le m \le 2K + 1.
\end{aligned}$$

(ii) The bandpass filters satisfy

$$\begin{aligned}
2-4K &\le 2K+3-2m \le 2K+1,\\
2 &\le 2m \le 6K+1,\\
1 &\le m \le 3K.
\end{aligned}$$

Thus

$$\begin{aligned}
\varphi_{j+1,2K+3} &= \sum_{m=2-K}^{2K+1} h_{2K+3-2m}\varphi_{j,m}(x) + \sum_{m=1}^{3K} g_{2K+3-2m}\psi_{j,m}(x)\\
&= h_{4K-1}\varphi_{j,2-K}(x) + h_{4K-3}\varphi_{j,3-K}(x)\\
&\quad + h_{4K-5}\varphi_{j,4-K}(x) + \cdots + h_{1-2K}\varphi_{j,2K+1}(x)\\
&\quad + g_{2K+1}\psi_{j,1}(x) + g_{2K-1}\psi_{j,2}(x)\\
&\quad + g_{2K-3}\psi_{j,3}(x) + \cdots + g_{3-4K}\psi_{j,3K}(x).
\end{aligned}$$

(5) Repeating the procedure, for $\varphi_{j+1,2K+4}$ we have

(i) The lowpass filters

$$\begin{aligned}
-2K &\le 2K+4-2m \le 4K-1,\\
5-2K &\le 2m \le 4K+4,\\
3-K &\le m \le 2K+2.
\end{aligned}$$

(ii) The bandpass filters

$$\begin{aligned}
2-4K &\le 2K+4-2m \le 2K+1,\\
3 &\le 2m \le 6K+2,\\
2 &\le m \le 3K+1.
\end{aligned}$$

Thus

$$\begin{aligned}
\varphi_{j+1,2K+4}(x) &= \sum_{m=3-K}^{2K+2} h_{2K+4-2m}\varphi_{j,m}(x) + \sum_{m=2}^{3K+1} g_{2K+4-2m}\psi_{j,m}(x)\\
&= h_{4K-2}\varphi_{j,3-K}(x) + h_{4K-4}\varphi_{j,4-K}(x)\\
&\quad + h_{4K-6}\varphi_{j,5-K}(x) + \cdots + h_{-2K}\varphi_{j,2K+2}(x)\\
&\quad + g_{2K}\psi_{j,2}(x) + g_{2K-2}\psi_{j,3}(x)\\
&\quad + g_{2K-4}\psi_{j,4}(x) + \cdots + g_{2-4K}\psi_{j,3K+1}(x).
\end{aligned}$$

This process can be conducted continuously. We skip the intermediate and jump to 5K and 8K shifts:

(6) For $\varphi_{j+1,5K+1}$,

(i) Lowpass

$$-2K \le 5K+1-2m \le 4K-1,$$

$$K + 2 \le 2m \le 7K + 1,$$
$$\frac{K}{2} + 1 \le m \le \frac{1}{2}(7K + 1).$$

(ii) Bandpass

$$2 - 4K \le 5K + 1 - 2m \le 2K + 1,$$
$$3K \le 2m \le 9K - 1,$$
$$\frac{3K}{2} \le m \le \frac{9K - 1}{2}.$$

Thus

$$\begin{aligned}
\varphi_{j,5K+1}(x) &= \sum_{m=(K/2)+1}^{(7K+1)/2} h_{5K+1-2m}\varphi_{j,m}(x) + \sum_{m=(3/2)K}^{(9K-1)/2} g_{5K+1-2m}\psi_{j,m}(x) \\
&= h_{4K-1}\varphi_{j,\frac{K}{2}+1}(x) \\
&\quad + h_{4K-3}\varphi_{j,\frac{K}{2}+2}(x) + \cdots + h_{-2K}\varphi_{j,\frac{1}{2}(7K+1)}(x) \\
&\quad + g_{2K+1}\psi_{j,\frac{3}{2}K}(x) \\
&\quad + g_{2K-1}\psi_{j,\frac{3K}{2}+1}(x) + \cdots + g_{2-4K}\psi_{j,\frac{1}{2}(9K-1)}(x).
\end{aligned}$$

(7) For $\varphi_{j+1,8K+1}$,

(i) Lowpass

$$-2K \le 8K + 1 - 2m \le 4K - 1,$$
$$4K + 2 \le 2m \le 10K + 1,$$
$$2K + 1 \le m \le 5K + \tfrac{1}{2}.$$

(ii) Bandpass

$$2 - 4K \le 3K + 1 - 2m \le 2K + 1,$$
$$6K \le 2m \le 12K - 1,$$
$$3K \le m \le 6K - \tfrac{1}{2}.$$

Thus

$$\begin{aligned}
\varphi_{j+1,8K+1}(x) &= \sum_{m=2K+1}^{5K} h_{8K+1-2m}\varphi_{j,m}(x) + \sum_{m=3K}^{6K-1} g_{8K+1-2m}\psi_{j,m}(x) \\
&= h_{4K-1}\varphi_{j,2K_+1}(x) \\
&\quad + h_{4K-3}\varphi_{j,2K-1}(x) + \cdots + h_{1-2K}\varphi_{j,5K}(x) \\
&\quad + g_{2K+1}\psi_{j,3K}(x) \\
&\quad + g_{2K-1}\psi_{j,3K+1}(x) + \cdots + g_{3-4K}\psi_{j,6K-1}(x).
\end{aligned}$$

Noticing that

$$\begin{aligned}
&\varphi_{j+1,8K+1}(x) \in S_{j+1}, \\
&8K + 1 \le 2^{j+1} - 4K + 1, \\
&12K \le 2^{j+1} \qquad \text{for } K = 2,\ j \ge 4.
\end{aligned}$$

As a result $\varphi_{j+1,8K+1}(x)$ can be represented as the linear combination of $\varphi_{j,m}(x) \in S_j$ and $\psi_{j,m}(x) \in W_j[0, 1]$.

(8) Furthermore

$$\varphi_{j+1,8K} = \sum_m h_{8K-2m}\varphi_{j,m}(x) + \sum_m g_{8K-2m}\psi_{j,m}(x).$$

(i) Lowpass

$$\begin{aligned}
-2K &\le 8K - 2m \le 4K - 1, \\
4K + 1 &\le 2m \le 10K, \\
2K + \tfrac{1}{2} &\le m \le 5K.
\end{aligned}$$

(ii) Bandpass

$$\begin{aligned}
2 - 4K &\le 8K - 2m \le 2K + 1, \\
6K - 1 &\le 2m \le 12K - 2, \\
3K - \tfrac{1}{2} &\le m \le 6K - 1.
\end{aligned}$$

Thus

$$\begin{aligned}
\varphi_{j+1,8K} &= \sum_{m=2K+1}^{5K} h_{8K-2m}\varphi_{j,m}(x) + \sum_{m=3K}^{6K-1} g_{8K-2m}\psi_{j,m}(x) \\
&= h_{4K-2}\varphi_{j,2K+1}(x) + h_{4K-3}\varphi_{j,2K+2}(x) + \cdots + h_{-2K}\varphi_{j,5K}(x) \\
&\quad + g_{2K}\psi_{j,3K}(x) + g_{2K-2}\psi_{j,3K+1}(x) + \cdots + g_{2-4K}\psi_{j,6K-1}(x).
\end{aligned}$$

(9) For $\varphi_{j+1,8K-1}$,

(i) Lowpass

$$\begin{aligned}
-2K &\le 8K - 1 - 2m \le 4K - 1, \\
4K &\le 2m \le 10K - 1, \\
2K &\le m \le 5K - \tfrac{1}{2}.
\end{aligned}$$

(ii) Bandpass

$$\begin{aligned}
2 - 4K &\le 8K - 1 - 2m \le 2K + 1, \\
6K - 2 &\le 2m \le 12K - 3, \\
3K - 1 &\le m \le 6K - \tfrac{3}{2}.
\end{aligned}$$

Thus

$$\begin{aligned}\varphi_{j,8K-1}(x) &= \sum_{m=2K}^{5K-1} h_{8K-1-2m}\varphi_{j,m}(x) + \sum_{m=3K-1}^{6K-2} g_{8K-1-2m}\psi_{j,m}(x) \\ &= h_{4K-1}\varphi_{j,2K}(x) + h_{4K-3}\varphi_{j,2K+1}(x) \\ &\quad + \cdots + h_{1-2K}\varphi_{j,5K-1}(x) \\ &\quad + g_{2K+1}\psi_{j,3K-1}(x) + g_{2K-1}\psi_{j,3K}(x) \\ &\quad + \cdots + g_{3-4K}\psi_{j,6K-2}(x).\end{aligned}$$

(10) For $\varphi_{j+1,8K-2}(x)$,

(i) Lowpass

$$\begin{aligned}&-2K \le 8K - 2 - 2m \le 4K - 1, \\ &4K - 1 \le 2m \le 10K - 2, \\ &2K - \tfrac{1}{2} \le m \le 5K - 1.\end{aligned}$$

(ii) Bandpass

$$\begin{aligned}&2 - 4K \le 8K - 2 - 2m \le 2K + 1, \\ &6K - 3 \le 2m \le 12K - 4, \\ &3K - 1.5 \le m \le 6K - 2.\end{aligned}$$

$$\begin{aligned}\varphi_{j+1,8K-2}(x) &= h_{4K-2}\varphi_{j,2K} + h_{4K-4}\varphi_{j,2K+1} + h_{4K-6}\varphi_{j,2K+2} + \cdots \\ &\quad + g_{2K-2}\psi_{j,3K} + h_{2K-4}\psi_{j,3K+1} + h_{2K-6}\psi_{j,3K+1} + \cdots\end{aligned}$$

In the same manner

$$\begin{aligned}\varphi_{j+1,8K-3}(x) &= \sum_m h_{8K-3-2m}\varphi_{j,m}(x) + \sum_m g_{8K-3-2m}\psi_{j,m}(x) \\ &= \sum_m h_{8K-1-2(m+1)}\varphi_{j,m}(x) + \sum_m g_{8K-1-2(m+1)}\psi_{j,m}(x) \\ &= \sum_m h_{8K-1-2m}\varphi_{j,m-1}(x) + \sum_m g_{8K-1-2m}\psi_{j,m-1}(x) \\ &= h_{4K-1}\varphi_{j,2K-1} + h_{4K-3}\varphi_{j,2K} + h_{4K-5}\varphi_{j,2K+1} + \cdots \\ \varphi_{j+1,8K-4}(x) &= \sum_m h_{8K-4-2m}\varphi_{j,m}(x) + \sum_m g_{8K-4-2m}\psi_{j,m}(x) \\ &= \sum_m h_{8K-2-2(m+1)}\varphi_{j,m}(x) + \sum_m g_{8K-2-2(m+1)}\psi_{j,m}(x) \\ &= \sum_m h_{8K-2-2m}\varphi_{j,m-1}(x) + \sum_m g_{8K-2-2m}\psi_{j,m-1}(x) \\ &= h_{4K-2}\varphi_{j,2K-1} + h_{4K-4}\varphi_{j,2K} + h_{4K-6}\varphi_{j,2K+1}\end{aligned}$$

In this sequence of functions, every second function is linearly dependent on the previous functions (modulo functions in $V_j[0, 1]$).

Proof. According to MRA, we know that

$$\sum_k h_k h_{k-2l} = \langle \varphi_{j,0}, \varphi_{j,l} \rangle = \delta_{0,l}.$$

It was even shown in Section 3.6 that

$$\begin{aligned}
\delta_{0,l} = \langle \varphi_{j,0}, \varphi_{j,l} \rangle &= \langle \varphi_{j,0}, 2^{j/2}, \varphi(2^j x - l) \rangle \\
&= \langle \varphi_{j,0}, 2^{j/2} \sum h_k 2^{1/2} \varphi 2(2^j x - l) - k \rangle \\
&= \langle \varphi_{j,0}, \sum_k h_k \varphi_{j,l+k} \rangle \\
&= \langle \varphi_{j,0}, \sum_{k'} h_{k'-2l} \varphi_{j+1,k'} \rangle \\
&= \sum_k \langle \varphi_{j,0}, \varphi_{j+1,k'} \rangle h_{k-2l} \\
&= \sum_k \left(\sum_k h_{k'} \langle \varphi_{j+1,k'}, \varphi_{j+1,k} \rangle \right) h_{k-2l} \\
&= \sum_k h_k h_{k-2l}.
\end{aligned}$$

Similarly

$$\sum_k h_k g_{k-2l} = \langle \varphi_{j,0}, \psi_{j,l} \rangle = 0.$$

As a consequence, for Coifman wavelet, $k = -2N, -2N+1, \ldots, 4N-1$. Let $l = 1-3N \Rightarrow 2l = -6N + 2 \Rightarrow -2l = 6N - 2 \Rightarrow \sum_{k=-2N}^{4N-1} h_{k+6N-2} = 0$. Thus

$$\begin{aligned}
&h_{-2N} h_{4N-2} + h_{-2N+1} h_{4N-1} + 0 = 0, \\
&h_{-2N} h_{4N-2} + h_{-2N+1} h_{4N-1} = 0.
\end{aligned}$$

In (1) and (2) except $\varphi_{j,2N}(x)$ all $\varphi_{j,k}(x)$ and $\psi_{j,k}(x) \in S_j^I$. We have

$$\begin{aligned}
&h_{-2N+1} * (1) + h_{-2N} * (2) \\
&\quad = \left(h_{-2N+1} \varphi_{j+1,8N-1}(x) + h_{-2N} \varphi_{j+1,8N-2}(x) \right) |_{[0,1]} \\
&\quad = \left(h_{-2N} h_{4N-2} + h_{-2N+1} h_{4N-1} \right) \varphi_{j,2N}(x) |_{[0,1]} + \cdots \\
&\quad = \left(h_{-2N} h_{4N-2} + h_{-2N+1} h_{4N-1} \right) \varphi_{j,2N}(x) |_{[0,1]} \mod V_j[0,1] \\
&\quad = 0 \operatorname{mod}_{V_j[0,1]}
\end{aligned}$$

Every second one is linearly dependent on the previous one (modulo functions in $V_j[0, 1]$).

Similarly

$$\begin{aligned}&\Big(h_{-2N}\varphi_{j+1,8N-4}(x) + h_{-2N+1}\varphi_{j+1,8N-3}(x) + h_{-2N+2}\varphi_{j+1,8N-2}(x)\\&\quad + h_{-2N+3}\varphi_{j+1,8N-1}(x)\Big)\mid_{[0,1]}\\&= 0 \bmod_{V_j[0,1]},\end{aligned}$$

and so on.

The missing functions in $W_j[0, 1]$ at the left endpoint are

$$\begin{aligned}\psi_{j,L}^{1} &= \varphi_{j+1,8N-1}(x)\mid_{[0,1]} - \mathrm{Proj}_{V_j[0,1]}\varphi_{j+1,8N-1}(x)\\ \psi_{j,L}^{2} &= \varphi_{j+1,8N-3}(x)\mid_{[0,1]} - \mathrm{Proj}_{V_j[0,1]}\varphi_{j+1,8N-3}(x)\\ &\ddots\\ \psi_{j,L}^{\alpha} &= \varphi_{j+1,8N-(2\alpha-1)}(x) - \mid_{[0,1]} - \mathrm{Proj}_{V_j[0,1]}\varphi_{j+1,8N-(2\alpha-1)}(x)\\ &= \varphi_{j+1,8N-2\alpha+1}(x) - \mid_{[0,1]} - \mathrm{Proj}_{V_j[0,1]}\varphi_{j+1,8N-2\alpha+1}(x)\\ &\ddots\\ \psi_{j,L}^{3N} &= \varphi_{j+1,8N-6N+1}(x) - \mid_{[0,1]} - \mathrm{Proj}_{V_j[0,1]}\varphi_{j+1,8N-6N+1}(x)\\ &= \varphi_{j+1,2N+1}(x) - \mid_{[0,1]} - \mathrm{Proj}_{V_j[0,1]}\varphi_{j+1,2N+1}(x).\end{aligned}$$

In the same way, we set $3N$ functions at the end of rightpoint:

$$\begin{aligned}\psi_{j,R}^{3N} &= \varphi_{j+1,2^{j+1}-4N}(x) - \mid_{[0,1]} - \mathrm{Proj}_{V_j[0,1]}\varphi_{j+1,2^{j+1}-4N}(x)\\ \psi_{j,R}^{3N-1} &= \varphi_{j+1,2^{j+1}-4N-2}(x) - \mid_{[0,1]} - \mathrm{Proj}_{V_j[0,1]}\varphi_{j+1,2^{j+1}-4N-2}(x)\\ \psi_{j,R}^{3N-2} &= \varphi_{j+1,2^{j+1}-4N-4}(x) - \mid_{[0,1]} - \mathrm{Proj}_{V_j[0,1]}\varphi_{j+1,2^{j+1}-4N-4}(x)\\ &\ddots\\ \psi_{j,R}^{1} &= \varphi_{j+1,2^{j+1}-10N+2}(x) - \mid_{[0,1]} - \mathrm{Proj}_{V_j[0,1]}\varphi_{j+1,2^{j+1}-10N+2}(x)\\ &\ddots\\ \psi_{j,R}^{\alpha} &= \varphi_{j+1,2^{j+1}-10N-2\alpha}(x) - \mid_{[0,1]} - \mathrm{Proj}_{V_j[0,1]}\varphi_{j+1,2^{j+1}-10-2\alpha}(x).\end{aligned}$$

In summary,

$$\begin{aligned}\psi_{j,L}^{\alpha} &= \varphi_{j+1,8N-2\alpha+1} - \sum_{l}\langle\varphi_{j+1,8N-2\alpha+1}, \phi_{j,l}\rangle\phi_{j,l}\\ \psi_{j,R}^{\alpha} &= \varphi_{j+1,2^{j+1}+2N-2\alpha} - \sum_{l}\langle\varphi_{j+1,2^{j+1}+2N-2\alpha}, \phi_{j,l}\rangle\phi_{j,l}.\end{aligned}$$

□

4.14 PROBLEMS

4.14.1 Exercise 5

1. Find the normalized capacitance of a square conducting plate using the MoM.

2. Expand in terms of Daubechies wavelets the following function

$$y(x) = \begin{cases} x+1, & -1 < x \le 0 \\ 2 - e^{-x^2/2}, & 0 \le x < 1. \end{cases}$$

4.14.2 Exercise 6

1. Show that $\sum_l 2^{j/2}\varphi(2^j x - k + 2^j l) = 2^{-j/2}$

2. Show that $\langle \psi_{j,k}^{per}, \psi_{j',k'}^{per} \rangle = \delta_{j,j'}\, \delta_{k,k'}$

3. Construct and plot the eight basis functions of V_2,

$$\varphi_{0,0}^{p},\ \psi_{0,0}^{p},\ \psi_{1,0}^{p},\ \psi_{1,1}^{p},\ \psi_{2,0}^{p},\ \psi_{2,1}^{p},\ \psi_{2,2}^{p},\ \psi_{2,3}^{p}$$

   (a) For the Franklin wavelets.

   (b) For the Coifman wavelets, $L = 4$.

4. Expand the following in terms of periodic wavelets

$$f(x) = \begin{cases} -x, & -1 \le x < 0 \\ 2x - x^2, & 0 \le x < 1 \end{cases}.$$

4.14.3 Exercise 7

1. For the Coifman scalets $\varphi(x)$, the nonzero support is $x = [-2K, 4K-1]$. Verify that for $\varphi(2^j x - k)$ by finding:

   (a) That the incomplete basis function, $\varphi(2^j x - k)$, beyond 0 of [0, 1], is for $-4K + 1 \le k \le 2K$.

   (b) That the incomplete basis function, $\varphi(2^j x - k)$, beyond 1 of [0, 1], are for $2^j - 4K + 1 \le k \le 2^j + 2K$.

   (c) That the basis functions, $\varphi(2^j x - k)$, completely within [0, 1], are for $2K + 1 \le k \le 2^j - 4K$.

2. Construct and plot the orthonormal edge basis functions $\varphi_{j,L}^{r}$, $r = 0, 1, 2, \ldots, 2K-1$ for the Coiflets when $K = 2$.

3. Show that $\langle \varphi_{j,L}^{n}, \varphi_{j,L}^{m} \rangle = \delta_{n,m}$, where $\varphi_{j,L}^{r}$ is constructed using the matrix approach in Section 4.10.

4.14.4 Exercise 8

1. Show that

$$\frac{\partial J_z}{\partial n} = i\omega\sigma \frac{\partial A_z}{\partial n},$$
$$\frac{\partial J_z}{\partial \ell} = i\omega\sigma \frac{\partial A_z}{\partial \ell}.$$

2. Use Green's vector identity

$$\int_V (\mathbf{P}\cdot\nabla\times\nabla\times\mathbf{Q} - \mathbf{Q}\cdot\nabla\times\nabla\times\mathbf{P})\,dv = \oint_S (\mathbf{Q}\times\nabla\times\mathbf{P} - \mathbf{P}\times\nabla\times\mathbf{Q})\cdot\hat{n}\,ds$$

to show that

$$\int \mathbf{H}\cdot\mathbf{H}^*\,ds = \left(\frac{1}{\omega\mu\sigma}\right)^2 \oint \mathrm{Re}\left\{ J_z^* \frac{\partial J_z}{\partial n}\right\} dl,$$

and that consequently

$$L = -\mu \frac{\oint_{all} dl \mathrm{Re}\{I_q/S_q\}}{\partial J_z^*/\partial n} \left| \oint_{\text{wire}} dl \frac{\partial J_z}{\partial n} \right|^2.$$

3. The nonzero support of the wavelets $\psi(x)$ is

$$1 - 3K \le x \le 3K.$$

Find k for $\psi_{j,k}(x)$ such that $\psi_{j,k}(x)$ is completely within [0, 1].

4. Show that the mutual inductance L_{12} of a two-wire system is

$$L_{12} = \frac{1}{2}\left(L_{11} + L_{22} - 4\frac{W_m}{I_q^2}\right),$$

where W_m is the stored magnetic energy and I_q is the current on wire 1, while $-I_q$ is the current on wire 2, L_{11} and L_{22} are the self-inductances.

5. For the periodic wavelets, show that

$$\sum_l \psi\left(x + \frac{l}{2}\right) = 0.$$

Hint: You may need to use $\sum h_{2m} = \sum h_{2m+1}$.

6. For the Coifman intervallic wavelets of order $2K$, it is known that for any monomial x^r, $r \le 2K - 1$,

$$x^r = \sum \langle x^r, \varphi_{j,k}\rangle \varphi(x).$$

Show that

(a) $x^r = 2^{-j[r+(1/2)]} \sum k^r \varphi_{j,k}(x)$,

(b) $x^r_{j+1,L} = 2^{r+(1/2)} x^r_{j,L}$,
where

$$x^r_{j,L} = 2^{j(r+\frac{1}{2})} \sum_{k=-4K+2}^{2K} \langle x^r, \varphi_{j,k}(x) \mid [0, 1].$$

4.14.5 Project 1

An alternative way of applying wavelets to the surface integral equations is to generate a dense MoM matrix by the standard pulse bases and point matching, and then to employ wavelets to sparsify the dense matrix. This way the dense matrix is treated as an image, and wavelets are employed to compress the image as if working on image processing.

(1) Read papers [20, 21, 22], of which all are in the imaging processing category.
(2) Follow the numerical example of the L-shaped scatterer in [21], and compare your results with Fig. 12 in [21].
(3) Write your report in terms of numerical accuracy, matrix filling time, memory consumption, operational counts, and so on.
(4) Compare the image processing approach with the Coiflet approach and the smooth local cosine (SLC) approach in Chapter 7. List major advantages of the Coiflet and/or SLC method over the imaging processing approach.

BIBLIOGRAPHY

[1] C. Baker, *The Numerical Treatment of Integral Equations*, Cambridge University Press, Cambridge, England, 1977.

[2] H. Brunner and P. van der Houwen, *The Numerical Solution of Volterra Equations*, North-Holland, Amsterdam, 1986.

[3] G. Beylkin, R. Coifman, and V. Rokhlin, "Fast wavelet transforms and numerical algorithms I," *Comm. Pure Appl. Math.*, **44**, 141–183, 1991.

[4] G. Wang and G. Pan, "Full wave analysis of microstrip floating line structures by wavelet expansion method," *IEEE Trans. Microw. Theory Tech.*, **43**, 131–142, Jan. 1995.

[5] B. Steinberg and Y. Leviatan, "On the use of wavelet expansions in the method of moments," *IEEE Trans. Ant. Propg.*, **41**(5), 610–619, May 1993.

[6] G. Pan, "Orthogonal wavelets with applications in electromagnetics," *IEEE Trans. Magn.*, **32**(3), 975–983, May 1996.

[7] G. Pan, M. Toupikov, and B. Gilbert, "On the use of Coifman intervallic wavelets in the method of moments for fast construction of wavelet sparsified matrices," *IEEE Trans. Ant. Prop.*, **47**(7), 1189–1200, 1999.

[8] D. Pozar, "Improved computational efficiency for the moment method solution of printed dipoles and patches," *Electromagn.*, **3**, 229–309, 1983.

[9] G. Pan, J. Tan and J. Murphy, "Full-wave analysis of microstrip floating-line discontinuities," *IEEE Trans. Electromagn. Compatibility*, **36**(1), 49–59, 1994.

[10] J. Mosig and F. Gardiol, "A dynamical radiation model for microstrip structures,"*Ad. Electronics Electron Phys.*, **59**, 138–236, 1982.

[11] R. Harrington, *Field Computation by Moment Methods*, Macmillan, New York, 1968.

[12] B. Alpert, "Hybrid Gauss-trapezoidal quadrature rules," *SIAM J. Sci. Comput.*, **20**(5), 1551–1584, 1999.

[13] R. Wu and J. Yang, "Boundary integral equation formulation of skin effect problems in multiconductor transmission lines," *IEEE Trans. Magn.*, **25**, 3013–3015, July 1989.

[14] A. Cohen, I. Daubechies, and P. Vial, "Wavelets on the interval and fast wavelet transforms," *Appl. Comput. Harmon. Anal.*, **I**, 54–81, 1993.

[15] L. Andersson, N. Hall, B. Jawerth, and G. Peters, "Wavelets on closed subsets of the real line," in *Topics in the Theory and Applications of Wavelets*, L. Schumaker and G. Webb, Eds., Academic Press, New York, 1994.

[16] W. Sweldens, "The lifting scheme: a construction of second generation wavelets," *SIAM J. Math. Anul.*, **29**(2), 511–546, 1997.

[17] M. Toupikov, G. Pan and B. Gilbert, "Weighted wavelet expansion in the method of moments," *IEEE Trans. Magn.*, **35**(3), 1550–1553, May 1999.

[18] A. Baghai-Wadji and G. Walter, "Green's function based wavelets: Selected properties," *Digest, IEEE Ultrason. Symp.*, 2000, IEEE Press, 537–546.

[19] I. Daubechies, *Ten Lectures on Wavelets*. SIAM, Philadelphia, 1992.

[20] B. Alpert, G. Beylkin, R. Coifman, and V. Rokhlin, "Wavelet-like bases for the fast solution of second-kind integral equations," *SIAM J. Sci. Comput.*, **14**(1), 159–184, Jan. 1993.

[21] R. Wagner and W. Chew, "A study of wavelets for the solution of electromagnetic integral equations," *IEEE Trans. Ant. Propg.*, **43**(8), 802–810, Aug. 1995.

[22] H. Deng and H. Ling, "Fast solution of electromagnetic integral equations using adaptive wavelet packet transform," *IEEE Trans. Ant. Propg.*, **47**(4), 674–682, Apr. 1999.

CHAPTER FIVE

Sampling Biorthogonal Time Domain Method (SBTD)

The finite difference time domain (FDTD) method was proposed by K. Yee [1] in 1966. The simplicity of the FDTD method in mathematics has proved to be its great advantage. The method does not involve any integral equations, Green's functions, singularities, nor matrix equations. Neither does it involve functional or variational principles. In addition the FDTD proves to be versatile when used in complicated geometries. The computational issues associated with the FDTD are the radiation boundary conditions or absorption boundary conditions for open structures, numerical dispersion, and stability conditions. Its major drawbacks include its massive memory consumption and huge computational time.

In these regard wavelets offer significant improvements to the FDTD. It will be shown that the Yee-based FDTD is identical to the Galerkin method using Haar wavelets. Since the Haar bases are discontinuous, the slow decay of the frequency components and the Gibbs phenomena of the Haar basis prevent the use of a coarse mesh in the FDTD. In contrast, the Daubechies-based sampling functions are continuous basis functions with fast decay in both the spatial and spectral domains. Thus a more efficient time domain method can be derived: the sampling biorthogonal time domain (SBTD) algorithm.

5.1 BASIS FDTD FORMULATION

For a lossy medium with a conductivity σ, we begin with Maxwell's two curl equations

$$\mu \frac{\partial \mathbf{H}}{\partial t} = -\nabla \times \mathbf{E}, \tag{5.1.1}$$

$$\epsilon \frac{\partial \mathbf{E}}{\partial t} + \sigma \mathbf{E} = \nabla \times \mathbf{H}. \tag{5.1.2}$$

We obtain by the leapfrog method [2] a set of finite difference equations

$$
\begin{aligned}
&{}_{k+(1/2)}H^x\left(\ell, m+\frac{1}{2}, n+\frac{1}{2}\right) \\
&\quad = {}_{k-(1/2)}H^x\left(\ell, m+\frac{1}{2}, n+\frac{1}{2}\right) \\
&\qquad + \frac{\Delta t}{\mu \Delta z}\left[{}_kE^y\left(\ell, m+\frac{1}{2}, (n+1)\right) - {}_kE^y\left(\ell, m+\frac{1}{2}, n\right)\right] \\
&\qquad - \frac{\Delta t}{\mu \Delta y}\left[{}_kE^z\left(\ell, (m+1), n+\frac{1}{2}\right) - {}_kE^z\left(\ell, m, n+\frac{1}{2}\right)\right], \\
&{}_{k+(1/2)}H^y\left(\ell+\frac{1}{2}, m, n+\frac{1}{2}\right) \\
&\quad = {}_{k-(1/2)}H^y\left(\ell+\frac{1}{2}, m, n+\frac{1}{2}\right) \\
&\qquad + \frac{\Delta t}{\mu \Delta x}\left[{}_kE^z\left(\ell+1, m, n+\frac{1}{2}\right) - {}_kE^z\left(\ell, m, n+\frac{1}{2}\right)\right] \\
&\qquad - \frac{\Delta t}{\mu \Delta z}\left[{}_kE^x\left(\ell+\frac{1}{2}, m, n+1\right) - {}_kE^x\left(\ell+\frac{1}{2}, m, n\right)\right], \qquad (5.1.3) \\
&{}_{k+(1/2)}H^z\left(\ell+\frac{1}{2}, m+\frac{1}{2}, n\right) \\
&\quad = {}_{k-(1/2)}H^z\left(\ell+\frac{1}{2}, m+\frac{1}{2}, n\right) \\
&\qquad + \frac{\Delta t}{\mu \Delta y}\left[{}_kE^x\left(\ell+\frac{1}{2}, m+1, n\right) - {}_kE^x\left(\ell+\frac{1}{2}, m, n\right)\right] \\
&\qquad - \frac{\Delta t}{\mu \Delta x}\left[{}_kE^y\left(\ell+1, m+\frac{1}{2}, n\right) - {}_kE^y\left(\ell, m+\frac{1}{2}, n\right)\right], \\
&{}_{k+1}E^x\left(\ell+\frac{1}{2}, m, n\right) \\
&\quad = \left(\frac{1-(\sigma\Delta t/2\epsilon)}{1+(\sigma\Delta t/2\epsilon)}\right){}_kE^x\left(\ell+\frac{1}{2}, m, n\right) \\
&\qquad + \left(\frac{1}{1+(\sigma\Delta t/2\epsilon)}\right)\left\{\frac{\Delta t}{\epsilon\Delta y}\left[{}_{k+(1/2)}H^z\left(\ell+\frac{1}{2}, m+\frac{1}{2}, n\right)\right.\right. \\
&\qquad \left. - {}_{k+(1/2)}H^z\left(\ell+\frac{1}{2}, m-\frac{1}{2}, n\right)\right]
\end{aligned}
$$

$$
\begin{aligned}
&- \frac{\Delta t}{\epsilon \Delta z}\left[{}_{k+(1/2)}H^{y}\left(\ell+\frac{1}{2}, m, n+\frac{1}{2}\right)\right.\\
&\left.\left.-{}_{k+(1/2)}H^{y}\left(\ell+\frac{1}{2}, m, n-\frac{1}{2}\right)\right]\right\},\\
{}_{k+1}E^{y}&\left(\ell, m+\frac{1}{2}, n\right)\\
&=\left(\frac{1-\frac{\sigma\Delta t}{2\epsilon}}{1+\frac{\sigma\Delta t}{2\epsilon}}\right){}_{k}E^{y}\left(\ell, m+\frac{1}{2}, n\right)\\
&+\left(\frac{1}{1+\frac{\sigma\Delta t}{2\epsilon}}\right)\left\{\frac{\Delta t}{\epsilon\Delta z}\left[{}_{k+(1/2)}H^{x}\left(\ell, m+\frac{1}{2}, n+\frac{1}{2}\right)\right.\right.\\
&\left.-{}_{k+(1/2)}H^{x}\left(\ell, m+\frac{1}{2}, n-\frac{1}{2}\right)\right]\\
&-\frac{\Delta t}{\epsilon\Delta x}\left[{}_{k+(1/2)}H^{z}\left(\ell+\frac{1}{2}, m+\frac{1}{2}, n\right)\right.\\
&\left.\left.-{}_{k+(1/2)}H^{z}\left(\ell-\frac{1}{2}, m+\frac{1}{2}, n\right)\right]\right\}, \qquad (5.1.4)\\
{}_{k+1}E^{z}&\left(\ell, m, n+\frac{1}{2}\right)\\
&=\left(\frac{1-\frac{\sigma\Delta t}{2\epsilon}}{1+\frac{\sigma\Delta t}{2\epsilon}}\right){}_{k}E^{z}\left(\ell, m, n+\frac{1}{2}\right)\\
&+\left(\frac{1}{1+\frac{\sigma\Delta t}{2\epsilon}}\right)\left\{\frac{\Delta t}{\epsilon\Delta x}\left[{}_{k+(1/2)}H^{y}\left(\ell+\frac{1}{2}, m, n+\frac{1}{2}\right)\right.\right.\\
&\left.-{}_{k+(1/2)}H^{y}\left(\ell-\frac{1}{2}, m, n+\frac{1}{2}\right)\right]\\
&-\frac{\Delta t}{\epsilon\Delta y}\left[{}_{k+(1/2)}H^{x}\left(\ell, m+\frac{1}{2}, n+\frac{1}{2}\right)\right.\\
&\left.\left.-\,{}_{k+(1/2)}H^{x}\left(\ell, m-\frac{1}{2}, n+\frac{1}{2}\right)\right]\right\}.
\end{aligned}
$$

In the equations above, indexes l, m, and n are the node numbers in the x, y, and z directions, respectively, and the leftscript k denotes the time step. Note that a central difference scheme has been used in all of the finite difference equations. Figure 5.1 illustrates a unit cell in the FDTD lattice where the electric and magnetic fields are spaced apart by a half-grid in each dimension. At the interface of two media (e.g., at the boundary between a conductor and a dielectric), the average values of ϵ and

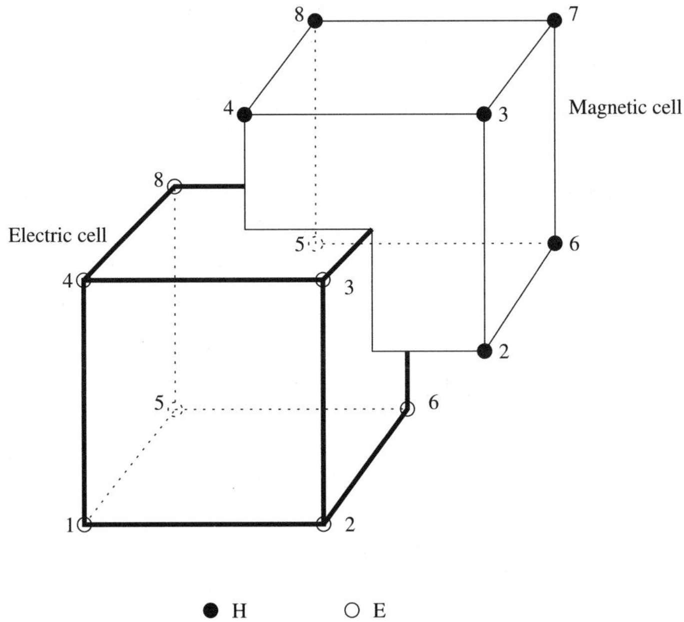

FIGURE 5.1 Standard Yee-FDTD lattice.

σ are used: $\epsilon = (\epsilon_1 + \epsilon_2)/2$ and $\sigma = (\sigma_1 + \sigma_2)/2$. To ensure the stability of the time-stepping algorithm of (5.1.1) and (5.1.2), a time increment is chosen to satisfy the inequality

$$c\Delta t \leq \frac{1}{\sqrt{1/\Delta x^2 + 1/\Delta y^2 + 1/\Delta z^2}}, \tag{5.1.5}$$

where c is the velocity of light in the computational space. Equation (5.1.5) will be derived in the next section. A Gaussian pulse

$$E_z = e^{-(t-t_0)^2/T^2}$$

is chosen as the excitation pulse and is imposed upon the rectangular region under the port to be excited.

The finite difference mesh must be truncated because of the finite ability of computers to solve across very large or even infinite 3D volumes. The field components tangential to the truncation planes cannot be evaluated from the FDTD equations above since they would require for their evaluation the values of field components outside the mesh. The tangential electric field components on the truncation planes must be specified in such a way that outgoing waves are not reflected; this is known as an *absorbing boundary condition* (ABC). There are many ABCs, including the Mur absorbing boundary conditions [3] and the perfectly matched layer absorbing boundary conditions (PML) of Berenger [4], among others. Here we have specified the boundary values of the fields according to the Engquist-Majda unconditionally stable, absorbing boundary condition [5]

$$\phi_0^{k+1} = \phi_1^k + \frac{c\Delta t - \Delta x}{c\Delta t + \Delta x}(\phi_1^{k+1} - \phi_0^k),$$

where ϕ_0 and ϕ_1 are the tangential electric field components at the mesh wall and at the first node within the wall, respectively.

Figure 5.2 depicts a system consisting of three coupled microstrip lines. In the direction of signal propagation, the y direction, we have chosen the parameters Δt, Δx, Δy, and Δz such that the wave travels one spatial step in approximately five temporal steps; this choice in turn requires a priori calculation in order to obtain the approximate wave velocity in the direction of propagation. At the top and side boundaries, the local velocity of light at the calculated node is used as the approximate wave velocity. Without a loss of generality, the time domain solution for this six-port system is obtained by means of the following procedures:

(1) Initialize (at $t = k\Delta t = 0$) all fields to 0.

(2) Impose Gaussian excitation on port 1:
- $\mathbf{H}^{k+(1/2)}$ is calculated from the FDTD equations.
- $\mathbf{E}^{k+1}$ is calculated from the FDTD equations.
- The tangential **E** field is set to 0 on the ground plane and the absorbing boundary condition is used on the truncation planes.
- Store port voltages $V_i^{(1)}(k\Delta t)$ at the reference plane of port i ($i = 1, 2, 3, 4, 5, 6$), where a port voltage V_i has been obtained by numerically integrating the vertical electric field beneath the center of port i.

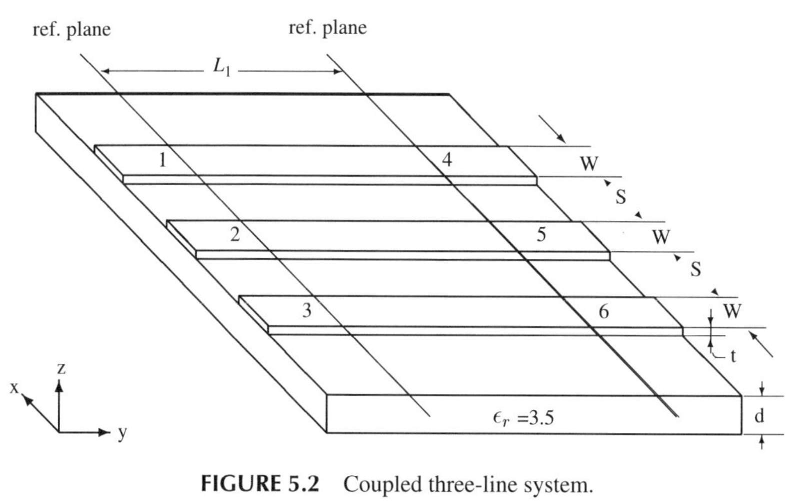

FIGURE 5.2 Coupled three-line system.

- Store port currents $I_i^{(1)}(k\Delta t)$ at the reference plane of port i ($i = 1, 2, 3$), where a port current I_i has been obtained by numerical integration of the magnetic field around the strip surface of port i in the reference plane.
- $k \to k+1$, repeat the previous steps 1 through 5 until the pulse and induced waves pass through the reference plane of ports (4, 5, 6) completely.

(3) Impose a Gaussian excitation on port 2 and repeat the above six procedures. Store all of the port voltages $V_i^{(2)}(i = 1, 2, 3, 4, 5, 6)$ and port currents $I_i^{(2)}(i = 1, 2, 3)$.

In the previous items the superscripts (1) and (2) represented port 1 excitation and port 2 excitation, respectively.

The Yee algorithm has been modified and extended into many versions and derivatives, including the nonuniform mesh FDTD, and the finite volume time domain (FVTD) method [6], nonorthogonal mesh [7], and the like. The transmission line matrix (TLM) method was proposed by Peter Johns [8] in 1971 independently of Yee's work. Nonetheless, it was proven that the TLM is equivalent to FDTD method. In handling lossy structures, the TLM needs to use artificial "stubs"; this necessity is inconvenient. Because of its simplicity and popularity, only the standard FDTD will be discussed in the text.

5.2 STABILITY ANALYSIS FOR THE FDTD

An unstable solution may occur owing to an improper choice of the time step Δt for the space intervals Δx, Δy, and Δz. The instability is not due to an accumulation of errors, but to causality. The analysis is conducted on plane waves, and is quite general, since any wave may be expressed as a superposition of plane waves. Let us write FDTD in terms of time–space eigenvalue problems. Space eigenvalues must be located in stable regions.

The two curl equations in a lossless medium are written in their component forms

$$\nabla \times \mathbf{H} = \epsilon \frac{\partial \mathbf{E}}{\partial t},$$

$$\frac{\partial E_z}{\partial t} = \frac{1}{\epsilon}\left(\frac{\partial H_y}{\partial x} - \frac{\partial H_x}{\partial y}\right),$$

$$\frac{\partial E_y}{\partial t} = \frac{1}{\epsilon}\left(\frac{\partial H_x}{\partial z} - \frac{\partial H_z}{\partial x}\right),$$

$$\frac{\partial E_x}{\partial t} = \frac{1}{\epsilon}\left(\frac{\partial H_z}{\partial y} - \frac{\partial H_y}{\partial z}\right),$$

and

$$\nabla \times \mathbf{E} = -\mu \frac{\partial \mathbf{H}}{\partial t},$$

$$\frac{\partial H_x}{\partial t} = \frac{1}{\mu}\left(\frac{\partial E_y}{\partial z} - \frac{\partial E_z}{\partial y}\right),$$

$$\frac{\partial H_y}{\partial t} = \frac{1}{\mu}\left(\frac{\partial E_z}{\partial x} - \frac{\partial E_x}{\partial z}\right),$$

$$\frac{\partial H_z}{\partial t} = \frac{1}{\mu}\left(\frac{\partial E_x}{\partial y} - \frac{\partial E_y}{\partial x}\right).$$

In the rest of this section, we will only attack 2D problems. In doing so, we will be able to capture the essence of the algorithms without spending too much time and effort on tedious details. In 2D problems we will deal with only three rather than six equations. The extension of the 2D formulation into 3D problems is straightforward, but time-consuming. Consider a 2D $TM^{(z)}$ wave, namely

$$\frac{\partial}{\partial z} = 0, \quad H_z = 0.$$

The remaining three equations are

$$\frac{\partial E_z}{\partial t} = \frac{1}{\epsilon}\left(\frac{\partial H_y}{\partial x} - \frac{\partial H_x}{\partial y}\right),$$

$$\frac{\partial H_x}{\partial t} = -\frac{1}{\mu}\frac{\partial E_z}{\partial y},$$

$$\frac{\partial H_y}{\partial t} = \frac{1}{\mu}\frac{\partial E_z}{\partial x}.$$

Using the center difference Yee scheme and simplified notations, we obtain

$$\frac{{}_{n+1}E^z_{i,j} - {}_{n}E^z_{i,j}}{\Delta t} = \frac{1}{\epsilon}\left[\frac{{}_{n+(1/2)}H^y_{i+(1/2),j} - {}_{n+(1/2)}H^y_{i-(1/2),j}}{\Delta x} - \frac{{}_{n+(1/2)}H^x_{i,j+(1/2)} - {}_{n+(1/2)}H^x_{i,j-(1/2)}}{\Delta y}\right],$$

$$\frac{{}_{n+(1/2)}H^x_{i,j+(1/2)} - {}_{n-(1/2)}H^x_{i,j+(1/2)}}{\Delta t} = -\frac{1}{\mu}\frac{{}_{n}E^z_{i,j+1} - {}_{n}E^z_{i,j}}{\Delta y},$$

$$\frac{{}_{n+(1/2)}H^y_{i+(1/2),j} - {}_{n-(1/2)}H^y_{i+(1/2),j}}{\Delta t} = \frac{1}{\mu}\frac{{}_{n}E^z_{i+1,j} - {}_{n}E^z_{i,j}}{\Delta x}. \tag{5.2.1}$$

CASE 1. TIME EIGENVALUE PROBLEM Separating the time derivatives in the preceding equations, we arrive at

$$\frac{{}_{n+1}E^z_{i,j} - {}_{n}E^z_{i,j}}{\Delta t} = \lambda\ {}_{n+(1/2)}E^z_{i,j}, \tag{5.2.2}$$

$$\frac{{}_{n+(1/2)}H^{x}_{i,j+(1/2)} - {}_{n-(1/2)}H^{x}_{i,j+(1/2)}}{\Delta t} = \lambda \, {}_{n}H^{x}_{i,j+(1/2)}, \tag{5.2.3}$$

$$\frac{{}_{n+(1/2)}H^{y}_{i+(1/2),j} - {}_{n-(1/2)}H^{y}_{i+(1/2),j}}{\Delta t} = \lambda \, {}_{n}H^{y}_{i+(1/2),j}. \tag{5.2.4}$$

The general form of (5.2.2) through (5.2.4) is

$$\frac{{}_{n+(1/2)}V_i - {}_{n-(1/2)}V_i}{\Delta t} = \lambda \, {}_{n}V_i. \tag{5.2.5}$$

Let us define a factor

$$q_i = \frac{{}_{n+(1/2)}V_i}{{}_{n}V_i}. \tag{5.2.6}$$

In order to have a stable solution of (5.2.2) through (5.2.4), we must meet the condition

$$|q_i| \leq 1.$$

Substituting (5.2.6) into (5.2.5), we have

$$\frac{{}_{n+(1/2)}V_i}{{}_{n}V_i} - \frac{{}_{n-(1/2)}V_i}{{}_{n}V_i} = \lambda \, \Delta t,$$

or equivalently

$$q_i - \frac{1}{q_i} = \lambda \, \Delta t.$$

Thus we obtain

$$q_i^2 - \lambda \, \Delta t q_i - 1 = 0,$$

$$q_i = \frac{\lambda \, \Delta t}{2} \pm \sqrt{1 + \left(\frac{\lambda \, \Delta t}{2}\right)^2}. \tag{5.2.7}$$

In order to have $|q_i| \leq 1$, we need

$$\begin{cases} \mathrm{Re}\{\lambda\} = 0 \\ -\frac{2}{\Delta t} \leq \mathrm{Im}\{\lambda\} \leq \frac{2}{\Delta t}. \end{cases}$$

Letting $\lambda = \mu + j\nu$, (5.2.7) gives

$$q_i = j\frac{\nu \Delta t}{2} \pm \sqrt{1 - \frac{(\nu \Delta t)^2}{4}}.$$

CASE 2. SPACE EIGENVALUE PROBLEM The right-hand side of (5.2.1) provides the following eigenvalue equations

$$\frac{H^y_{i+(1/2),j} - H^y_{i-(1/2),j}}{\Delta x} - \frac{H^x_{i,j+(1/2)} - H^x_{i,j-(1/2)}}{\Delta y} = \lambda \epsilon E^z_{i,j}, \tag{5.2.8}$$

$$\frac{E^z_{i,j+1} - E^z_{i,j}}{\Delta y} = -\lambda \mu H^x_{i,j+(1/2)}, \tag{5.2.9}$$

$$\frac{E^z_{i+1,j} - E^z_{i,j}}{\Delta x} = \lambda \mu H^y_{i+(1/2),j}. \tag{5.2.10}$$

Again, a nonplane wave can be expanded into a superposition of plane waves. Thus we may work with the following plane waves:

$$\begin{aligned} E^z_{I,J} &= E^z\, e^{j(k_x I\,\Delta x + k_y J\,\Delta y)}, \\ H^x_{I,J} &= H^x\, e^{j(k_x I\,\Delta x + k_y J\,\Delta y)}, \\ H^y_{I,J} &= H^y\, e^{j(k_x I\,\Delta x + k_y J\,\Delta y)}. \end{aligned} \tag{5.2.11}$$

Substitution of (5.2.11) into (5.2.8–5.2.10) leads to

$$\begin{aligned} E^z &= j\frac{2}{\lambda\epsilon}\left[\frac{H_y}{\Delta x}\sin\frac{k_x\,\Delta x}{2} - \frac{H_x}{\Delta y}\sin\frac{k_y\,\Delta y}{2}\right], \\ H^x &= -j\frac{2E_z}{\lambda\mu\,\Delta y}\sin\frac{k_y\,\Delta y}{2}, \\ H^y &= j\frac{2E_z}{\lambda\mu\,\Delta x}\sin\frac{k_x\,\Delta x}{2}, \end{aligned}$$

and

$$\lambda^2 = -\frac{4}{\epsilon\mu}\left[\left(\frac{\sin k_x\,\Delta x/2}{\Delta x}\right)^2 + \left(\frac{\sin k_y\,\Delta y/2}{\Delta y}\right)^2\right].$$

Note that $|\sin(\cdot)| \le 1$. Hence for $\forall\, k_x, k_y$,

$$\begin{aligned} &\mathrm{Re}\{\lambda\} = 0 \\ &\mathrm{Im}\{\lambda\} \le 2v\left[\left(\frac{1}{\Delta x}\right)^2 + \left(\frac{1}{\Delta y}\right)^2\right]^{1/2}. \end{aligned}$$

Case 3. Numerical Stability Relating the time eigenvalue problem to the space eigenvalue problem, we have the 2D stability condition

$$2v\left[\left(\frac{1}{\Delta x}\right)^2 + \left(\frac{1}{\Delta y}\right)^2\right]^{1/2} \le \frac{2}{\Delta t},$$

namely

$$\Delta t \leq \frac{1}{v\sqrt{(1/\Delta x)^2 + (1/\Delta y)^2}}.$$

For 3D cases, the stability condition is

$$\Delta t \leq \frac{1}{v\sqrt{(1/\Delta x)^2 + (1/\Delta y)^2 + (1/\Delta z)^2}}$$
$$= \frac{\Delta \ell}{v\sqrt{3}},$$

if $\Delta x = \Delta y = \Delta z = \Delta l$. For 1D, this condition reduces to

$$\Delta t \leq \frac{\Delta x}{v}.$$

5.3 FDTD AS MAXWELL'S EQUATIONS WITH HAAR EXPANSION

The finite difference time domain (FDTD) formulas of (5.1.3) to (5.1.4) are derived from the two Maxwell curl equations, using the finite difference to approximate the differential operators. In this section we will see that the FDTD can be derived as a special case of wavelet expansion using the Haar system.

To simplify our mathematical notation without losing generality, we consider the one-dimensional case, namely the telegraphers' equations in the frequency domain. The telegraphers' equations are

$$\begin{cases} -\dfrac{dI}{dx} = j\omega CV \\ -\dfrac{dV}{dx} = j\omega LI. \end{cases} \tag{5.3.1}$$

When the finite difference method is applied, the expected result is

$$\begin{cases} \dfrac{I_{n+1} - I_n}{\Delta x} = -j\omega C V_{n+(1/2)} \\ \dfrac{V_{n+(1/2)} - V_{n-(1/2)}}{\Delta x} = -j\omega L I_n. \end{cases} \tag{5.3.2}$$

Let us expand the unknown current and voltage in terms of Haar scalets, that is, pulse functions

$$\begin{cases} I = \sum_m I_m P_m(x) \\ V = \sum_m V_{m+(1/2)} P_{m+(1/2)}(x). \end{cases} \tag{5.3.3}$$

Notice that the voltage node and current node are offset by a half unit in space. The pulse function can be written in the form

$$P_k(x) = P\left(\frac{x}{\Delta x} - k\right),$$

where

$$P(x) = \begin{cases} 1 & \text{if} \quad |x| < \frac{1}{2} \\ \frac{1}{2} & \text{if} \quad |x| = \frac{1}{2} \\ 0 & \text{if} \quad |x| > \frac{1}{2}. \end{cases}$$

It can easily be seen that

$$\int_{-\infty}^{\infty} P_k(x) P_l(x)\, dx = (\Delta x)\, \delta_{k,l}, \tag{5.3.4}$$

which is analogous to the orthogonality for wavelets

$$\int_{-\infty}^{\infty} \varphi_{j,k}(x) \varphi_{j,l}(x)\, dx = \delta_{k,l}.$$

Show.

(1) If $k \neq l$, $P_k(x)$ and $P_l(x)$ have no overlaps; hence

$$\int_{-\infty}^{\infty} P_k(x) P_l(x)\, dx = 0. \tag{5.3.5}$$

(2) If $k = l$,

$$\begin{aligned} \int_{-\infty}^{\infty} P_k(x) P_l(x)\, dx &= \int P^2 \underbrace{\left(\frac{x}{\Delta x} - k\right)}_{u} dx \\ &= \Delta x \int P^2 \left(\frac{x}{\Delta x} - k\right) d\left(\frac{x}{\Delta x}\right) \\ &= \Delta x \int_{-1/2}^{1/2} P^2(u)\, du \\ &= \Delta x \int_{-1/2}^{1/2} 1\, du = \Delta x. \end{aligned} \tag{5.3.6}$$

Combining (5.3.5) and (5.3.6), we arrive at (5.3.4). □

Next we will show that

$$\int P_m(x) \frac{\partial}{\partial x} P_{m'+(1/2)}(x)\, dx = \delta_{m,m'} - \delta_{m,m'+1}. \tag{5.3.7}$$

In fact

$$\frac{\partial P(x)}{\partial x} = \delta\left(x + \frac{1}{2}\right) - \delta\left(x - \frac{1}{2}\right),$$

where the Dirac delta function $\delta(x - \tau)$ has been used. More rigorously, we can use the Heaviside step function $H(x - \tau)$

$$H(x - \tau) = \frac{1}{2\pi i} \int \frac{e^{(x-\tau)s}}{s}\, ds,$$

where we integrate along the imaginary axis. The Dirac delta function is defined by following integral

$$\int_a^b \delta(x - \tau)\, dx = \begin{cases} 0 & \text{if } b < \tau \text{ or } a > \tau \\ 1 & \text{if } a < \tau < b. \end{cases}$$

It is possible to write an integral representation that is similar to the step function

$$\delta(x - \tau) = \frac{1}{2\pi i} \int e^{(x-\tau)s}\, ds.$$

The Dirac delta function and the step function are related by the expression

$$H'(x - \tau) = \delta(x - \tau). \tag{5.3.8}$$

The pulse function can be written as

$$P(x) = H\left(x + \tfrac{1}{2}\right) - H\left(x - \tfrac{1}{2}\right). \tag{5.3.9}$$

Equations (5.3.9) and (5.3.8) lead to (5.3.7) as follows: using the previous results, we obtain

$$\begin{aligned}
&\int P_m(x) \frac{\partial}{\partial x} P_{m'+(1/2)}(x)\, dx \\
&\quad = \int P\left(\frac{x}{\Delta x} - m\right) \frac{\partial}{\partial x} P\left(\frac{x}{\Delta x} - \left(m' + \frac{1}{2}\right)\right) dx \\
&\quad = \int P\left(\frac{x}{\Delta x} - m\right) \frac{\partial}{\partial x} \left[H\left(\frac{x}{\Delta x} - \left(m' + \frac{1}{2}\right) + \frac{1}{2}\right)\right. \\
&\qquad \left. - H\left(\frac{x}{\Delta x} - \left(m' + \frac{1}{2}\right) - \frac{1}{2}\right)\right] dx \\
&\quad = \int P\left(\frac{x}{\Delta x} - m\right) \frac{1}{\Delta x} \left[\delta\left(\frac{x}{\Delta x} - m'\right) - \delta\left(\frac{x}{\Delta x} - (m' + 1)\right)\right] dx \\
&\quad = \delta_{m,m'} - \delta_{m,m'+1}.
\end{aligned}$$

Thus far we have sufficient knowledge to derive the finite difference equation (5.3.2). By substituting (5.3.3) into (5.3.1), we obtain

$$-\sum_m V_{m+(1/2)} \frac{d}{dx} P_{m+(1/2)}(x) = j\omega L \sum_m I_m P_m(x).$$

Multiplying both sides by $P_n(x)$ and integrating, we arrive at

$$\begin{aligned} \text{RHS} &= j\omega L \sum_m I_m \int P_m(x) P_n(x)\, dx \\ &= j\omega L I_n \,\Delta x, \end{aligned}$$

where the orthogonality of $P_m(x)$ and $P_n(x)$ has been employed in order to simplify the summation. In the meantime

$$\begin{aligned} \text{LHS} &= -\sum_m V_{m+(1/2)} \int P_n(x) \frac{d}{dx} P_{m+(1/2)}(x)\, dx \\ &= -\sum_m V_{m+(1/2)} [\delta_{n,m} - \delta_{n,m+1}] \\ &= -[V_{n+(1/2)} - V_{n-(1/2)}]. \end{aligned}$$

Equating the two sides, we finally have

$$\frac{V_{n+(1/2)} - V_{n-(1/2)}}{\Delta x} = -j\omega L I_n,$$

which is exactly the centralized finite difference expression of (5.3.2). Notice that the derivation is totally new and never makes use of the finite difference concept.

5.4 FDTD WITH BATTLE–LEMARIE WAVELETS

Battle–Lemarie wavelets possess better regularity than Haar wavelets. The Battle–Lemarie based time domain method, referred to as the multiresolution time domain (MRTD), improves numerical dispersion of the FDTD significantly [9]. However, the MRTD is not widespread in the field computation because of its high computational cost, complexity of its algorithm, CPU time required, and the difficulties in incorporating boundary conditions. For reasons of historical development and completeness, we will briefly discuss the scalet-based MRTD.

In the MRTD the time dependencies of the field quantities are still treated as pulse functions while the space dependencies are expanded in terms of the Battle–Lemarie (B-L) scalets instead of Haar scalets. The six components field equations are

$$
\begin{aligned}
E_x(\mathbf{r}, t) &= \sum_{k,l,m,n} {}_kE^{\varphi x}_{l+(1/2),m,n} P_k(t)\varphi_{l+(1/2)}(x)\varphi_m(y)\varphi_n(z),\\
E_y(\mathbf{r}, t) &= \sum_{k,l,m,n} {}_kE^{\varphi y}_{l,m+(1/2),n} P_k(t)\varphi_l(x)\varphi_{m+(1/2)}(y)\varphi_n(z),\\
E_z(\mathbf{r}, t) &= \sum_{k,l,m,n} {}_kE^{\varphi z}_{l,m,n+(1/2)} P_k(t)\varphi_l(x)\varphi_m(y)\varphi_{n+(1/2)}(z),\\
H_x(\mathbf{r}, t) &= \sum_{k,l,m,n} {}_{k+(1/2)}H^{\varphi x}_{l,m+(1/2),n+(1/2)} P_{k+(1/2)}(t)\varphi_l(x)\varphi_{m+(1/2)}(y)\varphi_{n+(1/2)}(z),\\
H_y(\mathbf{r}, t) &= \sum_{k,l,m,n} {}_{k+(1/2)}H^{\varphi y}_{l+(1/2),m,n+(1/2)} P_{k+(1/2)}(x)\varphi_{l+(1/2)}(x)\varphi_m(y)\varphi_{n+(1/2)}(z),\\
H_z(\mathbf{r}, t) &= \sum_{k,l,m,n} {}_{k+(1/2)}H^{\varphi z}_{l+(1/2),m+(1/2),n} P_{k+(1/2)}(x)\varphi_{l+(1/2)}(x)\varphi_{m+(1/2)}(y)\varphi_n(z).
\end{aligned}
\tag{5.4.1}
$$

The Fourier transform pair of the cubic spline Battle–Lemarie scalet is

$$\hat{\varphi}(\omega) = \int_{-\infty}^{\infty} \varphi(x)e^{-i\omega x}\, dx$$

and

$$\varphi(x) = \frac{1}{2\pi}\int_{-\infty}^{\infty} \hat{\varphi}(\omega)e^{i\omega x}\, d\omega.$$

It can be verified that the Fourier transform of the cubic Battle–Lemarie scalet is

$$\hat{\varphi}(\omega) = \left(\frac{\sin \omega/2}{\omega/2}\right)^4 \frac{1}{\sqrt{1-(4/3)\sin^2(\omega/2)+(2/5)\sin^4(\omega/2))-(4/315)\sin^6(\omega/2)}}. \tag{5.4.2}$$

Using properties of the Fourier integral, it is possible to write

$$
\begin{aligned}
&\int_{-\infty}^{\infty} \varphi_m(x)\frac{\partial}{\partial x}\varphi_{m'+(1/2)}(x)\, dx\\
&\quad= \int_{-\infty}^{\infty} dx \left[\left(\frac{1}{2\pi}\int d\omega\hat{\varphi}(\omega)e^{-i\omega m+i\omega x}\right)\right.\\
&\qquad \left.\frac{1}{2\pi}\int d\omega' \frac{\partial}{\partial x}\hat{\varphi}(\omega')e^{-i\omega'[m'+(1/2)]}e^{i\omega' x}\right]\\
&\quad= \iint d\omega\, d\omega' \int_{-\infty}^{\infty} dx \left[\frac{1}{2\pi}e^{ix(\omega+\omega')}\right]\frac{i\omega'}{2\pi}\hat{\varphi}(\omega')\hat{\varphi}(\omega)e^{-i(\omega m)-i\omega'[m'+(1/2)]}\\
&\quad= \frac{1}{2\pi}\int_{-\infty}^{\infty} d\omega\hat{\varphi}(\omega)e^{-i\omega m}\int d\omega'\,\delta(\omega+\omega')(i\omega')\hat{\varphi}(\omega')e^{-i\omega'[m'+(1/2)]}
\end{aligned}
$$

$$= \frac{1}{2\pi}\int_{-\infty}^{\infty} d\omega \hat{\varphi}(\omega) e^{-i(\omega m)}(-i\omega)\hat{\varphi}(-\omega) e^{i\omega[m'+(1/2)]}$$

$$= \frac{1}{\pi}\int_{0}^{\infty} \omega|\hat{\varphi}(\omega)|^2 \sin\left[\omega\left(m' - m + \frac{1}{2}\right)\right] d\omega.$$

This integral can be evaluated numerically. We can rewrite the expression above as

$$\int_{-\infty}^{\infty} \varphi_m(x)\frac{\partial \varphi_{m'+(1/2)}(x)}{\partial x}\, dx = \sum_{-\infty}^{\infty} a_i\, \delta_{m+i,m'}. \qquad (5.4.3)$$

It can be seen that

$$a_0 = \frac{1}{\pi}\int_0^{\infty} |\hat{\varphi}(\omega)|^2 \omega \sin \tfrac{\omega}{2} dw$$

$$a_1 = \frac{1}{\pi}\int_0^{\infty} |\hat{\varphi}(\omega)|^2 \omega \sin \tfrac{3}{2}\omega\, d\omega$$

$$a_2 = \frac{1}{\pi}\int_0^{\infty} |\hat{\varphi}(\omega)|^2 \omega \sin \tfrac{5}{2}\omega\, d\omega$$

$$\vdots$$

The Battle–Lemarie scalets decay rapidly, and the coefficients a_i are negligible for $i > 8$ and $i < -9$. Thus the summation can be truncated as

$$\int_{-\infty}^{\infty} \varphi_m(x)\frac{\partial \varphi_{m'+(1/2)}(x)}{\partial x}\, dx = \sum_{-9}^{8} a_i\, \delta_{m+i,m'}. \qquad (5.4.4)$$

For negative indexes

$$a_{-1-i} = -a_i, \qquad i = 0, 1, \ldots, 8.$$

Table 5.1 provides the values of a_i, $i = 0, 1, \ldots, 8$. Consider the x-component of Ampere's law

$$\frac{\partial H_z}{\partial y} - \frac{\partial H_y}{\partial z} = \epsilon \frac{\partial E_x}{\partial t}. \qquad (5.4.5)$$

We approximate the right-hand side as

$$\frac{\partial E_x}{\partial t} = \sum_{k',l',m',n'} {}_{k'}E^{\varphi x}_{l'+(1/2),m',n'} \varphi_{l'+(1/2)}(x)\varphi_{m'}(y)\varphi_{n'}(z)\frac{\partial}{\partial t}P_{k'}(t).$$

After sampling the right-hand side of (5.4.5), we obtain $\partial E_x/\partial t$, in space and time,

$$\iiiint dx\, dy\, dz\, dt \varphi_{l+(1/2)}(x)\varphi_m(y)\varphi_n(z)P_{k+(1/2)}(t)\frac{\partial E_x}{\partial t}$$

TABLE 5.1. Coefficients a_i

i	a_i
0	1.2918462
1	−0.1560761
2	0.0596391
3	−0.0293099
4	0.0153716
5	−0.0081892
6	0.0043788
7	−0.0023433
8	0.0012542

$$
\begin{aligned}
&= \sum_{k',l',m',n'} {}_{k'}E^{\varphi x}_{l'+(1/2),m',n'} \int \varphi_{l+(1/2)}(x)\varphi_{l'+(1/2)}(x)\,dx \int \varphi_m(y)\varphi_{m'}(y)\,dy \\
&\quad \int \varphi_n(z)\varphi_{n'}(z)dz \int P_{k+(1/2)}(t)\frac{\partial}{\partial t}P_{k'}(t) \\
&= \sum_{k',l',m',n} {}_{k'}E^{\varphi x}_{l'+(1/2),m',n'}\ \Delta x\ \delta_{l,l'}\ \Delta y\ \delta_{m,m'}\ \Delta z\ \delta_{n,n'}(\delta_{k+1,k'} - \delta_{k,k'}) \\
&= \left({}_{k+1}E^{\varphi x}_{l+(1/2),m,n} - {}_{k}E^{\varphi x}_{l+(1/2),m,n}\right)\ \Delta x\ \Delta y\ \Delta z.
\end{aligned}
$$

Then, sampling the first term of the left-hand side, $\partial H_z/\partial y$, and using the same testing functions as for the RHS, we have

$$
\begin{aligned}
&\iiiint dt\,dx\,dy\,dz \sum_{k',l',m',n'} {}_{k'+(1/2)}H^{\varphi z}_{l'+(1/2),m'+(1/2),n'} P_{k'+(1/2)}(t)\varphi_{l'+(1/2)}(x) \\
&\qquad \frac{\partial}{\partial y}\varphi_{m'+(1/2)}(y)\varphi_{n'}(z)P_{k+(1/2)}(t)\varphi_{l+(1/2)}(x)\varphi_m(y)\varphi_n(z) \\
&\qquad = \sum_{k',l',m',n'} {}_{k'+(1/2)}H^{\varphi z}_{l'+(1/2),m'+(1/2),n'}\ \Delta t\ \delta_{k,k'}\ \Delta x\ \delta_{l,l'}\ \Delta z\ \delta_{n,n'} \\
&\qquad \int dy\varphi_m(y)\frac{\partial}{\partial y}\varphi_{m'+(1/2)}(y) \\
&\qquad \approx \left(\sum_{i=-9}^{8} a_i\ {}_{k+(1/2)}H^{\varphi z}_{l+(1/2),m+(1/2)+i,n}\right)\ \Delta t\ \Delta x\ \Delta z.
\end{aligned}
$$

Applying the same procedure to the term $\partial H_y/\partial z$, we finally obtain a difference equation

$$\frac{\epsilon}{\Delta t}\left({}_{k+1}E^{\varphi x}_{l+(1/2),m,n} - {}_{k}E^{\varphi x}_{l+(1/2),m,n}\right) = \frac{1}{\Delta y}\sum_{i=-9}^{8} a_i \; {}_{k+(1/2)}H^{\varphi z}_{l+(1/2),m+(1/2)+i,n}$$
$$- \frac{1}{\Delta z}\sum_{i=-9}^{8} a_i \; {}_{k+(1/2)}H^{\varphi y}_{l+(1/2),m,n+(1/2)+i}. \tag{5.4.6}$$

Note that the space differential operator is approximated by an 18-term summation in the MRTD versus a 2-term summation in the traditional FDTD. The other five equations can be derived in the same manner.

5.5 POSITIVE SAMPLING AND BIORTHOGONAL TESTING FUNCTIONS

Recall that in communication theory Shannon's sampling theorem [10] is given by

$$x(t) = \sum_{k=-\infty}^{\infty} x(kT)\frac{\sin \sigma(t-kT)}{\sigma(t-kT)}, \; T = \frac{\pi}{\sigma}$$

for σ-band limited signals. For these signals $x(t) \in L^2(R)$, and the Fourier transform $\mathcal{F}(x(t))$ has finite support $[-\sigma, \sigma]$. Often we use the notation sinc as

$$\varphi(t) = \text{sinc}\,(t) := \frac{\sin \pi t}{\pi t}.$$

As studied in Chapter 3 that the Shannon sinc function is a scalet satisfying the sampling property

$$\varphi(n) = \delta_{0,n}.$$

In addition the sinc forms an orthogonal system

$$\int_{-\infty}^{\infty} \varphi(t)\varphi(t-n)\,dt = \delta_{0,n}.$$

Now we will construct sampling functions using the Daubechies scalets. Letting $\varphi(x)$ be the Daubechies scalet of $N = 2$, we can write a positive sampling function

$$S(x) = \frac{2\nu}{\nu - 1}\sum_{k=0}^{\infty}\left(\frac{1+\nu}{1-\nu}\right)^k \varphi(x-k+1), \tag{5.5.1}$$

where $\nu = -1/\sqrt{3}$. $S(x)$ was used to eliminate the Gibbs phenomenon [11]. We will demonstrate that $S(x)$ has a sampling property similar to the sinc function. By the factor $\varphi(1) = (\nu - 1)/2\nu$, $\varphi(2) = (\nu + 1)/2\nu$ (3.8.2), (5.5.1) may be rewritten in a more specific form as

$$S(x) = \frac{1}{\varphi(1)} \sum_{k=0}^{\infty} \left(\frac{|\varphi(2)|}{\varphi(1)} \right)^k \varphi(x - k + 1).$$

Introducing the notation

$$S_m(x) := S(x - m) = \frac{1}{\varphi(1)} \sum_{k=0}^{\infty} \left(\frac{|\varphi(2)|}{\varphi(1)} \right)^k \varphi(x - m - k + 1), \tag{5.5.2}$$

we will show the sampling property

$$S_m(n) = \delta_{m,n}. \tag{5.5.3}$$

Show. From the definition of $S_m(x)$, we have

$$S_m(n) = \frac{1}{\varphi(1)} \sum_{k=0}^{\infty} \left(\frac{|\varphi(2)|}{\varphi(1)} \right)^k \varphi(n - m - k + 1). \tag{5.5.4}$$

Notice that for the Daubechies scalet of $N = 2$, only two terms on the RHS of (5.5.4) are nonzero because $\text{supp}\{\varphi\} = [0, 3]$. Therefore

$$n - m - k + 1 = \begin{cases} 1 \\ 2 \end{cases}$$

or

$$k = \begin{cases} n - m \\ n - m - 1. \end{cases}$$

Hence we may write (5.5.4) explicitly as

$$S_m(n) = \frac{1}{\varphi(1)} \left[\left(\frac{|\varphi(2)|}{\varphi(1)} \right)^{n-m-1} \varphi(2) + \left(\frac{|\varphi(2)|}{\varphi(1)} \right)^{n-m} \varphi(1) \right]. \tag{5.5.5}$$

From (5.5.5) we immediately see that $S_m(n) = \delta_{m,n}$. In fact we can verify this property:

(1) When $n = m$, $k = n - m - 1 = -1$. However, the summation in (5.5.4) begins with $k = 0$. Thus the first term in (5.5.5) must be dropped, yielding

$$S_m(m) = \frac{1}{\varphi(1)} \left[\left(\frac{|\varphi(2)|}{\varphi(1)} \right)^0 \varphi(1) \right] = 1.$$

(2) When $n \neq m$, we can use the fact that $\varphi(2)$ is negative and obtain from (5.5.5),

$$\begin{aligned} S_m(n) &= \frac{1}{\varphi(1)} \left[-\frac{|\varphi(2)|^{n-m}}{\varphi(1)^{n-m-1}} + \frac{|\varphi(2)|^{n-m}}{\varphi(1)^{n-m-1}} \right] \\ &= 0. \end{aligned}$$

□

As $x \to \infty$, the D_2 (Daubechies scalet of $N = 2$) based sampling function $S_m(x)$ decays much faster than the sinc function and is compactly supported on the left endpoint of -1. As a matter of fact, $\text{supp}\{S(x)\} \approx [-1, 3]$ or $[-1, 4]$. Unfortunately, $S_m(x)$ is not orthogonal to its shifted versions, namely

$$\int_{-\infty}^{\infty} S_m(x) S_n(x)\, dx \neq \delta_{m,n}.$$

The biorthogonal testing functions $Q_n(x)$ were introduced by Walter [12] as the reproducing kernel

$$Q_n(x) = \sum_{p \in Z} \varphi(n-p)\varphi(x-p).$$

It has been shown that $\{Q_n(x)\}$ forms a Riesz basis [11]. We will now demonstrate that $\{Q_n(x)\}$ is biorthogonal to $\{S_m(x)\}$, namely

$$\int_{-\infty}^{\infty} S_m(x) Q_n(x)\, dx = \delta_{m,n}. \tag{5.5.6}$$

Owing to the finite support of D_2, the testing functions have a closed-form expression

$$Q_n(x) = \varphi(1)\varphi(x-n+1) + \varphi(2)\varphi(x-n+2). \tag{5.5.7}$$

From the previous equation and the support of D_2, we find immediately that

$$\text{supp}\{Q_n(x)\} = [n-2, n+2]. \tag{5.5.8}$$

Let us show the biorthogonality of (5.5.6).

Show.

$$\int_{-\infty}^{\infty} S_m(x) Q_n(x)\, dx = \frac{1}{\varphi(1)} \sum_{k=0}^{+\infty} \left(\frac{|\varphi(2)|}{\varphi(1)} \right)^k$$

$$\int_{-\infty}^{\infty} [\varphi(1)\varphi(x-n+1) + \varphi(2)\varphi(x-n+2)]\varphi(x-m-k+1)\, dx$$

$$= \frac{1}{\varphi(1)} \left[\left(\frac{|\varphi(2)|}{\varphi(1)} \right)^{n-m} \varphi(1) + \left(\frac{|\varphi(2)|}{\varphi(1)} \right)^{n-m-1} \varphi(2) \right], \tag{5.5.9}$$

where we have used the orthogonality of the Daubechies scalets

$$\langle \varphi(x-k), \varphi(x-\ell) \rangle = \delta_{k,\ell}.$$

Note that the right-hand side of (5.5.9) is identical to the expression (5.5.5), which is equal to $\delta_{m,n}$. Therefore

$$\int_{-\infty}^{\infty} S_m(x) Q_n(x)\, dx = \delta_{m,n}.$$

□

The sampling function $S(x)$ and testing function $Q(x)$ are plotted in Figs. 5.3 and 5.4.

In the Battle–Lemarie based MRTD, one must compute ahead of time the coefficients

$$a_i = \int_{-\infty}^{\infty} \varphi_{-i}(x) \frac{d\varphi_{1/2}(x)}{dx} \, dx.$$

In a similar fashion we need to evaluate the coefficients

$$c_i = \int_{-\infty}^{\infty} Q_{-i}(x) \frac{dS_{1/2}(x)}{dx} \, dx. \tag{5.5.10}$$

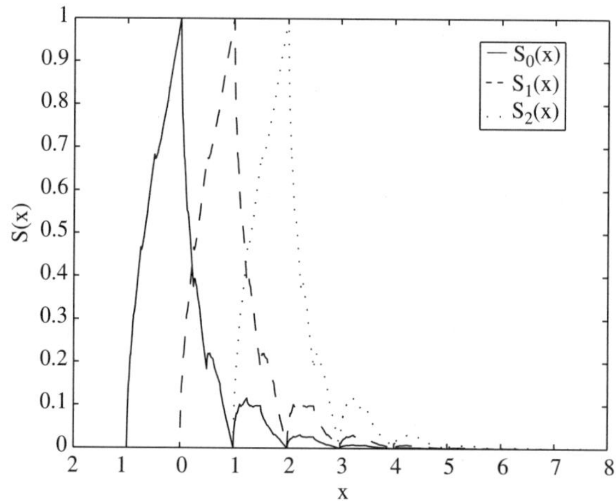

FIGURE 5.3 Daubechies-based positive sampling functions $S(x)$.

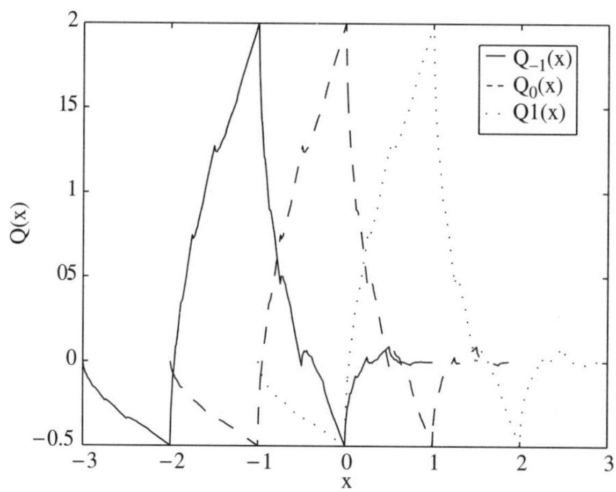

FIGURE 5.4 Reproducing kernel $Q(x)$ as the biorthogonal dual of $S(x)$.

TABLE 5.2. Coefficients c_i

i	c_i
0	1.22916661202745
1	−0.09374997764764
2	0.01041666418309

(1) For $i \geq 3$, $c_i = 0$ exactly, due to the finite support of D_2.

(2) For $i \leq -4$, the values of c_i evaluated according to (5.5.10) are identically zero, as remains to be shown later.

(3) For $-3 \leq i \leq 2$, it can be proved analytically that

$$c_i := \int_{-\infty}^{\infty} Q_{-i}(x) \frac{dS_{1/2}(x)}{dx}\, dx = \int_{-\infty}^{\infty} \varphi_{-i}(x) \frac{d\varphi_{1/2}(x)}{dx}\, dx \tag{5.5.11}$$

Thus we can compute the inner product

$$\langle Q_\ell(x), \frac{d}{dx} S_{\ell'+(1/2)}(x) \rangle = \sum_{i=-3}^{2} c_i\, \delta_{\ell+i,\ell'},$$

where the values of c_i have been evaluated numerically, similar to (5.4.4), as

$$c_i = \left\langle \varphi_{-i}, \frac{d}{dx}\varphi_{1/2}(x) \right\rangle = \frac{1}{\pi} \int_{-\infty}^{\infty} \omega |\hat{\varphi}(\omega)|^2 \sin\left[\omega\left(i + \frac{1}{2}\right)\right] d\omega. \tag{5.5.12}$$

The values of c_i are tabulated in Table 5.2.

For negative indexes, the symmetry holds, that is,

$$c_{-1-i} = -c_i, \qquad i = 0, 1, 2,$$

although $\varphi(x)$ is not symmetric. Before presenting an elegant proof in the transform domain, let us examine (5.5.11) in the spatial domain for $i = 1$. Other cases will follow the same procedure. The following proofs are lengthy, but they provide some physical insight.

Show. Using (5.5.7) for $Q_n(x)$ and (5.5.2) for $S_m(x)$, we have

$$c_i|_{i=1} = \int_{-\infty}^{\infty} Q_{-1}(x) \frac{d}{dx} S_{1/2}(x)\, dx = \frac{1}{\varphi(1)} \sum_{k=0}^{+\infty} \left(\frac{|\varphi(2)|}{\varphi(1)}\right)^k \int_{-\infty}^{\infty} dx[\varphi(1)\varphi(x+2)$$

$$+\varphi(2)\varphi(x+3)]\frac{d\varphi\left(x+\frac{1}{2}-k\right)}{dx}$$

$$=\frac{1}{\varphi(1)}[I_1+I_2]. \tag{5.5.13}$$

In the evaluation of I_1, and I_2 of the equation above, we need to utilize the fact that for the Daubechies scalet of $N=2$, $\text{supp}\{\varphi(x)\}=[0,3]$; only $k=0$ and $k=1$ remain in the infinite summation. For $k\geq 2$ the integrand of (5.5.13) is zero because the two scalets $\varphi(x+2)$ or $\varphi(x+3)$ and $\varphi(x+\frac{1}{2}-k)$ do not overlap. Hence

$$I_1=\varphi(1)\left[\left(\frac{|\varphi(2)|}{\varphi(1)}\right)^0\int_{-\infty}^{\infty}\varphi(x+2)\frac{d\varphi\left(x+\frac{1}{2}\right)}{dx}dx\right.$$
$$\left.+\left(\frac{|\varphi(2)|}{\varphi(1)}\right)^1\int_{-\infty}^{\infty}\varphi(x+2)\frac{d\varphi\left(x-\frac{1}{2}\right)}{dx}dx\right]$$
$$=\varphi(1)\int_{-\infty}^{\infty}\varphi(x+2)\frac{d\varphi\left(x+\frac{1}{2}\right)}{dx}dx$$
$$+|\varphi(2)|\int_{-\infty}^{\infty}\varphi(x+2)\frac{d\varphi\left(x-\frac{1}{2}\right)}{dx}dx \tag{5.5.14}$$

and

$$I_2=\varphi(2)\left[\left(\frac{|\varphi(2)|}{\varphi(1)}\right)^0\int_{-\infty}^{\infty}\varphi(x+3)\frac{d\varphi\left(x+\frac{1}{2}\right)}{dx}dx\right]$$
$$=\varphi(2)\left[\int_{-\infty}^{\infty}\varphi(y+2)\frac{d\varphi\left(y-\frac{1}{2}\right)}{dy}dy\right]$$
$$=-|\varphi(2)|\int_{-\infty}^{\infty}\varphi(x+2)\frac{d\varphi\left(x-\frac{1}{2}\right)}{dx}dx, \tag{5.5.15}$$

where we have used the substitution $y=x+1$ and the fact that $\varphi(2)=-|\varphi(2)|$ because $\varphi(2)<0$.

Combining (5.5.14) and (5.5.15), we obtain from (5.5.13) that

$$c_1=\frac{1}{\varphi(1)}[I_1+I_2]$$
$$=\int_{-\infty}^{\infty}\varphi(x+2)\frac{d\varphi\left(x+\frac{1}{2}\right)}{dx}dx$$
$$=\int_{-\infty}^{\infty}\varphi(x+1)\frac{d\varphi\left(x-\frac{1}{2}\right)}{dx}dx$$
$$=\int_{-\infty}^{\infty}\varphi_{-1}(x)\frac{d\varphi_{1/2}(x)}{dx}dx. \tag{5.5.16}$$

The last integral in (5.5.16) is exactly equal to c_i $(i=1)$ given by (5.5.11). □

The coefficients c_i are identically zero for $i > 2$ or $i < -3$, in contrast to the MRTD where a_i are approximately zero for $i > 8$ or $i < -9$. The verification of $c_{-4} = 0$ is provided below.

Show. We begin with

$$c_{-4} = \int_{-\infty}^{\infty} Q_4(x) \frac{dS_{1/2}(x)}{dx} \, dx.$$

Notice that

$$\mathrm{supp}\{S_{1/2}(x)\} = \left(-\tfrac{1}{2}, +\infty\right)$$

and that from (5.5.8),

$$\mathrm{supp}\{Q_4(x)\} = [2, 6].$$

Following the procedure in (5.5.13), we have

$$\begin{aligned} c_{-4} &= \frac{1}{\varphi(1)} \sum_{k=0}^{+\infty} \left(\frac{|\varphi(2)|}{\varphi(1)} \right)^k \int_{-\infty}^{\infty} [\varphi(1)\varphi(x-3) + \varphi(2)\varphi(x-2)] \frac{d\varphi\left(x + \frac{1}{2} - k\right)}{dx} \, dx \\ &= \frac{1}{\varphi(1)} [I_1 + I_2]. \end{aligned}$$

In the equation above the nonzero terms are $k = 1, 2, \ldots, 7$ for I_1 and $k = 0, 1, \ldots, 6$ for I_2. We see that

$$\begin{aligned} I_1 = \varphi(1) \Bigg[&\left(\frac{|\varphi(2)}{\varphi(1)} \right)^1 \int_{-\infty}^{\infty} \varphi(x-3) \frac{d\varphi\left(x - \frac{1}{2}\right)}{dx} \, dx \\ &+ \left(\frac{|\varphi(2)}{\varphi(1)} \right)^2 \int_{-\infty}^{\infty} \varphi(x-3) \frac{d\varphi\left(x - \frac{3}{2}\right)}{dx} \, dx \\ &+ \left(\frac{|\varphi(2)}{\varphi(1)} \right)^3 \int_{-\infty}^{\infty} \varphi(x-3) \frac{d\varphi\left(x - \frac{5}{2}\right)}{dx} \, dx \\ &+ \left(\frac{|\varphi(2)}{\varphi(1)} \right)^4 \int_{-\infty}^{\infty} \varphi(x-3) \frac{d\varphi\left(x - \frac{7}{2}\right)}{dx} \, dx \\ &+ \left(\frac{|\varphi(2)}{\varphi(1)} \right)^5 \int_{-\infty}^{\infty} \varphi(x-3) \frac{d\varphi\left(x - \frac{9}{2}\right)}{dx} \, dx \\ &+ \left(\frac{|\varphi(2)}{\varphi(1)} \right)^6 \int_{-\infty}^{\infty} \varphi(x-3) \frac{d\varphi(x - \frac{11}{2})}{dx} \, dx \\ &+ \left(\frac{|\varphi(2)}{\varphi(1)} \right)^7 \int_{-\infty}^{\infty} \varphi(x-3) \frac{d\varphi\left(x - \frac{13}{2}\right)}{dx} \, dx] \Bigg] \end{aligned}$$

and

$$I_2 = \varphi(2)\left[\left(\frac{|\varphi(2)}{\varphi(1)}\right)^0 \int_{-\infty}^{\infty} \varphi(x-2)\frac{d\varphi\left(x+\frac{1}{2}\right)}{dx}\,dx\right.$$

$$+\left(\frac{|\varphi(2)}{\varphi(1)}\right)^1 \int_{-\infty}^{\infty} \varphi(x-2)\frac{d\varphi\left(x-\frac{1}{2}\right)}{dx}\,dx$$

$$+\left(\frac{|\varphi(2)}{\varphi(1)}\right)^2 \int_{-\infty}^{\infty} \varphi(x-2)\frac{d\varphi\left(x-\frac{3}{2}\right)}{dx}\,dx$$

$$+\left(\frac{|\varphi(2)}{\varphi(1)}\right)^3 \int_{-\infty}^{\infty} \varphi(x-2)\frac{d\varphi\left(x-\frac{5}{2}\right)}{dx}\,dx$$

$$+\left(\frac{|\varphi(2)}{\varphi(1)}\right)^4 \int_{-\infty}^{\infty} \varphi(x-2)\frac{d\varphi\left(x-\frac{7}{2}\right)}{dx}\,dx$$

$$+\left(\frac{|\varphi(2)}{\varphi(1)}\right)^5 \int_{-\infty}^{\infty} \varphi(x-2)\frac{d\varphi\left(x-\frac{9}{2}\right)}{dx}\,dx$$

$$\left.+\left(\frac{|\varphi(2)}{\varphi(1)}\right)^6 \int_{-\infty}^{\infty} \varphi(x-2)\frac{d\varphi\left(x-\frac{11}{2}\right)}{dx}\,dx\right].$$

Further simplification yields

$$I_1 = |\varphi(2)| \int_{-\infty}^{\infty} \varphi(x-3)\frac{d\varphi\left(x-\frac{1}{2}\right)}{dx}\,dx$$

$$+\frac{|\varphi(2)|^2}{\varphi(1)} \int_{-\infty}^{\infty} \varphi(x-3)\frac{d\varphi\left(x-\frac{3}{2}\right)}{dx}$$

$$+\frac{|\varphi(2)|^3}{\varphi(1)^2} \int_{-\infty}^{\infty} \varphi(x-3)\frac{d\varphi\left(x-\frac{5}{2}\right)}{dx}$$

$$+\frac{|\varphi(2)|^4}{\varphi(1)^3} \int_{-\infty}^{\infty} \varphi(x-3)\frac{d\varphi\left(x-\frac{7}{2}\right)}{dx}$$

$$+\frac{|\varphi(2)|^5}{\varphi(1)^4} \int_{-\infty}^{\infty} \varphi(x-3)\frac{d\varphi\left(x-\frac{9}{2}\right)}{dx}$$

$$+\frac{|\varphi(2)|^6}{\varphi(1)^5} \int_{-\infty}^{\infty} \varphi(x-3)\frac{d\varphi\left(x-\frac{11}{2}\right)}{dx}$$

$$I_2 = -|\varphi(2)| \int_{-\infty}^{\infty} \varphi(x-2)\frac{d}{dx}\varphi\left(x+\frac{1}{2}\right)dx$$

$$-\frac{|\varphi(2)|^2}{\varphi(1)} \int_{-\infty}^{\infty} \varphi(x-2)\frac{d}{dx}\varphi\left(x-\frac{1}{2}\right)dx$$

$$
\begin{aligned}
&-\frac{|\varphi(2)|^3}{\varphi(1)^2}\int_{-\infty}^{\infty}\varphi(x-2)\frac{d}{dx}\varphi\left(x-\frac{3}{2}\right)dx \\
&-\frac{|\varphi(2)|^4}{\varphi(1)^3}\int_{-\infty}^{\infty}\varphi(x-2)\frac{d}{dx}\varphi\left(x-\frac{5}{2}\right)dx \\
&-\frac{|\varphi(2)|^5}{\varphi(1)^4}\int_{-\infty}^{\infty}\varphi(x-2)\frac{d}{dx}\varphi\left(x-\frac{7}{2}\right)dx \\
&-\frac{|\varphi(2)|^6}{\varphi(1)^5}\int_{-\infty}^{\infty}\varphi(x-2)\frac{d}{dx}\varphi\left(x-\frac{9}{2}\right)dx.
\end{aligned}
$$

I_1 and I_2 cancel each other out exactly item by item due to the fact that

$$
\begin{aligned}
\int_{-\infty}^{\infty}\varphi(x-3)\frac{d\varphi\left(x-\frac{1}{2}\right)}{dx}dx &= \int_{-\infty}^{\infty}\varphi(x-2)\frac{d\varphi\left(x+\frac{1}{2}\right)}{dx}dx \\
\int_{-\infty}^{\infty}\varphi(x-3)\frac{d\varphi\left(x-\frac{3}{2}\right)}{dx}dx &= \int_{-\infty}^{\infty}\varphi(x-2)\frac{d\varphi\left(x-\frac{1}{2}\right)}{dx}dx. \\
\cdots &= \cdots
\end{aligned}
$$

Proof. To begin with, we rewrite (5.5.12) below

$$
\begin{aligned}
\int_{-\infty}^{\infty}\varphi_{-l}(x)\frac{d}{dx}\varphi_{1/2}(x)\,dx &= \frac{1}{2\pi}\int_{-\infty}^{\infty}d\omega\hat{\varphi}(\omega)\hat{\varphi}(-\omega)e^{i\omega(l+(1/2))}(-i\omega) \\
&= \frac{1}{\pi}\int_{0}^{\infty}\omega|\hat{\varphi}(\omega)|^2\sin\left[\omega\left(l+\tfrac{1}{2}\right)\right]d\omega.
\end{aligned}
$$

On the other hand,

$$
\begin{aligned}
&\int_{-\infty}^{\infty}Q_{-l}(x)\frac{d}{dx}S_{1/2}(x)\,dx \\
&= \int_{-\infty}^{\infty}dx\left[\frac{1}{2\pi}\int_{-\infty}^{\infty}d\omega\hat{Q}(\omega)e^{(i\omega l+i\omega x)}\frac{1}{2\pi}\int_{-\infty}^{\infty}d\omega'\hat{S}(\omega')e^{-i\omega'(1/2)+i\omega' x}\right] \\
&= \iint d\omega\,d\omega'\frac{1}{2\pi}\int_{-\infty}^{\infty}dx e^{ix(\omega'+i\omega')}\frac{i\omega'}{2\pi}\hat{Q}(\omega)\hat{S}(\omega')e^{i\omega l-i\omega'(1/2)} \\
&= \frac{1}{2}\int_{-\infty}^{\infty}d\omega\hat{Q}(\omega)e^{i\omega l}\int(d\omega')\delta(\omega+\omega')(i\omega')\hat{S}(\omega')e^{i\omega'(1/2)} \\
&= \frac{1}{2}\int_{-\infty}^{\infty}d\omega\hat{Q}(\omega)\hat{S}(-\omega)e^{i\omega(l+(1/2))}(-i\omega) \qquad (5.5.17)
\end{aligned}
$$

where $\hat{Q}(\omega)$ is the Fourier transform of $Q_0(x)$, and $\hat{S}(\omega)$ is the Fourier transform of $S(x)$ given by (5.5.1). Since $Q_0(x)$ and $\hat{S}(\omega)$ are real, their Fourier transform $\hat{Q}(\omega)$ and $\hat{S}(\omega)$, are conjugate symmetric.

According to (5.5.2)

$$
S(x) = \frac{1}{\varphi(1)}\sum_{k=0}^{\infty}r^k\varphi(x-k+1)
$$

where $r := |\varphi(2)|/\varphi(1)$. Its Fourier transform is

$$\begin{aligned}\hat{S}(\omega) &= \frac{1}{\varphi(1)} \sum_{k=0}^{\infty} r^k \hat{\varphi}(\omega) e^{i\omega(-k+1)} \\ &= \frac{e^{i\omega}}{\varphi(1)} \hat{\varphi}(\omega) \sum_{k=0}^{\infty} (re^{-i\omega})^k \\ &= \frac{e^{i\omega}}{\varphi(1)} \hat{\varphi}(\omega) \frac{1}{1 - re^{-i\omega}}. \end{aligned} \tag{5.5.18}$$

The testing function

$$Q_0(x) = \varphi(1) \left[\varphi(x+1) - \frac{|\varphi(2)|}{\varphi(1)} \varphi(x+2) \right].$$

Thus its Fourier transform is

$$\begin{aligned}\hat{Q}(\omega) &= \varphi(1) \left[\hat{\varphi}(\omega) e^{i\omega} - r\hat{\varphi}(\omega) e^{2i\omega} \right] \\ &= \varphi(1) e^{i\omega} \hat{\varphi}(\omega) \left[1 - re^{i\omega} \right]. \end{aligned} \tag{5.5.19}$$

From (5.5.18) and (5.5.19) we have immediately

$$\hat{Q}(\omega)\hat{S}(-\omega) = \hat{\varphi}(\omega)\hat{\varphi}(-\omega).$$

Therefore (5.5.17) reduces to

$$\begin{aligned}\int_{-\infty}^{\infty} Q_{-l}(x) \frac{d}{dx} S_{1/2}(x)\, dx &= \frac{1}{2\pi} \int_{-\infty}^{\infty} d\omega \hat{\varphi}(\omega) e^{i\omega(l+(1/2))} \hat{\varphi}(-\omega)(-i\omega) \\ &= \int_{-\infty}^{\infty} \varphi_{-l}(x) \frac{d}{dx} \varphi_{1/2}(x)\, dx. \end{aligned}$$ □

We want to emphasize that this interpolation property is exact for our biorthogonal system. Interestingly the coefficients c_i in Table 5.2 were derived in a paper by Y. W. Cheong et al. [13]. Cheong and colleagues employed the approximate interpolation property of the Daubechies scalet D_2 from [14], namely

$$\varphi(\mathcal{M}_1 + k) \approx \delta_{0,k},$$

where $\mathcal{M}_1$ is the first-order moment such that

$$\mathcal{M}_1 = \int x\varphi(x)\, dx.$$

Numerically, it was reported in [14] that

$$\begin{aligned}&\mathcal{M}_1 \approx 0.683, \\ &\varphi(\mathcal{M}_1) \approx 1.00020859077,\end{aligned}$$

$$\varphi(\mathcal{M}_1 + 1) \approx -4.17181539384E - 04,$$

$$\varphi(\mathcal{M}_1 + 2) \approx 2.08590769692E - 04.$$

The approximate sampling of D_2 is reserved for the reader in Exercise 4. As mathematicians have pointed out, noone knows the exact values of $\varphi(\sqrt{2})$, nor how to solve for the exact value of $\mathcal{M}$ from $\varphi(\mathcal{M} + k) = \delta_{0,k}$.

From our derivation of the positive sampling function and its dual biorthogonal testing function, the sampling (interpolation) property is exact. Yet the coefficients c_i from the exact sampling system are identical to those of the shifted D_2. In consequence we have proved indirectly the existence theorem below:

Theorem. There exists a point $\mathcal{M} \in (0, 3)$ for the Daubechies scalet D_2 such that the following equation holds exactly

$$\varphi(\mathcal{M} + k) = \delta_{0,k}.$$

Note the significance of this theorem. In the shifted D_2 scheme, the authors have claimed that $\varphi(\mathcal{M}_1)$, $\varphi(\mathcal{M}_1{+}1)$, and $(\mathcal{M}_1{+}2)$ are only approximately interpolating. Therefore the resulting field equations are only approximations. Nonetheless, the amount of shift $\mathcal{M}$ does not appear in the equations for conducting the biorthogonal sampling time domain (BSTD) procedure in the next section. The existence theorem here guarantees that the field equations (5.6.3) through (5.6.8) are exact, and that the only possible error is attributable to the precision of the coefficients c_i.

5.6 SAMPLING BIORTHOGONAL TIME DOMAIN METHOD

5.6.1 SBTD versus MRTD

The multiresolution time domain (MRTD) method in Section 5.4 shows an excellent capacity to approximate a precise solution, even at a rate near the Nyquist sampling limit. However, in the MRTD the nonsampling properties of the Battle–Lemarie wavelets make the formulation difficult to compute. For instance, the two-term finite difference expression in the FDTD has been extended to 18 terms in the MRTD. The field quantity at a given node is the sum of the partial values at all related neighboring nodes. Such a distribution makes it very cumbersome to implement the radiation or absorption boundary conditions. The sampling property of $S_m(x)$ and the compact support of the D_2 have overcome these shortcomings or alleviated the burden of the MRTD. As a matter of fact, the biorthogonal sampling time domain (SBTD) has inherited all advantages of the MRTD but is a much simpler algorithm.

5.6.2 Formulation

For the SBTD scheme, we use the basis for expansion

$$s_m(x) = S\left(\frac{x}{\Delta x} - m\right)$$

and the biorthogonal testing functions

$$q_n(x) = Q\left(\frac{x}{\Delta x} - n\right).$$

The time discretization still occurs in pulses as in the FDTD:

$$P_k(t) = P\left(\frac{t}{\Delta t} - k + \frac{1}{2}\right),$$

where

$$P(t) = \begin{cases} 1, & |t| < \frac{1}{2} \\ \frac{1}{2}, & |t| = \frac{1}{2} \\ 0, & |t| > \frac{1}{2} \end{cases}.$$

Thus the two Maxwell equations become, after discretization by the SBTD, six component equations. More specifically, let us consider the x-component equation from $\epsilon(\partial \mathbf{E}/\partial t) = \nabla \times \mathbf{H}$, namely

$$\frac{\partial E_x}{\partial t} = \frac{1}{\epsilon}\left(\frac{\partial H_z}{\partial y} - \frac{\partial H_y}{\partial z}\right). \tag{5.6.1}$$

We can expand the fields in terms of the basis functions

$$\begin{aligned}
E_x &= \sum_{k',\ell',m',n'} {}_kE^x_{\ell'+(1/2),m',n'} P_{k'}(t) s_{\ell'+(1/2)}(x) s_{m'}(y) s_{n'}(z), \\
H_y &= \sum_{k',\ell',m',n'} {}_{k'+(1/2)}H^y_{\ell'+(1/2),m',n'+(1/2)} P_{k'+(1/2)}(t) s_{\ell'+(1/2)}(x) s_{m'}(y) s_{n'+(1/2)}(z), \\
H_z &= \sum_{k',\ell',m',n'} {}_{k'+(1/2)}H^z_{\ell'+(1/2),m'+(1/2),n'} P_{k'+(1/2)}(t) s_{\ell'+(1/2)}(x) s_{m'+(1/2)}(y) s_{n'}(z).
\end{aligned} \tag{5.6.2}$$

Substituting (5.6.2) into (5.6.1) and testing with $q_{\ell+(1/2)}(x) q_m(y) q_n(z) P_{k+(1/2)}(t)$, we arrive at

$$\begin{aligned}
\text{LHS} = &\sum_{k',\ell',m',n'} {}_{k'}E^x_{\ell'+(1/2),m',n'} \int dt\, P_{k+(1/2)}(t) \frac{\partial P_{k'}(t)}{\partial t} \\
&\int dx q_{\ell+(1/2)}(x) s_{\ell'+(1/2)}(x) \int dy q_m(y) s_{m'}(y) \int dz q_n(z) s_{n'}(z) \\
= &\left({}_{k+1}E^x_{\ell+(1/2),m,n} - {}_kE^x_{\ell+(1/2),m,n}\right) \Delta x\, \Delta y\, \Delta z,
\end{aligned}$$

where we have used

$$\int dt\, P_{k+(1/2)}(t) \frac{\partial P_{k'}(t)}{\partial t} = \delta_{k',k+1} - \delta_{k',k}$$

$$\int dyq_m(y)s_{m'}(y) = \int dyQ\left(\frac{y}{\Delta y} - m\right) S\left(\frac{y}{\Delta y} - m'\right) = \delta_{m,m'}\,\Delta y$$
$$\cdots = \cdots$$

In the same way, the first term on the RHS is

$$\iiiint dt\,dx\,dy\,dz \frac{\partial H_z}{\partial y} q_{\ell+(1/2)}(x) q_m(y) q_n(z) P_{k+(1/2)}(t)$$
$$= \sum_{k',\ell',m',n'} {}_{k'+(1/2)}H^z_{\ell'+(1/2),m'+(1/2),n'} \int dt\,P_{k'+(1/2)}(t) P_{k+(1/2)}(t)$$
$$\int dxq_{\ell+(1/2)}(x) s_{\ell'+(1/2)}(x) \int dyq_m(y) \frac{\partial s_{m'+(1/2)}(y)}{\partial y} \int dzq_n(z) s_{n'}(z)$$
$$\sum_{k',\ell',m',n'} {}_{k'+(1/2)}H^z_{\ell'+(1/2),m'+(1/2),n'}\,\Delta t\,\delta_{k,k'}\,\Delta x\,\delta_{\ell,\ell'}\,\Delta z\,\delta_{n,n'}$$
$$\int dyq_m(y) \frac{\partial s_{m'+(1/2)}(y)}{\partial y}.$$
$$= \sum_{i=-3}^{2} c_i\ {}_{k+(1/2)}H^z_{\ell+(1/2),(m+i)+(1/2),n}\,\Delta t\,\Delta x\,\Delta z,$$

where we have used the property from (5.5.11) that $\int dyQ_m(y)\partial S_{m'+(1/2)}(y)/\partial y = \int dy\varphi_m(y)\partial\varphi_{m'+(1/2)}/\partial y$. The second term on the RHS yields a similar result. Equating both sides, we arrive at

$$\begin{aligned} {}_{k+1}E^x_{\ell+(1/2),m,n} &= {}_kE^x_{\ell+(1/2),m,n} \\ &\quad + \frac{\Delta t}{\epsilon_{\ell+(1/2),m,n}} \left[\frac{1}{\Delta y} \sum_{i=-3}^{2} c_i \cdot {}_{k+\frac{1}{2}}H^z_{\ell+\frac{1}{2},m+i+(1/2),n} \right. \\ &\quad \left. - \frac{1}{\Delta z} \sum_{i=-3}^{2} c_i \cdot {}_{k+(1/2)}H^y_{\ell+(1/2),m,n+i+(1/2)} \right]. \end{aligned} \tag{5.6.3}$$

We can derive other component equations in the same fashion, yielding

$$\begin{aligned} {}_{k+1}E^y_{l,m+(1/2),n} &= {}_kE^y_{l,m+(1/2),n} \\ &\quad + \frac{\Delta t}{\epsilon_{l,m+(1/2),n}} \left[\frac{1}{\Delta z} \sum_{i=-3}^{2} c_i \cdot {}_{k+(1/2)}H^x_{l,m+(1/2),n+(1/2)+i} \right. \\ &\quad \left. - \frac{1}{\Delta x} \sum_{i=-3}^{2} c_i \cdot {}_{k+(1/2)}H^z_{l+(1/2)+i,m+(1/2),n} \right], \end{aligned} \tag{5.6.4}$$

$$
\begin{aligned}
{}_{k+1}E^{z}_{l,m,n+(1/2)} &= {}_{k}E^{z}_{l,m,n+(1/2)} \\
&\quad + \frac{\Delta t}{\epsilon_{l,m,n+(1/2)}}\left[\frac{1}{\Delta x}\sum_{i=-3}^{2} c_i \cdot {}_{k+(1/2)}H^{y}_{l+(1/2)+i,m,n+(1/2)}\right. \\
&\quad \left. - \frac{1}{\Delta y}\sum_{i=-3}^{2} c_i \cdot {}_{k+(1/2)}H^{x}_{l,m+(1/2)+i,n+(1/2)}\right], \qquad (5.6.5)
\end{aligned}
$$

$$
\begin{aligned}
&{}_{k+(1/2)}H^{x}_{l,m+(1/2),n+(1/2)} \\
&\quad = {}_{k-(1/2)}H^{x}_{l,m+(1/2),n+(1/2)} \\
&\qquad + \frac{\Delta t}{\mu_{l,m+(1/2),n+(1/2)}}\left[\frac{1}{\Delta z}\sum_{i=-3}^{2} c_i \cdot {}_{k}E^{y}_{l,m+(1/2),n+i+1}\right. \\
&\qquad \left. - \frac{1}{\Delta y}\sum_{i=-3}^{2} c_i \cdot {}_{k}E^{z}_{l,m+i+1,n}\right]. \qquad (5.6.6)
\end{aligned}
$$

$$
\begin{aligned}
&{}_{k+(1/2)}H^{y}_{l+(1/2),m,n+(1/2)} \\
&\quad = {}_{k-(1/2)}H^{y}_{l+(1/2),m,n+(1/2)} \\
&\qquad + \frac{\Delta t}{\mu_{l+(1/2),m,n+(1/2)}}\left[\frac{1}{\Delta x}\sum_{i=-3}^{2} c_i \cdot {}_{k}E^{z}_{l+i+1,m+(1/2),n}\right. \\
&\qquad \left. - \frac{1}{\Delta z}\sum_{i=-3}^{2} c_i \cdot {}_{k}E^{x}_{l+(1/2),m,n+i+1}\right], \qquad (5.6.7)
\end{aligned}
$$

$$
\begin{aligned}
&{}_{k+(1/2)}H^{z}_{l+(1/2),m+(1/2),n} \\
&\quad = {}_{k-(1/2)}H^{z}_{l+(1/2),m+(1/2),n} \\
&\qquad + \frac{\Delta t}{\mu_{l+(1/2),m+(1/2),n}}\left[\frac{1}{\Delta y}\sum_{i=-3}^{2} c_i \cdot {}_{k}E^{x}_{l+(1/2),m+1+i,n}\right. \\
&\qquad \left. - \frac{1}{\Delta x}\sum_{i=-3}^{2} c_i \cdot {}_{k}E^{y}_{l+1+i,m+(1/2),n}\right]. \qquad (5.6.8)
\end{aligned}
$$

For the 2D $TM^{(z)}$ case, $\partial/\partial z = 0$ and $H_z = 0$. Maxwell's curl equations reduce to three equations:

$$
\frac{\partial E_z}{\partial t} = \frac{1}{\epsilon}\left(\frac{\partial H_y}{\partial x} - \frac{\partial H_x}{\partial y}\right),
$$

$$\frac{\partial H_x}{\partial t} = -\frac{1}{\mu}\frac{\partial E_z}{\partial y},$$
$$\frac{\partial H_y}{\partial t} = -\frac{1}{\mu}\frac{\partial E_z}{\partial x}.$$

The expansions from (5.4.1) are

$$E_z(\boldsymbol{\rho}, t) = \sum_{k',\ell',m'} {}_{k'}E^z_{\ell',m'} P_{k'}(t) s_{\ell'}(x) s_{m'}(y),$$
$$H_x(\boldsymbol{\rho}, t) = \sum_{k',\ell',m'} {}_{k'+(1/2)}H^x_{\ell',m'+(1/2)} P_{k'+(1/2)}(t) s_{\ell'}(x) s_{m'+(1/2)}(y),$$
$$H_y(\boldsymbol{\rho}, t) = \sum_{k',\ell',m'} {}_{k'+(1/2)}H^y_{\ell'+(1/2),m'} P_{k'+(1/2)}(t) s_{\ell'+(1/2)}(x) s_{m'}(y).$$

The corresponding discretized equations are thus

$$
\begin{aligned}
{}_{k+1}E^z_{\ell,m} = {}_kE^z_{\ell,m} + \frac{\Delta t}{\epsilon_{\ell,m}}\Bigg[\frac{1}{\Delta x}\sum_{i=-3}^{2} c_i \, {}_{k+(1/2)}H^y_{\ell+(1/2)+i,m} \\
- \frac{1}{\Delta y}\sum_{i=-3}^{2} c_i \, {}_{k+(1/2)}H^x_{\ell,m+(1/2)+i}\Bigg],
\end{aligned}
$$

$${}_{k+(1/2)}H^x_{\ell,m+(1/2)} = {}_{k-(1/2)}H^x_{\ell,m+(1/2)} - \frac{\Delta t}{\mu_{\ell,m+(1/2)}}\frac{1}{\Delta y}\sum_{i=-3}^{2} c_i \, {}_kE^z_{\ell,m+i},$$

$${}_{k+(1/2)}H^y_{\ell+(1/2),m} = {}_{k-(1/2)}H^y_{\ell+(1/2),m} + \frac{\Delta t}{\mu_{\ell+(1/2),m}}\frac{1}{\Delta x}\sum_{i=-3}^{2} c_i \, {}_kE^z_{\ell+i,m}.$$

5.7 STABILITY CONDITIONS FOR WAVELET-BASED METHODS

The stability condition for the wavelet-based time domain method MRTD was derived in [9] for the Battle–Lemarie. This formulation applies to the Daubechies-based SBTD as well. Because of the sampling property and finite support of the SBTD system, no infinite summation nor summation of partial values is needed.

5.7.1 Dispersion Relation and Stability Analysis

The operator equation (5.10.5) in the Appendix is a homogeneous equation. To obtain the nontrivial solution, we must have

$$\det \tilde{W}(\tilde{T}_h, \tilde{X}_h, \tilde{Y}_h, \tilde{Z}_h) = 0. \tag{5.7.1}$$

If only plane waves are considered, (5.7.1) reduces to

$$\det \tilde{W}(e^{-i(\Omega/2)}, e^{-i(\chi/2)}, e^{-i(\eta/2)}, e^{-i(\xi/2)}) = 0, \tag{5.7.2}$$

where the dimensionless variables are

$$\begin{aligned}
\Omega &= \omega\,\Delta t, \\
\chi &= k_x\,\Delta x, \\
\eta &= k_y\,\Delta y, \\
\xi &= k_z\,\Delta z.
\end{aligned}$$

Equation (5.7.2) can be simplified as

$$d_t(\Omega) = 0 \tag{5.7.3}$$

and

$$\epsilon\mu(d_t(\Omega))^2 = (D_x^{\varphi}(\chi))^2 + (D_y^{\phi}(\eta))^2 + (D_z^{\varphi}(\xi))^2, \tag{5.7.4}$$

where the difference operator in the frequency domain is

$$d_t(\Omega) = \frac{1}{\Delta t}(e^{i(\Omega/2)} - e^{-i(\Omega/2)}) = \frac{1}{\Delta t}(\tilde{T}_h^{\dagger} - \tilde{T}_h) = \frac{2i}{\Delta t}\sin\frac{\Omega}{2}. \tag{5.7.5}$$

The difference operators in the wave vector domain for the MRTD are

$$\begin{aligned}
D_x^{\varphi}(\chi) &= \frac{2i}{\Delta x}\sum_{p=0}^{8} a_p \ \sin\chi\left(p + \frac{1}{2}\right), \\
D_y^{\varphi}(\eta) &= \frac{2i}{\Delta y}\sum_{p} a_p \ \sin\eta\left(p + \frac{1}{2}\right), \\
D_z^{\varphi}(\xi) &= \frac{2i}{\Delta z}\sum_{p} a_p \ \sin\xi\left(p + \frac{1}{2}\right).
\end{aligned} \tag{5.7.6}$$

Substituting (5.7.6) into (5.7.4), we obtain

$$\begin{aligned}
\epsilon\mu\left[\frac{2i}{\Delta t}\sin\frac{\Omega}{2}\right]^2 &= \left[\frac{2i}{\Delta x}\sum_{p=0}^{8} a_p \sin x\left(p + \frac{1}{2}\right)\right]^2 \\
&\quad + \left[\frac{2i}{\Delta y}\sum_{p=0}^{8} a_p \sin\eta\left(p + \frac{1}{2}\right)\right]^2 \\
&\quad + \left[\frac{2i}{\Delta z}\sum_{p=0}^{8} a_p \sin\xi\left(p + \frac{1}{2}\right)\right]^2.
\end{aligned} \tag{5.7.7}$$

If $\Delta x = \Delta y = \Delta z = \Delta\ell$ and $c = 1/\sqrt{\epsilon\mu}$, (5.7.7) reduces to

$$\sin^2\frac{\Omega}{2} = \left(\frac{c\,\Delta t}{\Delta\ell}\right)^2 \left\{ \left[\sum_{p=0}^{8} a_p \sin\chi\left(p+\frac{1}{2}\right)\right]^2 + \left[\sum_{p=0}^{8} a_p \sin\eta\left(p+\frac{1}{2}\right)\right]^2 + \left[\sum_{p=0}^{8} a_p \sin\xi\left(p+\frac{1}{2}\right)\right]^2 \right\}, \tag{5.7.8}$$

where

$$\begin{aligned}\sum_{p=0}^{8} a_p \sin\chi\left(p+\tfrac{1}{2}\right) &= a_0\sin\tfrac{1}{2}\chi + a_1\sin\tfrac{3}{2}\chi + a_2\sin\tfrac{5}{2}\chi + a_3\sin\tfrac{7}{2}\chi \\ &\quad + a_4\sin\tfrac{9}{2}\chi + a_5\sin\tfrac{11}{2}\chi + a_6\sin\tfrac{13}{2}\chi \\ &\quad + a_7\sin\tfrac{15}{2}\chi + a_8\sin\tfrac{17}{2}\chi \\ &\le |a_0| + |a_1| + |a_2| + |a_3| + |a_4| + |a_5| \\ &\quad + |a_6| + |a_7| + |a_8| \\ &= 1.2918462 + 0.1560761 + 0.0596391 + 0.0293099 \\ &\quad + 0.0153716 + 0.0081892 + 0.0043788 \\ &\quad + 0.0023433 + 0.0012542 \\ &= 1.5684084. \end{aligned} \tag{5.7.9}$$

Similarly

$$\sum_{p=0}^{8} a_p \sin\eta\left(p+\tfrac{1}{2}\right) \le 1.5684084$$

and

$$\sum_{p=0}^{8} a_p \sin\xi\left(p+\tfrac{1}{2}\right) \le 1.5684084.$$

Therefore (5.7.8) becomes

$$\sin^2\frac{\Omega}{2} \le \left(\frac{c\,\Delta t}{\Delta\ell}\right)^2 [1.5684084^2 + 1.5684084^2 + 1.5684084^2].$$

Finally, the stability condition of the MRTD is

$$\Delta t \le \frac{1}{\sqrt{3}\times 1.5684084}\cdot\frac{\Delta\ell}{c} = 0.368112201\frac{\Delta\ell}{c}. \tag{5.7.10}$$

For small arguments, $\sin\alpha \approx \alpha$. Hence the dispersing relation, Eq. (5.7.4), yields

$$\frac{\omega^2}{c^2} \approx k_x^2 + k_y^2 + k_z^2.$$

The stability condition for Yee's FDTD of $\Delta x = \Delta y = \Delta z = \Delta l$ is

$$\Delta t \le \frac{1}{\sqrt{3}} \frac{\Delta l}{c} = 0.57735\,\frac{\Delta l}{c}.$$

5.7.2 Stability Analysis for the SBTD

The stability condition for SBTD can be derived in the same manner. For Daubechies scalets of $N = 2$, we have

$$D_x^{\phi}(x) = \frac{2i}{\Delta x} \sum_{p=0}^{2} c_p \sin\chi\left(p + \frac{1}{2}\right),$$

$$D_y^{\phi}(y) = \frac{2i}{\Delta y} \sum_{p=0}^{2} c_p \sin\eta\left(p + \frac{1}{2}\right),$$

$$D_z^{\phi}(z) = \frac{2i}{\Delta z} \sum_{p=0}^{2} c_p \sin\xi\left(p + \frac{1}{2}\right).$$

Similar to (5.7.9), we obtain

$$\begin{aligned}\sum_{p=0}^{2} c_p \sin\chi\left(p + \tfrac{1}{2}\right) &= c_0 \sin\tfrac{1}{2}\chi + c_1 \sin\tfrac{3}{2}\chi + c_2 \sin\tfrac{5}{2}\chi \le |c_0| + |c_1| + |c_2| \\ &= 1.22916661 + 0.09374998 + 0.01041666 \\ &= 1.3333333.\end{aligned}$$

The stability condition in this case is

$$\begin{aligned}\sin^2\frac{\Omega}{2} &\le \left(\frac{c\,\Delta t}{\Delta\ell}\right)^2 [1.3333333^2 + 1.3333333^2 + 1.3333333^2] \\ &= \left(\frac{c\,\Delta t}{\Delta\ell}\right)^2 \times 3 \times 1.3333333^2\end{aligned}$$

or

$$\begin{aligned}\Delta t &\le \frac{1}{\sqrt{3} \times 1.3333333} \cdot \frac{\Delta\ell}{c} \\ &= 0.433012712\frac{\Delta\ell}{c}. \qquad (5.7.11)\end{aligned}$$

For 2D case, the formula above should be

$$\Delta t \leq \frac{1}{\sqrt{2} \times 1.3333333} \cdot \frac{\Delta \ell}{c} = 0.530330099 \cdot \frac{\Delta \ell}{c}. \tag{5.7.12}$$

5.8 CONVERGENCE ANALYSIS AND NUMERICAL DISPERSION

5.8.1 Numerical Dispersion

In a dispersionless medium the phase velocity of electromagnetic waves should be independent of the frequency. However, the phase velocity of numerical wave modes in the discretized grid can differ from the vacuum speed of light c. It varies with model wavelength, propagation direction, and grid discretization. This artifect is referred to as numerical dispersion. Numerial dispersion produces nonphysical results, including spurious anisotropy, pulse widening, accumulated phase error, and unexpected refraction. The cause of numerical dispersion is from the approximation of differential equations (Maxwell's curl equations) with the finite difference equations.

To analyze the numerical dispersion of a finite difference scheme we use dispersion relation (5.7.8), which can be directly derived by substituting a numerical plane wave (5.2.11) into the updating equations (e.g., Eq. (5.4.6) for the MRTD and (5.6.3)–(5.6.8) for the SBTD). As a special case of (5.7.8), the 1D dispersion equation is

$$\left(\frac{1}{q}\sin\left(\frac{\pi q}{n_l}\right)\right)^2 = \left(\sum_{i=0}^{N-1} a_i \sin\left(\frac{\pi u}{n_l}(2i+1)\right)\right)^2 \tag{5.8.1}$$

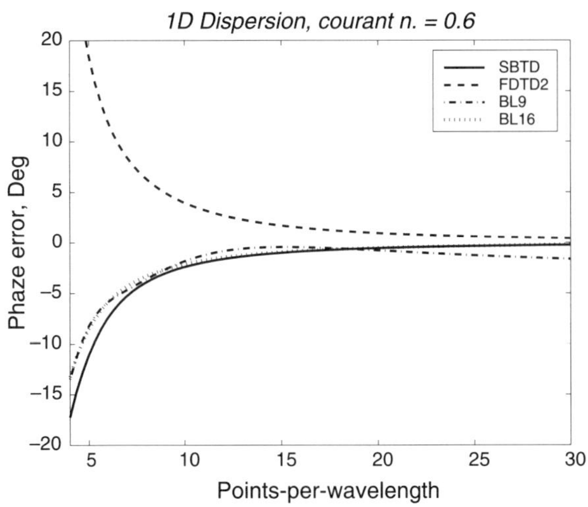

FIGURE 5.5 1D phase error versus discretization.

where $q = c\Delta t/\Delta x$ and $n_l = \lambda/\Delta x$, $u = \lambda/\lambda_{num}$. Figure 5.5 represents the 1D phase error versus sampling rate, which was computed by (5.8.1) with the Courant number $q = 0.6$. Four algorithms are compared, namely the SBTD, FDTD-2, MRTD-9, and MRTD-16.

For numerical plane waves propagating on angle ϕ in a square mesh, $\Delta x = \Delta y$, 2D dispersion equation becomes

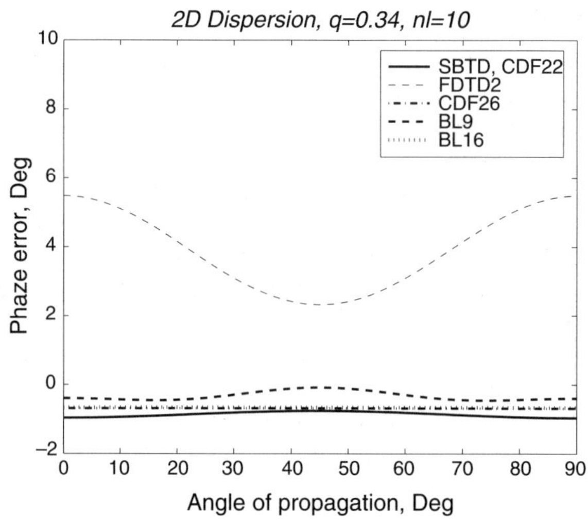

FIGURE 5.6 2D phase error versus angle of propagation.

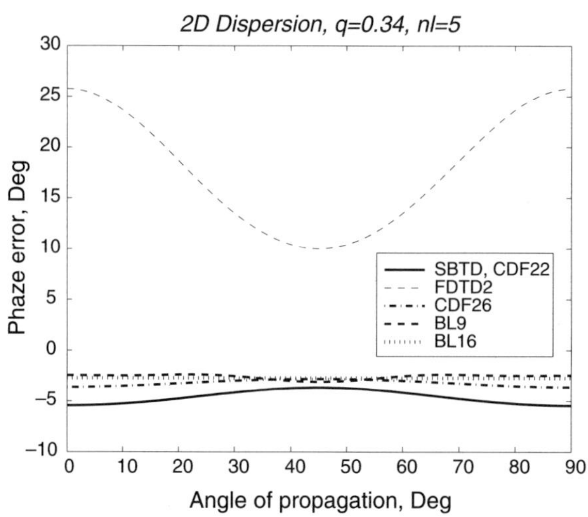

FIGURE 5.7 2D phase error versus angle of propagation.

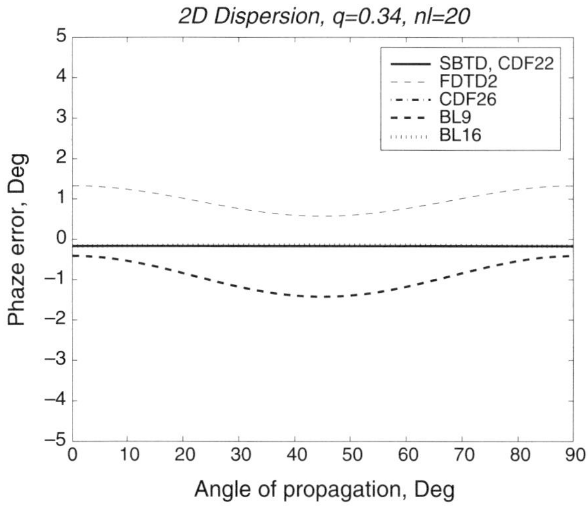

FIGURE 5.8 2D phase error versus angle of propagation.

$$\left(\frac{1}{q}\sin\left(\frac{\pi q}{n_l}\right)\right)^2 = \left(\sum_{i=0}^{N-1} a_i \sin\left(\frac{\pi u}{n_l}(2i+1)\cos\phi\right)\right)^2 + \left(\sum_{i=0}^{N-1} a_i \sin\left(\frac{\pi u}{n_l}(2i+1)\sin\phi\right)\right)^2. \qquad (5.8.2)$$

Figures 5.6–5.8 demonstrate results of numerical experiments for several finite difference schemes, including the SBTD, CDF2-2, FDTD2, CDF2-6, MRTD-9 and MRTD-16, with different sampling rates and Courant numbers. In creating these figures values of coefficients c_i for CDF2-6 and MRTD-16 are quoted from references [13, 16, and 17].

5.8.2 Convergence Analysis

From the MRTD and SBTD we see that the derivative can be approximated by

$$\frac{\partial f_n}{\partial x} := \left.\frac{\partial f}{\partial x}\right|_{x=x_n} \approx \frac{1}{h}\sum_{i=-N}^{N-1} c_i f\left(x_{n+i+1/2}\right), \qquad i = \begin{cases} 0, & 1, \ldots, N-1 \\ -1, & -2, \ldots, -N \end{cases} \qquad (5.8.3)$$

where h is step size and $x_{n+i+(1/2)} = x_n + (i + 1/2)h$. We will see that different schemes have different convergence rates. Notice also that for the SBTD and MRTD

$$c_{-1-i} = -c_i.$$

In Eq. (5.8.3), $i = 0, 1, 2, 3$ for the SBTD; $i = 0, 1, 2, \ldots$ and truncated at $N = 9$ for MRTD-9 and at $N = 16$ for MRTD-16.

Using symmetry of c_i, we may write the RHS of (5.8.3) as

$$P_N = \frac{1}{h}\sum_{i=0}^{N-1} c_i \left(f_{n+(i+1/2)} - f_{n-(i+1/2)}\right) \tag{5.8.4}$$

where $f_{n+i+(1/2)} := f|_{x=x_n+(i+(1/2))h}$. Taking the Taylor expansion about $x = x_n$, we have

$$\begin{aligned} f_{n\pm(i+(1/2))} = f_n &\pm \frac{h(i+1/2)}{1!}\frac{\partial f_n}{\partial x} + \frac{h^2(i+1/2)^2}{2!}\frac{\partial^2 f_n}{\partial x^2} \\ &\pm \frac{h^3(i+1/2)^3}{3!}\frac{\partial^3 f_n}{\partial x^3} \cdots + \cdots \end{aligned} \tag{5.8.5}$$

Combining (5.8.4) and (5.8.5), we arrive at

$$\begin{aligned} P_N = \frac{1}{h}\Bigg[&2c_0\left(\frac{h}{1!}\frac{1}{2}\frac{\partial f_n}{\partial x} + \frac{h^3}{3!}\left(\frac{1}{2}\right)^3\frac{\partial^3 f_n}{\partial x^3} + \frac{h^5}{5!}\left(\frac{1}{2}\right)^5\frac{\partial^5 f_n}{\partial x^5} + \cdots\right) \\ &+2c_1\left(\frac{h}{1!}\left(1+\frac{1}{2}\right)\frac{\partial f_n}{\partial x} + \frac{h^3}{3!}\left(1+\frac{1}{2}\right)^3\frac{\partial^3 f_n}{\partial x^3} + \frac{h^5}{5!}\left(1+\frac{1}{2}\right)^5\frac{\partial^5 f_n}{\partial x^5} + \cdots\right) \\ &+\cdots + 2c_{N-1}\left(\frac{h}{1!}\left(N-1+\frac{1}{2}\right)\frac{\partial f_n}{\partial x} + \frac{h^3}{3!}\left(N-1+\frac{1}{2}\right)^3\frac{\partial^3 f_n}{\partial x^3}\right. \\ &\left.+\frac{h^5}{5!}\left(N-1+\frac{1}{2}\right)^5\frac{\partial^5 f_n}{\partial x^5} + \cdots\right)\Bigg]. \end{aligned}$$

Grouping in powers of h, we obtain

$$\begin{aligned} P_N = \frac{2}{h}\Bigg[&\frac{h}{1!}\frac{\partial f_n}{\partial x}\left(c_0\frac{1}{2} + c_1\left(1+\frac{1}{2}\right) + \cdots + c_{N-1}\left(N-1+\frac{1}{2}\right)\right) \\ &+\frac{h^3}{3!}\frac{\partial^3 f_n}{\partial x^3}\left(c_0\left(\frac{1}{2}\right)^3 + c_1\left(1+\frac{1}{2}\right)^3 + \cdots + c_{N-1}\left(N-1+\frac{1}{2}\right)^3\right) \\ &+\cdots + \frac{h^{2k+1}}{(2k+1)!}\frac{\partial^{2k+1} f_n}{\partial x^{2k+1}}\left(c_0\left(\frac{1}{2}\right)^{2k+1} + c_1\left(1+\frac{1}{2}\right)^{2k+1}\right. \\ &\left.+\cdots + c_{N-1}\left(N-1+\frac{1}{2}\right)^{2k+1}\right) + \cdots\Bigg]. \end{aligned}$$

Now the residue between $\partial f_n/\partial x$ and its approximation P_N is

$$r_N := P_N - \frac{\partial f_n}{\partial x} = I_0 \frac{\partial f_n}{\partial x} + I_2 h^2 \frac{\partial^3 f_n}{\partial x^3} + I_4 h^4 \frac{\partial^5 f_n}{\partial x^5} + \cdots + I_{2k} h^{2k} \frac{\partial^{2k+1} f_n}{\partial x^{2k+1}} + \cdots \tag{5.8.6}$$

where

$$I_0 = \frac{2}{1!}\left[c_0\left(\frac{1}{2}\right) + c_1\left(1+\frac{1}{2}\right) + c_2\left(2+\frac{1}{2}\right) + \cdots + c_{N-1}\left(N-1+\frac{1}{2}\right) - \frac{1}{2}\right]$$

$$I_2 = \frac{2}{3!}\left[c_0\left(\frac{1}{2}\right)^3 + c_1\left(1+\frac{1}{2}\right)^3 + c_2\left(2+\frac{1}{2}\right)^3 + \cdots + c_{N-1}\left(N-1+\frac{1}{2}\right)^3\right]$$

$$\cdots$$

$$I_{2k} = \frac{2}{(2k+1)!}\left[c_0\left(\frac{1}{2}\right)^{2k+1} + c_1\left(1+\frac{1}{2}\right)^{2k+1} + c_2\left(2+\frac{1}{2}\right)^{2k+1} + \cdots + c_{N-1}\left(N-1+\frac{1}{2}\right)^{2k+1}\right]$$

$$\cdots$$

In Eq. (5.8.6) the first term on the RHS

$$T_1 := I_0 \frac{\partial f_n}{\partial x}$$

is h independent, which represents a systematic error of the approximation. The second term

$$T_2 := I_2 \frac{\partial^3 f_n}{\partial x^3} h^2$$

represents the error that is in $O(h^2)$. The third one is in $O(h^4)$, and so on.

Let us examine the central finite difference scheme (CFD), in which the nonzero coefficients in (5.8.3) are $c_0 = -c_{-1} = 1$. According to (5.8.6) the residues of the CFD are computed as follows. The systematic error is

$$T_1 \propto I_0 = 2\left(c_0 \tfrac{1}{2} - \tfrac{1}{2}\right) = 0.$$

This is to say that the CFD scheme produces no systematic error. Error in $O(h^2)$ is

$$T_2 = h^2 I_2 \frac{\partial^3 f_n}{\partial x^3} \propto I_2 = \frac{2}{3!} c_0 \left(\frac{1}{2}\right)^3 = \frac{1}{24} = 4.17e^{-2}.$$

TABLE 5.3. Coefficients I_{2k}

	SBTD, CDF2 − 2 $N = 3$	*FDTD* $N = 1$	*MRTD* $N = 9$	*MRTD* $N = 16$
I_0	$-2.0e-14$	0.0	**7.95e − 3**	**−4.53e − 4**
I_2	$9.93e-9$	**4.17e − 2**	0.11	−0.013
I_4	**5.73e − 3**	$5.21e-4$	0.45	−0.15
I_6	$1.89e-3$	$3.11e-6$	0.88	−0.85
I_8	$1.99e-4$	$1.08e-8$	0.97	−2.87
I_{10}	$1.20e-4$	$2.45e-10$	0.70	−6.44
I_{12}	$5.91e-5$	$4.70e-12$	3.85	−112
I_{14}	$2.48e-5$	$7.84e-14$	18.38	−1734

Following the same procedure, we estimated the errors for different approximation schemes, including the FDTD (based on CFD), MRTD-9, MRTD-16 (based on Battle–Lemarie wavelets of $N = 9$ and $N = 16$ in (5.8.3)), SBTD and CDF2-2 (based on CDF2-2 biorthogonal wavelets [16]). The results are listed in Table 5.3.

It can be seen from Table 5.3 that the SBTD essentially has no systematic error, and the SBTD is a scheme with convergence rate in $O(h^4)$.

5.9 NUMERICAL EXAMPLES

We provide several numerical examples to validate the biorthogonal sampling system.

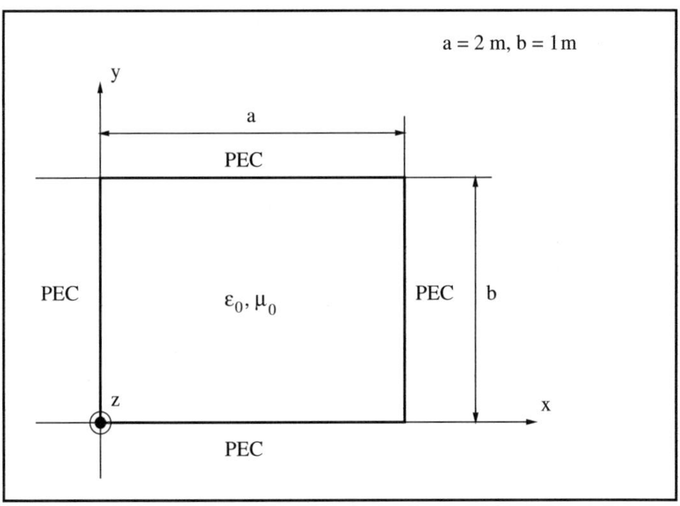

FIGURE 5.9 2D parallel plate resonator.

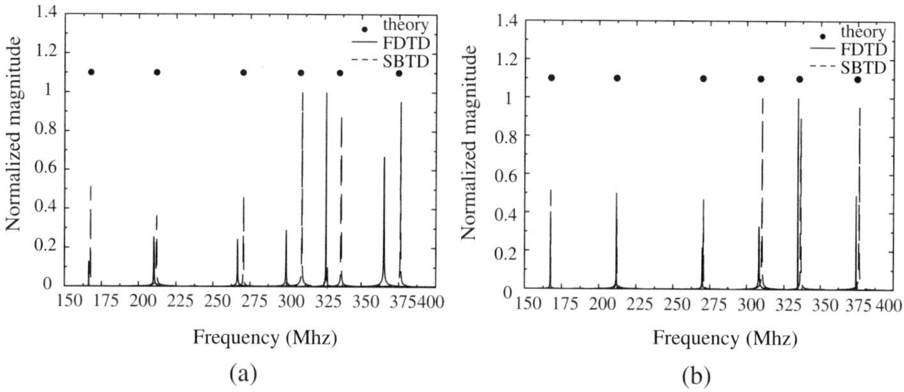

FIGURE 5.10 Magnitude of the electric field component E_z in the frequency domain, air-filled 2D resonator: (*a*) With 105 cells; (*b*) 800 cells for FDTD and 105 cells for SBTD.

Example 1 Resonance Frequency Problem. A 2D parallel plate resonator is depicted in Fig. 5.9. For simplicity, we analyze only the $TM^{(z)}$ polarization for which $E_x = 0$, $E_y = 0$, and $H_z = 0$. The dimensions are $a = 2$ m, $b = 1$ m and the time increment $\Delta t = 10^{-10}$ s. The electric field values E_z are sampled during the time period $T_s = 2^{16}\Delta t$ and the fast Fourier transform (FFT) is performed to obtain the spectrum of the sampled field E_z. Illustrated in Fig. 5.10*a* are the numerical results obtained with $15 \times 7 = 105$ Yee cells for both the FDTD and SBTD techniques, along with analytical values. It can be seen clearly that SBTD produces results in better agreement with the analytical solution, although it is slower than the FDTD approach. The computational time is 8.93 s for the FDTD method and 39.89 s for the SBTD.

To achieve the accuracy, we refined the mesh in the FDTD. As a result, $40 \times 20 = 800$ Yee cells were required by the FDTD in order to achieve the precision yielded by the SBTD with only 105 cells, as shown in Fig. 5.10*b*. The computational time for the FDTD increased to 66.36 s due to the increased number of Yee cells. As can be seen in the figure, both methods yield almost the same results for the resonance frequencies, but the SBTD approach is more efficient in terms of computational time and computer memory.

Example 2 Air-Filled 3D Cavity. An air-filled 3D cavity is shown in Fig. 5.11 with dimensions $a = 1.2$ m, $b = 0.6$ m, and $c = 0.8$ m. The time step was $\Delta t = 0.8 \cdot 10^{-10}$ s. The three electric field components were sampled during the time period $T_s = 2^{16}\,\Delta t$, and the FFT was performed to obtain the frequency spectrum of the sampled electric field. Fig. 5.12*a* displays the numerical results obtained with $6 \times 3 \times 4 = 72$ Yee cells for both FDTD and SBTD techniques, along with analytical values. One can see that the SBTD has better agreement with the theoretical results, though it is more time-consuming than FDTD. Namely the computational time is 23.8 s for the FDTD method and 125.7 s for the SBTD.

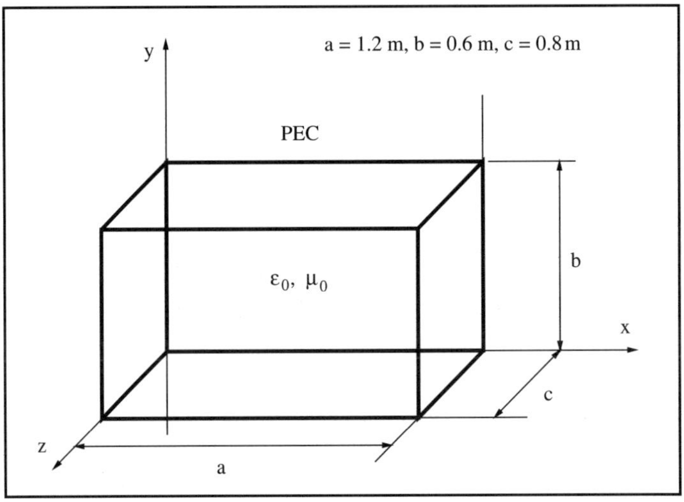

FIGURE 5.11 Air-filled 3D cavity.

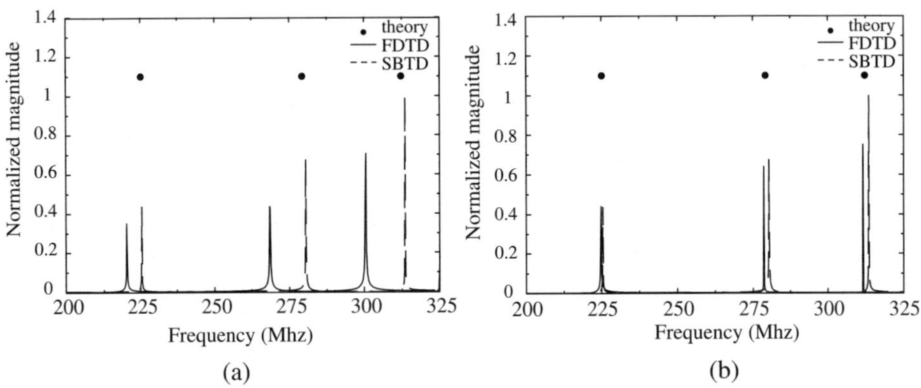

FIGURE 5.12 Magnitude of the electric field in the frequency domain, air-filled 3D cavity: (*a*) With 72 cells (*b*) 4608 cells for FDTD and 72 cells for SBTD.

To achieve more accuracy with the FDTD, we increased the number of Yee cells. The numerical results are shown in Fig. 5.12*b* where FDTD has $24 \times 12 \times 16 = 4608$ cells and SBTD has 72. The computational time for FDTD increased to 1608.9 s. It is obvious that the SBTD approach here is more efficient in terms of computational time and computer memory. To be more specific, we need $4608/72 = 64$ times less computer memory for the SBTD method than for the FDTD approach to obtain an accurate result. At the same time the SBTD technique will be also $1608.9/125.7 \approx 13$ times faster than the FDTD.

Table 5.4 summarizes the numerical results in terms of the lowest resonant frequency (TE_{101} mode), mesh size, computational time, and numerical error. Time

TABLE 5.4. Lowest Resonance Frequency (Air-Filled 3D Cavity)

Mesh	FDTD			SBTD		
	Frequency	Error	Time	Frequency	Error	Time
$x \times z \times y$	(MHz)	(%)	(s)	(MHz)	(%)	(s)
$6 \times 3 \times 4$	220.299	2.17251	23.8	225.449	0.11437	125.7
$12 \times 6 \times 8$	223.923	0.56322	193.9			
$24 \times 12 \times 16$	224.876	0.13972	1608.9			

is given in seconds, error in %, and the resonance frequency in MHz. The theoretical value for the lowest resonance frequency of the parameters in this example is 225.191 MHz.

Example 3 Partially Filled 3D Cavity. A 3D cavity partially filled with a dielectric is shown in Fig. 5.13. The parameters are $a = 1.2$ m, $b = 0.6$ m, $c = 0.8$ m, $h = 0.3$ m and $\epsilon_r = 2.0$.

Table 5.5 shows the numerical results in terms of the lowest resonant frequency, mesh size, computational time and numerical error. Time is given in seconds, error in %, and the resonant frequency in MHz. The theoretical value for the lowest resonant frequency is 224.364 MHz. It can be clearly seen from Table 5.5 that the $12 \times 8 \times 8$ FDTD and $6 \times 4 \times 4$ SBTD have about the same precision of less than 1%. But the FDTD needs $768/96 = 8$ times more computer memory than the SBTD and is $255.4/168.8 \approx 1.5$ times slower than the SBTD.

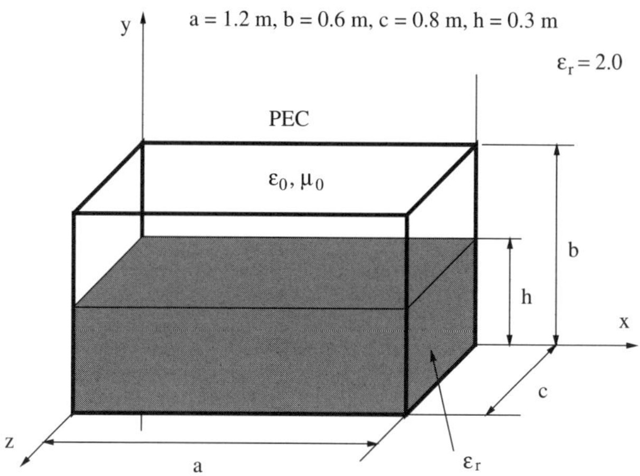

FIGURE 5.13 Partially filled 3D cavity.

TABLE 5.5. Lowest Resonance Frequency (Partially Filled 3D Cavity)

Mesh	FDTD			SBTD		
	Frequency	Error	Time	Frequency	Error	Time
$x \times z \times y$	(MHz)	(%)	(s)	(MHz)	(%)	(s)
$6 \times 4 \times 4$	219.272	2.06684	31.2	225.83	0.65352	168.8
$12 \times 8 \times 8$	223.161	0.53618	255.4	N/A	N/A	N/A
$24 \times 16 \times 16$	223.923	0.19656	2119.5	N/A	N/A	N/A

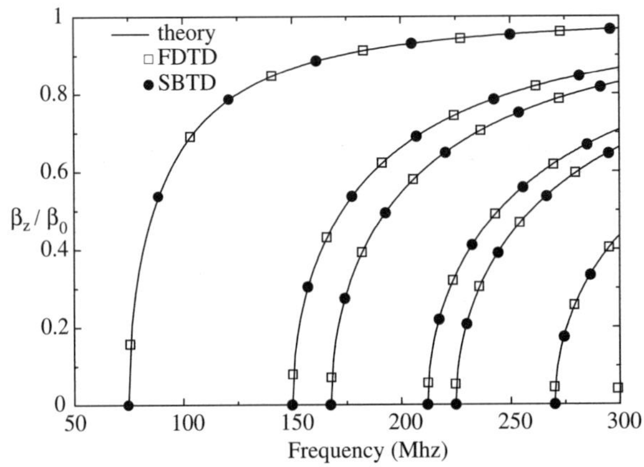

FIGURE 5.14 Normalized propagation constant β_z/k_0 versus frequency (MHz).

Example 4 Waveguide Problem. We model an air-filled rectangular waveguide using the technique described in [18] but with the SBTD method. The cross-sectional dimensions are $a = 2m,\ b = 1$ m. In Fig. 5.14 we plotted the normalized propagation constant β_z/k_0 versus frequency for a few eigenmodes, starting with the dominant mode TE^z_{10}. To verify our numerical SBTD results, we also plotted the dispersion curves from theoretical formulation and from the FDTD.

For the SBTD method we used a mesh with $20 \times 10 = 200$ cells. To reach a competitive precision, the FDTD mesh requires $44 \times 22 = 968$ cells. For each value of β_z the computational time was approximately equal to 150 seconds for the SBTD and 204 seconds for the FDTD.

Example 5 Rectangular Patch Antenna. We analyzed a rectangular microstrip patch antenna in Fig. 5.15. This structure had been previously analyzed in [19]. For the reference solution we used FDTD with space steps $\Delta x = 0.8120$ mm, $\Delta y = 0.8120$ mm, and $\Delta z = 0.3970$ mm. The mesh size is $60 \times 100 \times 16 = 96{,}000$ cells and the time step of 0.441 ps. For the SBTD technique we implemented a mesh with $30{\times}50{\times}8 = 12{,}000$ cells. We also used a smaller mesh size for the FDTD technique.

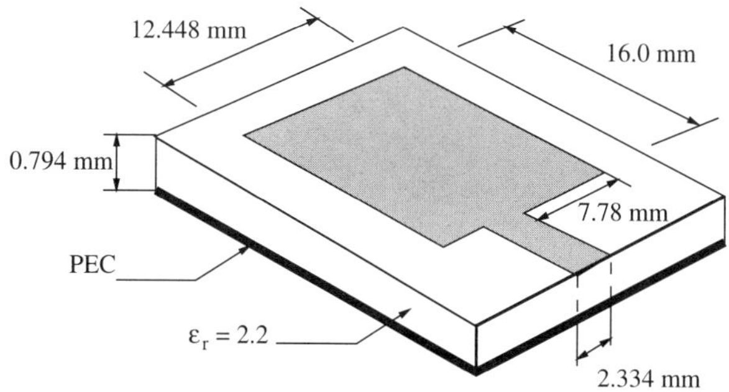

FIGURE 5.15 Rectangular patch antenna.

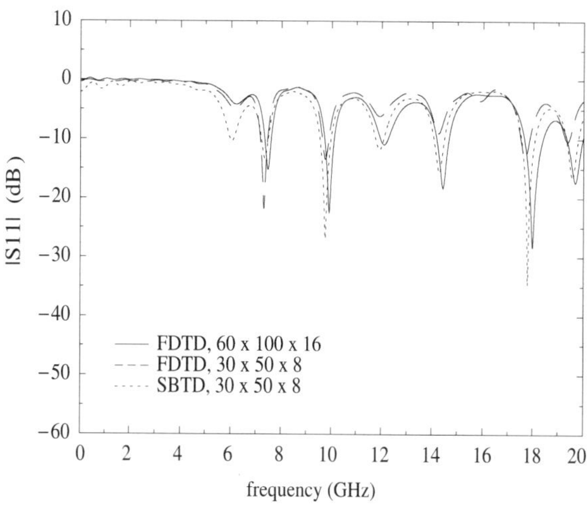

FIGURE 5.16 Return loss of the rectangular patch antenna.

The results in terms of the scattering parameter $|S_{11}|$ are shown in Fig. 5.16. The computation time for the most accurate solution was 4739.2 seconds. For the SBTD approach the CPU time was 2760.8 seconds. To close the open microwave patch antenna structure in the FDTD and SBTD techniques, the PML absorbing boundary conditions [20] were used.

From Fig. 5.16 we see that the SBTD technique provides a result with a coarser mesh size than the FDTD. For the results presented in Fig. 5.16, the SBTD method reduces by factors of 8 and $4739.2/2760.8 \approx 1.7$ for the computer memory and computational time required, respectively.

5.10 APPENDIX: OPERATOR FORM OF THE MRTD

The notation and procedure in this Appendix come from the Hilbert space representation in [9] and [15]. Let us introduce the component vectors

$$|E_{\varphi,\kappa}\rangle := \sum_{k,l,m,n=\infty}^{\infty} {}_kE^{\varphi\kappa}_{l,m,n}|k; l, m, n\rangle,$$

$$|H_{\varphi,\kappa}\rangle := \sum_{k,l,m,n=\infty}^{\infty} {}_kH^{\varphi\kappa}_{l,m,n}|k; l, m, n\rangle,$$

where $\kappa = x, y, z$. These vectors belong to the Hilbert product space $\mathcal{H}_{m\otimes t}$,

$$|H_{\varphi,\kappa}\rangle,\ |E_{\varphi,\kappa}\rangle \in \mathcal{H}_{m\otimes t} = \mathcal{H}_m \otimes \mathcal{H}_t.$$

The orthonormal basis vectors of $\mathcal{H}_{m\otimes t}$ are given by the tensor product

$$|k; l, m, n\rangle = |k\rangle \otimes |l, m, n\rangle.$$

The orthogonality relations are expressed as

$$\langle k_1; l_1, m_1, n_1|k_2; l_2, m_2, n_2\rangle = \delta_{k_1,k_2}\,\delta_{l_1,l_2}\,\delta_{m_1,m_2}\,\delta_{n_1,n_2}.$$

Let us define the half-shift operators

$$\tilde{X}_h\ |k; l, m, n\rangle = |k; l + \tfrac{1}{2}, m, n\rangle$$

and their Hermitian conjugates

$$\tilde{X}_h^\dagger\ |k, l, m, n\rangle = |k; l - \tfrac{1}{2}, m, n\rangle\ = \tilde{X}_h^{-1}.$$

The shift operator $\tilde{X}$ and its Hermitian conjugate $\tilde{X}^\dagger$ are

$$\begin{aligned}\tilde{X}\ |k; l, m, n\rangle &= |k; l + 1, m, n\rangle,\\ \tilde{X}^\dagger\ |k; l, m, n\rangle &= |k; l - 1, m, n\rangle\\ &= \tilde{X}^{-1}|k; l, m, n\rangle.\end{aligned}$$

In the same way the shift and half-shift operators for other spatial, and time domains are, respectively,

$$\begin{array}{ll}\tilde{Y}, & \tilde{Y}_h,\\ \tilde{Z}, & \tilde{Z}_h,\\ \tilde{T}, & \tilde{T}_h.\end{array}$$

Equation (5.4.6) can be written as

$$\epsilon\tilde{X}_h^\dagger\tilde{T}_h^\dagger\tilde{d}_t|E_{\varphi x}\rangle = \tilde{X}_h^\dagger\tilde{T}_h(\tilde{D}_y^\varphi|H_{\varphi z}\rangle - \tilde{D}_z^\varphi|H_{\varphi y}\rangle), \tag{5.10.1}$$

where

$$\begin{aligned}\tilde{d}_t &:= \frac{1}{\Delta t}(\tilde{T}_h^\dagger - \tilde{T}_h),\\ \tilde{D}_y^\varphi &:= \frac{1}{\Delta y}\tilde{Y}_h^\dagger\sum_i a_i\tilde{Y}^{-i},\\ \tilde{D}_z^\varphi &:= \frac{1}{\Delta z}Z_h^\dagger\sum_i a_i\tilde{Z}^{-i}.\end{aligned}$$

The detailed action of these operators can be written as

$$\tilde{d}t\colon\quad \frac{1}{\Delta t}\sum_{k,l,m,n}\left({}_kE^{\varphi x}_{l,m,n}|k'-\tfrac{1}{2},l',m'n'\rangle - {}_kE^{\varphi x}_{l,m,n}|k'+\tfrac{1}{2},l',m',n'\rangle\right)$$

$$\tilde{T}_h^{\dagger}\colon\quad \frac{1}{\Delta t}\sum_{k,l,m,n}\left({}_kE^{\varphi x}_{l,m,n}|k'-1,l',m',n'\rangle - {}_kE^{\varphi x}_{l,m,n}|k',l',m',n'\rangle\right)$$

$$\tilde{X}_h^{\dagger}\colon\quad \frac{\epsilon}{\Delta t}\sum_{k,l,m,n}\left({}_kE^{\varphi x}_{l,m,n}|k'-1,l'-\tfrac{1}{2},m'n'\rangle - {}_kE^{\varphi x}_{l,m,n}|k',l'-\tfrac{1}{2},m',n'\rangle\right).$$

Taking the inner product with $\langle k,l,m,n|$, and imposing orthogonality properties, we obtain

$$\frac{\epsilon}{\Delta t}\left({}_{k+1}E^{\varphi x}_{l+(1/2),m,n} - {}_kE^{\varphi x}_{l+(1/2),m,n}\right). \tag{5.10.2}$$

The rules for the shift operators are

$$\tilde{X}_h\colon\qquad E_l \leftarrow E_{l-(1/2)}$$

$$\tilde{T}^{\dagger}\colon\qquad {}_kE \leftarrow {}_{k+1}E.$$

Similarly we can show the equivalence of the RHS of (5.10.1) and (5.10.2) step by step

$$\tilde{D}_y^{\varphi}|H_{\varphi z}\rangle - \tilde{D}_z^{\varphi}|H_{\varphi y}$$

$$\Rightarrow \frac{1}{\Delta y}\tilde{Y}_h^{\dagger}\left(\sum_p c_p\tilde{Y}^{-p}\,{}_kH^{\varphi z}_{l,m,n}\right) - \frac{1}{\Delta z}\tilde{Z}_h^{\dagger}\left(\sum_p c_p\tilde{Z}^{-p}\,{}_kH^{\varphi y}_{l,m,n}\right)$$

$$\Rightarrow \frac{1}{\Delta y}\sum_p c_p\,{}_kH^{\varphi z}_{l,m+(1/2)+p,n} - \frac{1}{\Delta z}\sum_i c_p\,{}_kH^{\varphi y}_{l,m,n+(1/2)+p}$$

$$\tilde{X}_h^{\dagger}\tilde{T}_h^{\dagger}\colon\quad \frac{1}{\Delta y}\sum_p c_p\,{}_{k+(1/2)}H^{\varphi z}_{l+(1/2),m+(1/2)+p,n}$$

$$-\frac{1}{\Delta z}\sum_p c_p\,{}_{k+(1/2)}H^{\varphi y}_{l+(1/2),m,n+(1/2)+p}. \tag{5.10.3}$$

In equating (5.10.2) and (5.10.3), we arrive at (5.4.6). We recognize that the operator notation is very convenient, since we do not have to remember the half unit shifts in time interval and space for the **E** and **H** field components as they were given in (5.4.1). Introducing the field vector

$$|F_{\varphi}\rangle = \begin{pmatrix} |E_{\varphi x}\rangle \\ |E_{\varphi y}\rangle \\ |E_{\varphi z}\rangle \\ H_{\varphi x}\rangle \\ H_{\varphi y}\rangle \\ H_{\varphi z}\rangle \end{pmatrix}, \tag{5.10.4}$$

we can rewrite the six difference equations as

$$\tilde{W}|F_{\varphi}\rangle = 0, \tag{5.10.5}$$

where the operator $\tilde{\mathbf{W}}$ is given by

$$\tilde{\mathbf{W}} = \begin{bmatrix} \epsilon \mathbf{X}_h^\dagger \mathbf{T}_h^\dagger \mathbf{d}_t & 0 & 0 & 0 & \mathbf{T}_h^\dagger \mathbf{X}_h^\dagger \mathbf{D}_z^\phi & -\mathbf{T}_h^\dagger \mathbf{X}_h^\dagger \mathbf{D}_y^\phi \\ 0 & \epsilon \mathbf{Y}_h^\dagger \mathbf{T}_h^\dagger \mathbf{d}_t & 0 & -\mathbf{T}_h^\dagger \mathbf{Y}_h^\dagger \mathbf{D}_z^\phi & 0 & \mathbf{T}_h^\dagger \mathbf{Y}_h^\dagger \mathbf{D}_x^\phi \\ 0 & 0 & \epsilon \mathbf{Z}_h^\dagger \mathbf{T}_h^\dagger \mathbf{d}_t & \mathbf{T}_h^\dagger \mathbf{Z}_h^\dagger \mathbf{D}_y^\phi & -\mathbf{T}_h^\dagger \mathbf{Z}_h^\dagger \mathbf{D}_x^\phi & 0 \\ 0 & -\mathbf{Y}_h^\dagger \mathbf{Z}_h^\dagger \mathbf{D}_z^\phi & \mathbf{Y}_h^\dagger \mathbf{Z}_h^\dagger \mathbf{D}_y^\phi & \mu \mathbf{Y}_h^\dagger \mathbf{Z}_h^\dagger \mathbf{d}_t & 0 & 0 \\ \mathbf{X}_h^\dagger \mathbf{Z}_h^\dagger \mathbf{D}_z^\phi & 0 & -\mathbf{X}_h^\dagger \mathbf{Z}_h^\dagger \mathbf{D}_x^\phi & 0 & \mu \mathbf{X}_h^\dagger \mathbf{Z}_h^\dagger \mathbf{d}_t & 0 \\ -\mathbf{X}_h^\dagger \mathbf{Y}_h^\dagger \mathbf{D}_y^\phi & \mathbf{X}_h^\dagger \mathbf{Y}_h^\dagger \mathbf{D}_x^\phi & 0 & 0 & 0 & \mu \mathbf{X}_h^\dagger \mathbf{Y}_h^\dagger \mathbf{d}_t \end{bmatrix}.$$

Note that the field components of the wavelet-FDTD scheme are different from those of Yee's FDTD. For the wavelet MRTD, E_x at $[k - (1/2)]\,\Delta t < t < [k + (1/2)]\,\Delta t$ is given by

$$E_x(\mathbf{r}_0, t_0) = \sum_{l',m',n'=-\infty}^{\infty} {}_kE^{\varphi x}_{l'+(1/2),m',n'} \varphi_{l'+(1/2)}(x_0)\varphi_{m'}(y_0)\varphi_{n'}(z_0). \tag{5.10.6}$$

5.11 PROBLEMS

5.11.1 Exercise 9

1. Show that the center finite difference scheme converges in $O(h^2)$.

2. Show that

$$\int_{-\infty}^{\infty} P_m(x) \frac{\partial P_{m'+(1/2)}(x)}{\partial x}\, dx = \delta_{m,m'} - \delta_{m,m'+1},$$

where

$$P_m(x) = P\left(\frac{x}{\Delta x} - m\right)$$

and

$$P(x) = \begin{cases} 1 & \text{for } |x| < \frac{1}{2} \\ \frac{1}{2} & |x| = \frac{1}{2} \\ 0 & \text{otherwise.} \end{cases}$$

3. Show that the numerical stability condition for the uniform cubic Yee mesh is

$$\Delta t \le \frac{\Delta S}{v\sqrt{3}}.$$

4. Show that the 2D dispersion for the square Yee mesh is

$$\left(\frac{\Delta S}{v\,\Delta t}\right)^2 \sin^2\left(\frac{\omega\,\Delta t}{2}\right) = \sin^2\left(\frac{k\cos\alpha\,\Delta S}{2}\right) + \sin^2\left(\frac{k\sin\alpha\,\Delta S}{2}\right).$$

5. Evaluate the coefficients c_i, for $i = 0, 1, 2$, using the formula

$$c_i = \left\langle \varphi_{-i}, \frac{d}{dx}\varphi_{\frac{1}{2}}(x) \right\rangle = \frac{1}{\pi}\int_{-\infty}^{\infty} \omega|\hat{\varphi}(\omega)|^2 \sin\left[\omega\left(i + \frac{1}{2}\right)\right] dw.$$

5.11.2 Exercise 10

1. Construct the positive sampling function $S(x)$ by using the Daubechies scalet $\varphi(t-k)$ of order $N = 2$ (referred to as the D_2). Plot $S_m(x)$, $m = 0, 1, 2$.

2. Construct the biorthogonal testing function

$$Q_n(x) = \sum_{p \in \mathcal{Z}} \varphi(n-p)\varphi(x-p)$$

and plot $Q_n(x)$, $n = -1, 0, 1$.

3. Verify biorthogonality numerically by evaluating

(a) $\int S(x)Q(x)\,dx = ?$

(b) $\int S(x)Q_1(x)\,dx = ?$

5.11.3 Project 2

A two-dimensional rectangular resonator with dimensions $a = 2m$, $b = 1m$ is filled with dielectric $\epsilon_r = 2$. Find the first 4 to 6 resonant modes for the TM^z polarization. Compare the FDTD and SBTD results with your analytical answers.

In your work, you should verify that the stability condition is satisfied. Your submission should consist of four parts:

1. Analytic Solution. Derive expressions for the resonance frequencies.

2. FDTD Scheme

Formulation

(a) Three partial differential equations.

(b) Corresponding discretized equations.

Parameters

(c) $\Delta x = \Delta y = 0.05m \Rightarrow N_x = \frac{a}{\Delta x} = 40,\ N_y = \frac{b}{\Delta y} = 20,\ \Delta t = 10^{-10}s.$

Excitation

(d) Gaussian pulse in time, and located about (but not exactly) at the center of the cavity, $f(t) = e^{-(t-t_0)^2/(T^2)}$ with $T = 1.5\,\Delta t$ $t_0 = 5\,\Delta t$.

Boundary Conditions Radiation or Absorption.

(e) No need for closed boundary cavities.

Recording Time-Domain Signals

(f) The first $10\,\Delta t$, $E_z(t) = f(t)$. The FDTD-updated E_z at the exciting node must be dropped and replaced by $f(t)$.

(g) Afterward use the FDTD updated E_z at that node.

(h) First 5000 Δt are the relaxation time.

(i) Record the signal, $E_z(t)$, after relaxation until $t = 2^{16}\,\Delta t$.

Extraction of Frequency Parameters By the DFT, or FFT.

(j) $\Im\{E_z(t) : t = 5000\,\Delta t \text{ to } 2^{16}\,\Delta t\}$.

Report

(k) Plot $E_z(\omega)$ against frequency.

(l) Report the CPU time and memory usage.

3. SBTD

(m) Corresponding discretized equations.

(n) $N_x = \frac{z}{\Delta x} = 15$, $N_y = \frac{b}{\Delta y} = 7$.

(o) Boundary conditions must be applied.

4. Summary.

BIBLIOGRAPHY

[1] K. Yee, "Numerical solution of initial boundary value problems involving Maxwell's equation in isotropic media," *IEEE Trans. Ant. Propg.*, **14**, 302–307, May 1966.

[2] A. Taflove and K. Umashankar, "The finite-difference time-domain method for numerical modeling of electromagnetic wave interactions with arbitrary structures," in *Finite Element and Finite Difference Methods in Electromagnetic Scattering*, M. A. Morgan, Ed. Elsevier, New York, 1991.

[3] G. Mur, "Absorbing boundary conditions for the finite-difference approximation of the time-domain electromagnetic-field equations," *IEEE Trans. EMC*, **23**(4), 377–382, Nov. 1981.

[4] J. Berenger, "A perfectly matched layer for the absorption of electromagnetic waves," *J. Comput. Phys.*, **114**, 185–200, Oct. 1994.

[5] B. Engquist and A. Majda, "Absorbing boundary conditions for the numerical simulation of waves," *Math. Comput.*, **31**(139), 629–651, July 1977.

[6] M. Yang, Y. Chen, and R. Mittra, "Hybrid finite-difference/finite-volume time-domain analysis for microwave integrated circuits with curved PEC surface using nonuniform rectangular grid," *IEEE Trans. Microw. Theory Tech.*, **48**(2), 969–975, June 2000.

[7] G. Pan, D. Cheng, and B. Gilbert, "2D FDTD modeling of objects with curved boundaries, using embedded boundary orthogonal grids," *IEE Proc. Microw. Ant. Propg.*, **147**(5), 399–405, Oct. 2000.

[8] P. John and R. Beurle, "Numerical solutions of 2-dimensional scattering problems using a transmission-line matrix," *Proc. IEE*, **119**(8), 1086–1091, Aug. 1971.

[9] M. Krumpholz and L. Katehi, "MRTD: New time-domain schemes based on multiresolution analysis," *IEEE Trans. Microw. Theory Tech.*, **44**, 555–571, Apr. 1996.

[10] S. Shamugan, *Analog and Digital Communication Systems*, John Wiley, New York, 1978.

[11] G. Walter and X. Shen, *Wavelets and Other Orthogonal Systems*, 2nd ed., CRC Press, New York, 2001.

[12] G. Walter, *Wavelets and Other Orthogonal Systems with Applications*, CRC Press, Boca Raton, FL, 1994.

[13] Y. Cheong, Y. Less, K. Ra, J. Kang, and C. Shin, "Wavelet-Galerkin scheme of time-dependent inhomogeneous electromagnetic problems," *IEEE Microw. Guided Wave Lett.*, **9**, 297–299, Aug. 1999.

[14] W. Sweldens and R. Piessens, "Wavelet sampling techniques," in *1993 Proc. Statist. Comput. Sec.*, 20–29.

[15] P. Russer and M. Krumpholz, "The Hilbert space formulation of the TLM method," *Int. J. Num. Mod.: Electronic Networks, Devices Fields*, **6**(1), 29–45, Feb. 1993.

[16] T. Dogaru and L. Carin, "Multiresolution time-domain using CDF biorthogonal wavelets," *IEEE Trans. Microw. Theory Tech.*, **49**(5), 902–912, May 2001.

[17] E. Tentzeris, A. Langellaris, L. Katehi, and J. Harvey, "MRTD adaptive schemes using arbitrary resolution of wavelets," *IEEE Trans. Microw. Theory Tech.*, **50**(2), 501–515, Feb. 2002.

[18] E. Tentzeris, R. Robertson, M. Krumpholz, and L. Katehi, "Application of MRTD to printed transmission lines," in *Proc. Microw. Theory Tech. Soc.*, IEEE Press, 1996, 573–576.

[19] D. Sheen, S. Ali, M. Abouzahra, and J. Kong, "Application of the three-dimensional finite-difference time-domain method to the analysis of planar microstrip circuits," *IEEE Trans. Microw. Theory Tech.*, **38**, 849–857, July 1990.

[20] E. Tentzeris, R. Robertson, J. Harvey, and L. Katehi, "PML absorbing boundary conditions for the characterization of open microwave circuits using multiresolution time-domain techniques (MRTD)," *IEEE Trans. Ant. Propg.*, **47**(11), 1709–1715, November 1999.

CHAPTER SIX

Canonical Multiwavelets

As discussed in the previous chapters, wavelets have provided many beneficial features, including orthogonality, vanishing moments, regularity (continuity and smoothness), multiresolution analysis, among these features. Some wavelets are compactly supported in the time domain (Coifman, Daubechies) or in the frequency domain (Meyer), and some are symmetrical (Haar, Battle–Lemarie). On many occasions it would be very useful if the basis functions were symmetrical. For instance, it would be better to expand a symmetric object such as the human face using symmetric basis functions rather than asymmetric ones. In regard to boundary conditions, magnetic wall and electric wall are symmetric and antisymmetric boundaries, respectively. It might be ideal to create a wavelet basis that is symmetric, smooth, orthogonal, and compactly supported. Unfortunately, the previous four properties cannot be simultaneously possessed by any wavelets, as proved in [1].

To overcome the limitations of the regular (i.e., scalar) wavelets, mathematicians have proposed multiwavelets. There are two categories of multiwavelets, and both of them are defined on finite intervals. The first class is that of the canonical multiwavelets that are based upon the vector-matrix dilation equation [2–4]; this class will be studied in this chapter. The second class is based on the Lagrange or Legendre interpolating polynomials [5], which is similar in some respects to the pseudospectral domain method and as such facilitates MRA.

6.1 VECTOR-MATRIX DILATION EQUATION

Multiwavelets offer more flexibility than traditional wavelets by extending the scalar dilation equation

$$\varphi(t) = \sum h_k \varphi(2t - k)$$

into the matrix-vector version

$$|\phi(t)\rangle = \sum_k C_k\,|\phi(2t-k)\rangle,$$

where $C_k = [C_k]_{r\times r}$ is a matrix of $r \times r$, $|\phi(t)\rangle = (\phi_0(t)\cdots\phi_{r-1}(t))^T$ is a column vector of $r \times 1$, and r is the multiplicity of the multiwavelets. By taking the jth derivative, we have

$$|\phi^{(j)}(t)\rangle = \sum_k C_k 2^j\,|\phi^{(j)}(2t-k)\rangle.$$

Let us denote a matrix

$$\Phi(t) = \begin{bmatrix} \phi_0(t) & \phi_0'(t) & \cdots & \phi_0^{(r-1)}(t) \\ \phi_1(t) & \phi_1'(t) & & \vdots \\ \vdots & & & \\ \phi_{r-1}(t) & \vdots & & \phi_{r-1}^{(r-1)}(t) \end{bmatrix}. \tag{6.1.1}$$

Then

$$\begin{aligned} \Phi(t) &= \left[|\phi(t)\rangle \quad |\phi'(t)\rangle \quad \cdots \quad |\phi^{(r-1)}(t)\rangle\right] \\ &= \left[\sum C_k|\phi(2t-k)\rangle \;\; 2\sum C_k|\phi'(2t-k)\rangle \;\; \cdots \;\; 2^{r-1}\sum C_k|\phi^{(r-1)}(2t-k)\rangle\right] \\ &= \sum_k C_k|\phi(2t-k)\rangle \begin{bmatrix} 1 & & & \\ & 2 & & \\ & & \ddots & \\ & & & 2^{r-1} \end{bmatrix}_{r\times r}, \end{aligned}$$

or

$$\Phi(t) = \sum C_k \Phi(2t-k)\Lambda^{-1}, \tag{6.1.2}$$

where

$$\Lambda^{-1} = \text{diag}\{1, 2, \ldots, 2^{r-1}\}. \tag{6.1.3}$$

Equation (6.1.2) can be verified as follows:

Show.

$$\begin{aligned} \text{LHS} &= \Phi(t) \\ &= \begin{bmatrix} \phi_0(t) & & \phi_0^{(r-1)}(t) \\ \cdots & \cdots & \cdots \\ \phi_{r-1}(t) & & \phi_{r-1}^{(r-1)}(t) \end{bmatrix} \end{aligned}$$

$$= \left[|\phi(t)\rangle \quad |\phi^{(1)}(t)\rangle \quad \cdots \quad |\phi^{(r-1)}(t)\rangle \right]$$

$$= \left[\sum C_k |\phi(2t-k)\rangle \quad 2\sum C_k |\phi^{(1)}_{2t-k}\rangle \quad \cdots \quad 2^{(r-1)} \sum C_k^{(r-1)} |\phi(2t-k)\rangle \right].$$

$$\text{RHS} = \sum_k C_k \Phi(2t-k)\Lambda^{-1}$$

$$= \sum (C_k)_{r\times r} \left[|\phi(2t-2)\rangle \quad |\phi^{(1)}(2t-2)\rangle \quad \cdots \quad |\phi^{(r-1)}(2t-k)\rangle \right]$$

$$\cdot \begin{bmatrix} 1 & \cdots & \cdots & \cdots 0 \\ 0 & 2 & 0 & \cdots 0 \\ 0 & 0 & 2^2 & \cdots 0 \\ \cdots & \cdots & \cdots & \cdots \\ 0 & 0 & 0 & \cdots 2^{r-1} \end{bmatrix}_{r\times r}$$

$$= \sum_k (C_k)_{r\times r} \begin{bmatrix} \phi_0(2t-k) & \phi_0'(2t-k) & \cdots & \phi_0^{(r-1)}(2t-k) \\ \phi_1(2t-k) & \phi_1'(2t-k) & \cdots & \phi_1^{(r-1)}(2t-k) \\ \cdots & \cdots & \cdots & \cdots \\ \phi_{r-1}(2t-k) & \phi_{r-1}'(2t-k) & \cdots & \phi_{r-1}^{(r-1)}(2t-k) \end{bmatrix}$$

$$\cdot \begin{bmatrix} 1 & & & \\ & 2 & & \\ & & \cdots & \\ & & & 2^{r-1} \end{bmatrix}$$

$$= \sum_k (C_k)_{r\times r} \begin{bmatrix} \phi_0(2t-k) & 2\phi_0'(2t-k) & \cdots & 2^{(r-1)}\phi_0^{(r-1)}(2t-k) \\ \phi_1(2t-k) & 2\phi_1'(2t-k) & \cdots & \cdots \\ \cdots & \cdots & & \\ \phi_{r-1}(2t-k) & 2\phi_{r-1}'(2t-k) & \cdots & 2^{(r-1)}\phi_{r-1}^{(r-1)}(2t-k) \end{bmatrix}.$$

Hence

$$\Phi(t) = \sum C_k \Phi(2t-k)\Lambda^{-1}. \tag{6.1.4}$$

□

In the construction of the multiwavelets, we may use either the frequency domain approach or the time domain approach. The frequency approach is more elegant but requires more extensive mathematical background. We select the latter approach, which seems to be easier to follow despite being more cumbersome.

6.2 TIME DOMAIN APPROACH

We begin with the vector dilation equation

$$|\phi(t)\rangle = \sum_k C_k |\phi(2t-k)\rangle, \tag{6.2.1}$$

which has an explicit form of

$$\begin{bmatrix} \phi_0(t) \\ \phi_1(t) \\ \cdot \\ \cdot \\ \cdot \\ \phi_{r-1}(t) \end{bmatrix} = \sum_{k=0}^{n-1} [C_k]_{r\times r} \begin{bmatrix} \phi_0(2t-k) \\ \phi_1(2t-k) \\ \cdot \\ \cdot \\ \cdot \\ \phi_{r-1}(2t-k) \end{bmatrix},$$

where n is the order of approximation (see Eq. (6.2.6)).

Let us denote an infinite-dimensional matrix

$$L = \begin{bmatrix} \cdots & \cdots & & & & & & & \\ \cdots & C_3 & C_2 & C_1 & C_0 & & & & \\ & \cdots & \cdots & C_3 & C_2 & C_1 & C_0 & & \\ & & & & & C_3 & C_2 & C_1 & C_0 \\ & & & & & & & \cdots & \cdots \end{bmatrix}.$$

Then (6.2.1) becomes

$$|\,\Theta(t)\rangle = L\,|\Theta(2t)\rangle, \tag{6.2.2}$$

where $|\,\Theta(t)\rangle = [\cdots \langle\phi(t-1)\,|\,\langle\phi(t)|\,\langle\phi(t+1)\,|\cdots]^T$. The explicit form of (6.2.2) is

$$\begin{bmatrix} \cdots \\ |\,\phi(t-1)\rangle \\ |\,\phi(t)\rangle \\ |\,\phi(t+1)\rangle \\ \cdots \end{bmatrix} = \begin{bmatrix} \cdots & & & & & & & \\ C_3 & C_2 & C_1 & C_0 & & & & \\ & \cdots & C_3 & C_2 & C_1 & C_0 & & \\ & & \cdots & C_3 & C_2 & C_1 & C_0 \\ & & & & & & \cdots \end{bmatrix} \begin{bmatrix} \cdots \\ |\,\phi(2t-1)\rangle \\ |\,\phi(2t)\rangle \\ |\,\phi(2t+1)\rangle \\ \cdots \end{bmatrix},$$

or

$$\begin{bmatrix} \cdots & & & & & & & \\ C_3 & C_2 & C_1 & C_0 & & & & \\ & \cdots & C_3 & C_2 & C_1 & C_0 & & \\ & & \cdots & C_3 & C_2 & C_1 & C_0 \\ & & & & & & \cdots \end{bmatrix} \begin{bmatrix} \cdots \\ |\,\phi(2t-1)\rangle \\ |\,\phi(2t)\rangle \\ |\,\phi(2t+1)\rangle \\ \cdots \end{bmatrix} = \begin{bmatrix} \cdots \\ |\,\phi(t-1)\rangle \\ |\,\phi(t)\rangle \\ |\,\phi(t+1)\rangle \\ \cdots \end{bmatrix}. \tag{6.2.3}$$

Let us pick out the row that represents $|\,\phi(2t)\rangle$ and $|\,\phi(t)\rangle$ in (6.2.3), namely

$$\cdots + C_2|\,\phi(2t-2)\rangle + C_1|\,\phi(2t-1)\rangle + C_0|\,\phi(2t)\rangle = |\,\phi(t)\rangle.$$

If we replace t by $t-1$, then (6.2.1) becomes

$$\sum_k C_k|\,\phi(2t-k-2)\rangle = |\,\phi(t-1)\rangle.$$

Explicitly, the equation above is

$$\cdots + C_1 |\, \phi(2t-3)\rangle + C_0 |\, \phi(2t-2)\rangle = |\, \phi(t-1)\rangle,$$

which is one row above in (6.2.3). Notice the two-unit shift in the row of the matrix L that corresponds to the equation above.

Now consider the monomials t^j, $j = 0, 1, \ldots, r-1$, which span the scaling subspace. The $\phi(\cdot)$ are the basis function in V_r. Therefore

$$t^j := G_j(t) = \sum_{k=-\infty}^{\infty} \langle y_k^{[j]} |\, \phi(t-k)\rangle = \langle y^{[j]} |\, \Theta(t)\rangle, \tag{6.2.4}$$

where

$$\langle y^{[j]} \,| = \left[\cdots \langle y_0^{[j]} |\, \langle y_1^{[j]} \,| \langle y_2^{[j]} | \cdots \right]$$

and each piece $\langle y_k^{[j]} \,|$ is a row vector with r components that matches the vectors $|\, \phi(t-k)\rangle$. Substituting (6.2.2) into (6.2.4), we obtain

$$G_j(t) = \langle y^{[j]} |\, \Theta(t)\rangle = \langle y^{[j]} |\, L \,|\Theta(2t)\rangle.$$

On the other hand, we may rewrite this as

$$G_j(t) = t^j = 2^{-j}(2t)^j = 2^{-j}\langle y^{[j]} \,|\Theta(2t)\rangle.$$

Hence

$$\langle y^{[j]} |\, L \,|\Theta(2t)\rangle = 2^{-j}\langle y^{[j]} |\, \Theta(2t)\rangle,$$

and therefore

$$\langle y^{[j]} |\, L = 2^{-j}\langle y^{[j]} \,|. \tag{6.2.5}$$

The previous equation implies that L has eigenvalue 2^{-j} for the left eigenvector $\langle y^{[j]} \,|$. That is to say, if L has eigenvalues $1, 2^{-1}, 2^{-2}, \ldots, 2^{-(p-1)}$ with left eigenvectors $\langle y^{[j]} \,|$, then

$$G_j(t) = \sum_{k=-\infty}^{\infty} \langle y_k^{[j]} \,|\phi(t - k)\rangle.$$

A special and important case is $j = 0$, in which case

$$\sum_k \langle y_k^{[0]} \,|\phi(t-k)\rangle = 1 = t^0.$$

In the remainder of this section, we will list definitions, lemmas and theorems that will form a solid foundation of multiwavelets in the time domain.

Definition. A multiscalet $|\, \phi(t)\rangle$ has approximation order n if each monomial $t^j, j = 0, \ldots, n-1$ is a linear summation of integer translations $|\, \phi(t-k)\rangle$

such that

$$t^j = \sum_{k=-\infty}^{\infty} \langle y_k^{[j]} | \phi(t-k) \rangle, \qquad j = 0, 1, \ldots, n-1, \tag{6.2.6}$$

almost everywhere.

Lemma 1. Suppose that $\phi_j(t) \in L^1$ for $j = 0, \ldots, r-1$ and the translates $\phi_j(t-k)$, $k \in Z$, are linearly independent. Then $|\phi(t)\rangle$ provides an approximation of order n if and only if L has eigenvalues 2^{-j} corresponding to the left eigenvectors

$$\langle y^{[j]} | = \left[\cdots \langle y_0^{[j]} | \langle y_1^{[j]} \langle y_2^{[j]} | \cdots \right]$$

with a component

$$\langle y_k^{[j]} | = \sum_{\ell=0}^{j} \binom{j}{l} (-k)^{j-\ell} \langle u^{[\ell]} |, \qquad j = 0, 1, \ldots, n-1, \tag{6.2.7}$$

where $\langle u^{[\ell]} |$ are constant vectors that will be given in (6.12.12).

Lemma 2. Suppose that $\langle y^{[j]} |$ is given by (6.2.7) and that L corresponds to a multiscalet with an approximation order n. Then

$$\langle y^{[j]} | L = 2^{-j} \langle y^{[j]} |, \qquad j = 0, \ldots, n-1,$$

if and only if the following finite equations are held:

$$\sum_k \langle y_k^{[j]} | C_{2k+1} = 2^{-j} \langle u^{[j]} | \tag{6.2.8}$$

$$\sum_k \langle y_k^{[j]} | C_{2k} = 2^{-j} \langle y_1^{[j]} | = 2^{-j} \sum_{\ell=0}^{j} (-1)^{j-\ell} \binom{j}{\ell} \langle u^{[\ell]} | \quad \text{for } j = 0, 1, \ldots, n-1. \tag{6.2.9}$$

Equations (6.2.8) and (6.2.9) are referred to as the approximation conditions. The proofs of Lemma 1 and Lemma 2 are provided in the Appendix to this chapter.

6.3 CONSTRUCTION OF MULTISCALETS

We begin with the approximation conditions (6.2.8) and (6.2.9):

$$\sum_k \langle y_k^{[j]} | C_{2k+1} = 2^{-j} \langle u^{[\ell]} |, \tag{6.3.1}$$

$$\begin{aligned} \sum_k \langle y_k^{[j]} | C_{2k} &= 2^{-j} \langle y_1^{[j]} | \\ &= 2^{-j} \sum_{\ell=0}^{j} (-1)^{j-\ell} \binom{j}{\ell} \langle u^{[\ell]} |, \end{aligned} \tag{6.3.2}$$

which are a system of nonlinear equations in terms of matrix components and the starting vectors $\langle u^{[j]}|$. These equations can be solved effectively only for low approximation orders with a small number of dilation coefficients. Fortunately, in electromagnetics, the order is usually ≤ 4. An intervallic function of order r is a multiscalet

$$|\phi(t)\rangle = (\phi_0(t)\ldots\phi_{r-1}(t))^T \tag{6.3.3}$$

consisting of intervallic ϕ_j, which are piecewise polynomials of degree $2r-1$ with $r-1$ continuous derivatives. For all r, $\phi_j(t) \neq 0$ only on two intervals $[0, 1]$ and $[1, 2]$. The function value and its $r-1$ derivatives are specified at each integer node. If the intervallic functions are defined on $[0, 2]$, then they are alternatively symmetric and antisymmetric about $t = 1$. The translations of these functions span V_0.

The dilation equation may be written as

$$\begin{aligned} |\phi(t)\rangle &= \sum_k C_k|\phi(2t-k)\rangle \\ &= C_0|\phi(2t)\rangle + C_1|\phi(2t-1) + C_2|\phi(2t-2)\rangle. \end{aligned} \tag{6.3.4}$$

Since the support is $[0, 2]$, the only nonzero coefficients are C_0, C_1, and C_2. There are r basis functions at each node, and C_i are matrices of $r \times r$ ($i = 0, 1, 2$). The polynomials of degree $2r-1$ on $[0, 1]$ and $[1, 2]$ can be determined by

$$\left(\frac{d}{dt}\right)^k \phi_j(1) = \delta_{k,j}, \qquad k, j = 0, \ldots, r-1, \tag{6.3.5}$$

$$\left(\frac{d}{dt}\right)^k \phi_j(0) = 0 = \left(\frac{d}{dt}\right)^k \phi_j(2), \qquad k, j = 0, \ldots, r-1, \tag{6.3.6}$$

where $\delta_{k,j}$ is the Kronecker delta.

The symmetry and antisymmetry about $t = 1$ are given by

$$\phi_j(2-t) = (-1)^j \phi_j(t), \qquad j = 0, \ldots, r-1. \tag{6.3.7}$$

Notice that $C_0|\phi(2)\rangle = 0 = C_2|\phi(0)\rangle$ by (6.3.6). Equations (6.3.5) and (6.3.6) may be expressed compactly as

$$\Phi(n) = \delta_{1,n} I,$$

where $\delta_{1,n}$ is the Kronecker delta, I is the identity matrix of $r \times r$, and $\Phi(t)$ was defined in (6.1.4) as

$$\begin{aligned} \Phi(t) &= (|\phi(t)\rangle|\phi'(t)\rangle \cdots |\phi^{(r-1)}(t)\rangle) \\ &= \begin{bmatrix} \phi_0(t) & & \phi_0^{(r-1)}(t) \\ \cdots & \cdots & \cdots \\ \phi_{r-1}(t) & & \phi_{r-1}^{(r-1)}(t) \end{bmatrix} \end{aligned}$$

with $\phi_i^{(j)}(t) := (d/dt)^j \phi_i(t)$, $i, j = 0, \ldots, r-1$.

Example 1 The multiscalets for multiplicity $r = 2$ are

$$\phi_0(t) = (3t^2 - 2t^3), \quad \phi_1(t) = t^3 - t^2 \qquad \text{for } t \in [0, 1], \tag{6.3.8}$$

$$\phi_0(t) = \phi_0(2 - t), \quad \phi_1(t) = -\phi_1(2 - t) \qquad \text{for } t \in [1, 2]. \tag{6.3.9}$$

We can verify that

$$\Phi(t)|_{t=1} = \begin{bmatrix} \phi_0(t) & \phi_0'(t) \\ \phi_1(t) & \phi_1'(t) \end{bmatrix}\Bigg|_{t=1} = \begin{bmatrix} 1 & 0 \\ 0 & 1 \end{bmatrix}.$$

It is easy to find that

$$\begin{aligned}
&\phi_0(1) = 1, \\
&\phi_1(1) = 0, \\
&\phi_0'(t)|_{t=1} = [6t - 6t^2]_{t=1} = 0, \\
&\phi_1'(t)|_{t=1} = [3t^2 - 2t]|_{t=1} = 1.
\end{aligned}$$

The curves of $\phi_0(t)$ and $\phi_1(t)$ with explicit expressions are plotted in Fig. 6.1.

Recall from (6.1.2) and (6.1.3) that

$$\begin{aligned}
\Phi(t) &= \sum C_k \Phi(2t - k)\Lambda^{-1} \\
\Lambda^{-1} &= \text{diag}\{1, 2, \ldots, 2^{r-1}\}.
\end{aligned}$$

Let us evaluate the dilation coefficients by taking $t = m/2$, $m \in Z$ in (6.1.4),

$$\begin{aligned}
\Phi\left(\frac{m}{2}\right) &= \sum_k C_k \Phi(m - k)\Lambda^{-1} \\
&= \sum_k C_k \, \delta_{1,m-k} \Lambda^{-1} \\
&= C_{m-1}\Lambda^{-1}.
\end{aligned} \tag{6.3.10}$$

Since Φ has a support of $[0, 2]$, all $C_k = 0$ for $k \geq 3$. For the three nonzero coefficients, we have from (6.3.10) that

$$\begin{aligned}
C_0 &= \Phi\left(\tfrac{1}{2}\right)\Lambda, \\
C_1 &= \Phi(1)\Lambda = I\Lambda = \Lambda = \text{diag}\left\{1, \tfrac{1}{2}, \ldots, \left(\tfrac{1}{2}\right)^{r-1}\right\}, \\
C_2 &= \Phi\left(\tfrac{3}{2}\right)\Lambda.
\end{aligned} \tag{6.3.11}$$

While C_1 was given in (6.3.11) for any multiplicity r, C_0 and C_2 can be obtained for the case of $r = 2$ in the next example. For arbitrary r, the general expressions of C_0 and C_2 will be derived later in this section.

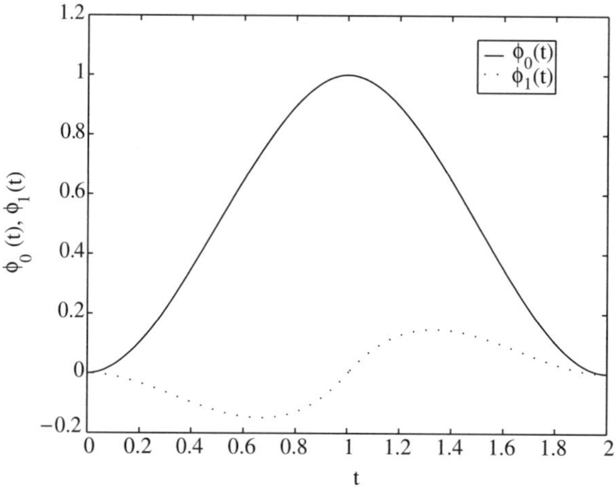

FIGURE 6.1 Multiscalets of $r = 2$ from analytic expression.

Example 2 Evaluate C_0 and C_2 for $r = 2$.

Solution

$$\Phi(t) = \begin{bmatrix} \phi_0(t) & \phi_0'(t) \\ \phi_1(t) & \phi_1'(t) \end{bmatrix}$$

$$= \begin{bmatrix} (3t^2 - 2t^3) & 6(t - t^2) \\ t^3 - t^2 & (3t^2 - 2t) \end{bmatrix} \qquad \text{for } t \le 1. \tag{6.3.12}$$

Hence

$$\Phi\left(\frac{1}{2}\right) = \begin{bmatrix} \frac{1}{2} & \frac{3}{2} \\ -\frac{1}{8} & -\frac{1}{4} \end{bmatrix},$$

$$C_0 = \Phi\left(\frac{1}{2}\right)\Lambda = \begin{bmatrix} \frac{1}{2} & \frac{3}{2} \\ -\frac{1}{8} & -\frac{1}{4} \end{bmatrix}\begin{bmatrix} 1 & 0 \\ 0 & \frac{1}{2} \end{bmatrix} = \begin{bmatrix} \frac{1}{2} & \frac{3}{4} \\ -\frac{1}{8} & -\frac{1}{8}. \end{bmatrix}. \tag{6.3.13}$$

To evaluate C_2, we need $\Phi(\frac{3}{2})$. However, we cannot set $t = \frac{3}{2}$ in (6.3.12). Instead, $\Phi(3/2)$ may be found from $\Phi(\frac{1}{2})$ by symmetry/antisymmetry about $t = 1$ (see Fig. 6.2), yielding

$$\Phi\left(\frac{3}{2}\right) = \begin{bmatrix} \frac{1}{2} & -\frac{3}{2} \\ \frac{1}{8} & -\frac{1}{4} \end{bmatrix}.$$

Therefore

$$C_2 = \Phi\left(\tfrac{3}{2}\right)\Lambda = \begin{bmatrix} \frac{1}{2} & -\frac{3}{2} \\ \frac{1}{8} & -\frac{1}{4} \end{bmatrix} \begin{bmatrix} 1 & 0 \\ 0 & \frac{1}{2} \end{bmatrix} = \begin{bmatrix} \frac{1}{2} & -\frac{3}{4} \\ \frac{1}{8} & -\frac{1}{8} \end{bmatrix}. \tag{6.3.14}$$

Next let us derive C_0 and C_2 for arbitrary r. The property (6.3.7) may be written in a matrix form as

$$|\phi((2-t)\rangle = S|\phi((t)\rangle, \tag{6.3.15}$$

where

$$S = \begin{bmatrix} 1 & & & \\ & -1 & & \\ & & \cdots & \\ & & & (-1)^{r-1} \end{bmatrix} = S^{-1}.$$

Applying the dilation equation (6.3.4) to (6.3.15), we obtain

$$\begin{aligned} \text{LHS} &= |\phi(2-t)\rangle \\ &= C_0|\phi(4-2t)\rangle + C_1|\phi(3-2t)\rangle + C_2|\phi(2-2t)\rangle \\ &= C_0|\phi[2-(2t-2)]\rangle + C_1|\phi[2-(2t-1)]\rangle + C_2|\phi(2-2t)\rangle \\ &= C_0S|\phi(2t-2)\rangle + C_1S|\phi(2t-1)\rangle + C_2S|\phi(2t)\rangle, \end{aligned}$$

where the last equality was arrived at by using the symmetry–antisymmetry property of (6.3.15).

Applying the dilation equation (6.3.4) to the right-hand side of (6.3.15), we have

$$\text{RHS} = S[C_0|\phi(2t)\rangle + C_1|\phi(2t-1)\rangle + C_2|\phi(2t-2)\rangle].$$

Equating both sides, we have

$$\begin{aligned} C_0|\phi(2t)\rangle + C_1|\phi(2t-1)\rangle| C_2\,|\phi(2t-2)\rangle = S^{-1}C_2S|\phi(2t)\rangle \\ + S^{-1}C_1S\,|\phi(2t-1)\rangle \\ + S^{-1}C_0S\,|\phi(2t-2)\rangle. \end{aligned}$$

By linear independence of translations $\phi(2t-k)$, we claim that

$$\begin{aligned} C_0 &= S^{-1}C_2S \\ &= SC_2S^{-1}. \end{aligned} \tag{6.3.16}$$

The component expression of (6.3.16) is

$$[C_0]_{ij} = (-1)^{i+j}[C_2]_{ij}.$$

As a result of (6.3.16), C_2 remains to be determined. The coefficient C_2 of arbitrary r can be obtained from the following theorem:

Theorem 1. The eigenvalues of C_2 are $(\frac{1}{2})^r, (\frac{1}{2})^{r+1}, \ldots, (\frac{1}{2})^{2r-1}$, and C_2 can be found from the similarity transformation of a diagonal matrix Γ by

$$C_2 = U^{-1}\Gamma U,$$

where the transformation matrix U is given by

$$[U]_{mn} = (-1)^{r+m-n} \frac{(r+m-1)!}{[r+m-n]!}. \tag{6.3.17}$$

Note that a similarity transform does not change eigenvalues. Therefore

$$\Gamma = \operatorname{diag}\left\{\left(\tfrac{1}{2}\right)^r, \left(\tfrac{1}{2}\right)^{r+1}, \ldots, \left(\tfrac{1}{2}\right)^{2r-1}\right\}.$$

The proof of this theorem is provided in the Appendix to this chapter.

Example 3 The piecewise cubic case $r = 2$.

$$\Gamma = \begin{bmatrix} \frac{1}{4} & 0 \\ 0 & \frac{1}{8} \end{bmatrix}$$

$$U = \begin{bmatrix} \frac{2!}{2!} & -\frac{2!}{1!} \\ -\frac{3!}{3!} & \frac{3!}{2!} \end{bmatrix} = \begin{bmatrix} 1 & -2 \\ -1 & 3 \end{bmatrix},$$

$$U^{-1} = \begin{bmatrix} 3 & 2 \\ 1 & 1 \end{bmatrix}.$$

Thus

$$C_2 = U^{-1}\Gamma U = \begin{bmatrix} \frac{1}{2} & -\frac{3}{4} \\ \frac{1}{8} & -\frac{1}{8} \end{bmatrix},$$

$$C_0 = SC_2S^{-1} = \begin{bmatrix} \frac{1}{2} & \frac{3}{4} \\ -\frac{1}{8} & -\frac{1}{8} \end{bmatrix}.$$

Recall that C_1 was given in Eq. (6.3.11) as

$$\begin{bmatrix} 1 & 0 \\ 0 & \frac{1}{2} \end{bmatrix}.$$

The resultant matrices C_0, C_1, and C_2 in this example agree exactly with (6.3.13) and (6.3.14) in Example 2. This implies that the multiscalets constructed by the analytic expressions and by the numerical (iterative or cascade) methods are identical.

Using the vector dilation equations, we obtain

$$|\phi(t)\rangle = \sum_k C_k |\phi(2t-k)\rangle$$

with given matrices C_0, C_1, and C_2, we may construct the scalets either by the iterative method or the cascade method, as in Chapter 3 for the Daubechies scalet. Figure 6.2 depicts the two multiscalets, ϕ_0 and ϕ_1; they are identical to those obtained from analytic expressions. For multiplicity $r = 3$, the corresponding lowpass matrices C_0, C_1, and C_2 can be calculated in the same manner outlined in Example 3 and are given below:

$$C_0 = \begin{bmatrix} \frac{1}{2} & \frac{15}{16} & 0 \\ -\frac{5}{32} & -\frac{7}{32} & \frac{3}{8} \\ \frac{1}{64} & \frac{1}{64} & -\frac{1}{16} \end{bmatrix},$$

$$C_1 = \begin{bmatrix} 1 & 0 & 0 \\ 0 & \frac{1}{2} & 0 \\ 0 & 0 & \frac{1}{4} \end{bmatrix}, \tag{6.3.18}$$

$$C_2 = \begin{bmatrix} \frac{1}{2} & -\frac{15}{16} & 0 \\ \frac{5}{32} & -\frac{7}{32} & -\frac{3}{8} \\ \frac{1}{64} & -\frac{1}{64} & -\frac{1}{16} \end{bmatrix}.$$

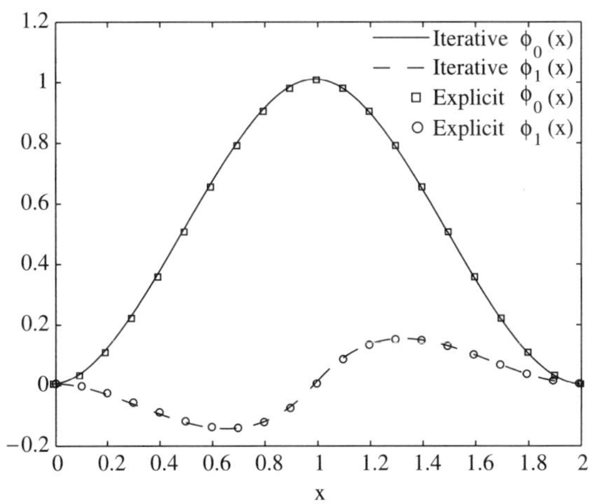

FIGURE 6.2 Multiscalets of $r = 2$, analytical and iterated.

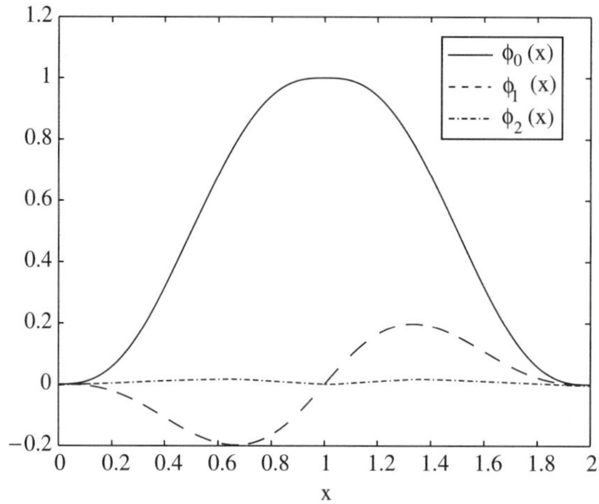

FIGURE 6.3 Multiscalets of $r = 3$.

The corresponding multiscalets are plotted in Fig. 6.3. The explicit polynomials of $\phi_0(t)$, $\phi_1(t)$ and $\phi_2(t)$ are

$$\phi_0(t) = 6t^5 - 15t^4 + 10t^3$$
$$\phi_1(t) = -3t^5 + 7t^4 - 4t^3$$
$$\phi_2(t) = \tfrac{1}{2}t^5 - t^4 + \tfrac{1}{2}t^3$$

on the interval [0, 1]. Using the symmetry/antisymmetry, we can obtain the closed form expressions on the interval [1, 2].

In general multiscalets with arbitrary r have the form

$$\phi_0(t) = p_{1,1}t^{2r-1} + p_{1,2}t^{2r-2} + \cdots + p_{1,r}t^r$$
$$\cdots = \cdots$$
$$\phi_{r-1}(t) = p_{r,1}t^{2r-1} + p_{r,2}t^{2r-2} + \cdots + p_{r,r}t^r$$

where the coefficients $p_{i,j}$ are obtained by inverting the matrix whose entries are

$$a(k,\ell) = \frac{(2r-k)!}{(2r-k-\ell+1)!}, \quad k,\ell = 1,2,\ldots,r. \tag{6.3.19}$$

Thus far we have constructed the multiscalets that are compactly supported on [0, 2]. These multiscalets do not satisfy orthogonality in the usual sense

$$\int \phi_i(t)\phi_i(t-n)\,dt = \delta_{0,n}.$$

Instead, condition (6.1.2) leads to another type of orthogonality of a Sobolev-type inner product. We define

$$\langle f, g\rangle_0 := \sum_{j=0}^{r-1}\sum_{k} \overline{f^{(j)}(k)g^{(j)}(k)}, \tag{6.3.20}$$

where the subscript 0 indicates scaling level 0, the overbar denotes the complex conjugate, and f and g are in C_0^{r-1}, which is $r-1$ times differentiable and satisfies zero boundary conditions. Then we have

$$\begin{aligned}
\langle \phi_i, \phi_k(\cdot - m)\rangle_0 &= \sum_{j=0}^{r-1}\sum_{p\in Z} \phi_i^{(j)}(p)\phi_k^{(j)}(p-m) \\
&= \sum_{j=0}^{r-1} \phi_i^{(j)}(1)\phi_k^{(j)}(1-m) \\
&= \sum_{j=0}^{r-1} \delta_{i,j}\,\delta_{k,j}\,\delta_{0,m} \\
&= \delta_{i,k}\,\delta_{0,m},
\end{aligned}$$

where we have used the property of (6.3.5) that

$$\phi_i^{(j)}(1) = \delta_{i,j} \qquad \text{for } j = 0, 1, \ldots, r-1,$$

and

$$\begin{aligned}
\phi(0) &= 0, \\
\phi(2) &= 0.
\end{aligned}$$

In order to simplify notation, we have denoted ϕ, without subscript, as a vector. The simplified notation of ϕ and ψ will be carried out throughout the chapter. The result above may be written in vector form as

$$\langle \phi, \phi^T(\cdot - m)\rangle_0 = \delta_{0,m} I,$$

or in matrix form as

$$\sum_k \Phi(k)\Phi^T(k-m) = \delta_{0,m} I. \tag{6.3.21}$$

In a similar manner we define at level p,

$$\langle f(t), g(t)\rangle_p := \sum_{j=0}^{r-1}\sum_{k} \overline{f^{(j)}(t)\, g^{(j)}(t)}|_{t=2^{-p}k}, \qquad p \in Z. \tag{6.3.22}$$

Unfortunately, we do not have orthogonality of $\{\phi(t-m)\}$ with regard to these inner products.

For $p = 1$, we obtain

$$\begin{aligned}\langle \phi, \phi^T(\cdot - m)\rangle_1 &= \sum_k \Phi\left(\frac{k}{2}\right)\Phi^T\left(\frac{k}{2}-m\right)\\ &= \sum_k C_{k-1}\Lambda^{-1}(C_{k-2m-1}\Lambda^{-1})^T\\ &= \sum_k C_{k-1}\Lambda^{-2}C_{k-2m-1}.\end{aligned}$$

Note that for $m = 1$,

$$\langle \phi, \phi^T(t-1)\rangle_1 = \sum_k C_{k-1}\Lambda^{-2}C_{k-3}^T = C_2\Lambda^{-2}C_0^T.$$

For $m = 0$,

$$\begin{aligned}\langle \phi, \phi^T\rangle_1 &= C_0\Lambda^{-2}C_0^T + C_1\Lambda^{-2}C_1 + C_2\Lambda^{-2}C_2\\ &= C_0\Lambda^{-2}C_0^T + I^2 + C_2\Lambda^{-2}C_2.\end{aligned}$$

For $m = -1$,

$$\langle \phi, \phi^T(t+1)\rangle = \sum_k C_{k-1}\Lambda^{-2}C_{k+1}^T.$$

Following the same derivation, we may show that

$$\langle \phi(2t), \phi(2t-m)\rangle_1 = \delta_{0,m}\Lambda^{-2}.$$

In fact

$$\begin{aligned}\langle \phi_p(2t), \phi_q(2t-m)\rangle_1 &= \sum_{j=0}^{r-1}\sum_{k\in Z} 2^j\phi_p^{(j)}\left(\tfrac{1}{2}\cdot 2k\right)2^j\phi_q^{(j)}\left(\tfrac{1}{2}\cdot 2k-m\right)\\ &= \sum_{j=0}^{r-1}\sum_{k\in Z} 2^{2j}\phi_p^{(j)}(k)\phi_q^{(j)}(k-m)\\ &= \sum_{j=0}^{r-1}\sum_{k\in Z} 2^{2j}\,\delta_{j,p}\,\delta_{1,k}\,\delta_{j,q}\,\delta_{k,m+1}\\ &= 2^{2p}\sum_{k\in Z}\delta_{p,q}\,\delta_{1,k}\,\delta_{k,m+1}.\end{aligned}$$

Hence

$$\langle \phi(2t), \phi(2t-m) \rangle_1 = \delta_{0,m} \Lambda^{-2}. \tag{6.3.23}$$

Equation (6.3.23) will be used in Section 6.5.

6.4 ORTHOGONAL MULTIWAVELETS $\breve{\psi}(t)$

In the previous section the orthogonal multiscalets were constructed. Naturally, one expects to build the corresponding multiwavelets. Multiwavelets $\breve{\psi}_0(t), \ldots, \breve{\psi}_{r-1}(t)$ are orthogonal to multiscalets $\phi_0(t), \ldots, \phi_{r-1}(t)$, and they satisfy the dilation equation

$$\begin{bmatrix} \breve{\psi}_0(t) \\ \vdots \\ \breve{\psi}_{r-1}(t) \end{bmatrix} = \sum_{k=0}^{4} [G_k] \begin{bmatrix} \phi_0(2t-k) \\ \vdots \\ \phi_{r-1}(2t-k) \end{bmatrix}. \tag{6.4.1}$$

Note that there are five nonzero matrices G. The support of $\Phi(2t-k)$ is $[k/2, (k+2)/2]$. With coefficients $G_0, \ldots, G_4$, the support of $\breve{\psi}(t)$ will be $[0, 3]$. In fact the left endpoint is for $k = 0$, and the right endpoint is for $k = 4$.

The orthogonality against $\Phi(t)$ and its translations provide equations for the G, and they are

$$\begin{aligned} \breve{\Psi}(t)\Phi^T(t+1) &= 0, \\ \breve{\Psi}(t)\Phi^T(t) &= 0, \\ \breve{\Psi}(t)\Phi^T(t-1) &= 0, \\ \breve{\Psi}(t)\Phi^T(t-2) &= 0. \end{aligned} \tag{6.4.2}$$

Note that

$$\begin{aligned} \text{supp}\{\Phi(t)\} &= [0, 2], \\ \text{supp}\{\breve{\Psi}(t)\} &= [0, 3]. \end{aligned}$$

The translations of $k > 2$ or $k < -1$ in (6.4.2) shift the functions so that there is no overlap between Φ and $\breve{\Psi}$. Therefore they are ruled out from (6.4.2).

Equation (6.4.2) involves integrals of $\breve{\psi}_i(t)\phi_j(t-k)$. By using the dilation equations of both scalets and wavelets, we integrate $\phi_i(2t-k)\phi_j(2t-m)$. A change of variable converts these inner product integrals into

$$\int \phi_i(2t-k)\phi_j(2t-m)\, dt = \tfrac{1}{2} \int \phi_i(t)\phi_j(t-m+k)\, dt, \tag{6.4.3}$$

where ϕ_i and ϕ_j are supported on $[0, 2]$. Hence the only inner products needed are the two matrices

$$
\begin{aligned}
X &= \Phi(t)\Phi^T(t), \\
Y &= \Phi(t)\Phi^T(t-1) = \Phi(t+1)\Phi^T(t).
\end{aligned} \tag{6.4.4}
$$

To avoid an explicit evaluation of polynomials $\phi_i(t)$ and their inner products, we substitute the dilation equations into (6.4.4) and impose (6.4.3) to convert all arguments involving $2t$ into t. The resultant two matrix equations are

$$
\begin{aligned}
2X &= C_0XC_0^T + C_1Y^TC_0^T + C_0YC_1^T + C_1XC_1^T \\
&\quad + C_2Y^TC_1^T + C_1YC_2^T + C_2XC_2^T, \\
2Y &= C_1YC_0^T + C_2XC_0^T + C_2YC_1^T.
\end{aligned}
$$

These equations determine X and Y up to a scalar factor, and the X and Y enter the orthogonality equation (6.4.2). Substituting in (6.4.2) for $\breve{\Psi}(t)$ introduces the unknown G, and substituting for $\Phi(t)$ links the known C. After some algebra we obtain

$$
\begin{aligned}
&G_0(Y^TC_1^T + XC_2^T) + G_1Y^TC_2^T = 0, \\
&G_0(XC_0^T + YC_1^T) + G_1(Y^TC_0^T + XC_1^T + YC_2^T) \\
&\quad + G_2(Y^TC_1^T + XC_2^T) + G_3Y^TC_2^T = 0, \\
&G_1YC_0^T + G_2(XC_0^T + YC_1^T) + G_3(Y^TC_0^T + XC_1^T + YC_2^T) \\
&\quad + G_4(Y^TC_1^T + XC_2^T) = 0, \\
&G_3YC_0^T + G_4(XC_0^T + YC_1^T) = 0.
\end{aligned} \tag{6.4.5}
$$

The equations above form a system of $4r^2$ homogeneous equations that consist of $5r^2$ entries in $G_0, \ldots, G_4$. We pick out the solution with $G_2 = I$. In this case, symmetry–antisymmetry is also held for the wavelets. The property $C_0 = SC_2S$ may be extended to the G as

$$
\begin{cases}
G_0 = \mathrm{SG_4S} \\
G_1 = \mathrm{SG_3S},
\end{cases}
$$

where $S = \mathrm{diag}\{1, -1, \ldots, (-1)^{r-1}\}$. As a result the first two equations in (6.4.5) become identical to the remaining two. Employing this pattern of the G and also

$$
\begin{cases}
X = \mathrm{SXS} \\
Y = \mathrm{SY^TS},
\end{cases}
$$

we obtain

$$
\begin{aligned}
[X]_{ij} = [X]_{ji} &= \int \phi_i(t)\phi_j(t)\,dt \\
&= \int \phi_i(2-t)\phi_j(2-t)\,dt
\end{aligned}
$$

$$= (-1)^{i+j} \int \phi_i(t)\phi_j(t)\,dt$$
$$= (-1)^{i+j}[X]_{ij}$$

and

$$\begin{aligned}
[Y]_{ij} &= \int \phi_i(t)\phi_j(t-1)\,dt \\
&= \int \phi_i(3-t)\phi(2-t)\,dt \\
&= (-1)^{i+j} \int \phi_i(t-1)\phi_j(t)\,dt \\
&= (-1)^{i+j}[Y]_{ji}.
\end{aligned}$$

Finally, Eq. (6.4.5) reduces to two matrix equations for two unknowns G_3 and G_4. Using Matlab, we have solved for the unknowns and listed them below

$$\begin{aligned}
G_0 &= \begin{bmatrix} -\frac{17}{98} & -\frac{89}{98} \\ \frac{79}{6438} & \frac{137}{2146} \end{bmatrix}, \\
G_1 &= \begin{bmatrix} -\frac{16}{49} & -\frac{286}{49} \\ \frac{152}{3219} & \frac{550}{1073} \end{bmatrix}, \\
G_2 &= \begin{bmatrix} 1 & 0 \\ 0 & 1 \end{bmatrix}, \\
G_3 &= \begin{bmatrix} -\frac{16}{49} & \frac{286}{49} \\ -\frac{152}{3219} & \frac{550}{1073} \end{bmatrix}, \\
G_4 &= \begin{bmatrix} -\frac{17}{98} & \frac{89}{98} \\ -\frac{79}{6438} & \frac{137}{2146} \end{bmatrix}.
\end{aligned} \tag{6.4.6}$$

The orthogonal wavelets are constructed according to Eq. (6.4.1). They are plotted in Fig. 6.4. By construction, the scalets are orthogonal to the wavelets and their integer translations. Hence

$$\begin{aligned}
V_1 &= V_0 \oplus W_0, \\
V_2 &= V_0 \oplus W_0 \oplus W_1, \\
W_i &\perp W_j \text{ for } i \neq j.
\end{aligned}$$

Unfortunately, a wavelet is not orthogonal to its translations, nor a scalet to its translations.

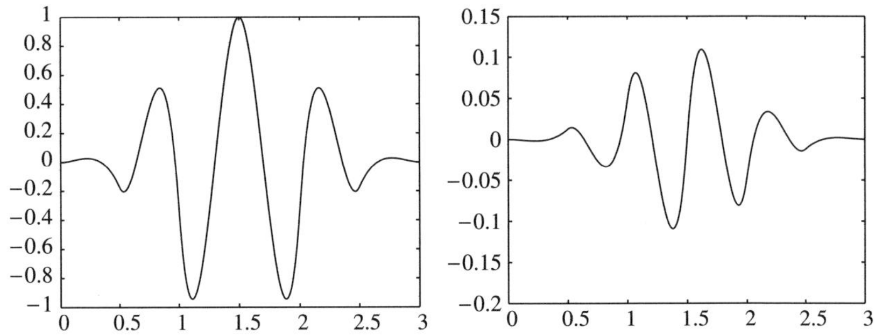

FIGURE 6.4 Intervallic multiwavelets $\breve{\psi}_0(t)$ and $\breve{\psi}_1(t)$ of $r = 2$.

6.5 INTERVALLIC MULTIWAVELETS $\psi(t)$

The orthogonal multiwavelets constructed in the previous section are orthogonal to the multiscalets in the standard L^2 sense. However, these multiwavelets are oscillatory and have relatively wide supports. Most inconveniently, they are not orthogonal to their translations. To improve the properties of the multiwavelet, Walter introduced the orthogonal finite element multiwavelets [4], which we referred to as the intervallic multiwavelets to avoid confusion with the finite element method (FEM) in electromagnetics. This multiwavelet family is comprised of the intervallic multiwavelet and its dual, namely the intervallic dual multiwavelets.

As usual, we denote the closed linear span V_p in $L^2(R)$ of

$$\{\phi_0(2^p t - k), \ldots, \phi_{r-1}(2^p t - k)\},$$

with an inner product $\langle , \rangle_p$. This is equivalent to the L^2 inner product in V_p. We now introduce a biorthogonal pair of wavelets $(\psi, \tilde{\psi})$, both of which belong to V_1. The first is in $V_0^{\perp}$ and is given by

$$\psi(t) = \sum_k D_k \phi(2t - k), \tag{6.5.1}$$

where $\psi(t) = [\psi_1(t), \psi_2(t), \ldots, \psi_n(t)]^T$.

Since $\psi_i \in V_1$, we use its inner product to determine the D_k so that $\psi_1(t)$ is orthogonal in the sense of V_1 to V_0. Namely we need

$$\langle \psi, \phi^T(\cdot - m) \rangle_1 = 0 \qquad \text{for all } m \in Z.$$

The LHS can be evaluated by using (6.5.1), and the weighted orthogonality of (6.3.23),

$$\langle \phi(2t), \phi^T(2t - m) \rangle_1 = \delta_{0,m} \Lambda^{-2}.$$

Derivation We had

$$\langle \phi(t), \phi^T(t - m) \rangle_0 = \delta_{0,m} I.$$

Thus

$$\langle \psi, \phi^T(\cdot - m)\rangle_1 = \sum_k D_k \langle \phi(2t-k), \phi^T(t-m)\rangle_1 .$$

Employing the dilation equation

$$\begin{aligned}\phi^T(t-m) &= \left(\sum_j C_j \phi(2t-2m-j)\right)^T \\ &= \sum_j \phi^T(2t-2m-j)_{1\times n} C_j^T ,\end{aligned}$$

we have

$$\begin{aligned}\langle \psi, \phi^T(\cdot - m)\rangle_1 &= \sum_k \sum_j D_k \underbrace{\langle \phi(2t-k), \phi^T(2t-2m-j)\rangle_1}_{\delta_{k,2m+j}\Lambda^{-2}} C_j^T \\ &= \sum_j D_{2m+j}\Lambda^{-2} C_j^T \\ &= 0, \qquad m \in Z.\end{aligned}$$

The previous inner products are identically zero for $m < -1$, or $m > 4$, because there will be no overlap between ψ and ϕ. For $-1 \le m \le 4$, we end with the following equations:

$$\begin{aligned}
m = -1: &\quad D_0\Lambda^{-2}C_2^T = 0 \\
m = 0: &\quad D_0\Lambda^{-2}C_0^T + D_1\Lambda^{-2}C_1^T + D_2\Lambda^{-2}C_2^T = 0 \\
m = 1: &\quad D_2\Lambda^{-2}C_0^T + D_3\Lambda^{-2}C_1^T + D_4^T\Lambda^{-2}C_2^T = 0, \\
m = 2: &\quad D_4\Lambda^{-2}C_0^T = 0, \\
m = 3: &\quad D_6\Lambda^{-2}C_0^T = 0.
\end{aligned}$$

Since both C_0 and C_2 are nonsingular, it follows that $D_0 = 0 = D_4 = D_6$. Hence only two equations are left

$$\begin{cases} D_1\Lambda^{-2}C_1^T = -D_2\Lambda^{-2}C_2^T \\ D_2\Lambda^{-2}C_0^T = -D_3\Lambda^{-2}C_1^T . \end{cases}$$

Also $C_1 = \Lambda$ from (6.3.11). One solution is to take

$$D_2 = C_1 = \Lambda,$$

which makes

$$\begin{cases} D_1\Lambda^{-1} = -\Lambda^{-1}C_2^T \to D_1 = -\Lambda^{-1}C_2^T\Lambda \\ \Lambda^{-1}C_0^T = -D_3\Lambda^{-1}. \end{cases}$$

In general,

$$D_m = (-1)^m \Lambda^{-1} C_{3-m}^T \Lambda, \qquad m \in Z. \tag{6.5.2}$$

Verification. One can verify from (6.5.2) that

$$D_2 = \Lambda^{-1} C_1^T \Lambda,$$
$$D_0 = \Lambda^{-1} C_3^T \Lambda = 0 \qquad \text{because } C_3 = 0.$$

Since only $C_0, C_1, C_2 \neq 0$, ($m = 3, 2, 1$ in (6.5.2)), we obtain

$$\begin{aligned}\psi(t) &= D_1\phi(2t-1) + D_2\phi(2t-2) + D_3\phi(2t-3)\\ &= -\Lambda^{-1}C_2^T\Lambda\phi(2t-1) + \Lambda\phi(2t-2) - \Lambda^{-1}C_0^T\Lambda\phi(2t-3).\end{aligned} \tag{6.5.3}$$

Figure 6.5 illustrates the two multiwavelets, ψ_0 and ψ_1, obtained by the iteration method and by the explicit polynomial expressions of Example 2 in Section 6.3. Noticing that supp $\psi(t) = [\frac{1}{2}, \frac{5}{2}]$, and $\psi(t) = \sum D_k\phi(2t-k)$, we arrive at

$$\begin{aligned}\psi^{(j)}(t)\,|_{n/2} = \psi^{(j)}\left(\frac{n}{2}\right) &= \sum_k D_k\phi^{(j)}(n-k)2^j\\ &= D_{n-1}\phi^{(j)}(1)2^j; \qquad j = 0, 1, \ldots, r-1;\; n = 2, 3, 4.\end{aligned} \tag{6.5.4}$$

For integer values of t only $\phi^{(j)}(1)$ may be nonzero, that is to have from (6.5.4) a sampling property

$$\psi_q^{(j)}\left(p + \tfrac{3}{2}\right) = \delta_{q,j}\,\delta_{p,0}. \tag{6.5.5}$$

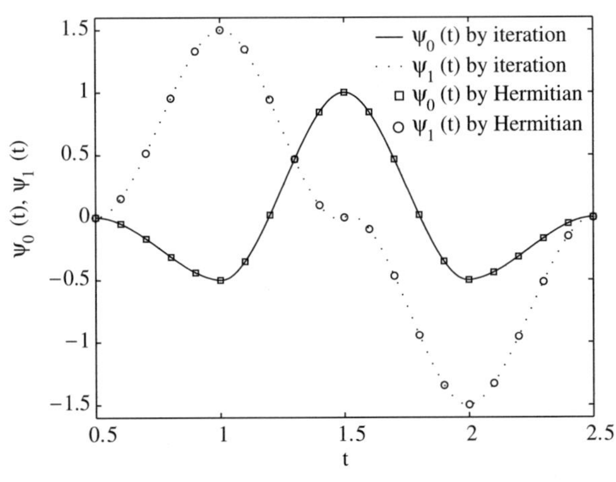

FIGURE 6.5 Intervallic multiwavelets of $r = 2$.

Show.

(1) For $n = 3$ in (6.5.4), namely $p = 0$ in (6.5.5), we have from (6.5.4),

$$\psi^{(j)}\left(\tfrac{3}{2}\right) = D_2\phi^{(j)}(1)2^j, \quad D_2 = \Lambda$$

that is,

$$\begin{bmatrix} \psi_0^{(j)}(t) \\ \cdots \\ \psi_{r-1}^{(j)}(t) \end{bmatrix}_{t=\frac{3}{2}} = \begin{bmatrix} 1 & & & \\ & 2^{-1} & & \\ & & \cdots & \\ & & & 2^{-(r-1)} \end{bmatrix} \begin{bmatrix} \phi_0^{(j)}(1) \\ \cdots \\ \phi_{r-1}^{(j)}(1) \end{bmatrix} 2^j.$$

The $(q+1)$th element

$$\psi_q^{(j)}(1) = 2^{-q}\phi_q^{(j)}(1)2^j = 2^{-q}\,\delta_{q,j}2^j = \delta_{q,j}, \qquad q = 0, 1, \ldots, (r-1).$$

(2) For $p \neq 0$, say $p = 1$ in (6.5.5), we have

$$\psi_q^{(j)}\left(\tfrac{5}{2}\right), \qquad n = 5.$$

$D_{n-1} = D_4 = 0$ in (6.5.4) which makes $\psi_q^{(j)}(\frac{5}{2}) = 0$. This is in agreement with (6.5.5), that is,

$$\psi_q^{(j)}\left(p + \tfrac{3}{2}\right) = \delta_{q,j}\,\delta_{p,0} = 0. \qquad \square$$

Regrettably, the wavelet $\psi(t)$ is not orthogonal to its translations $\psi(t - \ell)$ in the Sobolev sampling sense. Hence we introduce the dual multiwavelets. The dual multiwavelets $\tilde{\psi}$ are related to ϕ by

$$\begin{bmatrix} \phi_0(2t-2) \\ \cdots \\ \cdots \\ \cdots \\ \phi_{r-1}(2t-2) \end{bmatrix} = \begin{bmatrix} 1 & & & \\ & 2 & & \\ & & \cdots & \\ & & & 2^{(r-1)} \end{bmatrix} \begin{bmatrix} \tilde{\psi}_0(t) \\ \cdots \\ \cdots \\ \tilde{\psi}_{r-1}(t) \end{bmatrix} \to \phi_j(2t-2) = 2^j\tilde{\psi}_j(t),$$

$$j = 0, 1, \ldots, r-1.$$

Detailed study of the dual multiwavelets is deferred to Section 6.7.

6.6 MULTIWAVELET EXPANSION

Let us expand $f(t)$ in terms of $\phi_p(t-k)$:

$$f(t) = \sum_{p=0}^{r-1}\sum_{k\in Z} a_{k,p}\phi_p(t-k), \tag{6.6.1}$$

where the coefficients

$$a_{k,p} = f^{(p)}(k+1). \tag{6.6.2}$$

Show. Multiplying both sides of (6.6.1) by $\phi_q(t)$ and taking the inner product $\langle\ ,\ \rangle_0$, we arrive at

$$\langle \phi_q(t), f(t)\rangle_0 = \sum_{p=0}^{r-1}\sum_{k\in Z} a_{k,p}\langle \phi_q(t), \phi_p(t-k)\rangle_0, \tag{6.6.3}$$

where

$$\langle \phi_q(t), \phi_p(t-k)\rangle_0 = \sum_{j=0}^{r-1}\sum_{\alpha\in Z} \phi_q^{(j)}(\alpha)\phi_p^{(j)}(\alpha-k).$$

Taking a close look of (6.6.3), we have the following:

$$\begin{aligned}
\text{RHS} &= \sum_{p=0}^{r-1}\sum_{k\in Z} a_{k,p}\sum_{j=0}^{r-1}\sum_{\alpha\in Z}\phi_q^{(j)}(\alpha)\phi^{(j)}(\alpha-k)\\
&= \sum_{k\in Z}\sum_{p=0}^{r-1} a_{k,p}\sum_{j=0}^{r-1}\phi_q^{(j)}(1)\phi_p^{(j)}(1-k)\\
&= \sum_{k\in Z}\sum_{p=0}^{r-1} a_{k,p}\,\delta_{p,q}\,\delta_{0,k}\\
&= \sum_{k\in Z} a_{k,q}\,\delta_{0,k}
\end{aligned} \tag{6.6.4}$$

$$\begin{aligned}
\text{LHS} &= \sum_{j=0}^{r-1}\sum_{\ell\in Z}\phi_q^{(j)}(\ell)f^{(j)}(\ell)\\
&= \sum_{\ell\in Z}\sum_{j=0}^{r-1}\delta_{j,q}\,\delta_{1,\ell}f^{(j)}(\ell)\\
&= \sum_{\ell\in Z} f^{(q)}(\ell)\,\delta_{1,\ell}\\
&= \sum_{k\in Z} f^{(q)}(k+1)\,\delta_{0,k},
\end{aligned} \tag{6.6.5}$$

where we have used $\ell = k+1$. Comparing (6.6.4) and (6.6.5), we obtain

$$a_{k,q} = f^{(q)}(k+1). \qquad \square$$

Next, if we expand $f(t)$ in terms of $\phi_p(2t-k)$ as

$$f(t) = \sum_{p=0}^{r-1}\sum_{k} a_{k,p}^{1}\phi_p(2t-k), \tag{6.6.6}$$

then the coefficients

$$a^1_{k,p} = 2^{-p} f^{(p)}(2^{-1}(k+1)). \tag{6.6.7}$$

Show. Multiplying both sides of the expansion (6.6.6) by $\phi_q(2t)$ and performing the inner product operation $\langle\ ,\ \rangle_1$, we obtain

$$\langle \phi_q(2t), f(t)\rangle_1 = \sum_{p=0}^{r-1} \sum_{k\in Z} a^1_{k,p} \langle \phi_q(2t), \phi_p(2t-k)\rangle_1 \tag{6.6.8}$$

whereby, from (6.3.22),

$$\langle \phi_q(2t), \phi_p(2t-k)\rangle_1 = \sum_{j=0}^{r-1} \sum_{\alpha\in Z} 2^{2j} \phi_q^{(j)}(\alpha)\phi_p^{(j)}(\alpha-k).$$

From (6.6.8) we have

$$\begin{aligned}
\text{RHS} &= \sum_{p=0}^{r-1} \sum_{k\in Z} a^1_{k,p} \sum_{j=0}^{r-1} 2^{2j} \phi_q^{(j)}(1)\phi_p^{(j)}(1-k) \\
&= \sum_{k\in Z} \sum_{p=0}^{r-1} 2^{p+q} a^1_{k,p}\, \delta_{p,q}\, \delta_{0,k} \\
&= \sum_{k\in Z} 2^{2q} a^1_{k,q}\, \delta_{0,k}
\end{aligned} \tag{6.6.9}$$

$$\begin{aligned}
\text{LHS} &= \langle \phi_q(2t), f(t)\rangle_1 \\
&= \sum_{j=0}^{r-1} \sum_{\ell\in Z} 2^j \phi_q^{(j)}\left(\frac{2\ell}{2}\right) f^{(j)}\left(\frac{1}{2}\ell\right) \\
&= \sum_{j=0}^{r-1} \sum_{\ell\in Z} 2^j\, \delta_{j,q}\, \delta_{1,\ell} f^{(j)}\left(\frac{\ell}{2}\right) \\
&= \sum_{\ell\in Z} 2^q f^{(q)}\left(\frac{\ell}{2}\right) \delta_{1,\ell} \\
&= \sum_{k\in Z} 2^q f^{(q)}\left(\frac{k+1}{2}\right) \delta_{0,k}.
\end{aligned} \tag{6.6.10}$$

Comparing (6.6.10) with (6.6.9), we find that

$$a^1_{k,q} = 2^{-q} f^{(q)}\left(\frac{k+1}{2}\right). \qquad \square$$

Let us denote as $f^0 \in V_0$ the projection of f onto V_0, namely

$$f^0(t) = \sum_{p=0}^{r-1} \sum_{k\in Z} f^{(p)}(k+1)\phi_p(t-k),$$

and let $f \in V_1$ with expansion (6.6.6). Hence the difference

$$f(t) - f^0(t) = \sum_{j=0}^{r-1} \left\{ \sum_k f^{(j)}\left(k + \tfrac{1}{2}\right) 2^{-j} \phi_j(2t - 2k) \right.$$
$$\left. + \sum_k f^{(j)}(k+1)[2^{-j}\phi(2t - 2k - 1) - \phi_j(t - k)] \right\}.$$

The first summation on the RHS of the previous equation is related to the intervallic dual multiwavelet.

6.7 INTERVALLIC DUAL MULTIWAVELETS $\tilde{\psi}(t)$

The intervallic dual wavelet is defined as

$$\tilde{\psi}(t) = \Lambda\phi(2t - 2). \tag{6.7.1}$$

Lemma 3. Let ψ and $\tilde{\psi}$ be defined by (6.5.3) and (6.7.1) respectively. Then:

(i) $\psi_p^{(j)}(k + \frac{3}{2}) = \delta_{p,j}\,\delta_{k,0},\ j, p = 0, \ldots, r-1,\ k \in Z.$
(ii) $\tilde{\psi}_p^{(j)}(k + \frac{3}{2}) = \delta_{p,j}\,\delta_{k,0},\ j, p = 0, \ldots, r-1,\ k \in Z.$
(iii) $\tilde{\psi}_p^{(j)}(k) = 0,\ j, p = 0, \ldots, r-1,\ k \in Z.$
(iv) $\langle \psi_j(\cdot - k), \tilde{\psi}_j(\cdot - \ell)\rangle_1 = \delta_{k,\ell},\ k, \ell \in Z.$

We have proved (i) as (6.5.5). Property (ii) can be verified in the same manner. Let us verify Property (iii).

Proof. From (6.7.1)

$$\tilde{\psi}^{(j)}(t) = (\Lambda\phi(2t-2))^{(j)} = 2^j \begin{bmatrix} 1 & & & \\ & 2^{-1} & & \\ & & \cdots & \\ & & & 2^{-(r-1)} \end{bmatrix} \begin{bmatrix} \phi_0^{(j)}(2t-2) \\ \phi_1^{(j)}(2t-2) \\ \cdots \\ \phi_{r-1}^{(j)}(2t-2) \end{bmatrix}.$$

For $2t - 2 = k$, the $(p+1)$th element

$$2^j \cdot 2^{-p}\phi_p^{(j)}(2t-2)|_{(2t-2)=k} = 2^{j-p}\phi_p^{(j)}(k) = 2^{j-p}\,\delta_{j,p}\,\delta_{1,k}.$$

For $t \in Z$, $2t - 2 \neq 1$. Therefore $\tilde{\psi}(k) = 0$. □

We now summarize the multiwavelet properties into a theorem.

Theorem 2. If $W_0 = \text{closure}\{\psi_j(t-k)\}$, $\tilde{W}_0 = \text{closure}\{\tilde{\psi}_j(t-k)\}$, then $W_0 \perp V_0$, $\tilde{W}_0 \bigcap V_0 = \{0\}$, and $V_1 = \tilde{W}_0 \bigoplus V_0$.

Proof. Space W_0 takes $\{\psi_i(t-k)\}$ as a Riesz basis. Since $\psi_j(t-k) \in V_0^{\perp}$, we have $W_0 \perp V_0$. From (iii), each element $f \in \tilde{W}_0$ and its $(r-1)$ derivatives must be zero. Now, if $f \in V_0$ as well, then

$$f(t) = \sum_{j=0}^{r-1} \sum_k a_{k,j} \phi_j(t-k),$$

where

$$a_{k,j} = f^{(j)}(k+1) = 0.$$

Hence

$$f \equiv 0.$$

The last statement of the theorem remains to be proved. Assume that $f \in V_1$; then

$$f(t) = \sum_{j=0}^{r-1} \sum_k a^1_{k,j} \phi_j(2t-k),$$

where

$$a^1_{k,j} = f^{(j)}\left(\frac{k+1}{2}\right) 2^{-j}. \tag{6.7.2}$$

Let $f^0 \in V_0$ be the projection of f onto V_0, namely

$$f^0(t) = \sum_{j=0}^{r-1} \sum_k f^{(j)}(k+1) \phi_j(t-k). \tag{6.7.3}$$

Note that $f(t)$ in (6.7.2) has twice as many points as $f^0(t)$ in (6.7.3). Hence the difference

$$\begin{aligned} f(t) - f^0(t) = \sum_{j=0}^{r-1} \Bigg\{ & \sum_m f^{(j)}\left(m + \tfrac{1}{2}\right) 2^{-j} \phi_j(2t-2m) \\ & + \sum_m f^{(j)}(m+1)[2^{-j}\phi_j(2t-2m-1) - \phi_j(t-m)] \Bigg\}, \end{aligned} \tag{6.7.4}$$

where we have used $k = \{^{2m}_{2m+1}$ in the $f(t)$ expansion, and $k = m$ in the $f^0(t)$ expansion. Notice from (6.7.1) that

$$\phi(2t-2) = \Lambda^{-1} \tilde{\psi}(t)$$

that is,

$$\phi_j(2t-2) = 2^j \tilde{\psi}_j(t).$$

Hence we may simplify the first summation over m on the right-hand side of (6.7.4) as

$$\begin{aligned}
&\sum_m f^{(j)}\left((m-1)+\tfrac{3}{2}\right)2^{-j}\phi_j(2t-2-2(m-1)) \\
&\quad=\sum_k f^{(j)}\left(k+\tfrac{3}{2}\right)2^{-j}\phi_j(2t-2-2k) \\
&\quad=\sum_k f^{(j)}\left(k+\tfrac{3}{2}\right)2^{-j}2^j\tilde{\psi}_j(t-k) \\
&\quad=\sum_k f^{(j)}\left(k+\tfrac{3}{2}\right)\tilde{\psi}_j(t-k).
\end{aligned}$$

As a result (6.7.4) can be written in a vector form as

$$f(t)-f^0(t)=\sum_k \mathbf{f}^T\left(k+\frac{3}{2}\right)\tilde{\psi}(t-k)+\mathbf{f}^T(k+1)[\Lambda\phi(2t-2k-1)-\phi(t-k)],$$

where superscript T denotes the transpose. On the other hand, by using (6.3.4), we have

$$\begin{aligned}
\Lambda\phi(2t-1)-\phi(t)&=\Lambda\phi(2t-1)-[C_0\phi(2t)+C_1\phi(2t-1)+C_2\phi(2t-2)] \\
&=\Lambda\phi(2t-1)-C_0\phi(2t)-\Lambda\phi(2t-1)-C_2\phi(2t-2) \\
&=-C_0\phi(2t)-C_2\phi(2t-2) \\
&=-C_0\phi(2(t+1)-2)-C_2\phi(2t-2) \\
&=-C_0\Lambda^{-1}\tilde{\psi}(t+1)-C_2\Lambda^{-1}\tilde{\psi}(t).
\end{aligned}$$

That is,

$$\Lambda\phi(2(t-k)-1)-\phi(t-k)=-C_0\Lambda^{-1}\tilde{\psi}(t-k+1)-C_2\Lambda^{-1}\tilde{\psi}(t-k).$$

Finally

$$\begin{aligned}
f^1(t):&=f(t)-f^0(t) \\
&=\sum_k \mathbf{f}^T\left(k+\tfrac{3}{2}\right)\tilde{\psi}(t-k)-\mathbf{f}^T(k+1) \\
&\qquad\times\left[C_0\Lambda^{-1}\tilde{\psi}(t-k+1)+C_2\Lambda^{-1}\tilde{\psi}(t-k)\right] \\
&=\sum_k\left\{\mathbf{f}^T\left(k+\tfrac{3}{2}\right)-\mathbf{f}^T(k+2)C_0\Lambda^{-1}-\mathbf{f}^T(k+1)C_2\Lambda^{-1}\right\}\tilde{\psi}(t-k),
\end{aligned}\tag{6.7.5}$$

where $\mathbf{f}(\cdot)$ is a vector and $\mathbf{f}^T(\cdot):=[f^{(0)}(\cdot),f^{(1)}(\cdot),\ldots,f^{(r-1)}(\cdot)]$. This equation indicates clearly that

$$f^0\in V_0,\quad f^1\in\tilde{W}_0,$$

and therefore

$$f=f^0+f^1. \qquad \square$$

Expression (6.7.5) can be written in terms of $\psi(t-k)$. As a matter of fact, we know from (6.5.3) that

$$\begin{aligned}\langle \psi(t-k), f\rangle_1 &= \langle D_1\phi(2t-2k-1), f\rangle_1 + \langle D_2\phi(2t-2k-2), f\rangle_1 \\ &\quad + \langle D_3\phi(2t-2k-3), f\rangle_1 \\ &= D_1\Lambda^{-1}\mathbf{f}(k+1) + D_2\Lambda^{-1}\mathbf{f}\left(k+\tfrac{3}{2}\right) + D_3\Lambda^{-1}\mathbf{f}(k+2) \\ &= (-1)\Lambda^{-1}C_2^T\mathbf{f}(k+1) + \Lambda\Lambda^{-1}\mathbf{f}\left(k+\tfrac{3}{2}\right) \\ &\quad - \Lambda^{-1}C_0^T\Lambda\Lambda^{-1}\mathbf{f}(k+2) \\ &= -\Lambda^{-1}C_2^T\mathbf{f}(k+1) + \mathbf{f}\left(k+\tfrac{3}{2}\right) - \Lambda^{-1}C_0^T\mathbf{f}(k+2) \qquad (6.7.6)\end{aligned}$$

which agrees with (6.7.5). In the derivation of (6.7.6) we have used (6.3.23), namely

$$\langle \phi(2t), \phi(2t-m)\rangle_1 = \delta_{0,m}\Lambda^{-2},$$

in particular, its component form

$$\langle \phi_q(2t), \phi_p(2t-m)\rangle_1 = \delta_{q,p}\,\delta_{0,m}2^{q+p}.$$

Hence

$$\begin{aligned}\langle \phi_q(2t), f(t)\rangle_1 &= \sum_{p=0}^{r-1}\sum_{k\in Z} 2^{-p} f^{(p)}\left(\frac{k+1}{2}\right)\langle \phi_q(2t), \phi_p(2t-k)\rangle_1 \\ &= \sum_{p=0}^{r-1}\sum_{k\in Z} 2^{p} f^{(p)}\left(\frac{k+1}{2}\right)\delta_{0,k},\end{aligned}$$

that is,

$$\begin{aligned}\langle \phi(2t), f(t)\rangle_1 &= \Lambda^{-1}\mathbf{f}\left(\frac{k+1}{2}\right)|_{k=0} \\ &= \Lambda^{-1}\mathbf{f}\left(\frac{1}{2}\right).\end{aligned}$$

As a consequence

$$\begin{aligned}\langle \phi(2t-(2m+1)), f\rangle_1 &= \Lambda^{-1}\mathbf{f}\left(\frac{(2m+1)+1}{2}\right) = \Lambda^{-1}\mathbf{f}(m+1), \\ \langle \phi(2t-(2m+2)), f\rangle_1 &= \Lambda^{-1}\mathbf{f}\left(m+\frac{3}{2}\right), \\ \langle \phi(2t-(2m+3)), f\rangle_1 &= \Lambda^{-1}\mathbf{f}(m+2).\end{aligned}$$

Lemma 4. Let f, f^0, and f^1 be a function and its projections onto V_0 and W_0, respectively. Let

$$a^1_{k,j} = \langle f, \phi_j(2t-k)\rangle_1,$$

$$a^0_{k,j} = \langle f, \phi_j(t-k)\rangle_0,$$

$$b^0_{k,j} = \langle f, \psi_j(t-k)\rangle_1.$$

Then the decomposition algorithm has as its vector form

$$\mathbf{a}^0_k = \Lambda^{-1}\mathbf{a}^1_{2k+1}, \tag{6.7.7}$$

$$\mathbf{b}^0_k = \Lambda^{-1}(-C_2^T\Lambda^{-1}\mathbf{a}^1_{2k+1} + \mathbf{a}^1_{2k+2} - C_0^T\Lambda^{-1}\mathbf{a}^1_{2k+3}) \tag{6.7.8}$$

while the reconstruction algorithm works as

$$\mathbf{a}^1_{2k+1} = \Lambda\mathbf{a}^0_k,$$

$$\mathbf{a}^1_{2k+2} = \Lambda\mathbf{b}^0_k + C_2^T\Lambda^{-1}\mathbf{a}^0_k + C_0^T\Lambda^{-1}\mathbf{a}^0_{k+1}.$$

Derivation Equation (6.7.7) may be written explicitly as

$$(j+1)\text{th} \rightarrow \begin{bmatrix} a^0_{k,0} \\ a^0_{k,1} \\ \cdots \\ a^0_{k,j} \\ \cdots \\ a^0_{k,r-1} \end{bmatrix} = \begin{bmatrix} 1 & & & & & \\ & 2 & & & & \\ & & \cdots & & & \\ & & & 2^j & & \\ & & & & \cdots & \\ & & & & & 2^{r-1} \end{bmatrix} \begin{bmatrix} a^1_{2k+1,0} \\ a^1_{2k+1,1} \\ \cdots \\ a^1_{2k+1,j} \\ \cdots \\ a^1_{2k+1,r-1} \end{bmatrix}.$$

Its $(j+1)$th row satisfies the relation

$$a^0_{k,j} = 2^j a^1_{2k+1,j}.$$

The expression above is true because of (6.7.2),

$$a^1_{m,j} = 2^{-j} f^{(j)}\left(\frac{m+1}{2}\right),$$

or equivalently

$$a^1_{2k+1,j} = 2^{-j} f^{(j)}\left(\frac{2k+2}{2}\right) = 2^{-j} f^{(j)}(k+1). \tag{6.7.9}$$

However, from (6.6.2),

$$a^0_{k,j} = f^{(j)}(k+1). \tag{6.7.10}$$

Comparing (6.7.10) and (6.7.9), we arrive at

$$a^0_{k,j} = 2^j a_{2k+1,j}$$

which completes (6.7.7). □

Equation (6.7.8) is, in fact, a restatement of (6.7.6). The reconstruction algorithm is simply a rearrangement of the same formulas. Thus far we have completed the basis theory of the intervallic multiwavelets, and in the next section we will present some examples.

6.8 WORKING EXAMPLES

CASE 1. $r = 1$ Multiscalets $\phi_0(t), \ldots, \phi_{r-1}(t)$ have only one element in the vector, and $\phi_0^{(r-1)}(t)$ has no requirement on derivatives. From (6.2.1),

$$\phi(t) = \sum C_k \phi(2t - k)$$

and the nonzero C_k, $k = 0, 1, 2$, we have

$$\phi(t) = C_0\phi(2t) + C_1\phi(2t - 1) + C_2\phi(2t - 2)$$

with

$$\begin{aligned} C_1 &= \Lambda = 1, \\ C_0 &= \phi\left(\tfrac{1}{2}\right)\Lambda = \phi\left(\tfrac{1}{2}\right) = \tfrac{1}{2} \\ C_2 &= \phi\left(\tfrac{3}{2}\right)\Lambda = \phi\left(\tfrac{3}{2}\right) = \tfrac{1}{2}. \end{aligned}$$

C_2 can also be obtained from Theorem 1. Hence

$$\phi(t) = \tfrac{1}{2}\phi(2t) + \phi(2t - 1) + \tfrac{1}{2}\phi(2t - 2). \tag{6.8.1}$$

It follows from (6.5.3) that

$$\begin{aligned} \psi(t) &= -C_2\phi(2t - 1) + \phi(2t - 2) - C_0\phi(2t - 3) \\ &= -\tfrac{1}{2}\phi(2t - 1) + \phi(2t - 2) - \tfrac{1}{2}(2t - 3) \end{aligned} \tag{6.8.2}$$

and

$$\tilde{\psi}(t) = \phi(2t - 2).$$

The functions $\phi_0(t)$ and $\psi_0(t)$ are plotted in Fig. 6.6. The orthogonality can be expressed as follows:

(1) Sobolev-like inner product of multiscalets at level 0.

$$\begin{aligned} \langle \phi(t - n), \phi(t) \rangle_{0}, &= \sum_k \phi(k - n)\phi(k) \\ &= \phi(1 - n) \cdot 1 = \delta_{0,n}, \end{aligned}$$

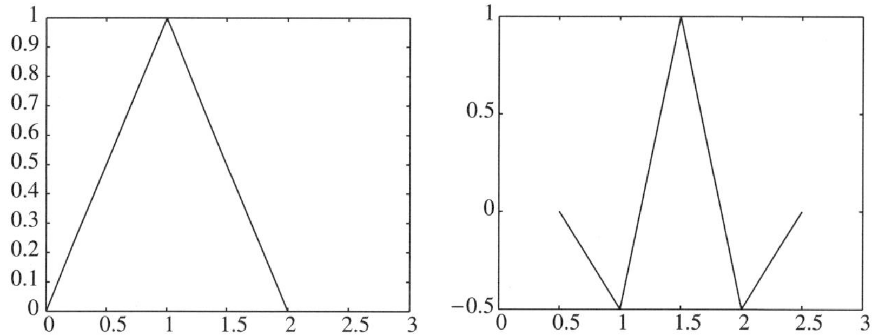

FIGURE 6.6 Multiscalet $\phi_0(t)$ and orthogonal multiwavelet $\psi_0(t)$ of $r = 1$.

where we have applied

$$\phi(k) = \delta_{1,k} = \begin{cases} 1, & k = 1 \\ 0, & k = \text{integer other than } 1. \end{cases}$$

(2) Sobolev-like inner product of multiscalet and intervallic multiwavelet at level 1. According to $\langle\ ,\ \rangle$, and (6.8.2)

$$\langle \phi(t-n), \psi(t) \rangle_1 = \sum_k \phi\left(\frac{k}{2} - n\right) \psi\left(\frac{k}{2}\right) = \sum_k \phi\left(\frac{k}{2} - n\right) \times \left[-\frac{1}{2}\phi(k-1) + \phi(k-2) - \frac{1}{2}\phi(k-3)\right].$$

Notice that $\phi(m) = \delta_{0,m}$. As a result the summation over k provides contributions only for $k = 2, 3, 4$. Therefore

$$\begin{aligned} \langle \phi(t-n), \psi(t) \rangle_1 &= -\tfrac{1}{2}\phi(1-n) + \phi\left(\tfrac{3}{2} - n\right) - \tfrac{1}{2}\phi(2-n) \\ &= -\tfrac{1}{2}\,\delta_{0,n} + \left(\tfrac{1}{2}\,\delta_{1,n} + \tfrac{1}{2}\,\delta_{0,n}\right) - \tfrac{1}{2}\,\delta_{1,n} \\ &= 0. \end{aligned}$$

This is a verification that $W_0 \perp V_0$.

(3) Sobolev-like inner product of multiwavelet and dual wavelet at 1.

$$\begin{aligned} \langle \tilde{\psi}(t-n), \psi(t) \rangle_1 &= \sum_k \tilde{\psi}\left(\frac{k}{2} - n\right) \psi\left(\frac{k}{2}\right) \\ &= \sum_k \phi\left(2\left(\frac{k}{2} - n\right) - 2\right) \psi\left(\frac{k}{2}\right) \\ &= \sum_k \phi(k - 2n - 2)\psi\left(\frac{k}{2}\right) \end{aligned}$$

$$= \sum_k \delta_{k-2n-2,1} \psi \left(\frac{k}{2} \right)$$
$$= \psi \left(n + \frac{3}{2} \right)$$
$$= \delta_{0,n}.$$

The last equality is held because of (i) in Lemma 3.

The decomposition algorithm is

$$a_k^0 = f(k+1) = f\left(\frac{(2k+1)+1}{2} \right) = a_{2k+1}^1,$$

which is $j = 0$ in (6.7.9), and

$$\begin{aligned} b_k^0 &= \langle f, \psi(t-k) \rangle_1 \\ &= \sum_m f\left(\frac{m}{2}\right) \psi\left(\frac{m}{2} - k\right) \\ &= \sum_m f\left(\frac{m}{2}\right) \left[-\frac{1}{2}\phi(m-2k-1) + \phi(m-2k-2) - \frac{1}{2}\phi(m-2k-3) \right] \\ &= -\frac{1}{2} f(k+1) + f\left(k + \frac{3}{2}\right) - \frac{1}{2} f(k+2) \\ &= -\frac{1}{2} a_{2k+1}^1 + a_{2k+2}^1 - \frac{1}{2} a_{2k+3}^1. \end{aligned}$$

Of course, the same results can be obtained directly from the general formulas (6.7.7) and (6.7.8).

The reconstruction algorithm is

$$\begin{aligned} a_{2k+1}^1 &= a_k^0, \\ a_{2k+2}^1 &= \tfrac{1}{2} a_k^0 + b_k^0 + \tfrac{1}{2} a_{k+1}^0. \end{aligned}$$

In programming, Case 1 needs the values of the function $f(t)$ at the finest scale, namely $f(2^{-M}k), k \in Z$.

CASE 2. $r = 2$ It can be derived as in Exercise 11 that the dilation equation is

$$\phi(t) = C_0 \phi(2t) + C_1 \phi(2t-1) + C_2 \phi(2t-2),$$

namely

$$\begin{pmatrix} \phi_0(t) \\ \phi_1(t) \end{pmatrix} = \begin{pmatrix} \frac{1}{2} & \frac{3}{4} \\ -\frac{1}{8} & -\frac{1}{8} \end{pmatrix} \begin{pmatrix} \phi_0(2t) \\ \phi_1(2t) \end{pmatrix} + \Lambda \begin{pmatrix} \phi_0(2t-1) \\ \phi_1(2t-1) \end{pmatrix} + \begin{pmatrix} \frac{1}{2} & -\frac{3}{4} \\ \frac{1}{8} & -\frac{1}{8} \end{pmatrix} \begin{pmatrix} \phi_0(2t-2) \\ \phi_1(2t-2) \end{pmatrix},$$

where

$$\Lambda = \begin{pmatrix} 1 & 0 \\ 0 & \frac{1}{2} \end{pmatrix},$$

$$\begin{aligned}
\begin{pmatrix} \psi_0(t) \\ \psi_1(t) \end{pmatrix} &= -\begin{pmatrix} 1 & 0 \\ 0 & 2 \end{pmatrix}\begin{pmatrix} \frac{1}{2} & -\frac{3}{4} \\ \frac{1}{8} & -\frac{1}{8} \end{pmatrix}\begin{pmatrix} 1 & 0 \\ 0 & \frac{1}{2} \end{pmatrix}\begin{pmatrix} \phi_0(2t-1) \\ \phi_1(2t-2) \end{pmatrix} + \Lambda \begin{pmatrix} \phi_0(2t-2) \\ \phi_1(2t-2) \end{pmatrix} \\
&\quad - \begin{pmatrix} 1 & 0 \\ 0 & 2 \end{pmatrix}\begin{pmatrix} \frac{1}{2} & \frac{3}{4} \\ -\frac{1}{8} & -\frac{1}{8} \end{pmatrix}\begin{pmatrix} 1 & 0 \\ 0 & \frac{1}{2} \end{pmatrix}\begin{pmatrix} \phi_0(2t-3) \\ \phi_1(2t-3) \end{pmatrix} \\
&= -\begin{pmatrix} -\frac{1}{2} & -\frac{3}{8} \\ -\frac{1}{4} & -\frac{1}{8} \end{pmatrix}\begin{pmatrix} \phi_0(2t-1) \\ \phi_1(2t-1) \end{pmatrix} + \Lambda \begin{pmatrix} \phi_0(2t-2) \\ \phi_1(2t-2) \end{pmatrix} \\
&\quad - \begin{pmatrix} \frac{1}{2} & \frac{3}{8} \\ -\frac{1}{4} & -\frac{1}{8} \end{pmatrix}\begin{pmatrix} \phi_0(2t-3) \\ \phi_1(2t-3) \end{pmatrix}.
\end{aligned}$$

Or explicitly

$$\begin{aligned}
\psi_0(t) &= -\tfrac{1}{2}\phi_0(2t-1) + \phi_0(2t-2) - \tfrac{1}{2}\phi_0(2t-3) \\
&\quad + \tfrac{3}{8}\phi_1(2t-1) - \tfrac{3}{8}\phi_1(2t-3), \\
\psi_1(t) &= -\tfrac{1}{4}\phi_0(2t-1) + \tfrac{1}{4}\phi_0(2t-3) + \tfrac{1}{8}\phi_1(2t-1) \\
&\quad + \tfrac{1}{2}\phi_1(2t-2) + \tfrac{1}{8}\phi_1(2t-3),
\end{aligned}$$

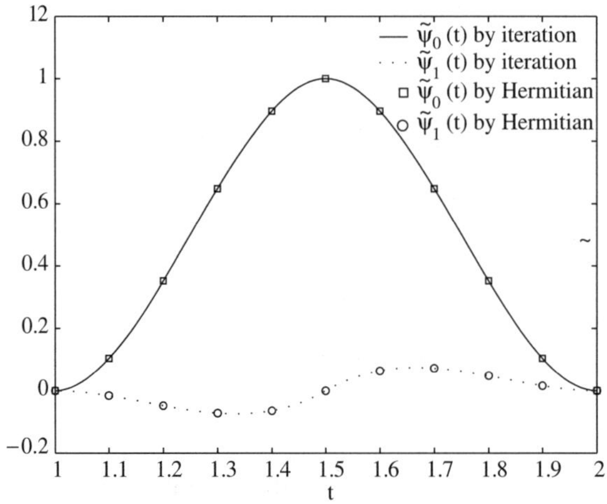

FIGURE 6.7 Intervallic dual multiwavelets of $r = 2$.

and

$$\begin{pmatrix} \tilde{\psi}_0(t) \\ \tilde{\psi}_1(t) \end{pmatrix} = \begin{pmatrix} \phi_0(2t-2) \\ \phi_1(2t-2) \end{pmatrix}.$$

The intervallic dual multiwavelets $\tilde{\psi}_0$ and $\tilde{\psi}_1$ are plotted in Fig. 6.7.

Example 4 Derive explicit expressions of the multiwavelets and dual multiwavelets for $r = 2$.

Solution

$$D_1 = \begin{pmatrix} -\frac{1}{2} & -\frac{1}{16} \\ \frac{3}{2} & \frac{1}{8} \end{pmatrix}, \quad D_2 = \begin{pmatrix} 1 & 0 \\ 0 & \frac{1}{2} \end{pmatrix}, \quad D_3 = \begin{pmatrix} -\frac{1}{2} & \frac{1}{16} \\ -\frac{3}{2} & \frac{1}{8} \end{pmatrix};$$

$$\psi(t) = D_1\phi(2t-1) + D_2\phi(2t-2) + D_3\phi(2t-3);$$

$$\psi(t) = \begin{pmatrix} \psi_0(t) \\ \psi_1(t) \end{pmatrix}.$$

The explicit forms of the multiscalets are

$$\begin{cases} \phi_0(t) = 2t^2 - 2t^3 & t \in [0,1] \\ \phi_0(t) = \phi_0(2-t) & t \in [1,2] \end{cases},$$

$$\begin{cases} \phi_1(t) = t^3 - t^2 & t \in [0,1] \\ \phi_1(t) = \phi_1(2-t) & t \in [1,2]. \end{cases}$$

Notice that

$$\begin{cases} t \in [0.5,1] & t \in [1,1.5] & t \in [1.5,2] & t \in [2,2.5] \\ 2t-1 \in [0,1] & 2t-1 \in [1,2] & 2t-1 \in [2,3] & 2t-1 \in [3,4] \\ 2t-2 \in [-1,10 & 2t-2 \in [0,1] & 2t-2 \in [1,2] & 2t-2 \in [2,3] \\ 2t-3 \in [-2,-1] & 2t-3 \in [-1,0] & 2t-3 \in [0,1] & 2t-3 \in [1,2]. \end{cases}$$

Hence, the intervallic multiwavelets are

CASE 1. $t \in [0.5, 1]$

$$\psi(t) = \begin{pmatrix} -\frac{1}{2} & -\frac{1}{16} \\ \frac{3}{2} & \frac{1}{8} \end{pmatrix} \begin{pmatrix} 3(2t-1)^2 - 2(2t-1)^3 \\ (2t-1)^3 - (2t-1)^2 \end{pmatrix}$$

$$= \begin{pmatrix} (2t-2)^2(\frac{15}{8}t - \frac{19}{8}) \\ (2t-1)^2(-\frac{23}{4}t + \frac{29}{4}) \end{pmatrix}.$$

CASE 2. $t \in [1, 1.5]$

$$\psi(t) = \begin{pmatrix} -\frac{1}{2} & -\frac{1}{16} \\ \frac{3}{2} & \frac{1}{8} \end{pmatrix} \begin{pmatrix} 3(-2t+3)^2 - 2(-2t+3)^3 \\ -(-2t+3)^3 + (-2t+3)^2 \end{pmatrix}$$
$$+ \begin{pmatrix} 1 & 0 \\ 0 & \frac{1}{2} \end{pmatrix} \begin{pmatrix} 3(2t-2)^2 - 2(2t-1)^3 \\ (2t-2)^3 - (2t-2)^2 \end{pmatrix}$$
$$= \begin{pmatrix} (-2t+3)^2(-\frac{17}{8}t + \frac{13}{8}) + (2t-2)^2(-4t+7) \\ (-2t+3)^2(\frac{25}{4}t - \frac{19}{4}) + (2t-2)^2(t - \frac{3}{2}) \end{pmatrix}.$$

CASE 3. $t \in [1.5, 2]$

$$\psi(t) = \begin{pmatrix} 1 & 0 \\ 0 & \frac{1}{2} \end{pmatrix} \begin{pmatrix} 3(-2t+4)^2 - 2(-2t+4)^3 \\ -(-2t+4)^3 + (-2t+4)^2 \end{pmatrix}$$
$$+ \begin{pmatrix} -\frac{1}{2} & \frac{1}{16} \\ -\frac{3}{2} & \frac{1}{8} \end{pmatrix} \begin{pmatrix} 3(2t-3)^2 - 2(2t-3)^3 \\ (2t-3)^3 - (2t-3)^2 \end{pmatrix}$$
$$= \begin{pmatrix} (-2t+4)^2(4t-5) + (2t-3)^2(\frac{17}{8}t - \frac{19}{4}) \\ (-2t+4)^2(t - \frac{3}{2}) + (2t-3)^2(\frac{25}{4}t - 14) \end{pmatrix}.$$

CASE 4. $t \in [2, 2.5]$

$$\psi(t) = \begin{pmatrix} -\frac{1}{2} & \frac{1}{16} \\ -\frac{3}{2} & \frac{1}{8} \end{pmatrix} \begin{pmatrix} 3(-2t+5)^2 - 2(-2t+5)^3 \\ -(-2t+5)^3 + (-2t+5)^2 \end{pmatrix} = \begin{pmatrix} (-2t+5)^2(-\frac{15}{8}t + \frac{13}{14}) \\ (-2t+5)^2(-\frac{23}{4}t + 10) \end{pmatrix};$$

$$\Lambda = \begin{pmatrix} 1 & 0 \\ 0 & \frac{1}{2} \end{pmatrix};$$

$$\tilde{\psi}(t) = \Lambda\phi(2t-2);$$

$$t \in [1, 1.5],$$
$$2t - 2 \in [0, 1],$$

$$t \in [1.5, 2];$$
$$2t - 2 \in [1, 2].$$

The intervallic dual multiwavelets are

CASE 1. $t \in [1, 1.5]$

$$\tilde{\psi}(t) = \begin{pmatrix} 1 & 0 \\ 0 & \frac{1}{2} \end{pmatrix} \begin{pmatrix} 3(2t-2)^2 - 2(2t-2)^3 \\ (2t-2)^3 + (2t-2)^2 \end{pmatrix} = \begin{pmatrix} (2t-2)^2(-4t+7) \\ \frac{1}{2}(2t-2)^2(2t-3) \end{pmatrix}.$$

CASE 2. $t \in [1.5, 2]$

$$\tilde{\psi}(t) = \begin{pmatrix} 1 & 0 \\ 0 & \frac{1}{2} \end{pmatrix} \begin{pmatrix} 3(-2t+4)^2 - 2(2t+4)^3 \\ -(-2t+4)^3 + (-2t+4)^2 \end{pmatrix}$$

$$= \begin{pmatrix} (-2t+4)^2(4t-5) \\ \frac{1}{2}(-2t+4)^2(2t-3) \end{pmatrix}.$$

CASE 3. OTHERWISE

$$\begin{bmatrix} \tilde{\psi}_0(t) \\ \tilde{\psi}_1(t) \end{bmatrix} = \begin{bmatrix} 0 \\ 0 \end{bmatrix}.$$

Example 5 Evaluate $\tilde{\psi}_0(\sqrt{2})$ and $\tilde{\psi}_1(\sqrt{3})$ from the explicit expressions of the dual multiwavelets of $r = 2$, namely $\tilde{\psi}_0(t)$ and $\tilde{\psi}_1(t)$.

Solution By definition,

$$\tilde{\psi}(t) = \Lambda\phi(2t-2),$$

or explicitly

$$\begin{pmatrix} \tilde{\psi}_0(t) \\ \tilde{\psi}_1(t) \end{pmatrix} = \begin{pmatrix} 1 & 0 \\ 0 & \frac{1}{2} \end{pmatrix} \begin{pmatrix} \phi_0(2t-2) \\ \phi_1(2t-2) \end{pmatrix}.$$

From the previous example

(1) $t \in [1, 1.5]$,

$$\begin{pmatrix} \tilde{\psi}_0(t) \\ \tilde{\psi}_1(t) \end{pmatrix} = \begin{pmatrix} 1 & 0 \\ 0 & \frac{1}{2} \end{pmatrix} \begin{pmatrix} 3(2t-2)^2 - 2(2t-2)^3 \\ (2t-2)^3 - (2t-2)^2 \end{pmatrix}$$

$$= \begin{pmatrix} 3(2t-2)^2 - 2(2t-2)^3 \\ \frac{1}{2}(2t-2)^3 - \frac{1}{2}(2t-2)^2 \end{pmatrix}$$

$$= \begin{pmatrix} 4(t-1)^2(7-4t) \\ 2(t-1)^2(2t-3) \end{pmatrix}.$$

(2) $t \in [1.5, 2]$,

$$\begin{pmatrix} \tilde{\psi}_0(t) \\ \tilde{\psi}_1(t) \end{pmatrix} = \begin{pmatrix} 1 & 0 \\ 0 & \frac{1}{2} \end{pmatrix} \begin{pmatrix} 3[2-(2t-2)]^2 - 2[2-(2t-2)]^2 \\ 2[2-(2t-2)]^2 - [2-(2t-2)]^3 \end{pmatrix}$$

$$= \begin{pmatrix} 3[(2-(2t-2)]^2 - 2[2-(2t-2)]^3 \\ \frac{1}{2}[2-(2t-2)]^2 - \frac{1}{2}[2-(2t-2)]^4 \end{pmatrix}.$$

Finally

(i) For $1 < \sqrt{2} < 1.5$. By using (1), we obtain

$$\begin{aligned}\tilde{\psi}_0(\sqrt{2}) &= 3(2\sqrt{2}-2)^2 - 2(2\sqrt{2}-2)^3 \\ &\approx 0.9218.\end{aligned}$$

(ii) For $1.5 < \sqrt{3} < 2$. By using (2), we have

$$\begin{aligned}\tilde{\psi}_1(\sqrt{3}) &= \tfrac{1}{2}[2-(2\sqrt{3}-2)]^2 - \tfrac{1}{2}[2-(2\sqrt{3}-2)]^3 \\ &\approx 0.06665.\end{aligned}$$

6.9 MULTISCALET-BASED 1D FINITE ELEMENT METHOD (FEM)

A typical boundary-value problem can be defined by a governing differential equation in a domain Ω as

$$L\phi = f, \tag{6.9.1}$$

together with conditions on the boundary $\partial\Omega$ that encloses the domain. Here L is a differential operator, f is the excitation and ϕ is the unknown function.

The multiscalets are employed as the shape functions in the FEM to substitute the Lagrange linear interpolation functions. Because of the interpolatory properties of the multiscalets and their derivatives, fast convergence in approximating a function is achieved. The new shape functions are $\in C^1$, meaning that the first derivatives are continuous on the connecting nodes. Thus the divergence-free condition is satisfied at the endpoints. The multiscalets along with their derivatives are orthonormal, as defined by (6.3.21) in the discrete sampling nodes. Therefore no coupled system of equations is involved in terms of the function and its derivative, resulting in a simple and efficient algorithm.

These shape functions are high-order interpolation functions:

$$\begin{aligned}N_1^e &= 3\left(\frac{x_2^e - x}{l_e}\right)^2 - 2\left(\frac{x_2^e - x}{l_e}\right)^3, \\ N_2^e &= 3\left(\frac{x - x_1^e}{l_e}\right)^2 - 2\left(\frac{x - x_1^e}{l_e}\right)^3, \\ D_1^e &= -\left[\left(\frac{x_2^e - x}{l_e}\right)^3 - \left(\frac{x_2^e - x}{l_e}\right)^2\right]\cdot l_e, \\ D_2^e &= \left[\left(\frac{x - x_1^e}{l_e}\right)^3 - \left(\frac{x - x_1^e}{l_e}\right)^2\right]\cdot l_e,\end{aligned} \tag{6.9.2}$$

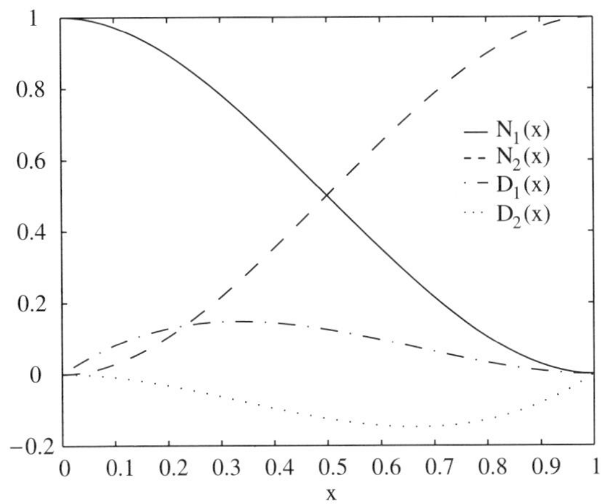

FIGURE 6.8 Four basis functions.

where the superscript e denotes element. Figure 6.8 depicts the four basis functions. In comparing (6.9.2) against (6.3.8) and (6.3.9), we find that N_2 is ϕ_0 given in (6.3.8), but is shifted by x_1^e and scaled by l_e. N_1 is ϕ_0 given in (6.3.9) but shifted by x_2^e and is scaled by l_e. Similarly D_1 and D_2 are the shifted and scaled versions of ϕ_1.

One feature of the multiwavelet basis is that the values of D vanish at the two nodes x_1^e and x_2^e, while the derivatives of N vanish at the two nodes. Therefore the unknown function can be written as

$$\phi^e(x) = \sum_{j=1}^{2} \phi_j^e N_j^e(x) + \phi_j'^e D_j^e(x), \tag{6.9.3}$$

$$\phi'^e(x) = \sum_{j=1}^{2} \phi_j^e N_j'^e(x) + \phi_j'^e D_j'^e(x). \tag{6.9.4}$$

Let us consider a 1D Sturm-Liouville problem

$$-\frac{d}{dx}\left(\alpha \frac{d\phi}{dx}\right) + \beta\phi = f.$$

Applying Galerkin's procedure to the shape function N_i^e and using integration by parts, we have

$$\left\langle N_i^e, -\frac{d}{dx}(\alpha \frac{d\phi}{dx}) + \beta\phi - f \right\rangle$$

$$= \int_{x_1^e}^{x_2^e} N_i^e \left[-\frac{d}{dx}\left(\alpha \frac{d\phi}{dx}\right) + \beta\phi\right] dx - \int_{x_1^e}^{x_2^e} N_i^e f \, dx$$

$$= \int_{x_1^e}^{x_2^e} \left(\alpha \frac{dN_i^e}{dx} \frac{d\phi}{dx} + \beta N_i^e \phi \right) dx - \int_{x_1^e}^{x_2^e} N_i^e f \, dx - \alpha N_i^e \frac{d\phi}{dx} = \Big|_{x_1^e}^{x_2^e}$$

$$= \sum_{j=1}^{2} \phi_j^e \underbrace{\int_{x_1^e}^{x_2^e} (\alpha N_i'^e N_j'^e + \beta N_i^e N_j^e) \, dx}_{K_{ij}} + \phi_j'^e \underbrace{\int_{x_1^e}^{x_2^e} (\alpha N_i'^e D_j'^e + \beta N_i^e D_j^e) \, dx}_{L_{ij}}$$

$$- \int_{x_1^e}^{x_2^e} N_i^e f \, dx - \underbrace{\alpha N_i^e (\phi_2'^e - \phi_1'^e)}_{\text{boundary}},$$

where we have employed the expansions (6.9.3) and (6.9.4). The boundary term $N_i^e(\phi_2'^e - \phi_1'^e)$ will be canceled, leaving only in the leftmost and rightmost elements. These are subjected to the boundary conditions of the given problem. In the same manner

$$\left\langle D_i^e, -\frac{d}{dx}\left(\alpha \frac{d\phi}{dx}\right) + \beta\phi - f \right\rangle$$

$$= \int_{x_1^e}^{x_2^e} D_i^e \left[-\frac{d}{dx}\left(\alpha \frac{d\phi}{dx}\right) + \beta\phi \right] dx - \int_{x_1^e}^{x_2^e} D_i^e f \, dx$$

$$= \int_{x_1^e}^{x_2^e} \left(\alpha \frac{dD_i^e}{dx} \frac{d\phi}{dx} + \beta D_i^e \phi \right) dx - \int_{x_1^e}^{x_2^e} D_i^e f \, dx - \alpha D_i^e \frac{d\phi}{dx} \Big|_{x_1^e}^{x_2^e}$$

$$= \sum_{j=1}^{2} \phi_j^e \underbrace{\int_{x_1^e}^{x_2^e} (\alpha D_i'^e N_j'^e + \beta D_i^e N_j^e) \, dx}_{P_{ij}} + \phi_j'^e \underbrace{\int_{x_1^e}^{x_2^e} (\alpha D_i'^e D_j'^e + \beta D_i^e D_j^e) \, dx}_{Q_{ij}}$$

$$- \int_{x_1^e}^{x_2^e} D_i^e f \, dx - \underbrace{\alpha D_i^e (\phi_2'^e - \phi_1'^e)}_{=0}.$$

The last term is zero because D_i^e are zero at the ends of the element. Expressing the previous equations in matrix form, we arrive at a system equations of the problem

$$\begin{bmatrix} K & L \\ P & Q \end{bmatrix} \begin{bmatrix} \phi \\ \phi' \end{bmatrix} = \begin{bmatrix} g \\ h \end{bmatrix}.$$

In the equation above

$$K_{ij} = \int_{x_1^e}^{x_2^e} (\alpha N_i'^e N_j'^e + \beta N_i^e N_j^e) \, dx$$

$$L_{ij} = \int_{x_1^e}^{x_2^e} (\alpha N_i'^e D_j'^e + \beta N_i^e D_j^e) \, dx$$

$$P_{ij} = \int_{x_1^e}^{x_2^e} (\alpha D_i'^e N_j'^e + \beta D_i^e N_j^e)\, dx$$

$$Q_{ij} = \int_{x_1^e}^{x_2^e} (\alpha D_i'^e D_j'^e + \beta D_i^e D_j^e)\, dx$$

$$g_i^e = \int_{x_1^e}^{x_2^e} N_i^e f\, dx$$

$$h_i^e = \int_{x_1^e}^{x_2^e} D_i^e f\, dx.$$

Example 6 Consider a simple problem

$$\frac{d^2\phi}{dx^2} = x + 1, \qquad 0 < x < 1,$$
$$\phi|_{x=0} = 0, \qquad \phi|_{x=1} = 1.$$

The analytical solution is given in [6] as

$$\phi(x) = \tfrac{1}{6}x^3 + \tfrac{1}{2}x^2 + \tfrac{1}{3}x.$$

The comparison of the analytical solution with the result obtained by multiscalet-based FEM is plotted in Fig. 6.9. From this picture we can see that the numerical solution matches the function and its derivative values very well.

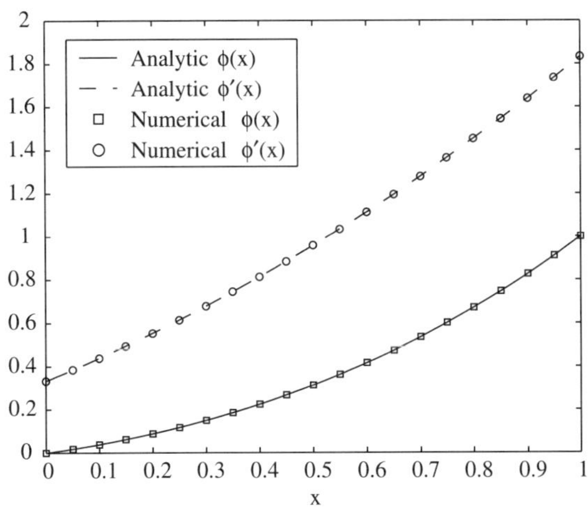

FIGURE 6.9 Multiwavelets based FEM in 1D case.

6.10 MULTISCALET-BASED EDGE ELEMENT METHOD

Electromagnetic fields are vectors. It is reported that edge-based finite element method, referred to as the edge element method (EEM), handles vector fields better than the node based FEM. The EEM is a popular and powerful numerical approach in computational electromagnetics [7–9]. It allows the normal component of the vector field to be discontinuous across the adjacent elements and handles field singularities better than the node based finite element method [10, 11]. While higher-order basis functions in the EEM improve the convergence and numerical accuracy, they increase the complexity of the algorithm and bandwidth of the system matrix dramatically. The Lagrange-based interpolation matches the function being approximated at the discrete points (nodes) by linear, quadratic, or cubic polynomials, depending on the interpolation order. Nonetheless, the slope (derivative) and curvature (second derivative) of the function has never been matched at the nodes, regardless of the order of the polynomials. Attempts were made to address the slope by using the splines because of the short support and beneficial features of the splines [12]. Unfortunately, simultaneous system equations in terms of the function and its derivative values must be solved in order to employ the splines. This complicity has rendered the interpolatory spline unpopular in the finite element method (FEM).

To avoid tedious manipulations of an excessive number of individual elements, only two-dimensional problems will be formulated in this section, and the element is rectangular shape. We selected a waveguide problem to demonstrate this concept. The methodology can be easily extended to 3D problems if the rectangular element is replaced by a brick box. For open boundaries, one needs to truncate the computation domain with absorption or radiation boundary conditions (ABC) or including the Mur ABC, Beranger's perfectly matched layers (PML), among others. The boundary-value problem in the full wave analysis of an inhomogeneous waveguide is governed by the vector wave equation

$$\nabla \times \left(\frac{1}{\mu_r} \nabla \times \mathbf{E} \right) - k_0^2 \epsilon_r \mathbf{E} = 0 \qquad \text{in } \Omega$$

with the boundary conditions

$$\begin{aligned} \hat{n} \times \mathbf{E} &= 0 \qquad \text{on } \Gamma_1, \\ \hat{n} \times (\nabla \times \mathbf{E}) &= 0 \qquad \text{on } \Gamma_2. \end{aligned}$$

In the previous equations Ω denotes the cross section of the structure whose boundary is comprised by the electrical wall Γ_1 and the magnetic wall Γ_2. The equivalent variational problem with real ϵ_r and μ_r is given by

$$\begin{cases} \delta F(\mathbf{E}) = 0 \\ \hat{n} \times \mathbf{E} = 0 \text{ on } \Gamma_1, \end{cases}$$

where

$$F(\mathbf{E}) = \frac{1}{2}\iint_{\Omega}\left[\frac{1}{\mu_r}(\nabla \times \mathbf{E}) \cdot (\nabla \times \mathbf{E})^* - k_0^2 \epsilon_r \mathbf{E} \cdot \mathbf{E}^*\right] d\Omega.$$

Assuming a known z-dependence of $\mathbf{E}(x, y, z) = \mathbf{E}(x, y)e^{-jk_z z}$, the functional can be written as

$$F(\mathbf{E}) = \frac{1}{2}\iint_{\Omega}\left[\frac{1}{\mu_r}(\nabla_t \times \mathbf{E}_t) \cdot (\nabla_t \times \mathbf{E}_t)^* - k_0^2 \epsilon_r \mathbf{E} \cdot \mathbf{E}^* + \frac{1}{\mu_r}(\nabla_t E_z + jk_z \mathbf{E}_t) \cdot (\nabla_t E_z + jk_z \mathbf{E}_t)^*\right] d\Omega.$$

The functional is discretized to yield an eigenvalue system that can be solved for k_0^2 of a given k_z. However, in engineering practice it is usually preferable to specify the operating frequency, and then solve for propagation constant k_z.

To alleviate the difficulty, we adopt the following transformation [6]:

$$\mathbf{e}_t = k_z \mathbf{E}_t, \quad e_z = -jE_z.$$

The normalized version of the functional is

$$F(\mathbf{e}) = \frac{1}{2}\iint_{\Omega}\left\{\frac{1}{\mu_r}(\nabla_t \times \mathbf{e}_t) \cdot (\nabla_t \times \mathbf{e}_t)^* - k_0^2 \epsilon_r \mathbf{e}_t \cdot \mathbf{e}_t^* + k_z^2\left[\frac{1}{\mu_r}(\nabla_t e_z + \mathbf{e}_t) \cdot (\nabla_t e_z + \mathbf{e}_t)^* - k_0^2 \epsilon_r e_z e_z^*\right]\right\} d\Omega.$$

Apparently the eigenvalue equation of the discretized functional for a given k_0 will result in a system with k_z^2 as its eigenvalue. To this end, the cross-sectional area Ω is subdivided into small rectangular or triangular elements. Within each element, the vector field can be expanded as

$$\mathbf{e}_t^e = \sum_{i=1}^{n} \mathbf{N}_i^e e_{ti}^e = \{\mathbf{N}^e\}^T \{e_t^e\} = \{e_t^e\}^T \{\mathbf{N}^e\}$$

and

$$e_z^e = \sum_{i=1}^{n} N_i^e e_{zi}^e = \{N^e\}^T \{e_z^e\} = \{e_z^e\}^T \{N^e\},$$

where $\mathbf{N}_i^e$ and N_i^e are vector and scalar interpolation functions, respectively.

The functional can then be discretized as

$$F = \frac{1}{2}\sum_{e=1}^{M}\left(\{e_t^e\}^T [A_{tt}^e]\{e_t^e\}^* + k_z^2 \begin{Bmatrix} e_t^e \\ e_z^e \end{Bmatrix}^T \begin{bmatrix} B_{tt}^e & B_{tz}^e \\ B_{zt}^e & B_{zz}^e \end{bmatrix} \begin{Bmatrix} e_t^e \\ e_z^e \end{Bmatrix}^*\right),$$

where A_{tt}^e, B_{tt}^e, B_{tz}^e, B_{zt}^e, and B_{zz}^e are all integrals in the corresponding elements, which can be evaluated analytically.

Adding all elements into a global matrix, we obtain the system matrix

$$[A_{tt}]\{e_t\} = k_z^2[B'_{tt}]\{e_t\},$$

where

$$[B'_{tt}] = [B_{tz}][B_{zz}]^{-1}[B_{zt}] - [B_{tt}].$$

Traditionally one employs the linear interpolation functions

$$\mathbf{N}_1^e = \left(\frac{y_2^e - y}{l_y}\right)\hat{x}, \qquad N_1^e = \frac{(x_2^e - x)(y_2^e - y)}{l_x l_y},$$

$$\mathbf{N}_2^e = \left(\frac{x - x_1^e}{l_x}\right)\hat{y}, \qquad N_2^e = \frac{(x - x_1^e)(y_2^e - y)}{l_x l_y},$$

$$\mathbf{N}_3^e = \left(\frac{y - y_1^e}{l_y}\right)\hat{x}, \qquad N_3^e = \frac{(x - x_1^e)(y - y_1^e)}{l_x l_y},$$

$$\mathbf{N}_4^e = \left(\frac{x_2^e - x}{l_x}\right)\hat{y}, \qquad N_4^e = \frac{(x_2^e - x)(y - y_1^e)}{l_x l_y}.$$

In contrast, we use multiwavelet interpolation functions as the edge bases

$$\mathbf{N}_1^e = \left[3\left(\frac{y_2^e - y}{l_y}\right)^2 - 2\left(\frac{y_2^e - y}{l_y}\right)^3\right]\hat{x},$$

$$\mathbf{N}_2^e = \left[3\left(\frac{x - x_1^e}{l_x}\right)^2 - 2\left(\frac{x - x_1^e}{l_x}\right)^3\right]\hat{y},$$

$$\mathbf{N}_3^e = \left[3\left(\frac{y - y_1^e}{l_y}\right)^2 - 2\left(\frac{y - y_1^e}{l_y}\right)^3\right]\hat{x},$$

$$\mathbf{N}_4^e = \left[3\left(\frac{x_2^e - x}{l_x}\right)^2 - 2\left(\frac{x_2^e - x}{l_x}\right)^3\right]\hat{y},$$

$$\mathbf{N}_5^e = -l_y\left[\left(\frac{y_2^e - y}{l_y}\right)^3 - \left(\frac{y_2^e - y}{l_y}\right)^2\right]\hat{x},$$

$$\mathbf{N}_6^e = l_x\left[\left(\frac{x - x_1^e}{l_x}\right)^3 - \left(\frac{x - x_1^e}{l_x}\right)^2\right]\hat{y},$$

$$\mathbf{N}_7^e = l_y\left[\left(\frac{y - y_1^e}{l_y}\right)^3 - \left(\frac{y - y_1^e}{l_y}\right)^2\right]\hat{x},$$

$$\mathbf{N}_8^e = -l_x\left[\left(\frac{x_2^e - x}{l_x}\right)^3 - \left(\frac{x_2^e - x}{l_x}\right)^2\right]\hat{y}.$$

The interpolation functions for z-components remain unchanged, and the transverse fields are written as

$$\mathbf{e}_t^e = \sum_{i=1}^{8} \mathbf{N}_i^e e_{ti}^e = \{\mathbf{N}^e\}^T \{e_t^e\},$$

where

$$e_t^e = [e_{t1}^e, e_{t2}^e, e_{t3}^e, e_{t4}^e, e_{t1}'^e, e_{t2}'^e, e_{t3}'^e, e_{t4}'^e]^T.$$

To demonstrate the fast convergence and high degree of accuracy of the new algorithm, we analyze the dispersion characteristics of a partially loaded waveguide.

Example 7 Figure 6.10 depicts the geometric dimensions and material properties of an inhomogeneous waveguide. The problem is attacked by the traditional EEM with the linear interpolation function and by the multiscalet EEM. To compare the results, we plot in Fig. 6.11 the first six propagation modes obtained from the traditional linear edge element method and from multiwavelet EEM, along with the analytical solutions.

While the wavelet FEM faithfully predicts the dominating mode, the linear FEM demonstrates substantial errors in the dominating mode. For the higher-order modes the wavelet FEM follows the trend of the analytic solutions closely. In contrast, the linear FEM lost tracking of 4th to 6th modes. To compare the two approaches quantitatively, we have created Table 6.1. It can be seen clearly that in terms of the L^2 error the wavelet FEM with 4×4 elements performs better than the linear FEM of 16×16 elements. This result reveals a saving in memory by a factor of 16 and a CPU time cut by a factor of 435. The fast convergence is achieved because of the smooth-

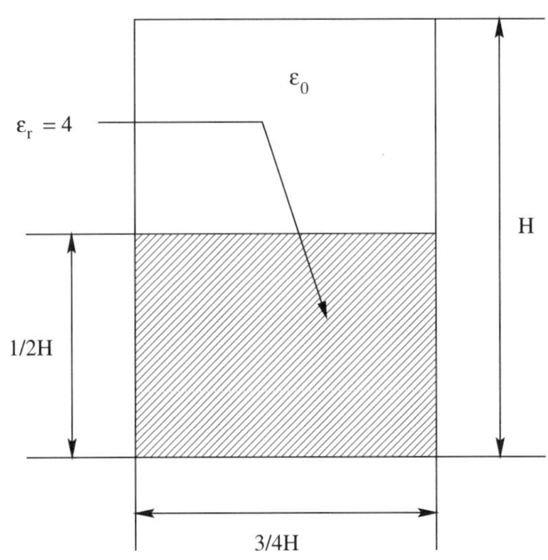

FIGURE 6.10 Configuration of a partially loaded waveguide.

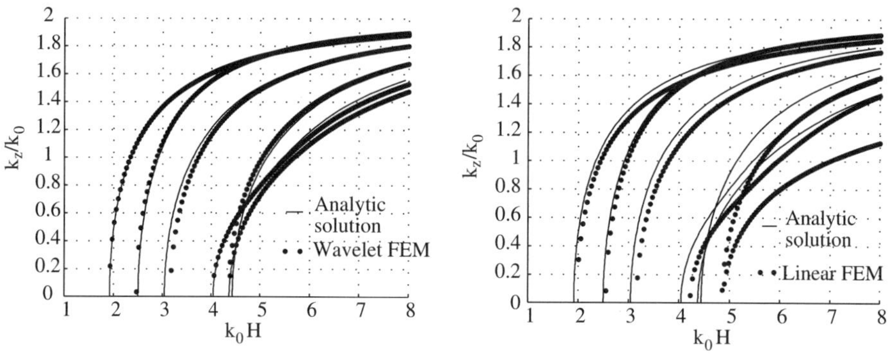

FIGURE 6.11 Results obtained by multiwavelet and linear FEM (4 × 4 elements).

TABLE 6.1. Performance Comparison for the Dominating Mode

Elements	L^2 Error (%)	CPU Time (s) on DEC-Alpha 433 (161 Frequency Points)
Multiwavelets-based FEM		
4 × 4	0.064	24
Linear-based FEM		
4 × 4	12.9	1.7
6 × 6	8.73	20
8 × 8	8.23	117
10 × 10	3.46	486
12 × 12	2.21	1539
16 × 16	1.15	10,459

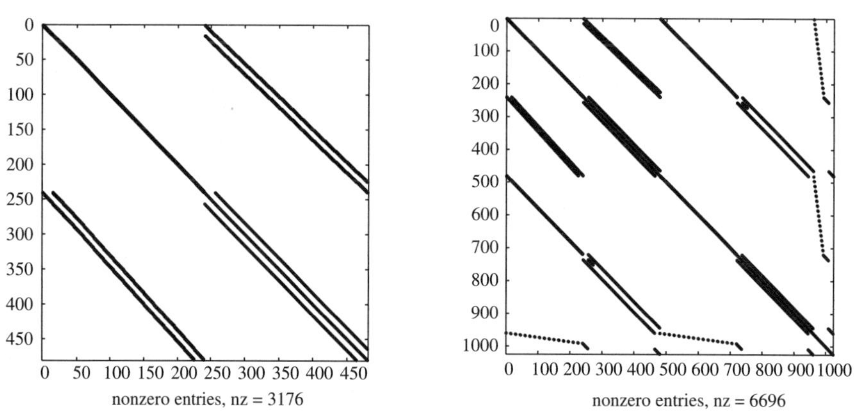

FIGURE 6.12 Linear and multiwavelet EEM matrix from inhomogeneous waveguide with 16 × 16 elements.

ness, completeness, compact support, and interpolation property of the multiscalets in terms of the basis function and its derivatives.

It is interesting to note that under the same discretization, the system matrix size of the multiwavelet EEM has increased roughly 2×2 with respect to that from the linear EEM. This is due to the fact that we have added the derivative bases into the multiwavelet EEM, in addition to the function bases. However, the nonzero entries in the multiwavelet matrix only increased by a factor of 2. Figure 6.12 illustrates the matrix pattern and nonzero elements from the inhomogeneous waveguide with 16×16 elements. Notice that the multiwavelet solution in Table 6.1 never used the 16×16 division, since the 4×4 multiwavelet scheme has already had superior performance to that of the 16×16 linear FEM.

6.11 SPURIOUS MODES

Spurious modes are numerical solutions to the vector wave equation that have no correspondence to physical reality. Sometimes the spurious modes are referred to as vector parasites, which occur in the FEM as wrong solutions. Many authors have observed the fact that spurious modes do not satisfy the condition $\nabla \cdot \epsilon \mathbf{E} = 0$ (or $\nabla \cdot \mu \mathbf{H} = 0$) that is required for physical solutions. It was reported in [13] that the true cause of spurious modes is the incorrect approximation of the null space of the curl operator, or inconsistent approximations of the static solutions to the wave equation. There are many articles discussing the spurious mode problem [14–17].

The edge elements introduced allow the normal component of the vector field to be discontinuous from one element to the next, while keeping the tangential component continuous [18, 19]. It was reported that the edge elements (sometimes also called tangential vector finite elements) eliminate spurious modes [18].

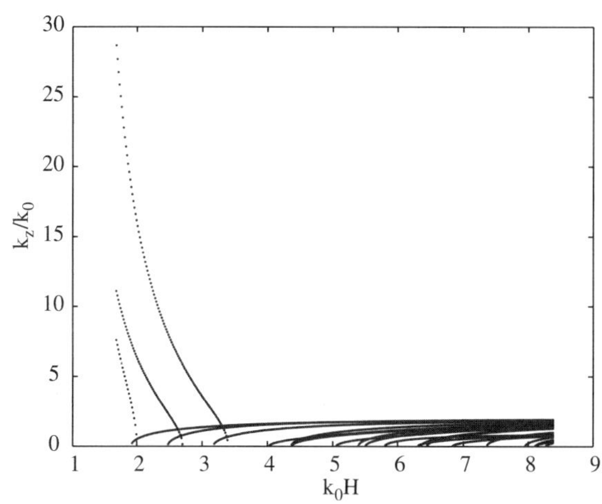

FIGURE 6.13 Spurious modes mixed with propagation modes in inhomogeneous waveguide.

In our work using multiwavelets, we have adopted the edge elements with our high-order basis of multiwavelets. However, spurious modes occurred, even though edge elements were employed. We will illustrate the method by which to detect and eliminate the spurious modes. Figure 6.13 shows the spurious modes in addition to the propagation modes for the inhomogeneous waveguide. Some of the spurious modes can be easily identified because they are beyond the meaningful region of the dispersion chart; some are difficult to distinguish. To make a decisive detection, we plotted the eigenvectors (electric field in the transverse plane corresponding to the spurious eigenvalue) in the waveguide. Figure 6.14 depicts the electric eigenfield in the transverse plane of the inhomogeneous waveguide. For an even spurious mode, the flux demonstrates source, or drain, nature as do star-star (or blackhole-blackhole) pairs in the left panel of Fig. 6.14. Shown in the right panel is for an odd spurious

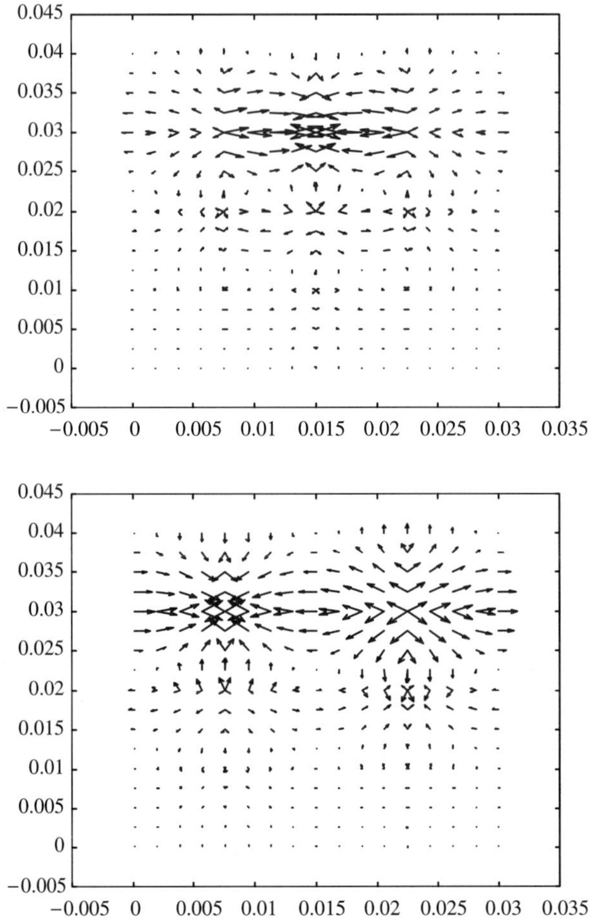

FIGURE 6.14 Electric eigenfield corresponding to even and odd spurious modes in inhomogeneous waveguide.

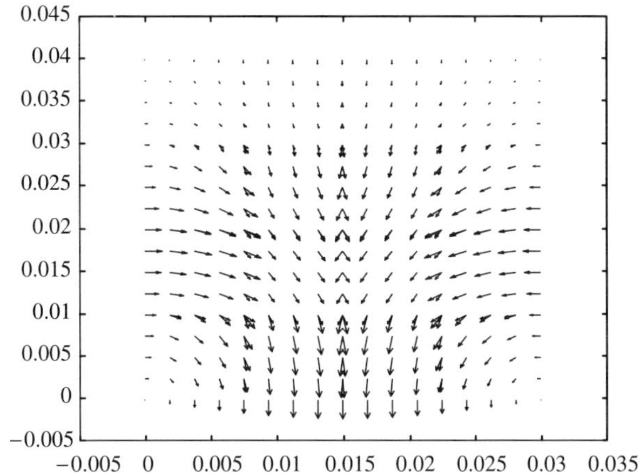

FIGURE 6.15 Electric eigenfield corresponding to physical mode in inhomogeneous waveguide.

mode, where the flux demonstrates source and drain behavior as star-blackhole pairs. In both events fictitious sources are generated, and they do not satisfy the zero divergence condition. We have carefully checked the edge element requirement and ensured the continuity of the tangential component across two adjacent elements. Nevertheless, for a rectangular element the tangent at a corner is undefined. The violation of $\nabla \cdot \epsilon \mathbf{E} = 0$ may happen at the element corner as seen the star or blackhole pattern. In contrast, for the eigenfield of a physical eigenvalue, an eigenvalue in which the mode exists physically, no star or blackhole ever occurs, as such the condition of $\nabla \cdot \epsilon \mathbf{E} = 0$ is strictly satisfied. Figure 6.15 illustrates a typical eigenfield pattern corresponding to the spurious free solution.

6.12 APPENDIX

Lemma 1. Suppose that $\phi(t) \in L^1$ (absolutely integrable) and $\phi_j(t-k)$, $j = 0, \ldots, r-1, k \in Z$, are linearly independent. The scalet $\phi(t)$ provides approximation of order m if and only if L has eigenvalues 2^{-j} corresponding to the left eigenvectors $\langle y^{[j]} | = (C \cdots \langle y_0^{[j]} |, \langle y_1^{[j]} |, \langle y_2^{[j]} |, \ldots), \langle y_k^{[j]} | = \sum_{\ell=0}^{j} \binom{j}{\ell} (-k)^{j-\ell} \langle u^{[\ell]} |$, $j = 0, \ldots, m-1$ where $\langle u_\ell |$ are constant row vectors.

Proof.

(1) *Necessity.* Approximation order m is defined by

$$t^j := G_j(t) = \sum_k \langle y_k^{[j]} | \phi(t-k) \rangle$$

$$= \langle y^{[j]} | \Theta(t) \rangle$$
$$= \langle y^{[j]} | L | \Theta(2t) \rangle, \qquad j = 0, 1, \ldots, m-1,$$

where $| \Theta(t) \rangle$ is given in (6.2.2). On the other hand,

$$\begin{aligned}(t^j) &= 2^{-j}(2t)^j \\ &= 2^{-j} G_j(2t) \\ &= 2^{-j} \langle y^{[j]} | \Theta(2t) \rangle.\end{aligned}$$

Thus

$$\langle y^{[j]} | L | \Theta(2t) \rangle = 2^{-j} \langle y^{[j]} | \Theta(2t) \rangle$$

and

$$\langle y^{[j]} | L = 2^{-j} \langle y^{[j]} |.$$

This means that L has eigenvalues 2^{-j} for left eigenvector $\langle y^{[j]} |$. In other words, if L has eigenvalues $1, \frac{1}{2}, \ldots, (\frac{1}{2})^{m-1}$ with eigenvector $\langle y^{[j]} |$, then

$$\begin{aligned}G_j(t) &= \sum_{-\infty}^{\infty} \langle y_k^{[j]} | \phi(t-k) \rangle \\ &= 2^{-j} G_j(2t). \end{aligned} \tag{6.12.1}$$

Recall from (6.2.4) that

$$\begin{aligned}G_j(t+1) &= (t+1)^j \\ &= \sum_k \langle y_k^{[j]} | \phi(t+1-k) \rangle \\ &= \langle y_{k-1}^{[j]} | \phi(t-k) \rangle\end{aligned}$$

and

$$\begin{aligned}G_j(t) - G_j(t+1) &= \sum_{k=-\infty}^{\infty} \langle (y_k^{[j]} - y_{k-1}^{[j]}) | \phi(t-k) \rangle \\ &= t^j - (t+1)^j \\ &= t^j - \sum_{\ell=0}^{j} \binom{j}{\ell} t^\ell \\ &= -\sum_{\ell=0}^{j-1} \binom{j}{\ell} t^\ell\end{aligned}$$

$$= -\sum_{\ell=0}^{j-1}\binom{j}{\ell}\sum_{k=-\infty}^{\infty}\langle y_k^{[\ell]}|\phi(t-k)\rangle$$

$$= -\sum_{k=-\infty}^{\infty}\sum_{\ell=0}^{j-1}\binom{j}{\ell}\langle y_k^{[\ell]}|\phi(t-k)\rangle.$$

The linear independence of the translations $\phi(t-k)$ gives

$$\langle y_k^{[j]}| - \langle y_{k-1}^{[j]}| = -\sum_{\ell=0}^{j-1}\binom{j}{\ell}\langle y_k^{[\ell]}|$$

or

$$\langle y_k^{[j]}| + \sum_{\ell=0}^{j-1}\binom{j}{\ell}\langle y_k^{[\ell]}| = \langle y_{k-1}^{[j]}|,$$

that is,

$$\sum_{\ell=0}^{j}\binom{j}{\ell}\langle y_k^{[\ell]}| = \langle y_{k-1}^{[j]}|. \tag{6.12.2}$$

It is easy to verify by direct substitution that (6.12.2) has a solution:

$$\langle y_k^{[j]}| = \sum_{\ell=0}^{j}(-1)^{\ell}\binom{j}{\ell}k^{\ell}\langle u^{[j-\ell]}| \tag{6.12.3}$$

$$= \sum_{\ell=0}^{j}(-k)^{j-\ell}\binom{j}{\ell}\langle u^{[\ell]}|, \tag{6.12.4}$$

where $\langle u^{[\ell]}|$ are some constant vectors. In fact, we may show that (6.12.3) and (6.12.4) are identical. Let

$$j-\ell = p, \qquad \text{i.e., } \ell = j-p.$$

$$\ell: 0 \to p \Rightarrow p := j \to 0.$$

From (6.12.3),

$$\sum_{\ell=0}^{j}(-1)^{\ell}\binom{j}{\ell}k^{\ell}\langle u^{[j-\ell]}| = \sum_{p=j}^{0}(-1)^{j-p}\binom{j}{j-p}k^{j-p}\langle u^{[p]}|$$

$$= \sum_{p=0}^{j}(-k)^{j-p}\binom{j}{p}\langle u^{[p]}|,$$

which is (6.12.4).

Next we show that (6.12.3) or (6.12.4) satisfies (6.12.2). Substitution of (6.12.3) into the RHS of (6.12.2) leads to

$$\begin{aligned}
\text{RHS} &= \langle y_{k-1}^{[j]} | \\
&= \sum_{\ell=0}^{j} (-(k-1))^{\ell} \binom{j}{\ell} \langle u^{[j-\ell]} | \\
&= \sum_{\ell=0}^{j} (1-k)^{\ell} \binom{j}{\ell} \langle u^{[j-\ell]} | \\
&= \sum_{\ell=0}^{j} \left\{ \sum_{m=0}^{\ell} (-k)^m \binom{\ell}{m} \langle u^{[\ell-m]} | \right\}.
\end{aligned}$$

The last equality comes from the fact that $\langle u^{[\ell-m]} |$ are constant vectors. The curly brackets in the previous equation are equal to $\langle y_k^{[\ell]} |$ by (6.12.3). Therefore

$$\langle y_{k-1}^{[\ell]} | = \sum_{\ell=0}^{j} \binom{j}{\ell} \langle y_k^{[\ell]} |.$$

(2) *Sufficiency.* If L has eigenvalues 2^{-j} corresponding to the eigenvectors

$$\begin{aligned}
\langle y_k^{[j]} | &= \sum_{\ell=0}^{j} \binom{j}{\ell} (-k)^{j-\ell} \langle u^{[\ell]} | \\
&= \sum_{\ell=0}^{j} \binom{j}{\ell} (-k)^{\ell} \langle u^{[j-\ell]} |.
\end{aligned}$$

We wish to show that for all $t^j,\ j < m,$

$$G_j(t) = t^j = \sum_k \langle y_k^{[j]} | \phi(t-k) \rangle = 2^{-j} G_j(2t)$$

Show. From (6.12.3) we have

$$\langle y_k^{[0]} | - \langle y_{k-1}^{[0]} | = \langle u_{k-1}^{[0]} | = \langle u^{[0]} | - \langle u^{[0]} | = 0,$$

that is

$$G_0(t) - G_0(t+1) = 0,$$

which implies that $G_0(t)$ is a periodic function. From (6.12.1), we obtain $G_0(t) = G_0(2t)$. Because $\phi(t) \in \mathrm{L}^1$, so is $G_0(t)$. Therefore we may apply the ergodic theorem that

$$G_0(t) = \sum_{k=-\infty}^{\infty} \langle u^{[0]} | \phi(t-k) \rangle = \text{constant}.$$

By proper normalization of $\langle u^{[0]} |$, we obtain

$$G_0(t) = 1.$$

Substituting $j = 1$ into (6.12.3), we have

$$\langle y_k^{[1]}| - \langle y_{k-1}^{[1]}| = -\langle u^{[0]}|.$$

Thus

$$\begin{aligned} G_1(t) - G_1(t+1) &= \sum_{k=-\infty}^{\infty} \left[\langle y_k^{[1]}| - \langle y_{k-1}^{[1]}\right] |\phi(t+k)\rangle \\ &= -\sum_k \langle u^{[0]}|\phi(t-k)\rangle \\ &= -1. \end{aligned}$$

This means that

$$F(t) := G_1(t) - t$$

is a periodic function, $F(t) = F(t+1)$. In fact,

$$\begin{aligned} &F(t) - F(t+1) \\ &\quad = [G_1(t) - t] - [G_1(t+1) - (t+1)] \\ &\quad = G_1(t) - G_1(t+1) + 1 \\ &\quad = -1 + 1 = 0. \end{aligned}$$

Again, using (6.12.1), we have

$$G_1(t) = 2^{-1} G_1(2t),$$

and the Birkhoff ergodic theorem gives

$$F_1(t) = 0$$

or

$$G_1(t) = \sum_{k=-\infty}^{\infty} \langle y_k^{[1]}|\phi(t+k)\rangle = t.$$

In general,

$$\begin{aligned} G_j(t) - G_j(t+1) &= \sum_{k=-\infty}^{\infty} [\langle y_k^{[j]}| - \langle y_{k-1}^{[j]}|]|\phi(t-k)\rangle \\ &= -\sum_{k=-\infty}^{\infty} \sum_{\ell=0}^{j-1} \binom{j}{\ell} \langle y_k^{[\ell]}|\phi(t-k)\rangle \\ &= -\sum_{\ell=0}^{j-1} \binom{j}{\ell} \sum_k \langle y_k^{[\ell]}|\phi(t-k)\rangle \\ &= -\sum_{\ell=0}^{j-1} \binom{j}{\ell} t^{\ell}. \end{aligned}$$

Again,

$$F_j(t) := G_j(t) - t^j$$

is a periodic function, which is verified as follows:

$$\begin{aligned} F_j(t+1) &= G_j(t+1) - (t+1)^j \\ &= G_j(t) + \sum_{\ell=0}^{j-1} \binom{j}{\ell} t^\ell - \sum_{\ell=0}^{j} \binom{j}{\ell} t^\ell \\ &= G_j(t) - t^j \\ &= F_j(t). \end{aligned}$$

Again,

$$G_j(t) = 2^{-j} G_j(2t).$$

Hence, as in the previous cases, we arrive at

$$G_j(t) = \sum_{k=-\infty}^{\infty} \langle y_k^{[j]} | \phi(t-k) = t^j$$

almost everywhere. □

Lemma 2. Suppose that $\langle y^{[j]} |$ is defined by

$$\langle y_k^{[j]} | = \sum_{\ell=0}^{j} \binom{j}{\ell} (-k)^{j-\ell} \langle u^{[\ell]} |$$

and L corresponds to a multiscalet with approximation order m. Then

$$\langle y^{[j]} | L = 2^{-j} \langle y^{[j]} |, \qquad j = 0, \ldots, m-1,$$

if and only if the following equations are held:

$$\sum_k \langle y_k^{[j]} | C_{2k+1} = 2^{-j} \langle u^{[j]} | \tag{6.12.5}$$

$$\sum_k \langle y_k^{[j]} | C_{2k} = 2^{-j} \langle y_1^{[j]} | \tag{6.12.6}$$

$$= 2^{-j} \sum_{\ell=0}^{j} (-1)^{j-\ell} \binom{j}{\ell} \langle u^{[\ell]} |. \tag{6.12.7}$$

Proof. The equation

$$\langle y^{[j]} | L = 2^{-j} \langle y[j] |$$

may be written explicitly as

$$\begin{bmatrix} \cdots & y_0^{[j]} & y_1^{[j]} & y_2^{[j]} \end{bmatrix} \begin{bmatrix} \cdots & & & & & & & & \\ \cdots & C_3 & C_2 & C_1 & C_0 & & & & \\ & & & C_3 & C_2 & C_1 & C_0 & & \\ & & & & & C_3 & C_2 & C_1 & C_0 \\ & & & & & & & & \cdots \end{bmatrix}$$

$$= 2^{-j} \begin{bmatrix} \cdots & y_0^{[j]} & y_1^{[j]} & y_2^{[j]} & \cdots \end{bmatrix}$$

Even though the infinite dimensional vectors and matrices are involved, the patterns are clear. We have either odd- or even-indexed equations

$$\left.\begin{aligned} y_0^{[j]} C_1 + y_1^{[j]} C_3 &= 2^{-j} y_0^{[j]} \\ y_1^{[j]} C_1 + y_2^{[j]} C_3 &= 2^{-j} y_2^{[j]} \end{aligned}\right\} \sum_k y_{k+\ell}^{[j]} C_{2k+1} = 2^{-j} y_{2\ell}^{[j]},$$

$$\left.\begin{aligned} y_0^{[j]} C_0 + y_1^{[j]} C_2 &= 2^{-j} y_1^{[j]} \\ y_1^{[j]} C_0 + y_2^{[j]} C_2 &= 2^{-j} y_3^{[j]} \end{aligned}\right\} \sum_k y_{k+\ell}^{[j]} C_{2k} = 2^{-j} y_{2\ell+1}^{[j]}. \tag{6.12.8}$$

Note that (6.12.6) is a special case of (6.12.8) with $\ell = 0$. Therefore let us prove that if (6.12.8) is held for $\ell = 0$, then it is true for all ℓ.

(1) For $j = 0$, $\langle y_k^{[0]} | = \langle u^{[0]} |$. Equation (6.12.8) becomes

$$\sum_k \langle y_0^{[0]} | C_{2k+1} = \langle u^{[0]} |,$$

$$\sum_k \langle y_0^{[0]} | C_{2k} = \langle u^{[0]} |,$$

which is independent of ℓ, and the two equations above are (6.12.6) for $j = 0$.

(2) Suppose that (6.12.6) is held for $j = 0, \ldots, n-1$. We show that the first equation in (6.12.8) is true for $j = n$ all $\ell \in Z$. Recall that from Theorem 1 we may have approximate order n, namely

$$\begin{aligned} \sum_K \langle y_{k+\ell}^{[n]} | \phi(t-k)\rangle &= \sum_m \langle y_m^{[n]} | \phi(t-\ell-m)\rangle \\ &= (x-\ell)^n \\ &= \sum_{j=0}^{n} \binom{n}{j} t^j (-\ell)^{n-j} \\ &= \sum_{j=0}^{n} \binom{n}{j} (-\ell)^{n-j} \sum_k \langle y_k^{[j]} | \phi(t-k)\rangle. \end{aligned}$$

The linear independence of $| \phi(t-k)\rangle$ requires that

$$\langle y_{k+\ell}^{[n]} | = \sum_{j=0}^{n} \binom{n}{j} (-\ell)^{n-j} \langle y_k^{[j]} |.$$

Using the induction hypothesis in conjunction with (6.12.6), we obtain

$$\begin{aligned}\sum_k \langle y_{k+\ell}^{[n]} | C_{2k+1} &= \sum_k \sum_{j=0}^{n} \binom{n}{j} (-\ell)^{n-j} \langle y_k^{[j]} | C_{2k+1} \\ &= \sum_{j=0}^{n} \binom{n}{j} (-\ell)^{n-j} \sum_k \langle y_k^{[j]} | C_{2k+1} \\ &= \sum_{j=0}^{n} \binom{n}{j} (-\ell)^{n-j} 2^{-j} \langle u^{[j]} | + \sum_k \\ &= 2^{-n} \sum_{j=0}^{n} \binom{n}{j} (-2\ell)^{n-j} \langle y_0^{[j]} | \\ &= 2^{-n} \langle y_{2\ell}^{[n-1]} |\end{aligned}$$

The second equation in (6.12.8) can be examined in the same manner. □

Lemmas 1 and 2 may be combined to give the following theorem.

Theorem 1. A multiscalet $\phi(t) \in L^1$ with linearly independent translations $|\phi(t-k)\rangle, k \in Z$ has approximation n if and only if there exist vectors $\langle u^{[j]}|, j = 0, \ldots, n-1$, such that (6.12.6) is satisfied.

Remark. For the scalar case, (6.12.6) reduces to the sum rules

$$\sum_{k=0}^{N} (-1)^k k^j C_k = 0, \qquad j = 0, \ldots, m-1,$$

which provide an approximation of order m.

Proof. Integer translations of intervallic functions of order r represent polynomials $1, \ldots, t^{2r-1}$; that is to say, $|\phi(t)\rangle$ provides approximation order $2r$. Hence, matrices C_0, C_1, and C_2 must satisfy requirements (6.3.1), with some starting vectors $\langle u^{[0]}|, \ldots, \langle u^{[r-1]}|$. In particular, for the intervallic multiwavelets, equations (6.3.1) and (6.3.2) reduce to

$$\langle u^{[j]} | C_1 = 2^{-j} \langle u^{[j]} |, \qquad j = 0, \ldots, 2r-1, \tag{6.12.9}$$

$$\begin{aligned}\langle u^{[j]} | C_0 + \langle y_1^{[j]} | C_2 &= 2^{-j} \langle y_1^{[j]} | \\ &= 2^{-j} \sum_{\ell=0}^{j} (-1)^{j-\ell} \langle u^{[j]} |, \qquad j = 0, \ldots, 2r-1.\end{aligned} \tag{6.12.10}$$

Since $C_1 = \operatorname{diag}\{1, \frac{1}{2}, \ldots, \frac{1}{2^{r-1}}\}$ has been found in (6.3.11), we may use (6.12.9) to find $\langle u^{[j]}|$. By simple algebra, we have

$$\langle u^{[j]} | = (u_0^{[j]}, u_1^{[j]}, \ldots, u_j^{[j]}, \ldots, u_{r-1}^{[j]})$$

with

$$u_0^{[j]} = u_j^{[j]} \delta_{j,\ell}, \qquad j, \ell = 0, 1, \ldots, r-1.$$

To determine $u_j^{[j]}$, $j = 0, \ldots, r-1$, we recall that

$$t^j = \sum_k \langle y_k^{[j]} | \phi(t-k) \rangle, \qquad j = 0, \ldots, r-1, \tag{6.12.11}$$

and that $\langle y_0^{[j]} | = \langle u^{[j]} |$. Differentiating (6.12.11) j times, we obtain

$$j! = \sum_k \langle y_k^{[j]} | \phi^{(j)}(t-k) \rangle, \qquad j = 0, \ldots, r-1.$$

Evaluating the previous equation at $t = 1$, and noticing that

$$\begin{aligned} \phi_\ell^{(k)}(1) &= \delta_{k,\ell}, \\ \phi_\ell^{(k)}(0) &= 0 = \phi^{(k)}(2), \end{aligned}$$

we have

$$\begin{aligned} j! &= \langle y_0^{[j]} | \phi^{(j)}(1) \rangle \\ &= \langle u^{[j]} | \phi^{(j)}(1) \rangle = [00 \cdots u_j^{[j]} \cdots 0][00 \cdots 1 \cdots 0]^T. \end{aligned}$$

Hence

$$u_j^{[j]} = j!$$

and

$$\langle u^{[j]} | = [0 \cdots j! \cdots 0], \qquad j = 0, \ldots, r-1. \tag{6.12.12}$$

Furthermore C_1 is $r \times r$, and has only r eigenvectors. We may extend the augment $\langle u^{[j]} |$ from $j = 0, \ldots, r-1$ to $2r-1$ by defining

$$\langle u^{[j]} | = 0 \qquad \text{for } j \geq r.$$

As a result (6.12.10) becomes

$$\langle y_1^{[j]} | C_2 = 2^{-j} \langle y_1^{[j]} |, \qquad j = r, \ldots, 2r-1,$$

which states that $\langle y_1^{[j]} |$ are the left eigenvectors of C_2 corresponding to the eigenvalues $\lambda_j = 2^{-j}$, $j = r, \ldots 2r-1$. Hence C_2 may be expressed as

$$C_2 = U^{-1} \Gamma U,$$

where

$$\Gamma = \text{diag}\{2^{-r}, \ldots, 2^{-2r-1}\}.$$

U is an $r \times r$ matrix with rows $\langle y_1^{[j]} |$, $j = r, \ldots, 2r-1$.

Recall from Lemma 1 that

$$\langle y_k^{[j]} | = \sum_{\ell=0}^{j} \binom{j}{\ell} (-k)^{j-\ell} \langle u^{[\ell]} |$$

or

$$\langle y_1^{[j]} | = \sum_{\ell=0}^{j} \binom{j}{\ell} (-1)^{j-\ell} \langle u^{\ell]} |.$$

As a result

$$U = \begin{bmatrix} \langle y_1^{[r]} | \\ \langle y_1^{[r+1]} | \\ \cdots \\ \langle y_1^{[2r-1]} | \end{bmatrix}.$$

The mnth entry of U is

$$U_{mn} = \{\langle y_1^{[r+m-1]} |\}_{n\text{th}} = \left\{ \sum_{\ell=0}^{r+m-1} \binom{r+m-1}{\ell} (-1)^{r+m-1-\ell} \langle u^{[\ell]} | \right\}_{n\text{th}}$$

$$= \{(-1)^{r+m-1} \binom{r-m-1}{0} [0!0 \cdots 0 \cdots 0] + \{(-1)^{r+m-2} \binom{r+m-1}{1} [01! \cdots 0 \cdots 0] + \cdots$$

$$+ (-1)^{r+m-n} \binom{r+m-1}{n-1} [00 \cdots (n-1)! \cdots 0] + \cdots \}_{n\text{th}}$$

$$= (-1)^{r+m-n} \binom{r+m-1}{n-1} (n-1)!$$

$$= (-1)^{r+m-n} \frac{(r+m-1)!}{(n-1)!(r+m-n)!} (n-1)!$$

$$= (-1)^{r+m-n} \frac{(r+m-1)!}{(r+m-n)!}$$

The equation above is in fact (6.3.17). □

6.13 PROBLEMS

6.13.1 Exercise 11

1. The lowpass filter matrices are given as

$$C_0 = \begin{bmatrix} \frac{1}{2} & \frac{3}{4} \\ -\frac{1}{8} & -\frac{1}{8} \end{bmatrix},$$

$$C_1 = \begin{bmatrix} 1 & 0 \\ 0 & \frac{1}{2} \end{bmatrix},$$

$$C_2 = \begin{bmatrix} \frac{1}{2} & -\frac{3}{4} \\ \frac{1}{8} & -\frac{1}{8} \end{bmatrix}.$$

Construct the multiscalets $\phi_0(t)$ and $\phi_1(t), t \in [0,2]$ by the dilation equation and iterative procedure (which is similar to that used in constructing the Daubechies D_2).

2. Compare the solutions $\phi_0(t)$ and $\phi_1(t)$ from the previous problems with the cubic Hermitian functions

$$H_0(t) = (3t^2 - 2t^3), \quad H_1(t) = t^3 - t^2, \qquad t \in [0,1],$$

and

$$H_0(t) = H_0(2-t), \quad H_1(t) = -H_1(2-t), \qquad t \in [1,2].$$

You may plot these functions in the same coordinates.

3. For multiplicity $r = 3$, compute the coefficient matrices C_0, C_1, and C_2, and then construct the multiscalets $\phi_0(t)$, $\phi_1(t)$, and $\phi_2(t)$. Verify numerically that

$$\Phi(1) = \begin{bmatrix} 1 & 0 & 0 \\ 0 & 1 & 0 \\ 0 & 0 & 1 \end{bmatrix}.$$

4. Compute the 2×2 matrices D_1, D_2, and D_3. Using these matrices, construct the multiwavelets ψ_1, ψ_2 and the dual multiwavelets $\tilde{\psi}_1$ and $\tilde{\psi}_2$ by

(a) Employing the iterative results of $\phi_0(t)$ and $\phi_1(t)$.

(b) Employing the cubic Hermitian functions.

Plot the results.

5. Show that the intervallic dual wavelet satisfies

$$\tilde{\psi}_p^{(j)}\left(k + \frac{3}{2}\right) = \delta_{p,j}\,\delta_{k,0}, \qquad j, p = 0, \ldots, r-1;\ k \in Z.$$

6. Derive the explicit expression for multiscalets of arbitrary r. Using the explicit expression of $r = 4$ construct the multiscalets and plot them.

BIBLIOGRAPHY

[1] I. Daubechies, *Ten Lectures on Wavelets*, SIAM, Philadelphia, 1992, at pp. 251–253.

[2] V. Strela and G. Strang, "Finite element multiwavelets," in *Approximation Theory, Wavelets and Applications*, Kluwer Academic, Dordrecht, 1995, pp. 485–496.

[3] V. Strela, "Multiwavelets: Theory and applications," Ph.D. dissertation, MIT, Cambridge, June 1996.

[4] G. Walter and X. Shen, *Wavelets and Orthogonal Systems*, 2nd ed., Chapman/CRC Press, Boca Raton, FL, pp. 242–252, 2001.

[5] B. Alpert, G. Beylkin, D. Gines, and L. Vozovoi, "Adaptive solution of partial differential equations in multiwavelet bases," PAM Report 409, 1999.

[6] J. Jin, *The Finite Element Method in Electromagnetics*, John Wiley, New York, 1993.

[7] R. Graglia, D. Wilton, and A. Peterson, "Higher order interpolatory vector bases for computational electromagnetics," *IEEE Trans. Ant. Propg.*, **45**(3), 329–342, Mar. 1997.

[8] G. Pan, K. Wang, and B. Gilbert, "Application of multiwavelet to the edge element method," *Microw. Optic. Tech. Letters*, **34**(2), 96–100, July 2002.

[9] J. Lee, D. Sun, and Z. Cendes, "Full-wave analysis of dielectric waveguides using tangential vector finite elements," *IEEE Trans. Microw. Theory Tech.*, **39**(8), 1262–1271, Aug. 1991.

[10] J. Tan and G. Pan, "A new edge element analysis of dispersive waveguide structures," *IEEE Trans. Microw. Theory Tech.*, **43**(11), 2600–2607, Nov. 1995.

[11] L. Andersen and J. Volakis, "Development and application of a novel class of hierarchical tangential vector finite elements for electromagnetics," *IEEE Trans. Ant. Propg.*, **47**(1), 112–120, Jan. 1999.

[12] X. Liang, B. Jian, and G. Ni, "The B-spline finite element method applied to axisymmetrical and nonlinear field problems," *IEEE Trans. Magn.*, **24**(1), 27–30, Jan. 1988.

[13] D. Sun, J. Manges, X. Yuan, and Z. Cendes, "Spurious modes in finite-element methods," *IEEE Ant. and Propg.*, **37**(5), 12–24, Oct. 1995.

[14] J. Nedelec, "Mixed finite elements in R^3," *Numer. Math.*, **35**, 315–341, 1980.

[15] J. Nedelec, "A new family of mixed finite elements in R^3," *Numer. Math.*, **50**, 57–81, 1986.

[16] D. Lynch and K. Paulsen, "Origin of vector parasites in numerical maxwell solutions," *IEEE Trans. Microw. Theory Tech.*, **39**(3), 383–394, Mar. 1991.

[17] S. Wong and Z. Cendes, "Combined finite element-modal solution of three-dimensional eddy current problems," *IEEE Trans. Magn.*, **24**, 2685–2687, 1988.

[18] J. Webb, "Edge elements and what they can do for you," *IEEE Trans. Magn.*, **29**(2), 1460–1465, Mar. 1993.

[19] J. Lee, "Tangential vector finite elements and their application to solving electromagnetic scattering problems," ECE Dept. WPI, Worcester, MA.

CHAPTER SEVEN

Wavelets in Scattering and Radiation

In this chapter we examine scattering from 2D grooves using standard Coiflets, scattering from 2D and 3D objects, scattering and radiation of curved wire antennas, and scatterers employing Coifman intervallic wavelets. We provide the error estimate and convergence rate of the single-point quadrature formula based on Coifman scalets. We also introduce the smooth local cosine (SLC), which is referred to as the Malvar wavelet [1], as an alternative to the intervallic wavelets in handling bounded intervals.

7.1 SCATTERING FROM A 2D GROOVE

The scattering of electromagnetic waves from a two-dimensional groove in an infinite conducting plane has been studied using a hybrid technique of physical optics and the method of moments (PO-MoM) [2], where pulses and Haar wavelets were employed to solve the integral equation.

In this section we apply the same formulation as in [2] but implement the Galerkin procedure with the Coifman wavelets. We first evaluate the physical optics (PO) current on an infinite conducting plane [3] and then apply the hybrid method, which solves for a local correction to the PO solution. In fact the unknown current is expressed by a superposition of the known PO current induced on an infinite conducting plane by the incident plane wave plus the local correction current in the vicinity of the groove. Because of its local nature the correction current decays rapidly and is essentially negligible several wavelengths away from a groove.

Because of the rapidly decaying nature of the unknown correction current, the Coiflets can be directly employed on a finite interval without any modification (periodizing or intervallic treatment). Hence all advantages of standard wavelets, including orthogonality, vanishing moments, MRA, single-point quadrature, and the like, are preserved. The localized correction current is numerically evaluated using the

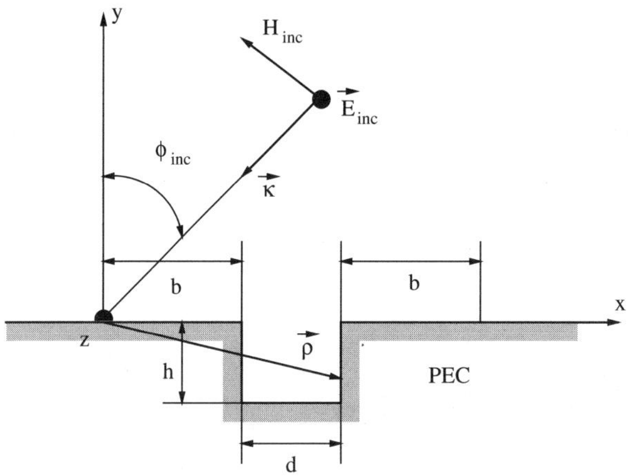

FIGURE 7.1 Geometry of the 2D groove in a conducting plane.

MoM with the Galerkin technique [4]. The hybrid PO-MoM formulation is implemented with the Coiflets of order $L = 4$, which are compactly supported and possess the one-point quadrature rule with a convergence of $O(h^5)$ in terms of the interval size h. This reduces the computational effort of filling the MoM impedance matrix entries from $O(n^2)$ to $O(n)$. As a result the Coiflet based method with twofold integration is faster than the traditional pulse-collocation algorithm. The obtained system of linear equations is solved using the standard LU decomposition [5] and iterative Bi-CGSTAB [6] methods. For an impedance matrix of large size, the Bi-CGSTAB method performs faster than the standard LU decomposition approach, especially when sparse matrices are involved.

7.1.1 Method of Moments (MoM) Formulation

In this section the Coifman wavelets are used on a finite interval without any modification. The scattering of the $TM^{(z)}$ and $TE^{(z)}$ time-harmonic electromagnetic plane waves by a groove in a conducting infinite plane is considered. The cross-sectional view of the 2D scattering problem is shown in Fig. 7.1.

The angle of incidence ϕ_{inc} is measured with respect to the y axis. The depth and width of the groove are h and d, respectively. For the $TM^{(z)}$ polarization of the incident plane wave, the induced current $\mathbf{J}_s$ is z-directed and independent of z, that is, $\mathbf{J}_s = \hat{z} \cdot J_z(x, y)$. For the $TE^{(z)}$ scattering case, the current $\mathbf{J}_s$ is also z-independent and lies in the (x, y) plane.

First, we consider the case of the $TM^{(z)}$ scattering. We split the geometry of our scattering problem into segments $\{l_s\}$, $s = 1, \ldots, 6$, as shown in Fig. 7.2. The segments l_1 and l_5 are semi-infinite. We write J_z in terms of four current distributions J^{PO}, J_L^{PO}, J_C, and $\tilde{J}_C$ as

$$J_z = J^{\text{PO}} - J_L^{\text{PO}} + J_C + \tilde{J}_C. \tag{7.1.1}$$

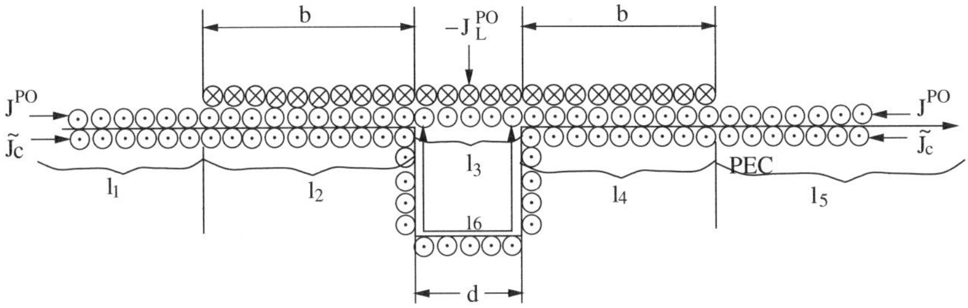

FIGURE 7.2 Partition of the induced current J_z.

In (7.1.1) we partitioned the induced current J_z into the following components:

- J^{PO} is the known physical optics current of the unperturbed problem (the current that would be induced on a perfectly infinite plane formed by $\bigcup_{s=1}^{5} l_s$).
- J_L^{PO} is the portion of the physical optics current J^{PO} residing on $\bigcup_{s=2}^{4} l_s$.
- J_C is the unknown surface *correction* current on the groove region l_6 and its vicinity l_2 and l_4.
- $\tilde{J}_C$ is the unknown surface correction current, defined on l_1 and l_5.

The widths of the segments l_2 and l_4 are chosen sufficiently large to ensure that the induced current on the segments l_1 and l_5 is almost equal to the physical J^{PO} optics current on an infinite plane.

To find the induced current J_z, we use the following boundary condition on the surface of the perfect conductor

$$L_z^s(J_z) + E_z^{\text{inc}} = 0 \qquad \text{on } l_1 \bigcup l_2 \bigcup l_6 \bigcup l_4 \bigcup l_5, \tag{7.1.2}$$

where the operator $L_z^s(\cdot)$ denotes the scattered electric field component which is tangential to the surface of the groove scatterer and caused by the current J_z. The electric field component E_z^{inc} is the tangential component of the incident electric field. From (7.1.1) and (7.1.2) we get the following:

$$L_z^s(J^{PO} - J_L^{PO} + J_C + \tilde{J}_C) + E_z^{\text{inc}} = 0 \qquad \text{on } l_1 \bigcup l_2 \bigcup l_6 \bigcup l_4 \bigcup l_5.$$

The operator $L_z^s(\cdot)$, which describes the scattered field, is a linear function of the induced current. Thus

$$L_z^s(J^{PO}) - L_z^s(J_L^{PO}) + L_z^s(J_C) + L_z^s(\tilde{J}_C) = -E_z^{\text{inc}}$$
$$\text{on } l_1 \bigcup l_2 \bigcup l_6 \bigcup l_4 \bigcup l_5. \tag{7.1.3}$$

We should note here that the sum of the incident field and scattered field evaluated beneath the interface is equal to zero, according to the extinction theorem [7]. This means that

$$L_z^s(J^{PO}) = -E_z^{\text{inc}} \qquad \text{on } l_1 \bigcup l_2 \bigcup l_6 \bigcup l_4 \bigcup l_5. \tag{7.1.4}$$

Combining (7.1.3) and (7.1.4), we obtain

$$L_z^s(J_C + \tilde{J}_C) = L_z^s(J_L^{\text{PO}}) \qquad \text{on } l_1 \bigcup l_2 \bigcup l_6 \bigcup l_4 \bigcup l_5. \tag{7.1.5}$$

We can further simplify equation (7.1.5) by recalling that the induced current on l_1 and l_5 is essentially equal to the physical optics current J^{PO}. This gives the following approximation:

$$\tilde{J}_C \approx 0. \tag{7.1.6}$$

From (7.1.6) and (7.1.5) it follows immediately that $L_z^s(\tilde{J}_C) \approx 0$, and hence

$$L_z^s(J_C) = L_z^s(J_L^{\text{PO}}) \qquad \text{on } l_1 \bigcup l_2 \bigcup l_6 \bigcup l_4 \bigcup l_5, \tag{7.1.7}$$

where the right-hand side is the known tangential electric field due to the current J_L^{PO}, while J_C is the unknown correction current. The correction current J_C is defined on $l_2 \bigcup l_6 \bigcup l_4$, and therefore (7.1.7) can be rewritten in the following way:

$$L_z^s(J_C) = L_z^s(J_L^{\text{PO}}) \qquad \text{on } l_2 \bigcup l_6 \bigcup l_4. \tag{7.1.8}$$

For the $TM^{(z)}$ scattering, the operator $L_z^s(\cdot)$ has the form

$$L_z^s(J(\boldsymbol{\rho}')) = -\frac{\kappa\eta}{4} \int_l J(\boldsymbol{\rho}') \cdot H_0^{(2)}(\kappa|\boldsymbol{\rho} - \boldsymbol{\rho}'|)\, dl'.$$

Therefore we can rewrite (7.1.8) as

$$\int_{l_2+l_6+l_4} J_C(\boldsymbol{\rho}') \cdot H_0^{(2)}(\kappa|\boldsymbol{\rho} - \boldsymbol{\rho}'|)\, dl' = \int_{l_2+l_3+l_4} J_L^{\text{PO}}(\boldsymbol{\rho}') \cdot H_0^{(2)}(\kappa|\boldsymbol{\rho} - \boldsymbol{\rho}'|)\, dl', \tag{7.1.9}$$

where $\boldsymbol{\rho} \in l_2 \bigcup l_6 \bigcup l_4$, J_L^{PO} is the known physical optics current, and $J_C(\boldsymbol{\rho}')$ is the unknown local current.

Equation (7.1.9) is sufficient for the determination of the local current J_C. The unknown current J_C is defined on the finite contour $l_2 \bigcup l_6 \bigcup l_4$ and is almost equal to the physical optics current J^{PO} at the starting and end points of the integral path.

The Coifman wavelets are defined on the real line. In order to apply the Coifman wavelets to the MoM on a finite interval, we change (7.1.9) into a slightly different form, such that the solution is almost equal to zero at the endpoints of the interval. This is due to the fact that the local current J_C is approximately equal to the physical optics current J_L^{PO} at the endpoints of the interval l_2 and l_4. We subtract the known current J^{PO}, defined on the intervals l_2 and l_4, from the unknown current J_L. Hence (7.1.9) becomes

$$\int_{l_2+l_6+l_4} J_C(\boldsymbol{\rho}') \cdot H_0^{(2)}(\kappa|\boldsymbol{\rho} - \boldsymbol{\rho}'|)\, dl' - \int_{l_2+l_4} J_L^{\text{PO}}(\boldsymbol{\rho}') \cdot H_0^{(2)}(\kappa|\boldsymbol{\rho} - \boldsymbol{\rho}'|)\, dl'$$
$$= \int_{l_3} J_L^{\text{PO}}(\boldsymbol{\rho}') \cdot H_0^{(2)}(\kappa|\boldsymbol{\rho} - \boldsymbol{\rho}'|)\, dl'. \tag{7.1.10}$$

We define the new unknown current

$$J_p = \begin{cases} J_C & \text{on } l_6 \\ J_C - J_L^{\mathrm{PO}} & \text{on } l_2 \bigcup l_4. \end{cases} \tag{7.1.11}$$

Using the new definition, we rewrite (7.1.10) in a compact form:

$$\int_{l_2+l_6+l_4} J_p(\boldsymbol{\rho}') \cdot H_0^{(2)}(\kappa|\boldsymbol{\rho}-\boldsymbol{\rho}'|)\,dl' = \int_{l_3} J_L^{\mathrm{PO}}(\boldsymbol{\rho}') \cdot H_0^{(2)}(\kappa|\boldsymbol{\rho}-\boldsymbol{\rho}'|)\,dl',$$
$$\boldsymbol{\rho} \in l_2 \bigcup l_6 \bigcup l_4. \tag{7.1.12}$$

The unknown current J_p in (7.1.12) is solved by the MoM with Galerkin's technique. First, we expand J_p in terms of the basis functions $\{q_i\}_{i=1}^N$ defined on $l_2 \bigcup l_6 \bigcup l_4$ as

$$J_p = \sum_{n=1}^{N} a_n q_n.$$

Then, we use the same basis as the testing functions to convert the integral equation (7.1.12) into a matrix equation

$$[Z][I] = [V], \tag{7.1.13}$$

where

$$\begin{aligned} Z_{m,n} &= \int_{S_m}\int_{S_n} q_m(l) q_n(l') H_0^{(2)}(\kappa|\boldsymbol{\rho}-\boldsymbol{\rho}'|)\,dl'\,dl, \\ I_n &= a_n, \\ V_m &= \int_{S_m}\int_{l_3} q_m(l) J_L^{\mathrm{PO}}(l') H_0^{(2)}(\kappa|\boldsymbol{\rho}-\boldsymbol{\rho}'|)\,dl'\,dl. \end{aligned} \tag{7.1.14}$$

In the previous equations, S_m denotes the support of the basis function q_m. By solving (7.1.13) numerically, we obtain the solution to the scattering problem of Fig. 7.1 with a finite number of unknowns.

To calculate V_m by using (7.1.14), we also need an expression for the physical optics current J^{PO}. For the $TM^{(z)}$ scattering we find J_L^{PO} [3]

$$\mathbf{J}^{\mathrm{PO}} = 2\hat{n} \times \mathbf{H}_{\mathrm{inc}}.$$

The incident electric and magnetic field components are given by

$$\begin{aligned} \mathbf{E}_{\mathrm{inc}} &= \hat{z} \cdot \eta \cdot e^{j\kappa(x \sin\phi_{\mathrm{inc}} + y\cos\phi_{\mathrm{inc}})}, \\ \mathbf{H}_{\mathrm{inc}} &= (-\hat{x} \cdot \cos\phi_{\mathrm{inc}} + \hat{y} \cdot \sin\phi_{\mathrm{inc}}) \cdot e^{j\kappa(x \sin\phi_{\mathrm{inc}} + y\cos\phi_{\mathrm{inc}})}. \end{aligned}$$

Upon substituting (7.1.15) into (7.1.1), we obtain

$$\mathbf{J}_L^{\mathrm{PO}} = \hat{z} \cdot 2\cos\phi_{\mathrm{inc}} \cdot e^{j\kappa x \sin\phi_{\mathrm{inc}}}.$$

The same approach is employed to construct the integral equation for the $TE^{(z)}$ scattering. For the sake of simplicity, we will omit the detailed derivation of the $TE^{(z)}$ case and present only numerical results.

7.1.2 Coiflet-Based MoM

The Coifman scalets of order $L = 2N$ and resolution level j_0 are employed as the basis functions to expand the unknown surface current J_p in (7.1.12) in the form

$$J_p(t') = \sum_n a_n \varphi_{j_0,n}(t'),$$

where we have employed the parametric representation $\boldsymbol{\rho} = \boldsymbol{\rho}(t)$ and $\boldsymbol{\rho}' = \boldsymbol{\rho}'(t')$, and $\varphi_{j_0,n}(t') = 2^{j_0/2}\varphi(2^{j_0}t' - n)$. Again, all equations are presented only for the $TM^{(z)}$ scattering, and the $TE^{(z)}$ case is treated in the same way.

After testing the integral equation (7.1.12) with the same Coifman scalets $\{\varphi_{j_0,m}(t)\}$, we arrive at the impedance matrix with the mnth entry

$$Z_{m,n} = \int_{S_m}\int_{S_n} H_0^{(2)}(\kappa|\boldsymbol{\rho} - \boldsymbol{\rho}'|)\varphi_{j_0,m}(t)\varphi_{j_0,n}(t')\,dt'\,dt \tag{7.1.15}$$

and

$$V_m = \int_{S_m}\int_{l_3} \varphi_{j_0,m}(t) J^{\mathrm{PO}}(t') H_0^{(2)}(\kappa|\boldsymbol{\rho} - \boldsymbol{\rho}'|)\,dt'\,dt, \tag{7.1.16}$$

where S_n and S_m are the support of the expansion and testing wavelets, respectively.

The following one-point equation rule [8]:

$$\int_{S_m}\int_{S_n} K(t,t')\varphi_{j_0,m}(t)\varphi_{j_0,n}(t')\,dt'\,dt \approx \frac{1}{2^{j_0}} K\left(\frac{m}{2^{j_0}}, \frac{n}{2^{j_0}}\right) \tag{7.1.17}$$

is used to evaluate the matrix elements for which $H_0^{(2)}(\kappa|\boldsymbol{\rho} - \boldsymbol{\rho}'|)$ is free of singularity within the interval of integration. To be more specific, the one-point quadrature formula (7.1.17) is used to calculate elements of the impedance matrix for which $|m - n| \geq 1$. In addition to that, it is also used to construct the right-hand side vector (7.1.16). The error estimate of (7.1.17) can be found in Section 7.2.3.

For all diagonal elements, the kernel of the integral (7.1.15) has a singularity at $t = t'$, where the diagonal elements are computed using standard Gauss–Legendre quadrature [5]. We used different number of Gaussian points with respect to t and t' in order to avoid the situation where $t = t'$. For the MoM with pulse basis, we used 4 and 6 Gaussian points for the integration with respect to t' and t. They are the minimum numbers of Gaussian points guaranteeing accurate and stable numerical results. For the Coiflet-based MoM, we split a support of each scalet into 5 small segments and used 4 and 6 points on each subinterval. In all numerical examples, the Coiflets are of order $L = 2N = 4$, this reflects a good trade off between accuracy and computation time.

It has also been noted that the accuracy of expression (7.1.17) depends on the resolution level j_0. The higher the resolution level is, the more accurate the results are. Here we mainly use the Coifman scalets with a resolution level $j_0 = 5$ to compute the MoM impedance matrix. We then perform the fast wavelet transform (FWT) of Section 4.8 to further sparsify the impedance matrix in standard form.

7.1.3 Bi-CGSTAB Algorithm

For the solution of the linear algebraic system (7.1.13), one could use the standard LU decomposition in combination with backsubstitution, numerically available in many books. When the size of the impedance matrix Z becomes large, it is better to use the iterative method to speed up the numerical computation. In our numerical calculations we use the standard LU decomposition technique as well as the stabilized variant of the bi-conjugate gradient (Bi-CG) iterative solver, named Bi-CGSTAB [6].

It is very important to note that the Bi-CGSTAB method does not involve the transpose matrix Z^T. The actual stopping criteria used in all numerical calculations is

$$||\, r_i \,||_{L_2} < EPS \cdot ||\, b - Ax_0 \,||_{L_2}$$

with $EPS = 10^{-5}$. It has been found from experiment that with this value of EPS we maintain accurate results in comparison with those of the LU decomposition.

We have also employed the sparse version of the Bi-CGSTAB algorithm for the wavelet solution with a sparse standard matrix form. The row-indexed sparse storage technique has been implemented [5] to store a given sparse matrix in the computer memory. To be more specific, we have also used the special fast algorithm for production of the sparse matrix with a given vector at every iteration step of the Bi-CGSTAB.

7.1.4 Numerical Results

We will first present the numerical results obtained from the $TM^{(z)}$ scattering with the following dimensions: $b = 3.09375\lambda$, $h = 0.5\lambda$, and $d = 0.5\lambda$. The number of unknowns for the pulse basis is 246. We used 256 Coifman scaling functions to expand the unknown current J_p. The order of the Coiflets is $L = 2N = 4$ with the resolution level $j_0 = 5$. The obtained numerical results for different incident angles are presented in Fig. 7.3. We plotted the normalized correction current J_c with respect to the length parameter (arclength) given in λ. The local current J_L was obtained from (7.1.1) after we found the unknown current J_p numerically. Numerical results for thc case of $TE^{(z)}$ scattering are shown in Fig. 7.4.

To demonstrate the advantage of the Coifman wavelets and Bi-CGSTAB algorithm, we present in Tables 7.1 and 7.2 the results of computation time. All numerical computations presented here were performed on a standard personal computer with 32-bit 400 MHz clock CPU from Advanced Micro Devices (AMD), 128 Mb RAM and Suse 6.3 Linux operational system. The public domain GNU g++ compiler was used to create executable codes. The following parameters were chosen to create the

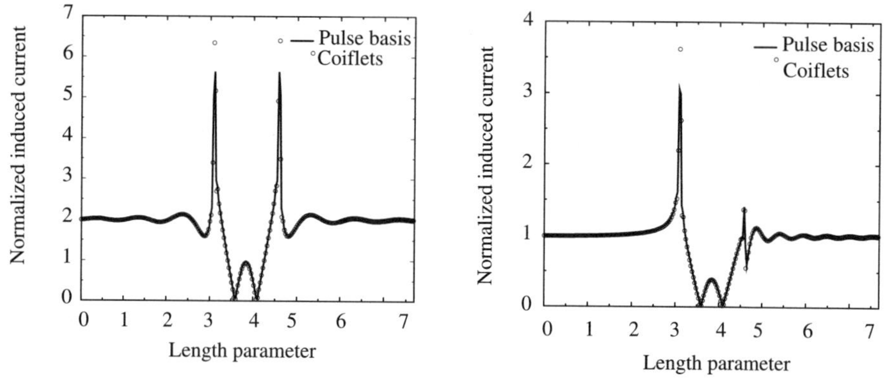

FIGURE 7.3 Normalized induced current versus length λ, $TM^{(z)}$ case with: $b = 3.09375\lambda$, $h = 0.5\lambda$, $d = 0.5\lambda$, $N_p = 246$, $N_c = 256$. *Left:* $\phi_{\text{inc}} = 0°$; *right:* $\phi_{\text{inc}} = 60°$.

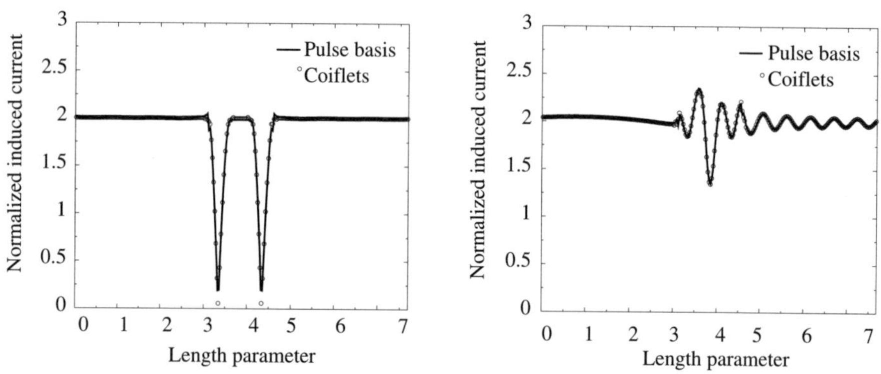

FIGURE 7.4 Normalized induced current versus length λ, $TE^{(z)}$ case with $b = 3.09375\lambda$, $h = 0.5\lambda$, $d = 0.5\lambda$, $N_p = 246$, $N_c = 256$, *Left:* $\phi_{\text{inc}} = 0°$; *right:* $\phi_{\text{inc}} = 60°$.

TABLE 7.1. Computation Time for the Pulse Basis, $TM^{(z)}$ Scattering

N_p	LU Time (s)	Bi-CGSTAB time (sec)	Iteration, N_{it}
1014	522.86	331.85	61
502	85.94	73.21	44
246	16.57	16.47	33

TABLE 7.2. Computation Time for the Coifman Wavelets, $TM^{(z)}$ Scattering

N_c	LU Time (s)	Bi-CGSTAB Time (s)	N_{it}	Sparse Bi-CGSTAB Time (s)	N_{it}	Sparsity (%)
1024	354.42	168.82	61	60.91	62	11.94
512	45.94	31.50	43	18.19	45	15.78
256	8.03	8.49	34	6.65	34	22.28

data presented in Tables 7.1 and 7.2:

$$b = 3.09375\lambda,\ h = 0.5\lambda,\ d = 0.5\lambda,\ \phi_{\text{inc}} = 60^\circ,\ N_p = 246,\ N_c = 256.$$

$$b = 6.34375\lambda,\ h = 1.0\lambda,\ d = 1.0\lambda,\ \phi_{\text{inc}} = 60^\circ,\ N_p = 502,\ N_c = 512.$$

$$b = 12.84375\lambda,\ h = 2.0\lambda,\ d = 2.0\lambda,\ \phi_{\text{inc}} = 60^\circ,\ N_p = 1014,\ N_c = 1024.$$

The numbers N_p and N_c denote the number of pulses and Coiflets in the MoM, N_{it} is the number of iterations in the Bi-CGSTAB algorithm. We implemented the LU and Bi-CGSTAB methods to solve the system of linear equations. We also decompose the system matrix of the Coifman-based MoM into the standard matrix. The sparse version of the Bi-CGSTAB is used to solve the system of linear equations. Then the threshold level of $10^{-4} \cdot p$ is selected to sparsify the system matrix, where parameter p is the maximum entry in magnitude. The relative error of 10^{-5} has been used as a stopping criterion for the Bi-CGSTAB. The sparsity of a matrix is defined as the percentage of the nonzero entries in the matrix.

From Tables 7.1 and 7.2 it can be seen that the use of Coifman wavelet-based MoM in combination with the standard form matrix achieves a factor of approximately 2.5 to 8.5 in the CPU time savings over the pulse-based MoM with the LU decomposition. This is due to the one-point quadrature formula, fast wavelet transform, and fast sparse matrix solver. Figure 7.6 illustrates the local current J_L obtained from

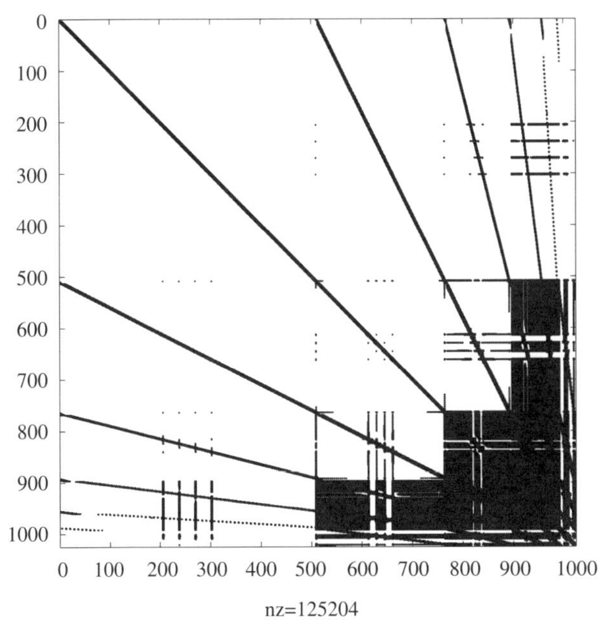

FIGURE 7.5 Standard form matrix, $TM^{(z)}$ scattering.

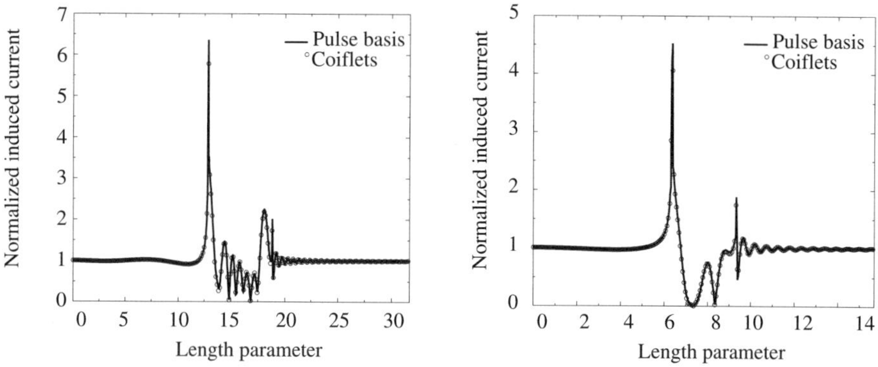

FIGURE 7.6 Normalized induced current versus length λ, $TM^{(z)}$ case. *Left:* $b = 12.84375\lambda$, $h = 2.0\lambda$, $d = 2.0\lambda$, $\phi_{\text{inc}} = 60°$, $N_p = 1014$, $N_c = 1024$; *right:* $b = 6.34375\lambda$, $h = 1.0\lambda$, $d = 1.0\lambda$, $\phi_{\text{inc}} = 60°$, $N_p = 502$, $N_c = 512$.

the $TM^{(z)}$ scattering with the parameters in Tables 7.1 and 7.2. Figure 7.5 shows the standard form matrix with 1024 unknowns and five resolution levels.

For all numerical results presented here, we made use of the Coiflets with resolution level $j_0 = 5$. This level has been chosen after a number of numerical trials indicating that this resolution level is the minimum at which there is good agreement between the pulse basis approach and wavelet technique. As the last numerical example we decrease the resolution level to $j_0 = 4$, thus obtaining fewer unknowns than in Fig. 7.3. Actually we used 123 pulse functions and 133 Coifman scalets to arrive at the results shown in Fig. 7.7. We can see that we still have good agreement

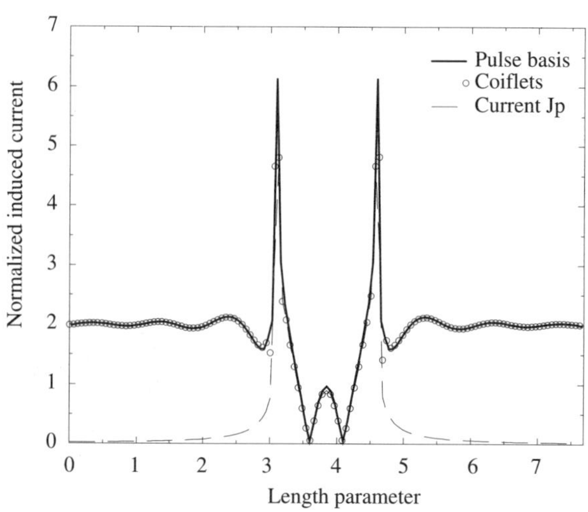

FIGURE 7.7 Normalized induced current versus length λ, $TM^{(z)}$ case: $b = 3.09375\lambda$, $h = 0.5\lambda$, $d = 0.5\lambda$, $\phi_{\text{inc}} = 0°$, $N_p = 123$, $N_c = 133$.

between the two approaches, though a small difference between the methods appears at the groove edges. The current J_p in (7.1.11) is also plotted in Fig. 7.7.

7.2 2D AND 3D SCATTERING USING INTERVALLIC COIFLETS

Periodic wavelets were applied to bounded intervals in Chapter 4. Nonetheless, the unknown functions must take on equal values at the endpoint of the bounded interval in order to apply periodic wavelets as the basis functions. The intervallic wavelets release the endpoints restrictions imposed on the periodic wavelets. The intervallic wavelets form an orthonormal basis and preserve the same multiresolution analysis (MRA) of other usual unbounded wavelets. The Coiflets possess a special property: their scalets have many vanishing moments. As a result the zero entries of the matrices are identifiable directly, without using a truncation scheme of an artificially established threshold. Furthermore the majority of matrix elements are evaluated directly, without performing numerical integration procedures such as Gaussian quadrature. For an $n \times n$ matrix the number of actual numerical integrations is reduced from n^2 to the order of $3n(2L-1)$ when the Coiflets of order L are employed.

7.2.1 Intervallic Scalets on $[0, 1]$

The basic concepts of intervallic wavelets were derived in Chapter 4. Here we will quickly review some major facts and then present the new material.

Starting from an orthogonal Coifman scalet with $3L$ nonzero coefficients (where $L = 2N$ is the order of the Coifman wavelets), we will assume that the scale is fine enough that the left- and right-edge bases are independent. Since the Coifman wavelets have vanishing moment properties in both scalets and wavelets, we have

$$\int \varphi(x)\,dx = 1, \tag{7.2.1}$$

$$\int x^p \varphi(x)\,dx = 0, \qquad p = 1, 2, \ldots, 2N-1, \tag{7.2.2}$$

$$\int x^p \psi(x)\,dx = 0, \qquad p = 0, 1, 2, \ldots, 2N-1. \tag{7.2.3}$$

Scalets under the L^2 norm exhibit the Dirac δ-like sampling property for smooth functions. Namely, if $\varphi(x)$ is supported in $[p, q]$, and we expand $f(x)$ at a point $0 \in [p, q]$, then

$$\int_p^q f(x)\varphi(x) = \int_p^q \left\{ f(0) + f'(0)x + \cdots + \frac{f^{2N-1}(0)x^{2N-1}}{(2N-1)!} + \cdots \right\} \varphi(x)\,dx$$
$$\approx f(0). \tag{7.2.4}$$

This property in a simple sense is similar to the Dirac δ function property

$$\int f(x)\,\delta(x)\,dx = f(0).$$

Of course, the Dirac δ-function is the extreme example of localization in the space domain, with an infinite number of vanishing moments. In all numerical examples we have chosen Coiflets of order $2N = 4$. From (7.2.4) the convergence rate is $O(h^4)$. Since the fourth moment is negligibly small in Table 7.3, we essentially have the convergence rate $O(h^5)$. This is in contrast to the MoM single-point quadrature, where only $O(h)$ is expected.

All polynomials of degree $< 2N$ can be written as linear combinations of $\varphi_{j,k}$ for $k \in Z$, with coefficients that are polynomials of degree $< 2N$. More precisely, if A is a polynomial of degree $p \leq 2N - 1$, then a polynomial B of the same degree exists such that

$$A(x) = \sum_k B(k)\varphi_{j,k}(x).$$

Since $\{\varphi_{j,k}\}$ is an orthonormal basis for V_j, any monomial x^α, $\alpha \leq 2N - 1$ can be seen by using equations (7.2.1) and (7.2.2) to have the representation (see (4.13.9))

$$\begin{aligned} x^\alpha &= \sum_k \langle x^\alpha, \varphi_{j,k}\rangle \varphi_{j,k}(x) \\ &= \sum_k \frac{k^\alpha}{2^{j(\alpha+1/2)}}\varphi_{j,k}(x), \end{aligned}$$

where j is the level of the Coifman wavelets. The restriction to $[0, 1]$ can be written as

$$x^\alpha \mid_{[0,1]} = \left(\sum_{k=-4N+2}^{2N-1} + \sum_{k=2N}^{2^j-4N+1} + \sum_{k=2^j-4N+2}^{2^j+2N-1} \right) \langle x^\alpha, \varphi_{j,k}\rangle \varphi_{j,k}(x) \mid_{[0,1]} .$$

Let

$$x^\alpha_{j,L} = 2^{j(\alpha+1/2)} \sum_{k=-4N+2}^{2N-1} \langle x^\alpha, \varphi_{j,k}\rangle \varphi_{j,k}(x) \mid_{[0,1]}$$

and

$$x^\alpha_{j,R} = 2^{j(\alpha+1/2)} \sum_{k=2^j-4N+2}^{2^j+2N-1} \langle x^\alpha, \varphi_{j,k}\rangle \varphi_{j,k}(x) \mid_{[0,1]},$$

where subscript L and R represent left and right, respectively. Hence

$$2^{j/2}(2^j x)^\alpha = x^\alpha_{j,L} + 2^{j(\alpha+1/2)} \sum_{k=2N}^{2^j-4N+1} \langle x^\alpha, \varphi_{j,k}\rangle \varphi_{j,k}(x) \mid_{[0,1]} + x^\alpha_{j,R}.$$

Define spaces

$$\{\tilde{V}_j, \quad j \geq j_0\},$$

to be linear spans of functions $\{x^\alpha_{j,L}\}_{\alpha\leq 2N-1}$, $\{x^\alpha_{j,R}\}_{\alpha\leq 2N-1}$, and $\{\varphi_{j,k} \mid_{[0,1]}\}_{k=2N}^{2^j-4N+1}$, namely

$$\tilde{V}_j = \overline{\{x^\alpha_{j,L}\}}_{\alpha\leq 2N-1} \cup \overline{\{\varphi_{j,k} \mid_{[0,1]}\}}_{k=2N}^{2^j-4N+1} \cup \overline{\{x^\alpha_{j,R}\}}_{\alpha\leq 2N-1}.$$

Collections $\{x^\alpha_{j,L}\}_{\alpha\leq 2N-1}$, $\{x^\alpha_{j,R}\}_{\alpha\leq 2N-1}$, and $\{\varphi_{j,k} \mid_{[0,1]}\}_{k=2N}^{2^j-4N+1}$ are mutually orthogonal.

As discussed in the previous paragraph, all polynomials of degree $\leq 2N-1$ are in $\tilde{V}_j$, and spaces $\tilde{V}_j$ form an increasing sequence

$$\tilde{V}_j \subset \tilde{V}_{j+1}.$$

It can be proved that $\tilde{V}_j$ form the MRA of $L^2([0,1])$. All of the functions in the collections are linearly independent and can be used as basis functions. In order to form an orthonormal basis, we only have to orthogonalize the functions $x^\alpha_{j,L}$ and $x^\alpha_{j,R}$.

Orthogonalization More specifically, let us consider the left endpoint, and set

$$\varphi^\alpha_{j,L} = \sum_{\beta=0}^{2N-1} a_{\alpha,\beta} x^\beta_{j,L}.$$

After defining

$$A = \{a_{\alpha,\beta}\},$$

$$X = \{\langle x^\alpha_{j,L}, x^\beta_{j,L}\rangle\},$$

we write the orthonormality condition as

$$I = AXA^*.$$

Now note that X is positive, definite, and symmetric; the Cholesky decomposition holds, namely $X = CC^*$. The selection of

$$A = C^{-1}$$

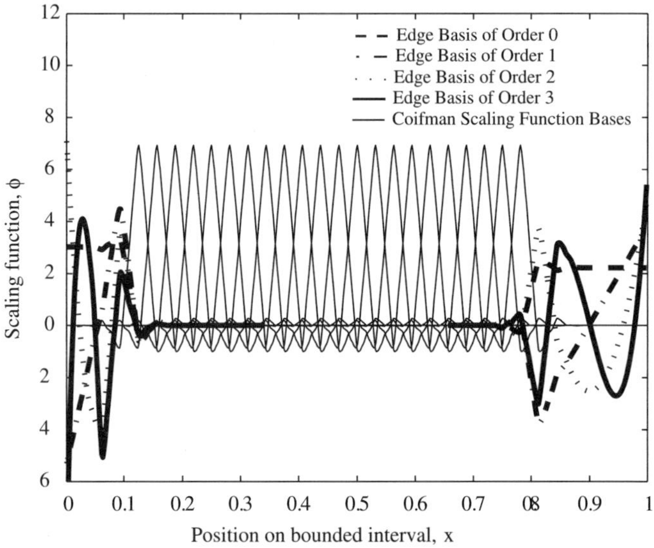

FIGURE 7.8 Coifman intervallic scalet at level 5 for use in solution of integral equations.

will be used to perform the orthogonalization process. That is, we have proved that the functions in $\{\varphi_{j,L}^{\alpha}\}_{\alpha=0}^{2N-1}$ are orthonormal. Similarly we can perform the orthogonalization of $x_{j,R}^{\alpha}$.

Let us order the basis elements of $V_j[0, 1]$ as follows

$$\phi_{j,k} = \begin{cases} \varphi_{j,L}^{k} & \text{if } k = 0, 1, \ldots, 2N-1 \\ \varphi_{j,k} & \text{if } k = 2N, \ldots, 2^j - 4N + 1 \\ \varphi_{j,R}^{k-(2^j-4N+2)} & \text{if } k = 2^j - 4N + 2, \ldots, k = 2^j - 2N + 1. \end{cases}$$

Figure 7.8 depicts the resultant scalets for $j = 5$ and $N = 2$. It can be seen in Fig. 7.8 that there are three kinds of basis functions, namely the left-edge functions, right-edge functions, and complete basis functions as indicated by thin solid lines.

7.2.2 Expansion in Coifman Intervallic Wavelets

In this section we apply the intervallic Coifman scalets to the solution of the integral equation

$$\int f(x')K(x, x')\,dx' + c(x)f(x) = g(x), \tag{7.2.5}$$

where $f(x)$ is the unknown and $c(x)$ is a known function. Equation (7.2.5) is an integral equation of the second kind if $c(x) \neq 0$, or of the first kind if $c(x) = 0$.

Within the integration domain [0, 1], let us expand the unknown function $f(x)$ in the integral equation in terms of scalets at the highest level J on the bounded interval as

$$f(x) = \sum_k \overline{f}_{J,k}\varphi_{J,k}(x), \qquad 1 \le k \le 2^J - 2N + 2.$$

We define

$$B_i(x) = \varphi_{J,i}(x),$$
$$a_i = \overline{f}_{J,i},$$

for $i = 1, 2, 3, \ldots, 2^J - 2N + 2$. The expansion of $f(x)$ is substituted into the integral equation (7.2.5), and the resultant equation is tested with the same set of expansion functions:

$$\sum_n a_n \left\{ c(x)B_n(x) + \int B_n(x')K(x,x')\,dx' \right\} = g(x), \tag{7.2.6}$$

$$\sum_n a_n \left\{ \int c(x)B_m(x)B_n(x)\,dx + \iint K(x,x')B_n(x')B_m(x)\,dx'\,dx \right\}$$
$$= \int g(x)B_m(x)\,dx. \tag{7.2.7}$$

As a result a set of linear equations is formed:

$$\mathbf{Ax} = \mathbf{g},$$

where

$$a_{m,n} = \int c(x)B_m(x)B_n(x)\,dx + \iint K(x,x')B_n(x')B_m(x)\,dx'\,dx, \tag{7.2.8}$$

$$g_m = \int g(x)B_m(x)\,dx. \tag{7.2.9}$$

7.2.3 Numerical Integration and Error Estimate

The evaluation of the coefficient matrix entries involves time-consuming numerical integrations. However, by taking advantage of vanishing moments and compact support of the Coiflets, many entries can be directly identified without performing numerical quadrature. Away from singular points of the kernel, the integrand behaves as a polynomial locally. Consequently the integral that contains at least one complete wavelet function, as the basis or testing function, will result in a zero value. On the other hand, the integral that contains only complete scalets as basis and testing functions will take a zero-order moment of the kernel. Even if supports of basis and testing functions overlap but do not coincide, it is still possible to impose the vanishing moment property and reduce partially the double integration to single integration for the nonsingular part.

Using the Taylor expansion of the integral kernel, we can approximate the nonsingular coefficient matrix entries in (7.2.8), which contain complete wavelets and scalets. For ease of reference, three basic cases are considered and relative errors are analyzed.

CASE 1. DOUBLE INTEGRAL, CONTAINING ONLY COIFMAN SCALETS Consider the second term of (7.2.8). The integral that contains only scalets as basis and testing functions

$$b_{n,m} = \int_{S_n} \int_{S_m} K(x, x')\varphi_{J,m}(x')\varphi_{J,n}(x)\, dx'\, dx$$

will take a zero-order moment of the kernel. It follows that for nonzero entries the error between the exact value and the Coiflet approximation is

$$\begin{aligned}\left|b_{m,n} - 2^{-J} K(2^{-J}n, 2^{-J}m)\right| &\leq 2^{-J} \left\{ \sum_{l \geq 2N} \left| 2^{-Jl} \frac{K_{x'}^{(l)}(2^{-J}n, 2^{-J}m)}{l!} \right| \left| \int_S y^l \varphi(y)\, dy \right| \right. \\ &+ \sum_{l \geq 2N} \left| 2^{-Jl} \frac{K_x^{(l)}(2^{-J}n, 2^{-J}m)}{l!} \right| \left| \int_S y^l \varphi(y)\, dy \right| \\ &+ \left. \sum_{l,p \geq 2N} \left| 2^{-J(l+p)} \frac{K_{x,x'}^{(l)(p)}(2^{-J}n, 2^{-J}m)}{l!p!} \right| \left| \int_S y^l \varphi(y)\, dy \right| \left| \int_S y^p \varphi(y)\, dy \right| \right\}, \end{aligned} \tag{7.2.10}$$

where S_m is a support of the mth scalet and S is the same support after a coordinate transform $x = 2^{-J}(y + m)$.

CASE 2. DOUBLE INTEGRAL, CONTAINING ONLY COIFMAN WAVELET FUNCTIONS ON LEVELS J_1 AND J_2

$$c_{n,m} = \int_{S_n}\int_{S_m} K(x; x')\psi_{J_1,m}(x')\psi_{J_2,n}(x)\, dx'\, dx.$$

It follows that for entries near zero the error between the exact value and the Coiflet approximation is

$$\begin{aligned}| c_{n,m} | \leq 2^{-(J_1+J_2)/2} &\left\{ \sum_{l,p \geq 2N} 2^{-(J_1 p + J_2 l)} \left| \frac{K_{x,x'}^{(l)(p)}(2^{-J_2}n, 2^{-J_1}m)}{l!p!} \right| \right. \\ &\left. \left| \int_S y^l \psi(y)\, dy \right| \left| \int_S y^p \psi(y)\, dy \right| \right\}.\end{aligned}$$

CASE 3. DOUBLE INTEGRAL, CONTAINING COIFMAN WAVELET AND SCALETS ON LEVELS J_1 AND J_2

$$d_{n,m} = \int_{S_n}\int_{S_m} K(x; x')\varphi_{J_1,m}(x')\psi_{J_2,n}(x)\, dx'\, g\, dx.$$

For zero entries the error between the exact value and the Coiflet approximation is

$$
\begin{aligned}
| d_{m,n} | \le 2^{-(J_1+J_2)/2} \Bigg\{ & \sum_{l \ge 2N} 2^{-J_2 l} \left| \frac{K_x^{(l)}(2^{-J_2}n, 2^{-J_1}m)}{l!} \right| \left| \int_S y^l \psi(y)\, dy \right| \\
& + \sum_{l,p \ge 2N} 2^{-(J_1 p + J_2 l)} \left| \frac{K_{x,x'}^{(l)(p)}(2^{-J_2}n, 2^{-J_1}m)}{l!p!} \right| \\
& \left| \int_S y^l \psi(y)\, dy \right| \left| \int_S y^p \varphi(y)\, dy \right| \Bigg\}.
\end{aligned}
\tag{7.2.11}
$$

Figure 7.9 shows the error introduced by the fast evaluation of the impedance matrix elements as will be discussed in Section 7.2.5, in an example where the basis and testing functions consist of ϕ and ψ that are both at level 7. In the Galerkin procedure the impedance matrix is given a block structure that involves combinations of basis and testing functions

$$
\begin{bmatrix} \langle \varphi, \varphi' \rangle & \langle \psi, \varphi' \rangle \\ \langle \varphi, \psi' \rangle & \langle \psi, \psi' \rangle \end{bmatrix}.
$$

Let us select a given row (e.g., row 96 at level 6, or row 192 at level 7) while varying the column number. This row crosses blocks $\langle \varphi, \psi' \rangle$ and $\langle \psi, \psi' \rangle$. The correspond-

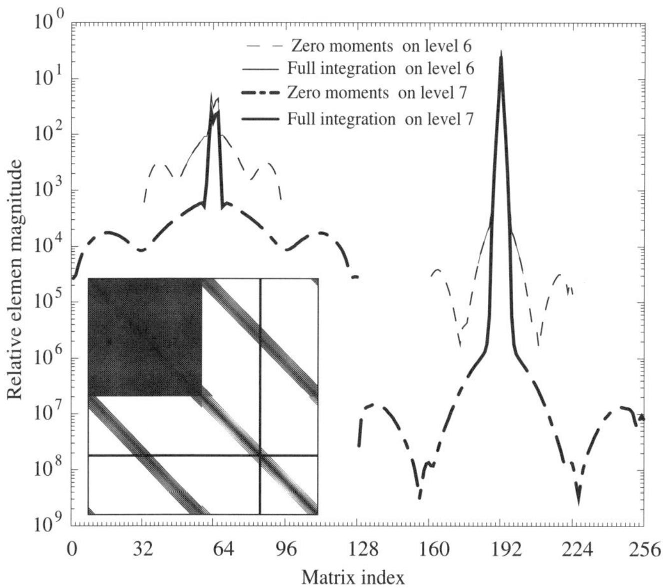

FIGURE 7.9 Error distribution induced by Coifman zero moment approach on resolution levels 6 and 7. (Source: G. Pan, M. Toupikov, and B. Gilbert, *IEEE Trans. Ant. Propg.*, **47**, 1189–1200, July 1999, ©1999 IEEE.)

ing entries are plotted in Fig. 7.9, where the solid lines are computed by Gaussian quadrature and the dashed/dashed-dotted lines are the error introduced by the zero moment property of Coiflets. To illustrate the effects of the resolution level on the error, we plotted two curves (bold versus thin) on levels 7 and 6 for the corresponding locations. It can be observed from the figure that at higher levels the error is reduced.

We need only a few items in each summation to estimate the order of the approximation error. Expressions that involve derivatives of the kernel can be estimated manually or by using symbolic derivation software such as *Maple*. The moment integrals

$$\int_S y^n \varphi(y)\,dy, \quad \int_S y^n \psi(y)\,dy, \qquad n \geq 2N$$

can be calculated directly using wavelet theory.

The nth moment integral for the scalet can be identified using the Fourier transform of the scalet

$$\int t^n \varphi(t)\,dt = \frac{\hat{\varphi}^{(n)}(0)}{(-i)^n}, \tag{7.2.12}$$

where $i = \sqrt{-1}$. Interestingly, the right-hand side of (7.2.12) has a closed form:

$$\hat{\varphi}^{(n)}(0) = \frac{\hat{h}^{(n)}(0)}{2^n - 1}, \qquad 2N \leq n \leq 4N - 1,$$

with

$$\hat{h}^{(n)}(0) = \frac{(-i)^n}{\sqrt{2}} \sum_k k^n h_k, \qquad n = 0, 1, 2, \ldots,$$

where h_k is the lowpass filter. The nth moment integral for the wavelet can be evaluated in a similar fashion.

The first two terms of the right-hand side in (7.2.10) are of the same order and represent the dominant portion of the error. The main part of the approximation error in (7.2.11) is also represented by the first term. Listed in Table 7.3 are the first nine moment integrals for the scalet $\varphi(y)$ and the associated error of expression (7.2.11) for the elliptic cylinder in the example of Section 7.2.5. It will be shown in the next section that for an $n \times n$ matrix, we need to perform numerical integration not on the order of n^2 separate twofold Gaussian quadrature operations, but only on the order of $3n(2L-1) - 7L(L-1) + 2L^2 - 2$ integrations, where $L = 2N$ is the order of the Coifman wavelets, as mentioned before. For a practical problem of $n = 10,000$ unknowns, instead of requiring one hundred million numerical integrations, we will need only 210,000.

From our experience, in most cases we can use the single-point quadrature everywhere except at the diagonal entries. For the Pocklington equation, where singularity seems to be more severe, the tri-diagonal elements are evaluated by standard Gaussian quadrature.

TABLE 7.3. First Nine Moment Integrals for Coifman Scalet of Order $2N = 4$

n	Moment Integral Value	Associated Error
0	1.0000000	N/A
1	0.0000000	N/A
2	0.0000000	N/A
3	0.0000000	N/A
4	4.9333e-11	0.00038057
5	−0.1348373	0.00013809
6	3.5308e-10	0.00004144
7	−3.2646135	0.00000960
8	−8.5859678	0.00000210

Note: The associated error is expressed by (7.2.11).

7.2.4 Fast Construction of Impedance Matrix

Consider a case where the set of basis functions consists of scalets only. The total number of basis functions in the set is $n = 2^j - L + 2$, where j is the level of resolution, and $L = 2N$ is the order of the Coiflets. The number of the left-edge basis functions is L and that of the right-edge basis functions is also L. As a result the number of the center (complete Coiflet) basis functions, which are complete Coifman scalets, is $2^j - 3L + 2 = n - 2L$. The Galerkin method suggests the following structure of the impedance matrix:

$$\begin{pmatrix} B_L B_L' & B_C B_L' & B_R B_L' \\ B_L B_C' & B_C B_C' & B_R B_C' \\ B_L B_R' & B_C B_R' & B_R B_R' \end{pmatrix}. \tag{7.2.13}$$

Specifically, we need to count the interactions of the left-edge basis functions with the left-edge testing functions, denoted as $B_L B_L'$; the left edge basis functions with the center basis functions are denoted as $B_L B_C'$, and so on. Note that only these items within $B_C B_C'$ may fully facilitate the Coiflet zero moments for a twofold integration, provided that the corresponding basis and testing functions do not overlap in their supports. If only one (basis or testing function) is complete, we may use a Coiflet zero moment for that function, and perform the other integration with Gaussian quadrature.

The Coifman scalets have a finite support length of $3L - 1$, namely $[-L, 2L - 1]$. The following derivation evaluates the number of double and single Gaussian quadrature operations, referring to Fig. 7.10.

CASE 1. DOUBLE GAUSSIAN QUADRATURE

- Edge functions react with edge functions. The edge basis functions are constructed from incomplete Coiflets; therefore the Coiflet vanishing moments can-

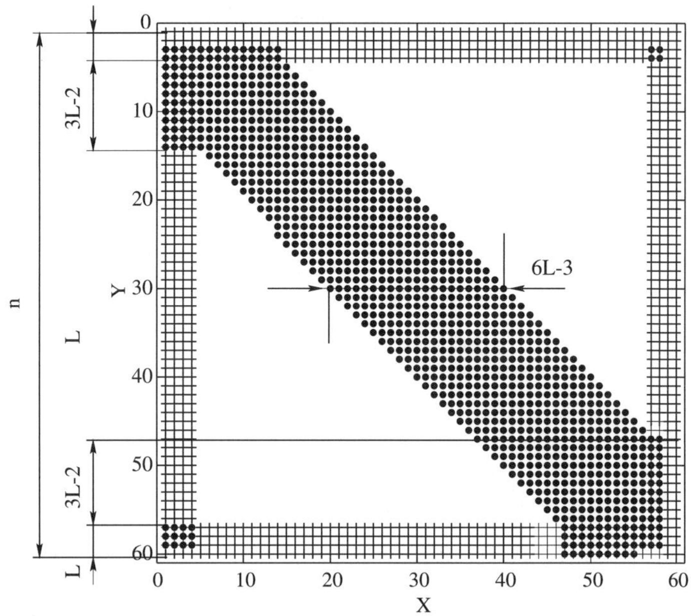

FIGURE 7.10 Impedance matrix structure of the intervallic Coiflet method.

not be imposed. The total number of elements is $4L^2$, as indicated by the four corner terms in Eq. (7.2.13), or the four corners in Fig. 7.10.

- The center functions react with left- (right-) edge functions. The support length of the edge functions is $3L - 2$, which is one unit shorter than the length of the complete scalets. Therefore each edge function overlaps with $3L - 2$ center functions. Since there are $2L$ edge functions, the total number of elements is $4L(3L - 2)$, where an additional factor of 2 is counted for the commutation between testing and expansion.
- Center basis functions are tested by center weighting functions.

(1) Incomplete diagonal (the number of complete testing functions to the left of the complete basis function does not equal the number of complete testing functions to its right). The leftmost complete center function overlaps with $(3L - 1)$ complete center functions, namely the leftmost with itself and $3L - 2$ to its right. The second left complete center function overlaps with $(3L - 1 + 1)$ complete center functions, the additional 1 is the overlap to its left neighbor. The 3rd left complete center function overlaps overlaps with $(3L - 1 + 2)$ complete center functions, the additional 2 are the overlaps to its left 2 neighbors. And so it goes until the last left complete center function overlaps with $(3L - 1 + 3L - 3)$ complete center functions. Summing up the preceding numbers, we obtain the number of total elements as $(3L - 2)(9L - 5)$, where a factor of

two has been multiplied, taking into account the reactions among right center functions.

(2) Complete diagonal (the number of complete testing functions to the left of the complete basis function equals the number of complete testing functions to its right). For these testing functions that may overlap with sufficient number of complete basis functions on both sides, the overlap width is $(6L-3)$. The number of such functions is $(n-2L-2(3L-2)) = (n-8L+4)$. Thus the number of complete overlap is $(6L-3)(n-8L+4)$.

The summation of all the items above gives us the total number that needs to be implemented in twofold Gaussian quadrature operations:

$$3n(2L-1) - 7L(L-1) + 2L^2 - 2 \approx 3n(2L-1).$$

These operations are indicated in Fig. 7.10 as dark regions.

CASE 2. SINGLE GAUSSIAN QUADRATURE In a similar but simpler fashion, we obtain the total number for single Gaussian quadrature operations as $4L(n-5L+2)$. These areas are marked in Fig. 7.10 with light shading.

CASE 3. THE DOUBLE COIFLET VANISHING MOMENT The remainder in Fig. 7.10 is the area where no numerical integration is needed. It is very clear that as the number n increases, the Coiflets becomes more efficient.

In Fig. 7.10 we created the impedance matrix for the scattering problem in which $j = 6$, $L = 4$, and the total number of unknown functions $n = 60$. The number of double Gaussian quadrature elements is reduced from 3600 to 1206, by a factor of 3. If the number of unknown function is 10^5, one may reduce the number of double Gaussian quadrature operations by a factor of 5000. Note that the conclusion we draw here is for the case where all basis functions are scalets. The number of $3n(2L-1)$ in twofold Gaussian quadratures does not represent nonzero entries (although it closely relates to nonzero elements). If both scalets and wavelets are employed, the matrix sparsity may be further improved, and the complexity of matrix construction may also be increased.

7.2.5 Conducting Cylinders, TM Case

Consider a perfectly conducting cylinder excited by an impressed electric field E_z^i. In the TM case, the impressed field induces current J_z on the conducting cylinder, which produces a scattered field E_z^s. By applying boundary conditions, we derive the integral equation as

$$E_z^i = \frac{k\eta}{4} \int_C J_z(\boldsymbol{\rho}') H_0^{(2)}(k \mid \boldsymbol{\rho} - \boldsymbol{\rho}' \mid)\, dl' \qquad \boldsymbol{\rho} \text{ on } C,$$

where $E_z^i(\boldsymbol{\rho})$ is known, J_z is to be determined, $H_0^{(2)}$ is the Hankel function of the second kind, zero order, $k = 2\pi/\lambda$, and $\eta \approx 120\pi$, and the incident field

$$E_z^i = e^{jk(x\cos(\phi_i)+y\sin(\phi_i))}.$$

After the current J_z is found, the scattered field and the scattering coefficient can be evaluated using the following formulas from [9]

$$E^s(\phi) = \eta k K \int_C J_z(x', y') e^{jk(x'\cos(\phi)+y'\sin(\phi))}\, dl',$$

where

$$K(\rho) = \frac{1}{\sqrt{8\pi k\rho}} e^{-j(\mathbf{k}\cdot\boldsymbol{\rho}+3\pi/4)}$$

and

$$\sigma(\phi) = \frac{k\eta^2}{4} \left| \int_C J_z(x', y') e^{jk(x'\cos(\phi)+y'\sin(\phi))}\, dl' \right|^2.$$

We will consider TM plane-wave scattering by an elliptic cylindrical surface, the geometric configuration for which is depicted in Fig. 7.11. In this case the impressed uniform plane wave is incident on the cylinder along the direction of the positive

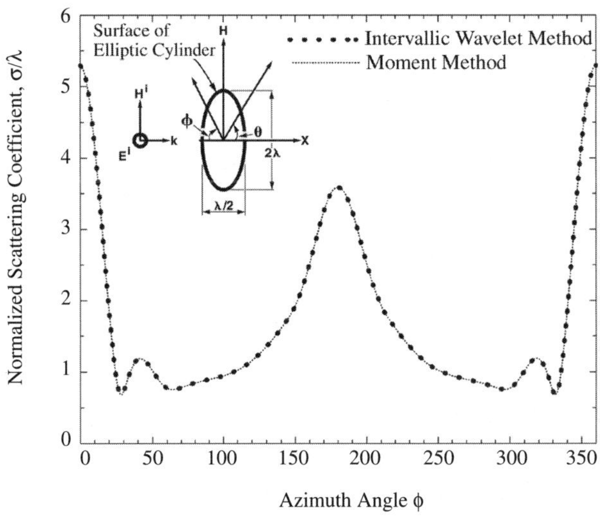

FIGURE 7.11 Radar cross section of a perfectly conducting elliptic cylindrical surface: Transverse magnetic (TM) case. (Source: G. Pan, M. Toupikov, and B. Gilbert, *IEEE Trans. Ant. Propg.*, **47**(7), 1189–1200, July 1999, ©1999 IEEE.)

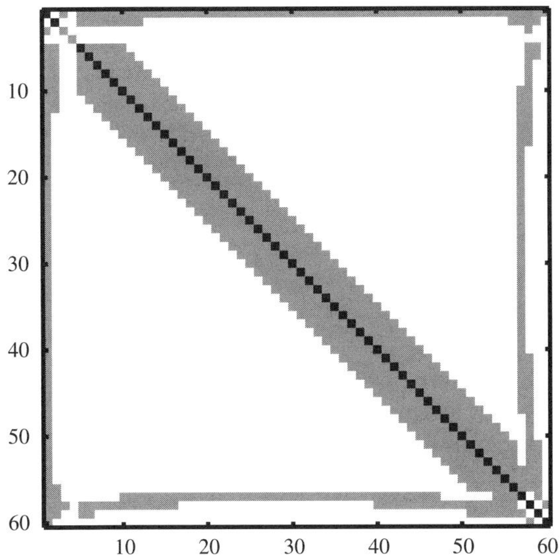

FIGURE 7.12 Magnitude of impedance matrix at level 6, generated by intervallic wavelets method.

x-axis. The procedures described in the solution for J_z are then used to expand the current to Coifman intervallic wavelets. Figure 7.12 shows the impedance matrix, which is produced by the intervallic Coifman scalet on level 6. In the figure the magnitudes of the entries have been digitized into 8-bit gray levels. Figure 7.13 shows the surface current density J_z that is produced by the vanishing moment properties of the Coifman wavelets. We compare it with the current found by using the Gaussian quadrature for the calculation of matrix elements. The magnitude of matrix elements, which are set to zero, does not exceed 0.1% of the largest element in the matrix. In this example the scalets and wavelets are both chosen on level 6 with a total of 60 basis functions. The circumference of the cylinder is approximately 5λ; thus we have 12 basis functions per wavelength. Figure 7.11 shows the radar cross section as computed by the conventional MoM and by this method. The results from the conventional MoM and this method agree very well.

We recall from Chapter 4 that as long as the boundary curve is a closed contour, there is no need to employ the intervallic wavelets, nor the periodic wavelets; instead, the standard wavelets are sufficient. At the left edge, portions of the wavelets that are beyond the interval are circularly shifted to the right edge, and vice versa. This procedure is similar to the circular convolution in the discrete Fourier transform. In this example we employed the intervallic Coifman wavelets, although we could have used the standard wavelets.

This example is a typical onefold wavelet expansion. It is mainly designed to demonstrate the fast construction of an impedance matrix for general problems in the confined interval.

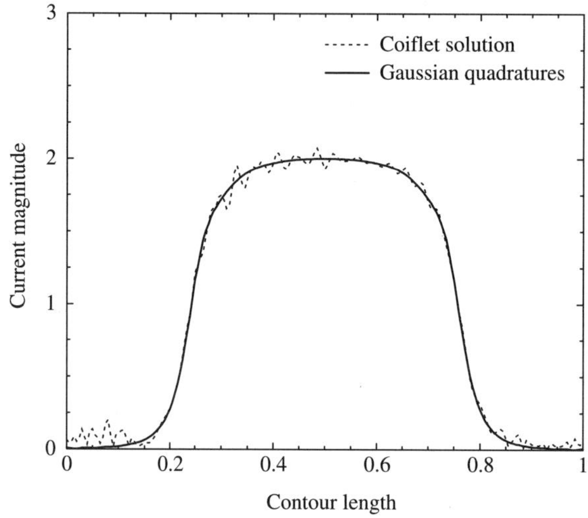

FIGURE 7.13 Current distribution on a 2D PEC elliptic cylinder, as computed by using Gaussian quadrature and vanishing moment wavelets.

7.2.6 Conducting Cylinders with Thin Magnetic Coating

The total fields in free space can be considered to be the sum of the incident fields and the scattered fields radiated by equivalent sources in the thin coating and electric currents on the surface of a perfect conductor. If the contribution of volume integration over all real sources is denoted by $\mathbf{E}^i$ and $\mathbf{H}^i$, based on the equivalence principles, the integral equations for the $\mathbf{E}$ and $\mathbf{H}$ fields can be established as

$$
\begin{aligned}
\mathbf{E}^{\text{tot}}(\mathbf{r}) = T\mathbf{E}^i + T \int_V \left[-j\omega\mu_0 \mathbf{J}_e^{\text{eq}} G - \mathbf{J}_m^{\text{eq}} \times \nabla' G + \frac{\rho_e^{\text{eq}}}{\epsilon_0} \nabla' G \right] dV' \\
+ T \int_S \big[-j\omega\mu_0 (\hat{n} \times \mathbf{H}) G + (\hat{n} \times \mathbf{E}) \times \nabla' G \\
+ (\hat{n} \cdot \mathbf{E}) \nabla' G \big] \, dS' \\
\mathbf{H}^{\text{tot}}(\mathbf{r}) = T\mathbf{H}^i + T \int_V \left[-j\omega\epsilon_0 \mathbf{J}_m^{\text{eq}} G + \mathbf{J}_e^{\text{eq}} \times \nabla' G + \frac{\rho_m^{\text{eq}}}{\mu_0} \nabla' G \right] dV' \\
+ T \int_S \big[-j\omega\epsilon_0 (\hat{n} \times \mathbf{E}) G + (\hat{n} \times \mathbf{H}) \times \nabla' G \\
+ (\hat{n} \cdot \mathbf{H}) \nabla' G \big] \, dS',
\end{aligned}
$$

where

$$
G(\mathbf{r}, \mathbf{r}') = \frac{e^{-jkR}}{4\pi R},
$$

$$R = |\mathbf{r} - \mathbf{r}'|,$$
$$\mathbf{J}_m^{\text{eq}} = j\omega(\mu - \mu_0)\mathbf{H},$$
$$\mathbf{J}_e^{\text{eq}} = j\omega(\epsilon - \epsilon_0)\mathbf{E},$$
$$\rho_e^{\text{eq}} = -\nabla \cdot ((\epsilon - \epsilon_0)\mathbf{E}),$$
$$\rho_m^{\text{eq}} = -\nabla \cdot ((\mu - \mu_0)\mathbf{H}),$$

and

$$T = \begin{cases} 2 & \text{if } r \in S \\ 1 & \text{otherwise,} \end{cases}$$

$\mathbf{J}_e^{\text{eq}}$ and $\mathbf{J}_m^{\text{eq}}$ are equivalent electric and magnetic current sources [10].

In the two-dimensional case, for the TM wave we have

$$-4\pi E_z^i(\boldsymbol{\rho}) = 2\pi\sigma_m t\mathbf{J}(\boldsymbol{\rho})\,|_{\tan} + \left\{ \int_C [(\sigma_m t)(\hat{n} \times \mathbf{J}(\boldsymbol{\rho}) \times (\nabla_t' + j\beta\hat{z}))G \right. \\ \left. - j\omega\mu_0\mathbf{J}(\boldsymbol{\rho})G + \frac{j}{\epsilon_0\omega}(\nabla_t' + j\beta\hat{z}) \cdot \mathbf{J}(\boldsymbol{\rho})(\nabla_t' + j\beta\hat{z})G]\, dl' \right\}_{\tan}, \tag{7.2.14}$$

where

$$G = \frac{\pi}{j} H_0^2 \left(\sqrt{(k^2 - \beta^2)}|\,\boldsymbol{\rho} - \boldsymbol{\rho}'\,| \right)$$

is the two-dimensional Green's function.

Equation (7.2.14) is an electric field integral equation for two-dimensional bodies with arbitrary cross sections. Compared to the case of the perfect conductor [10], an extra term is contributed by the equivalent magnetic current. The contribution from the magnetic current will give scattering that is different from that of a perfect conductor with a coating.

When the current density is known, the radar cross section can be evaluated by asymptotic expressions of Bessel functions. Here we are interested in the bistatic scattering cross section, which is defined by

$$\sigma(\phi) = \lim_{\rho \to \infty} 2\pi\rho \left| \frac{E_z^s}{E_z^i} \right|^2.$$

The normalized radar cross section of a circular cylinder excited by TM wave is given by

$$\frac{\sigma(\phi)}{\lambda} = \frac{(ka\eta)^2}{8\pi} \left| \int \left[1 - \frac{\sigma_m t}{\eta_0} \cos(\theta' - \phi) \right] J_z(\theta') e^{jka\cos(\theta' - \phi)}\, d\theta' \right|^2.$$

Based on the intervallic wavelet approach formulations, numerous numerical results have been obtained. To validate the new surface integral equation, the current distribution and the radar cross section of a circular cylinder were calculated using the intervallic wavelet approach.

Consider an infinitely long, perfectly conducting circular cylinder with $k_0 a = 2\pi$, where a is the radius of the circular cylinder. The perfectly conducting cylinder is assumed to be partially coated with a magnetic film which covers 25% of the circumference over the range $180° - 45° \leq \theta \leq 180° + 45°$. The normalized permeability is $\mu_r t/a = 0.01 - j0.03$. A uniform plane wave with an electric field E_z^i is assumed to be propagating at 135° (Fig. 7.15) in free space. Assuming TM excitation, the radar cross section and the current distribution on a fully coated, a partially coated, and a bare cylinder are plotted in Figs. 7.14 and 7.15.

The current distribution of a partially coated cylinder exhibits rapid variation at the edges of the coating. On the remaining portion of the cylinder without coating, the current is almost the same as that of an uncoated cylinder. The radar cross section of a partially coated cylinder is between that of a fully coated cylinder and that of a bare cylinder except near the edges of the coating. Again, for this example of the 2D cylinder with a closed contour, standard wavelets may be employed.

7.2.7 Perfect Electrically Conducting (PEC) Spheroids

To demonstrate the application of the 2D wavelet expansion to a 3D geometry, the generalized Mie scattering is considered, where the analytical solution and published results are available. We do not utilize the symmetry of revolution; otherwise, the 1D

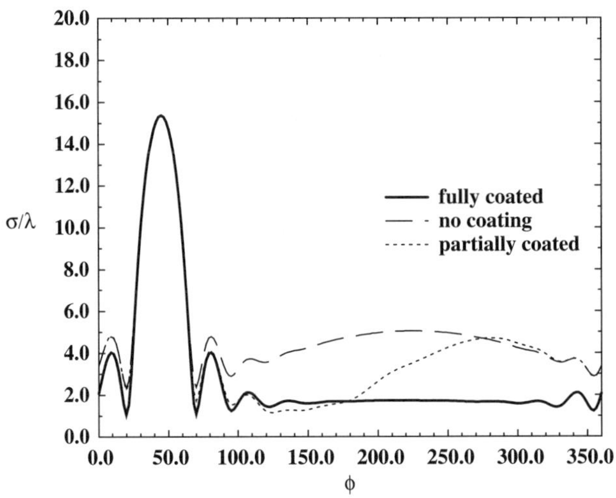

FIGURE 7.14 Radar cross section of a PEC right circular cylinder: transverse magnetic (TM) case, as computed by intervallic wavelet method for different amounts of surface coating, assuming asymmetric incident waves.

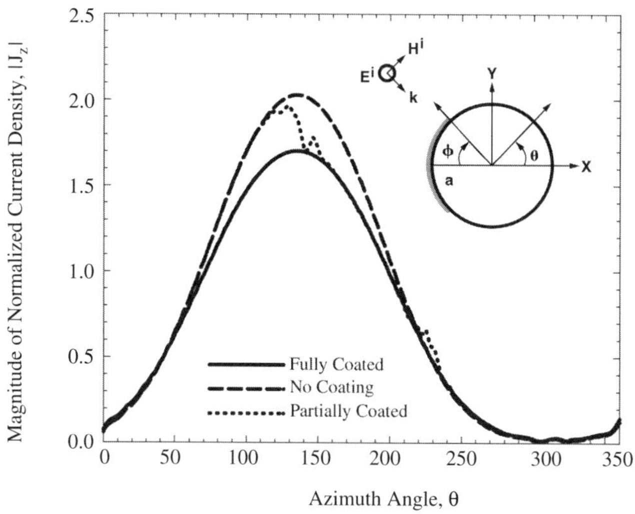

FIGURE 7.15 Current distribution on an infinitely long right circular cylinder for three different coating cases, assuming asymmetric incident waves. (Source: G. Pan, M. Toupikov, and B. Gilbert, *IEEE Trans. Ant. Propg.*, **47**, 1189–1200, July 1999; ©1999 IEEE.)

wavelet would be sufficient. A perfectly conducting prolate spheroid is excited by a uniform plane wave that is incident along the positive z-axis. The total electric current density $\mathbf{J}_s(\mathbf{r})$ induced at any point $\mathbf{r}$ on the surface of the spheroids can be found from the magnetic field integral equation (MFIE)

$$\mathbf{J} = 2\hat{n} \times \mathbf{H}^i + \frac{1}{2\pi}\hat{\mathbf{n}} \times \int_S \mathbf{J}(\mathbf{r}') \times \nabla' G(\mathbf{r}, \mathbf{r}')\, dS', \tag{7.2.15}$$

where ∇' is the surface gradient defined on the primed coordinates and $\hat{n}$ is the unit vector normal to the surface. The integral is interpreted in the Cauchy principal value sense. In a spherical coordinate system $\{r, \theta, \varphi\}$ the tangential electric current density on the spheroid surface can be described by its two components $\{J_\theta, J_\varphi\}$, where $0 \le \theta \le \pi$ and $0 \le \varphi \le 2\pi$. Formally, we can consider the coordinate θ on a bounded interval while the coordinate φ is on a closed contour.

Following the intervallic wavelet approach from Section 7.2.2, the unknown components of the surface current are expanded in the finite series of basis functions as

$$\begin{aligned} J_\theta(\theta, \varphi) &= \sum_k a_k^\theta B_k(\theta, \varphi), \\ J_\varphi(\theta, \varphi) &= \sum_k a_k^\varphi B_k(\theta, \varphi), \end{aligned} \tag{7.2.16}$$

where

$$B_k(\theta, \varphi) = \phi_{J_1,m}(\theta)\phi_{J_2,n}(\varphi)$$

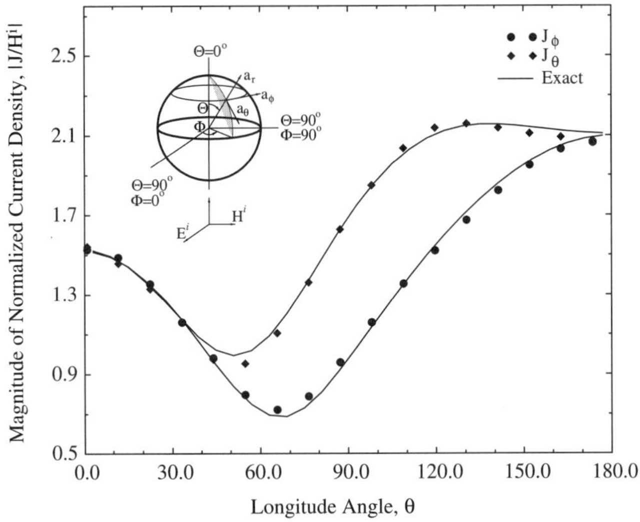

FIGURE 7.16 Current distribution along the principal cuts on a conducting sphere evaluated by using the Coifman scalets. (Source: G. Pan, M. Toupikov, and B. Gilbert, *IEEE Trans. Ant. Propg.*, **47**, 1189–1200, July 1999, ©1999 IEEE.)

Functions $\phi_{J_1,m}(\theta)$ are intervallic Coifman scalets of level J_1, functions $\phi_{J_2,n}(\varphi)$ are ordinary Coifman scalets of level J_2 that are defined on a closed contour, and $k = \{m, n\}$ is a double summation index.

PEC Sphere The surface current distribution of a sphere has been calculated for an incident plane wave with

$$\mathbf{E}^i = E_0\hat{\mathbf{x}}e^{-jkz}, \quad \mathbf{H}^i = H_0\hat{\mathbf{y}}e^{-jkz}.$$

Figure 7.16 shows the computed current distribution along the principal cuts for a sphere with radius 0.2λ, where the θ variation is discretized into 12 intervallic Coifman scalets and the φ variation is discretized into 32 standard Coifman scalets. These results are in good agreement with the exact solution.

PEC Spheroid Depicted in Fig. 7.17 is the configuration of the scattering of electromagnetic waves from a PEC spheroid with $b/a = 2$, where a and b are respectively the semi-minor axis and semi-major axis of the spheroid. Here we used 12 intervallic Coifman scalets in the θ and 32 regular Coifman scalets in the φ directions, respectively. Employing the MFIE formulation, we computed the bistatic radar cross section and plotted it into Fig. 7.17 with $ka = 1.7$. This solution agrees well with previously published data [11]. Figure 7.18 illustrates the backscattering coefficient versus the normalized wavenumber ka. Our numerical results agree well with the curve and data given by Moffat [12].

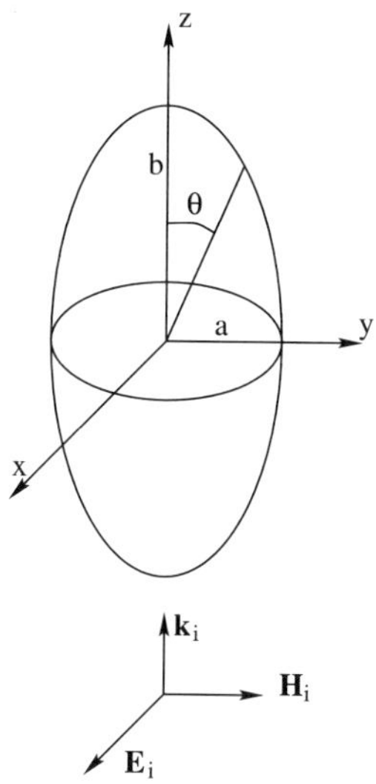

FIGURE 7.17 End-on plane-wave scattering by a prolate perfect electric conductor spheroid.

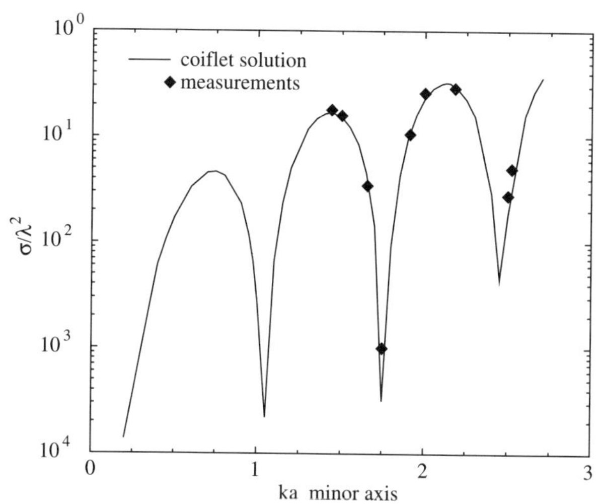

FIGURE 7.18 Normalized backscattering coefficient of a prolate spheroid for end-on plane wave incidence.

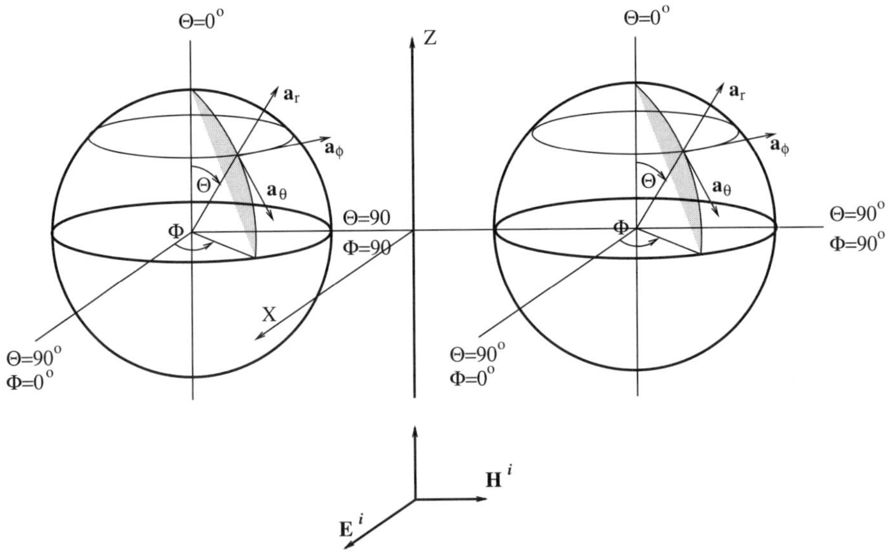

FIGURE 7.19 Scattering on two conducting spheres.

Two PEC Spheres Two perfectly conducting spheres (see Fig. 7.19) are excited by a uniform plane wave incident along the positive z-axis. In this case Eq. (7.2.15) can be written with respect to tangential electric currents as

$$J_\theta^1 \mathbf{E}_{\theta 1} + J_\phi^1 \mathbf{E}_{\phi 1} = 2\mathbf{E}_{r1} \times \mathbf{H}_1^{\text{inc}} + \frac{1}{2\pi}\mathbf{E}_{r1} \times \int_{S_1} (J_\theta^1 \mathbf{e}'_{\theta 1} + J_\phi^1 \mathbf{e}'_{\phi 1}) \times \nabla' G \, dS' + \frac{1}{2\pi}\mathbf{E}_{r1} \times \int_{S_2} (J_\theta^2 \mathbf{e}'_{\theta 2} + J_\phi^2 \mathbf{e}'_{\phi 2}) \times \nabla' G \, dS', \tag{7.2.17}$$

$$J_\theta^2 \mathbf{E}_{\theta 2} + J_\phi^2 \mathbf{E}_{\phi 2} = 2\mathbf{E}_{r2} \times \mathbf{H}_2^{\text{inc}} + \frac{1}{2\pi}\mathbf{E}_{r2} \times \int_{S_1} (J_\theta^1 \mathbf{e}'_{\theta 1} + J_\phi^1 \mathbf{e}'_{\phi 1}) \times \nabla' G \, dS' + \frac{1}{2\pi}\mathbf{E}_{r2} \times \int_{S_2} (J_\theta^2 \mathbf{e}'_{\theta 2} + J_\phi^2 \mathbf{e}'_{\phi 2}) \times \nabla' G \, dS'. \tag{7.2.18}$$

Following the intervallic wavelet approach the unknown components of the surface current are expanded in a finite series of basis functions as in (7.2.16):

$$J_\theta(\theta, \varphi) = \sum_k a_k^\theta B_k(\theta, \varphi), \quad J_\varphi(\theta, \varphi) = \sum_k a_k^\varphi B_k(\theta, \varphi),$$

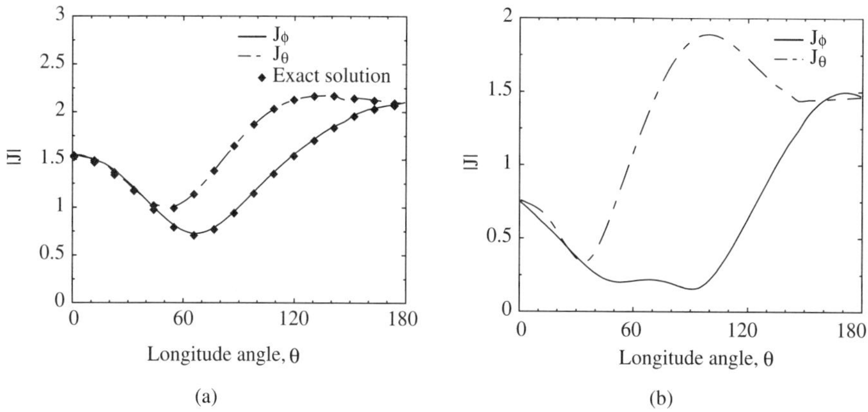

FIGURE 7.20 Surface currents: (a) 10λ separation; (b) 2λ separation.

where

$$B_k(\theta, \varphi) = \phi_{J_1,m}(\theta)\phi_{J_2,n}(\varphi).$$

Functions $\phi_{J_1,m}(\theta)$ are intervallic Coifman scalets of level J_1, functions $\phi_{J_2,n}(\varphi)$ are ordinary Coifman scalets of level J_2 that are defined on a closed contour, and $k = \{m, n\}$ is a double summation index. The expansion of $\mathbf{J}$ is substituted into the integral equation, and the resultant equation is tested with the same set of expansion functions.

For an incident plane wave with

$$\mathbf{E}^i = \hat{\mathbf{x}}e^{-jkz}, \quad \mathbf{H}^i = \hat{\mathbf{y}}e^{-jkz},$$

the surface current distribution for a sphere has been calculated. Figure 7.20 shows the computed current distribution along the principal cuts for two spheres with radius 0.2λ, where the θ variation is discretized with 12 scalets and the φ variation with 32 scalets. For the edge-to-edge separation of 10λ, the current on each sphere is close to that of a single sphere. For the separation of 2λ, the current shows the electrical interaction of two spheres.

7.3 SCATTERING AND RADIATION OF CURVED THIN WIRES

The current distribution on conducting wires are governed by Hallen's integral equation or Pocklington's integrodifferential equation. In this section we employ the Coifman intervallic scalets of $L = 4$ to solve Pocklington's integrodifferential equation. General geometry of a thin-wire problem is shown in Fig. 7.21, where the field point and source point are respectively on the surface and axis of the wire.

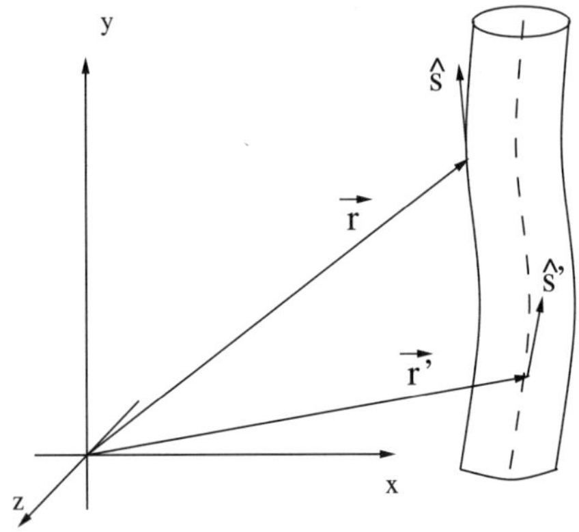

FIGURE 7.21 Thin-wire scatterer.

7.3.1 Integral Equation for Curved Thin-Wire Scatterers and Antennae

For general curved thin wires, we apply the generalized Pocklington's integral equation [13]

$$\int_c I(\mathbf{r}')\left(\frac{\partial^2}{\partial s\partial s'} - k^2\hat{s}\cdot\hat{s}'\right)G(\mathbf{r},\mathbf{r}')\,ds' = j\omega\epsilon\hat{s}\cdot\mathbf{E}^i(\mathbf{r}),$$

where I is the current on the wires, c is the path along the wire, $\mathbf{E}^i$ is the primary field, k is the wave number, s and s' are length variables at $\mathbf{r}$ and $\mathbf{r}'$, respectively, $\hat{s}$ and $\hat{s}'$ are the unit tangent vectors of the wires at $\mathbf{r}$ and $\mathbf{r}'$, respectively, and function $G(\mathbf{r},\mathbf{r}')$ is the free-space Green function given by

$$G(\mathbf{r},\mathbf{r}') = \frac{e^{-jk|\mathbf{r}-\mathbf{r}'|}}{4\pi|\mathbf{r}-\mathbf{r}'|}.$$

The Pocklington equation is an electric field integral equation (EFIE), which is the Fredholm integral equation of the first kind.

In order to avoid singularity in $G(\mathbf{r},\mathbf{r}')$, the observation point $\mathbf{r}$ is taken on the wire surface and the source point $\mathbf{r}'$ on the wire axis. Since the intervallic wavelets are defined in [0, 1], we need to map the integral path c onto [0, 1] such that

$$\mathbf{r} = \Omega(\xi), \tag{7.3.1}$$

where $\xi \in [0, 1]$. Through this mapping and by virtue of the wavelet expansion, the current over c can be expressed as

$$I(\mathbf{r}) = \sum_{n=0}^{N} I_n g_n[\Omega^{-1}(\mathbf{r})] = \sum_{n=0}^{N} I_n g_n(\xi), \tag{7.3.2}$$

where $\mathbf{r}$ is a point of c and $\Omega^{-1}(\mathbf{r})$ denotes the inverse mapping of Ω, I_n is the unknown coefficient to be determined, g_n is the orthogonal intervallic wavelet function which is defined in [0, 1]. Using (7.3.2) and applying Galerkin's method, we obtain a set of linear algebraic equations in matrix form

$$[Z_{m,n}][I_n] = [V_m], \tag{7.3.3}$$

where

$$Z_{m,n} = \int_c g_m(\Omega^{-1})(\mathbf{r}) \left\{ \int_c g_n(\Omega^{-1}(\mathbf{r}') \cdot \left(\frac{\partial^2}{\partial s \partial s'} - k^2 \hat{s} \cdot \hat{s}' \right) G(\mathbf{r}, \mathbf{r}')\, ds' \right\} ds,$$

$$V_m = \int_c g_m(\Omega^{-1}(\mathbf{r})[j\omega\epsilon \hat{s} \cdot \mathbf{E}^i(\mathbf{r})]\, ds.$$

From the map of (7.3.1), we rewrite (7.3.3) as

$$Z_{m,n} = \int_0^1 g_m(\xi)|\, D_\ell \,|\, d\xi \left\{ \int_0^1 g_n(\xi') G_d(\xi, \xi')|\, D'_\ell \,|\, d\xi' \right\},$$

$$V_m = \int_0^1 g_m(\xi)[j\omega\epsilon \hat{s} \cdot \mathbf{E}^i(\xi)|\, D_\ell \,|\, d\xi,$$

where $G_d(\xi, \xi') = (\partial^2/\partial s \partial s' - k^2 \hat{s} \cdot \hat{s}') G(\boldsymbol{\xi}, \boldsymbol{\xi}')$

$$|\, D_\ell \,| = \left| \frac{d\mathbf{r}}{d\xi} \right|,$$

$$|\, D'_\ell \,| = \left| \frac{d\mathbf{r}'}{d\xi'} \right|,$$

$$\hat{s} = \frac{d\mathbf{r}/d\xi}{|\, D_\ell \,|},$$

$$\hat{s}' = \frac{d\mathbf{r}'/d\xi'}{|\, D'_\ell \,|}.$$

7.3.2 Numerical Examples

Based on the former procedures and formulas, we worked through several examples of antennae and scatterers with complicated shapes. The intervallic Coifman wavelets with $L = 4$ were employed. The direct numerical integral algorithm has been implemented using Gaussian quadrature for the evaluation of the matrix elements. The following examples are selected from [14].

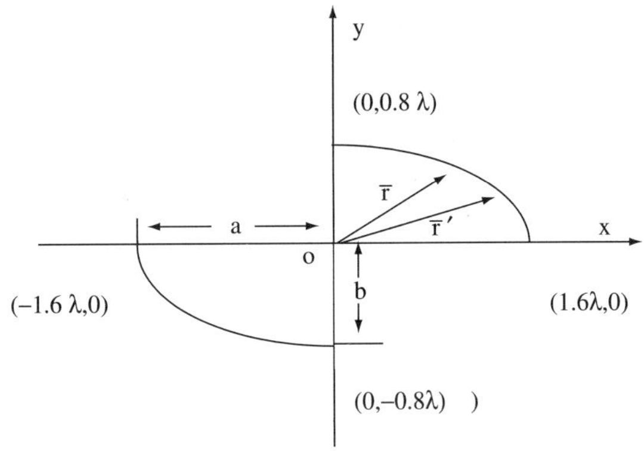

FIGURE 7.22 Segment wire consisting of two quarters of elliptic arc.

Example 1 Shown in Figure 7.22 is the broadside plane wave scattering from a pair of quarter ellipse placed antisymmetrically. Each of these wires is one-quarter of an entire ellipse whose major and minor axes are 3.2λ and 1.6λ, respectively. The electric field polarization of the broadside incident plane wave is parallel to the major axis. The normalized arc length variables of both wires start at the major axis and stop at the minor axis.

For this case, we specified the parameters in (7.3.3) as

$$G(\mathbf{r}, \mathbf{r}') = \frac{e^{-jk|\mathbf{r}-\mathbf{r}'|}}{4\pi|\mathbf{r}-\mathbf{r}'|} = \frac{e^{-jkR}}{4\pi R}.$$

$R = \sqrt{(x-x')^2 + (y-y')^2}$ is the distance between the observation point and the source point. The observation point and source point satisfy the parametric equations

$$\begin{cases} x = (a+d)\cos\theta \\ y = (b+d)\sin\theta, \end{cases} \qquad \begin{cases} x' = a\cos\theta' \\ y' = b\sin\theta', \end{cases}$$

and vectors $\mathbf{r}$ and $\mathbf{r}'$ are given by

$$\begin{cases} \mathbf{r} = x\hat{x} + y\hat{y} = (a+d)\cos\theta\hat{x} + (b+d)\sin\theta\hat{y} \\ \mathbf{r}' = x'\hat{x} + y'\hat{y} = a\cos\theta'\hat{x} + b\sin\theta'\hat{y}, \end{cases}$$

where d is the radius of the wire, and $d \ll b$.

Performing the following mapping

$$\theta = \begin{cases} \pi\xi & 0 \le \xi < 0.5 \\ \pi\xi + \frac{\pi}{2} & 0.5 < \xi \le 1, \end{cases} \qquad \theta' = \begin{cases} \pi\xi' & 0 \le \xi' < 0.5 \\ \pi\xi' + \frac{\pi}{2} & 0.5 < \xi \le 1, \end{cases}$$

we have

$$
\begin{aligned}
|D_\ell| &= \left|\frac{d\mathbf{r}}{d\xi}\right| = \left|\frac{d\mathbf{r}}{d\theta}\cdot\frac{d\theta}{d\xi}\right| \\
&= \pi\left|\frac{d\mathbf{r}}{d\theta}\right| = \pi\sqrt{(a+d)^2\sin^2\theta + (b+d)^2\cos^2\theta} = \pi\cdot A, \\
|D'_\ell| &= \left|\frac{d\mathbf{r}'}{d\xi'}\right| = \left|\frac{d\mathbf{r}}{d\theta'}\cdot\frac{d\theta'}{d\xi'}\right| \\
&= \pi\left|\frac{d\mathbf{r}'}{d\theta'}\right| = \pi\sqrt{a^2\sin^2\theta' + b^2\cos^2\theta'} = \pi\cdot B, \\
\hat{s} &= \frac{d\mathbf{r}/d\xi}{|D_\ell|} = \frac{-(a+d)\sin\theta}{\sqrt{(a+d)^2\sin^2\theta + (b+d)^2\cos^2\theta}}\hat{x} \\
&\quad + \frac{(b+d)\cos\theta}{\sqrt{(a+d)^2\sin^2\theta + (b+d)^2\cos^2\theta}}\hat{y} \\
&= -\frac{(a+d)y}{(b+d)A}\cdot\hat{x} + \frac{(b+d)x}{(a+d)A}\cdot\hat{y}, \\
\hat{s}' &= \frac{d\mathbf{r}'/d\xi'}{|d\mathbf{r}'/d\xi'|} = \frac{-a\sin\theta'}{\sqrt{a^2\sin^2\theta' + b^2\cos^2\theta'}}\hat{x} + \frac{b\cos\theta'}{\sqrt{a^2\sin^2\theta' + b^2\cos^2\theta'}}\hat{y} \\
&= -\frac{ay'}{bB}\cdot\hat{x} + \frac{bx'}{aB}\cdot\hat{y} \\
\hat{s}\cdot\hat{s}' &= \frac{[a(a+d)/b(b+d)]yy' + [b(b+d)/a(a+d)]xx'}{AB}, \\
G_d(\mathbf{r},\mathbf{r}') &= G_d(\xi,\xi') = \left(\frac{\partial^2}{\partial s\partial s'} - k^2\hat{s}\cdot\hat{s}'\right)G(\xi,\xi') \\
&= \frac{e^{-jkr_0}}{4\pi r_0^3 AB}\left\{\left[\frac{(x-x')(k^2r_0^2 - 3jkr_0 - 3)}{r_0^2}\right.\right. \\
&\quad \left.\times\left[-\frac{1}{\beta}y'(x-x') + \beta x'(y-y')\right] - \frac{y'}{\beta}(jkr_0+1)\right]\cdot\left(-\frac{(a+d)y}{(b+d)}\right) \\
&\quad + \left\{\frac{(y-y')(k^2r_0^2 - 3jkr_0 - 3)}{r_0^2}\cdot\left[-\frac{1}{\beta}y'(x-x') + \beta x'(y-y')\right]\right. \\
&\quad \left.\left. + \beta x'(jkr_0+1)\right]\cdot\left(-\frac{(b+d)x}{(a+d)}\right)\right\} \\
&\quad - \frac{k^2e^{-jkr_0}}{4\pi r_0}\cdot\frac{[a(a+d)/b(b+d)]\cdot yy' + [b(b+d)/a(a+d)]\cdot xx'}{AB}, \\
\mathbf{E}^i &= e^{-j\mathbf{k}\cdot\mathbf{r}}\cdot\hat{x}, \\
\hat{s}\cdot\hat{E}^i &= -\frac{(a+d)y}{(b+d)A}e^{-j\mathbf{k}\cdot\mathbf{r}},
\end{aligned}
$$

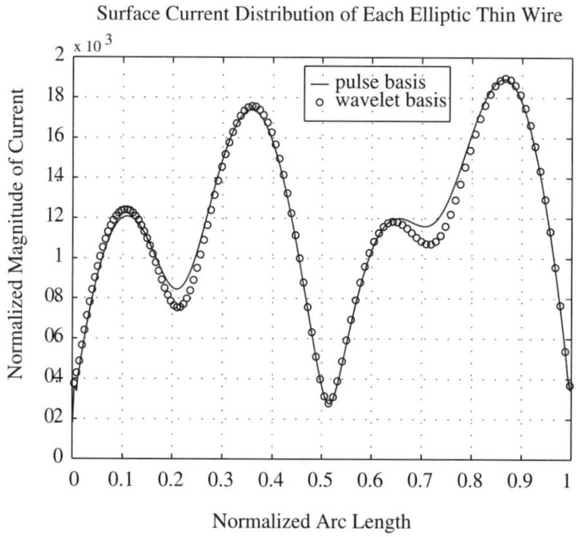

FIGURE 7.23 Current on elliptic wire segment.

where $\beta = (b + d)/(a + d)$. The results for current magnitude obtained by using this technique are shown in Fig. 7.23. The number of basis functions is $N = 252$. By comparison, the results from the MoM with pulse basis functions are also displayed in Fig. 7.23, where the number of basis functions is $N = 512$. We can see that two results agree well. Figure 7.24 shows the sparsity of the impedance matrix from the wavelet approach.

Example 2 A gull-shaped antenna is sketched in Fig. 7.25. The antenna has dimensions of 150 mm length and 0.5 mm radius and is excited by a center-fed voltage.

Solution The Pocklington's integral equation is employed here. The related parameters are

$$\hat{s}_1 = \hat{s}_1' = \hat{x},$$

$$\hat{s}_2 = \hat{s}_2' = \cos\alpha \cdot \hat{x} - \sin\alpha \cdot \hat{y},$$

$$\hat{s}_3 = \hat{s}_3'\hat{x},$$

$$\hat{s}_4 = \hat{s}_4' \cos\alpha \cdot \hat{x} + \sin\alpha \cdot \hat{y},$$

$$\hat{s}_5 = \hat{s}_5'\hat{x},$$

$$G(\mathbf{r}, \mathbf{r}') = \frac{e^{-jk|\mathbf{r}-\mathbf{r}'|}}{4\pi|\mathbf{r} - \mathbf{r}'|} = \frac{e^{-jkR}}{4\pi R},$$

$$G_d(\mathbf{r}, \mathbf{r}') = \frac{\partial G}{\partial s \partial s'} - k^2 \hat{s} \cdot \hat{s}' G(\mathbf{r}, \mathbf{r}')$$

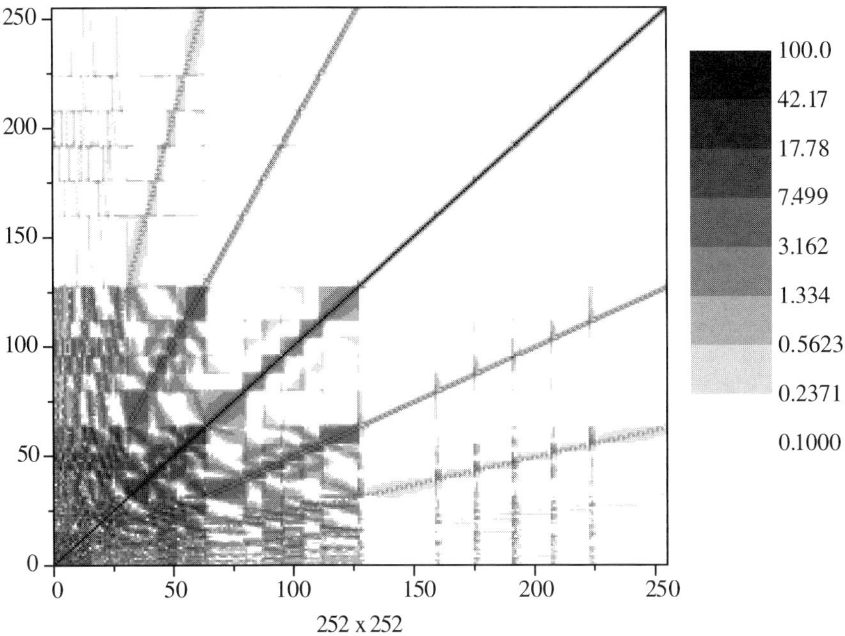

FIGURE 7.24 Sparse impedance matrix of elliptic wire.

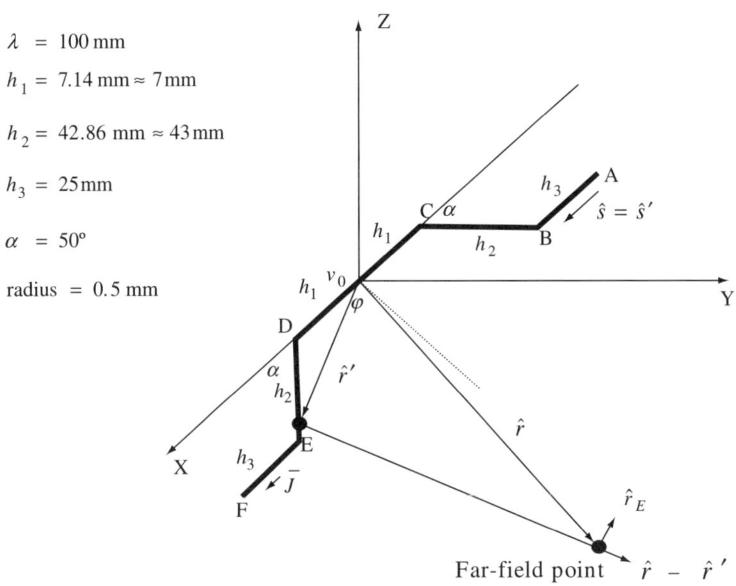

FIGURE 7.25 Gull antenna.

$$
\begin{aligned}
&= \frac{e^{-jkR}}{4\pi R^3}\left\{\left(k^2R - 3jk - \frac{3}{R}\right)[(x-x')s'_x + (y-y')s'_y]\right. \\
&\quad \times \left[\frac{(x-x')}{R}\cdot s_x + \frac{(y-y')}{R}\cdot s_y\right] \\
&\quad \left. + (jkR+1)\cdot[s'_x\cdot s_x + s'_y\cdot s_y)\right\} - k^2(s_x s'_x + s_y s'_y)\cdot\frac{e^{-jkR}}{4\pi R},
\end{aligned}
$$

where $\hat{s} = s_x \cdot \hat{x} + s_y \cdot \hat{y}$ and $\hat{s}' = s'_x \cdot \hat{x} + s'_y \cdot \hat{y}$. We apply the map

$$
\begin{cases} x = -\frac{\ell}{2} + \xi \cdot \ell \\ x' = -\frac{\ell}{2} + \xi' \cdot \ell \end{cases} \qquad \xi, \xi' \in [0, 1].
$$

On the right side of (7.3.3) we use the delta-gap model to yield

$$
V_m = \int_0^1 g_m(\xi)\cdot j\omega\epsilon\hat{s}\cdot \mathbf{E}^i(\xi) = g_m(0.5)\cdot j\omega\epsilon\cdot\frac{V_0}{\Delta d},
$$

where V_0 and Δd are the voltage and distance between two segments of the antenna.

Upon the solution of the surface current distribution over the antenna, the far field due to the current source $\mathbf{J}$ is obtained by [14]

$$
\mathbf{E}(\mathbf{r}) = -\frac{jk\eta e^{-jkr}}{4\pi r}\int_c \mathbf{J}\cdot e^{-jk\hat{r}\cdot\hat{r}'}\,ds' = -\frac{jk\eta e^{-jkr}}{4\pi r}\sum_{n=1}^{N} J\cdot(\hat{s}'\cdot\hat{r}_E)\,\Delta s',
$$

and the radiation pattern is

$$
P = 20\log\left|\frac{\mathbf{E}(\theta,\varphi)}{\mathbf{E}_{\max}}\right|.
$$

In Figs. 7.26 and 7.27 we plotted the current distribution and the radiation pattern. The number of basis functions is $N = 124$ for wavelets, and $N = 150$ for pulse MoM.

Example 3 Curl-Wire Scatterer. The plane-wave scattering of a 3D spiral wire with a relatively large electrical size is analyzed by use of this technique. Figure 7.28 shows its geometrical configuration. In this structure a pair of identical planar curl-wire segments are located on planes $z = -0.75\lambda$ and $z = 0.75\lambda$, respectively, and they are connected by a straight wire segment along the z-axis. Each of the curl-wire segments consists of two half-circular wire segments, which are described by

$$
x = \begin{cases} \lambda(\cos\phi - 1) & \text{if } 0 \le \phi \le \pi \\ 2\lambda\cos\phi & \text{if } \pi \le \phi \le 2\pi, \end{cases}
$$

$$
y = \begin{cases} \lambda\sin\phi & \text{if } 0 \le \phi \le \pi \\ 2\lambda\sin\phi & \text{if } \pi \le \phi \le 2\pi, \end{cases}
$$

$$
z = \mp 0.75\lambda.
$$

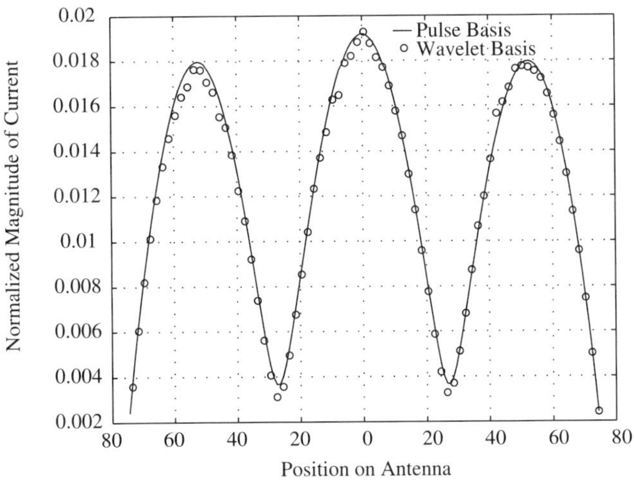

FIGURE 7.26 Current on gull antenna.

The total length of this wire is $(6\pi + 1.5)\lambda \approx 20.35\lambda$, and the radius of the wire is 0.05λ. This structure is illuminated by an incident plane wave that propagates along the positive z direction and its electrical field is parallel to the x-axis. The 3D parameters in the generalized Pocklington EFIE are given as

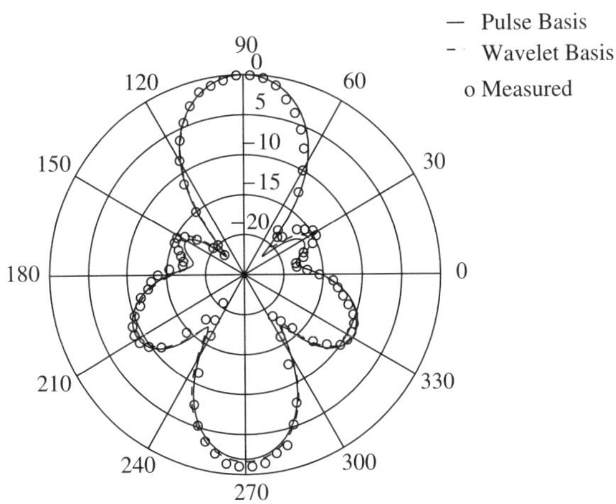

FIGURE 7.27 Radiation pattern of gull-shaped antenna.

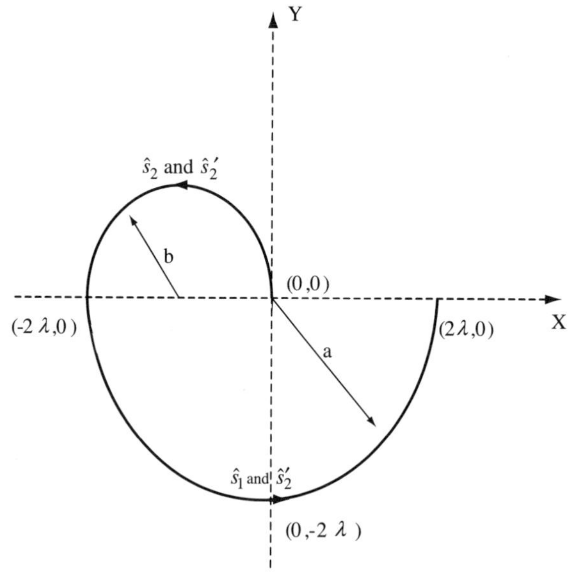

FIGURE 7.28 Spiral wire.

$$\begin{cases} \hat{s}_1 = \dfrac{y}{a+d} \cdot \hat{x} - \dfrac{x}{a+d} \cdot \hat{y} + 0 \cdot \hat{z} \\ \hat{s}_2 = \dfrac{y}{b+d} \cdot \hat{x} - \dfrac{x+b}{b+d} \cdot \hat{y} + 0 \cdot \hat{z} \\ \hat{s}_3 = 0 \cdot \hat{x} + 0 \cdot \hat{y} + 1 \cdot \hat{z} \\ \hat{s}_4 = -\dfrac{y}{b+d} \cdot \hat{x} + \dfrac{x+b}{b+d} \cdot \hat{y} + 0 \cdot \hat{z} \\ \hat{s}_5 = -\dfrac{y}{a+d} \cdot \hat{x} + \dfrac{x}{a+d} \cdot \hat{y} + 0 \cdot \hat{z}, \end{cases} \qquad \begin{cases} \hat{s}_1' = \dfrac{y}{a} \cdot \hat{x} - \dfrac{x}{a} \cdot \hat{y} + 0 \cdot \hat{z} \\ \hat{s}_2' = \dfrac{y}{b} \cdot \hat{x} - \dfrac{x+b}{b} \cdot \hat{y} + 0 \cdot \hat{z} \\ \hat{s}_3' = 0 \cdot \hat{x} + 0 \cdot \hat{y} + 1 \cdot \hat{z} \\ \hat{s}_4' = -\dfrac{y}{b} \cdot \hat{x} + \dfrac{x+b}{b} \cdot \hat{y} + 0 \cdot \hat{z} \\ \hat{s}_5' = -\dfrac{y}{a} \cdot \hat{x} + \dfrac{x}{a} \cdot \hat{y} + 0 \cdot \hat{z}, \end{cases}$$

$$G(\mathbf{r}, \mathbf{r}') = \frac{e^{-jk|\mathbf{r}-\mathbf{r}'|}}{4\pi|\mathbf{r}-\mathbf{r}'|} = \frac{e^{-jkR}}{4\pi R},$$

$$R = \sqrt{(x-x')^2 + (y-y')^2 + (z-z')^2},$$

$$\begin{aligned} G_d(\mathbf{r}, \mathbf{r}') &= \frac{\partial G}{\partial s \partial s'} - k^2 \hat{s} \cdot \hat{s}' G(\mathbf{r}, \mathbf{r}') \\ &= \frac{e^{-jkR}}{4\pi r_0^3} \left\{ \frac{k^2 r_0^2 - 3jkR - 3}{r_0^2} \left[s_x'(x-x') + s_y'(y-y') + s_z'(z-z') \right] \right. \\ &\quad \times \left[s_x(x-x') + s_y(y-y') + s_z(z-z') \right] + (jkR+1) \end{aligned}$$

$$\times \left[s_x s'_x + s_y s'_y + s_z s'_z + \frac{ds'_y}{dx}(y - y')s_x + \frac{ds'_x}{dy}(x - x')s_y \right] \Bigg\}$$

$$- k^2 \hat{s} \cdot \hat{s}' \frac{e^{-jkR}}{4\pi R},$$

and

$$\begin{cases} \hat{s} = s_x \hat{x} + s_y \hat{y} + s_z \hat{z} \\ \hat{s}' = s'_x \hat{x} + s'_y \hat{y} + s'_z \hat{z}. \end{cases}$$

We introduce the following map:

$$\begin{cases} \theta = 2\pi - \dfrac{l \cdot L}{a}, & 0 < l < l_1 \\ \theta = \pi - \dfrac{l \cdot L - 2\pi\lambda}{b}, & l_1 < l < l_2 \\ z = l \cdot L - 3\pi\lambda - 0.75\lambda, & l_2 < l < l_3 \\ \theta = \dfrac{l \cdot L - 3\pi\lambda - 1.5\lambda}{b}, & l_3 < l < l_4 \\ \theta = \dfrac{l \cdot L - 4\pi\lambda - 1.5\lambda}{a}, & l_4 < l < l_1, \end{cases} \qquad \begin{cases} \theta' = 2\pi - \dfrac{l' \cdot L}{a}, & 0 < l' < l_1 \\ \theta' = \pi - \dfrac{l' \cdot L - 2\pi\lambda}{b}, & l_1 < l' < l_2 \\ z' = l' \cdot L - 3\pi\lambda - 0.75\lambda, & l_2 < l' < l_3 \\ \theta' = \dfrac{l' \cdot L - 3\pi\lambda - 1.5\lambda}{b}, & l_3 < l' < l_4 \\ \theta' = \dfrac{l' \cdot L - 4\pi\lambda - 1.5\lambda}{a}, & l_4 < l' < l_1, \end{cases}$$

where $L = (6\pi + 1.5)\lambda$ is the total length of the curl wire, l, l' are the normalized wire length variables, and

$$\begin{cases} l_1 = \dfrac{2\pi\lambda}{L} \\ l_2 = \dfrac{3\pi\lambda}{L} \\ l_3 = \dfrac{(3\pi + 1.5)\lambda}{L} \\ l_4 = \dfrac{(4\pi + 1.5)\lambda}{L}. \end{cases}$$

The intervallic wavelets at resolution level $j = 8$ are used to expand the unknown current over this wire, yielding 252 unknown coefficients. Figure 7.29 shows the surface current distribution from this technique and standard pulse basis with $N = 390$. They agree very well with each other.

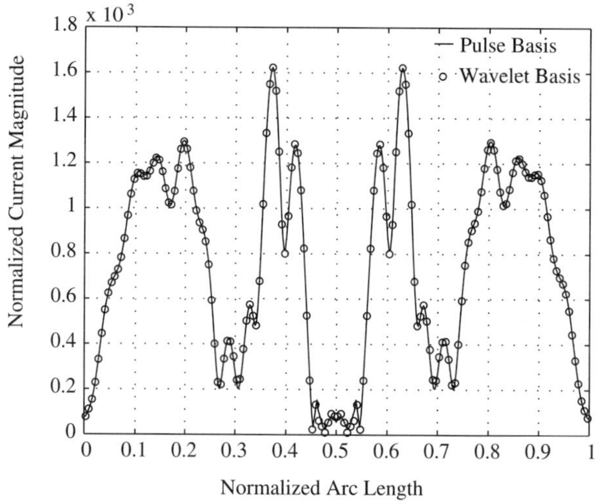

FIGURE 7.29 Current on spiral wire.

7.4 SMOOTH LOCAL COSINE (SLC) METHOD

Wavelets have been employed to solve integral equations, resulting in sparse impedance matrices [15–26]. This is due to the vanishing moment, orthogonality, and multiresolution analysis of wavelets. Despite these attractive features the standard wavelets are defined on the real line, while practical electromagnetic problems are often confined to a finite interval or domain. To incorporate structures with geometric constraints, modified wavelets, including periodic wavelets and intervallic wavelets, were introduced [8, 27]. Nevertheless, the modified wavelets, or the wavelet-like bases [28] have sacrificed some useful properties of wavelets. In this section we employ the smooth local trigonometric (SLT) bases for the method of moments, where the scatters are of finite dimensions.

The discovery of the smooth local trigonometric systems was accomplished by Malvar [29], followed by Coifman and Meyer [30]. The SLT bases are also called the Malvar wavelets. They are trigonometric functions multiplied by a smooth bell-shaped window, and they form an orthogonal basis in $L^2((n, n+1])$. Similar to wavelets, the SLT system constructs its basis functions utilizing both translation and dilation of a single function. However, the construction is accomplished in a more flexible manner, thereby overcoming the inconvenience of conventional wavelets in handling the end points of nonperiodic functions. The basic idea of SLT is to use smooth cutoff functions to split the function and to fold overlapping parts back into the intervals so that the orthogonality of the system is preserved. Moreover, by choosing the correct trigonometric basis, rapid convergence in the case of smooth functions is ensured. Intuitively, one can use a relatively small number of the SLT bases (in comparison to the number of pulse bases) to cover the dominating spectral components of the unknown spatial current of the scatterer. In addition, the folding operator allows the usage of the FFT-like fast numerical technique, such as the fast discrete cosine transform (DCT) for all numerical integrations. Hence accurate and fast algorithms can be developed. In a recent paper, leading mathematicians indicated that: "Classic wavelets seem to be good in computing low frequency scattering and antenna problems," but "high frequency oscillatory integral kernels need local cosines, not classic wavelets" [31].

In this section we construct a smooth cutoff function, cosine-IV, which is twice differentiable. We then apply the SLT to the integral equations to solve scattering and radiation problems, in which the scatterers and antennas are electrically large, resulting in highly oscillatory currents. In case the scatterer consists of several segments, we divide the contour into pieces according to the geometric and physical nature of the problem. The SLT bases are allocated to each segment and overlapping with the SLT bases of the neighboring segments so that the continuity of the solution is guaranteed.

Numerical examples of conductors with smooth contours and with sharp edges are presented for both the TM and TE cases, as well as wire antennas. The results are compared with those obtained by using the standard pulse basis approach as well as by the wavelet expansion technique in terms of computational speed and accuracy.

7.4.1 Construction of Smooth Local Cosine Basis

The most popular transform is the Fourier transform and its variations and modifications. The advantages of the Fourier transform are frequency localization, orthogonality, and the existence of fast algorithms, such as the fast Fourier transform (FFT). The main drawback of the FFT is that the basis functions of the Fourier transform are nonlocalized. Many applications need the use of basis functions that are localized in the time and frequency domains. The reason for this requirement is that most signals have both temporal and spectral correlation, and the use of basis functions that are local in time and frequency results in good approximation properties. Thus we may obtain a good approximation of the analyzed signal with only a small error using only a few basis functions.

One method of construction an orthogonal basis with time-frequency localization is to divide the real axis into disjoint intervals and use the Fourier series on each interval. However, such a local trigonometric basis has several disadvantages. First, Fourier series converge rapidly when the function to be approximated is smooth and periodic. Second, since each interval is handled separately, the approximations are, in general, discontinuous. An improvement has been proposed by Malvar [32] and Coifman and Meyer [30], called the smooth local cosine basis (SLC). The folding of the function is implemented by the introduction of a folding operator.

The SLC bases consist of sines or cosines multiplied by a smooth, compactly supported window. The smoothness of the selected window improves the convergence of the Fourier coefficients in the spectral domain without creating any discontinuities at the endpoints of the sampled signal. The finite support of the smooth window is particularly suitable for approximating functions that are restricted on an interval with an arbitrary smooth periodic basis set. In this section we employ a local cosine basis for the solution of the integral equation (7.4.8).

Let us begin with a smooth cutoff function introduced in [33]

$$r_{\sin}(t) = \begin{cases} 0 & \text{if } t \le -1 \\ \sin\left[\frac{\pi}{4}(1+t)\right] & \text{if } -1 < t < 1 \\ 1 & \text{if } t \ge 1. \end{cases} \tag{7.4.1}$$

From this function, we can obtain real-valued d-times continuously differentiable functions for arbitrary large fixed d by repeatedly replacing t with $\sin(\pi t/2)$, namely

$$r^{(0)}(t) = r_{\sin}(t), \quad r^{(k+1)}(t) = r^{(k)}\left[\sin\left(\frac{\pi}{2}t\right)\right]. \tag{7.4.2}$$

Using the cutoff function $r(t)$, we define the folding operator $U(r, \alpha, \epsilon)$ of which the folding action takes place on an interval $(\alpha - \epsilon, \alpha + \epsilon)$ in the following way:

$$U(r, \alpha, \epsilon) = \begin{cases} r\left(\frac{t-\alpha}{\epsilon}\right) f(t) + r\left(\frac{\alpha-t}{\epsilon}\right) f(2\alpha - t) & \text{if } \alpha < t < \alpha + \epsilon \\ \overline{r}\left(\frac{\alpha-t}{\epsilon}\right) f(t) - \overline{r}\left(\frac{t-\alpha}{\epsilon}\right) f(2\alpha - t) & \text{if } \alpha - \epsilon < t < \alpha \\ f(t) & \text{otherwise.} \end{cases} \tag{7.4.3}$$

In the previous equation the overbar $\overline{r}(\cdot)$ denotes the complex conjugate. Since the cutoff function here is real, $\overline{r}(\cdot) = r(\cdot)$. Immediately we may define the adjoint unfolding operator as

$$U^a(r, \alpha, \epsilon) = \begin{cases} r\left(\frac{t-\alpha}{\epsilon}\right) f(t) - r\left(\frac{\alpha-t}{\epsilon}\right) f(2\alpha - t) & \text{if } \alpha < t < \alpha + \epsilon \\ \overline{r}\left(\frac{\alpha-t}{\epsilon}\right) f(t) + \overline{r}\left(\frac{t-\alpha}{\epsilon}\right) f(2\alpha - t) & \text{if } \alpha - \epsilon < t < \alpha \\ f(t) & \text{otherwise.} \end{cases} \tag{7.4.4}$$

Observe that

$$U(r, \alpha, \epsilon) f(t) = f(t),$$
$$U^a(r, \alpha, \epsilon) f(t) = f(t),$$

if $t \ge \alpha + \epsilon$ or $t \le \alpha - \epsilon$. Also

$$U^a(r, \alpha, \epsilon) U(r, \alpha, \epsilon) f(t) = f(t) U(r, \alpha, \epsilon) U^a(r, \alpha, \epsilon) f(t) = f(t),$$
$$\left(\left| r\left(\frac{t-\alpha}{\epsilon}\right)\right|^2 + \left| r\left(-\frac{t-\alpha}{\epsilon}\right)\right|^2 \right) f(t) = f(t),$$

for all $t \ne \alpha$ and any $r(t) = r^{(k)}(t)$ given by (7.4.2). This statement means that $U(r, \alpha, \epsilon)$ and $U^a(r, \alpha, \epsilon)$ are unitary isomorphisms of L^2(R).

As an example, we have computed $U(r^{(1)}(t), \alpha, \epsilon) f(t)$ for the two functions $f(t) \equiv 1$ and $f(t) = \exp(t)$. The results are plotted in Fig. 7.30 for $\alpha = -2, -1, 0, \ldots, 2$ and $\epsilon = 0.25$.

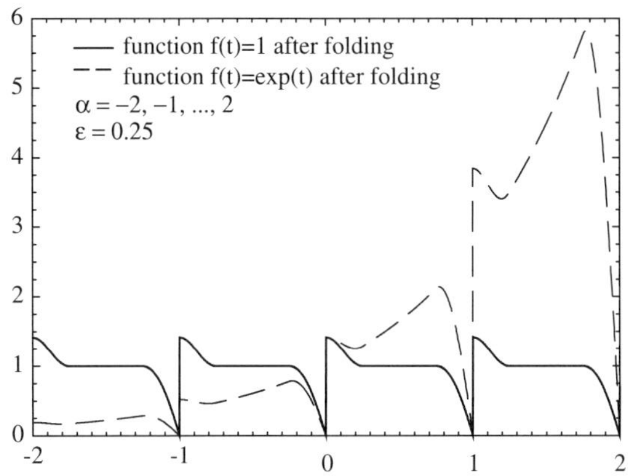

FIGURE 7.30 Action of $U(r_1(t), \alpha, \epsilon)$ on the constant function $f(t) \equiv 1$ and exponential function $f(t) = \exp(t)$.

Let k be an integer and $C_k(t) = \cos[\pi(k + \frac{1}{2})t]$ be a cosine function at half-integer frequency. We consider the block cosine function $1(t)C_n(t)$, which is this cosine function, to be restricted to the interval $[0, 1]$. The block cosine functions may be dilated, normalized, and translated to the interval $I_j = [\alpha_j, \alpha_{j+1}]$ by the formulas

$$C_{j,k}(t) = \sqrt{\frac{2}{|I_j|}} C_k\left(\frac{t - \alpha_j}{|I_j|}\right),$$

$$1_{I_j}(t) = 1\left(\frac{t - \alpha_j}{|I_j|}\right).$$

We can then unfold $1_{I_j}(t)C_{j,k}(t)$ with the active regions of radii ϵ_j and ϵ_{j+1}, respectively, possibly using different cutoffs $r_j(t)$ and $r_{j+1}(t)$ in each active region. This approach is equivalent to multiplying $C_{j,k}(t)$ by the window function $b_j(t)$, supported on $[\alpha_j - \epsilon_j, \alpha_{j+1} + \epsilon_{j+1}]$ and defined by

$$b_j(t) = r_j\left(\frac{t - \alpha_j}{\epsilon_j}\right) r_{j+1}\left(\frac{\alpha_{j+1} - t}{\epsilon_{j+1}}\right). \tag{7.4.5}$$

We may also write the window function $b_j(t)$ in another way as

$$b_j(t) = \begin{cases} r_j\left(\frac{t-\alpha_j}{\epsilon_j}\right), & t \in [\alpha_j - \epsilon_j, \alpha_j + \epsilon_j) \\ 1, & t \in [\alpha_j + \epsilon_j, \alpha_{j+1} - \epsilon_{j+1}) \\ r_{j+1}\left(\frac{\alpha_{j+1}-t}{\epsilon_{j+1}}\right), & t \in [\alpha_{j+1} - \epsilon_{j+1}, \alpha_{j+1} + \epsilon_{j+1}) \\ 0, & \text{otherwise.} \end{cases} \tag{7.4.6}$$

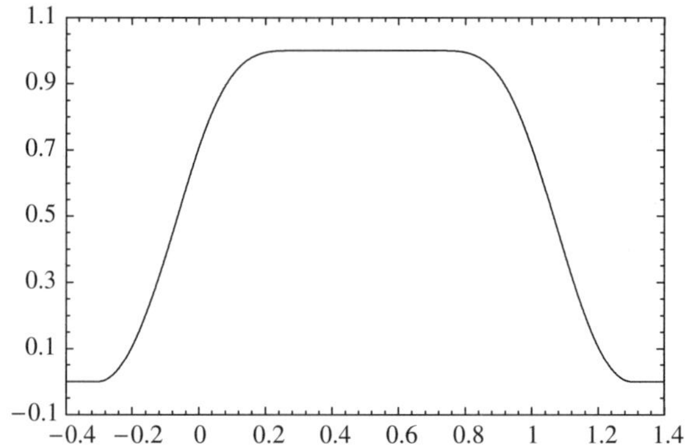

FIGURE 7.31 Bell-shaped window function.

In Fig. 7.31 we plotted a bell-shaped window function given by

$$b(t) = r^{(1)}\left(\frac{t}{0.3}\right) r^{(1)}\left(\frac{1.0 - t}{0.3}\right).$$

We refer to the windowed or unfolded block cosine as the smooth local cosine basis function. For integers $k \geq 0$ and j, the SLC function has the following general form:

$$\begin{aligned}\psi_{j,k}(t) &= U^a\left(r_j, \alpha_j, \epsilon_j\right) U^a\left(r_{j+1}, \alpha_{j+1}, \epsilon_{j+1}\right) 1_{I_j} C_{j,k}(t) \qquad (7.4.7)\\ &= b_j(t) C_{j,k}(t)\\ &= \sqrt{\frac{2}{|I_j|}} r_j\left(\frac{t-\alpha_j}{\epsilon_j}\right) r_{j+1}\left(\frac{\alpha_{j+1}-t}{\epsilon_{j+1}}\right) \cos\left[\frac{\pi\left(k+\frac{1}{2}\right)\left(t-\alpha_j\right)}{|I_j|}\right],\end{aligned}$$

where $I_j = \alpha_{j+1} - \alpha_j$. It should be noted here that the cosine function can be replaced by a sine function in order to get smooth, local sine functions. Other modifications are possible as well.

In Figs. 7.32 and 7.33 we plotted a few examples of the local cosine basis functions, defined on the same interval or on two adjacent intervals. For these examples we used $r_j(t) = \sin\left[\frac{\pi}{4}\left(1 + \sin\left(\frac{\pi}{2}t\right)\right)\right] = r^{(1)}(t)$ for all j.

7.4.2 Formulation of 2D Scattering Problems

The scattering problems are formulated by the integral equation and are numerically solved using the MoM [9]. A 2D perfect electrical conductor (PEC) scatterer is shown in Fig. 7.34, in which the total length ℓ is designated for the circumference of the scatterer in the (x, y) plane.

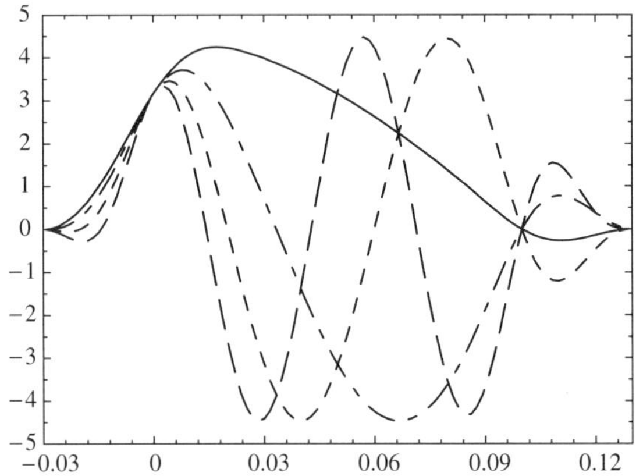

FIGURE 7.32 Local cosine basis functions defined on the same interval, for $\alpha_j = 0, \alpha_{j+1} = 0.1, \epsilon_j = \epsilon_{j+1} = 0.03, k = 0, 1, 2, 3$.

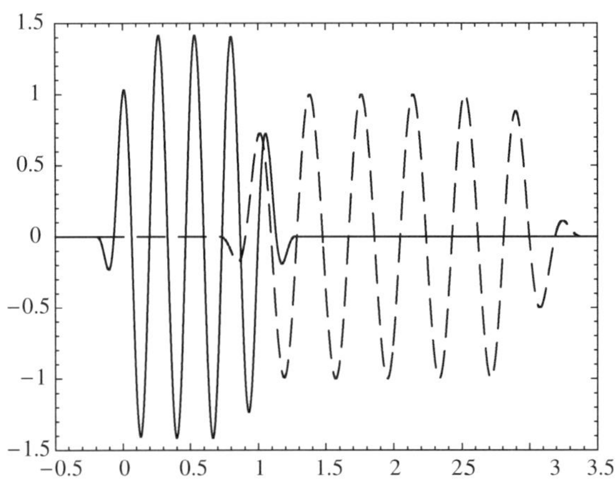

FIGURE 7.33 Two local cosine basis functions with different (I_j) and (ϵ_j), defined on adjacent intervals.

In the figure the impressed incident TM or TE field $\Psi^i(\boldsymbol{\rho})$ induces a surface current J_s on the surface of the scatterer. This current gives rise to the scattered fields. The corresponding integral equation is

$$\int_C G(\boldsymbol{\rho}, \boldsymbol{\rho}') J_s(\boldsymbol{\rho}')\, dl' = -\Psi^i(\boldsymbol{\rho}), \tag{7.4.8}$$

where $\boldsymbol{\rho}, \boldsymbol{\rho}' \in C$ and $G(\boldsymbol{\rho}, \boldsymbol{\rho}')$ is the Green's function.

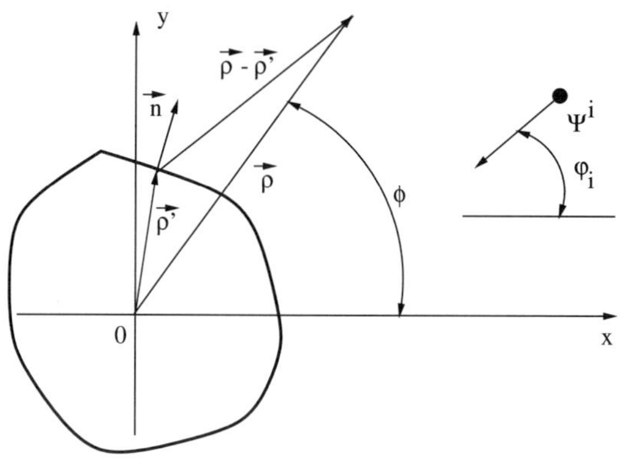

FIGURE 7.34 Geometry of a 2D scatterer.

In the TM case, $\Psi^i(\boldsymbol{\rho})$ is the incident electric field, and Green's function is given by

$$G(\boldsymbol{\rho}, \boldsymbol{\rho}') = -\frac{\kappa\eta}{4} H_0^{(2)}\left(\kappa\left|\boldsymbol{\rho} - \boldsymbol{\rho}'\right|\right). \tag{7.4.9}$$

In the TE case, $\Psi^i(\boldsymbol{\rho})$ is the incident magnetic field, and Green's function is

$$G(\boldsymbol{\rho}, \boldsymbol{\rho}') = \frac{1}{2}\delta(\boldsymbol{\rho} - \boldsymbol{\rho}') + \frac{j\kappa}{4} H_1^{(2)}\left(\kappa\left|\boldsymbol{\rho} - \boldsymbol{\rho}'\right|\right)\cos\left[\hat{n}(\boldsymbol{\rho}'), \boldsymbol{\rho} - \boldsymbol{\rho}'\right], \tag{7.4.10}$$

where $\hat{n}$ is the unit normal vector directed out of the scatterer and $\delta(\cdot)$ is the Dirac delta function. In the previous equations, η is the free-space impedance, κ is the wave number, and J_s is the induced current to be determined.

The scattering coefficient or radar cross section $\sigma(\phi)$ is given by

$$\sigma(\phi) = 2\pi\rho\left|\frac{\Psi^s(\phi)}{\Psi^i}\right|^2. \tag{7.4.11}$$

Upon the numerical solution of current J_s, the scattering coefficient in the far-field zone for the TM case can be evaluated using

$$\sigma(\phi) = \frac{k\eta^2}{4}\left|\int_C J_s(x', y')\, e^{jk(x'\cos(\phi)+y'\sin(\phi))}\, dl'\right|^2, \tag{7.4.12}$$

and for the TE case we can use

$$\sigma(\phi) = \frac{k}{4}\left|\int_C J_s(x', y')\left(\hat{n}\cdot\hat{R}\right) e^{jk(x'\cos(\phi)+y'\sin(\phi))}\, dl'\right|^2, \tag{7.4.13}$$

where $\hat{R}$ is the unit directed from the source point (x', y') to the observation point in the far-field zone.

To solve the integral equation (7.4.8), the unknown current J_s is first expressed as a function of the arclength, namely $J_s(x, y) = J_s(x(t), y(t))$, where $t \in [0, \ell]$. The current is then approximated as a summation of the given basis functions multiplied by unknown coefficients. The basis functions can be the standard pulse bases, triangular functions, or piecewise sinusoidal functions or wavelets. In this section we choose the smooth local cosine bases that have been used extensively in the discipline of signal processing to compress data. Upon solution for the current, (7.4.12) and (7.4.13) are employed to evaluate the normalized scattering coefficient $\sqrt{\sigma/\lambda}$ as a function of the scattering angle ϕ for TM and TE cases.

7.4.3 SLC-Based Galerkin Procedure and Numerical Results

The SLC basis (7.4.7) constructed in the previous section is applied to solve the integral equation (7.4.8). We expand the unknown current J_s using the SLC basis in the form

$$J_s(t) = J_s(x(t), y(t)) \simeq \sum_{j,k} s_{j,k}\psi_{j,k}(t), \tag{7.4.14}$$

where the basis functions (7.4.7) are copied here as

$$\psi_{j,k}(t) = b_j(t)C_{j,k}(t) = b_j(t)\sqrt{\frac{2}{|I_j|}}\cos\left[\frac{\pi}{|I_j|}\left(k+\frac{1}{2}\right)(t-\alpha_j)\right]. \tag{7.4.15}$$

In one particular case, Fig. 7.34, the unknown current can be considered as a periodic function of t with period ℓ, where t is the distance along the circumference of a 2D scatterer. The same functions (7.4.15) are chosen as the testing functions that are chosen for the Galerkin procedure in the MoM. The elements of the impedance matrix are given by the double integral

$$A_{k+j\cdot N_{j,k},k'+j'\cdot N_{j',k'}} = \iint K(t,t')\psi_{j',k'}(t')\psi_{j,k}(t)\,dt'\,dt, \tag{7.4.16}$$

where $N_{j,k}$ denotes the number of frequency components used on the interval $[\alpha_j, \alpha_{j+1}]$ to approximate the unknown current and $K(t, t')$ is the kernel of the integral equation (7.4.8).

The double integral (7.4.16) may be evaluated by double discrete cosine transforms. Let us consider the internal integral with respect to t' for a fixed t in $K(t, t')$. We note the following properties of the folding and unfolding in (7.4.3) and (7.4.4), defined in the previous discussion:

$$\begin{aligned} a_{j,k}(t) &= \left\langle K(t,t'), U^*\left(r,\alpha'_j,\epsilon'_j\right)U^*\left(r,\alpha'_{j+1},\epsilon'_{j+1}\right)1_{I_j}C_{j,k}\right\rangle \\ &= \left\langle U\left(r,\alpha'_j,\epsilon'_j\right)U\left(r,\alpha'_{j+1},\epsilon'_{j+1}\right)K(t,t'), 1_{I_j}C_{j,k}\right\rangle. \end{aligned} \tag{7.4.17}$$

The last expression can be evaluated numerically by using the discrete cosine transform, DCT-IV. Let us define a function

$$g(t,t') = U\left(r, \alpha'_j, \epsilon'_j\right) U\left(r, \alpha'_{j+1}, \epsilon'_{j+1}\right) K(t,t') \tag{7.4.18}$$

that is obtained after folding function $K(t,t')$ with respect to t' from the interval $[\alpha'_j - \epsilon'_j, \alpha'_{j+1} + \epsilon'_{j+1}]$ to the interval $[\alpha'_j, \alpha'_{j+1}]$. Then from (7.4.17) it follows that

$$\begin{aligned} a_{j,k}(t) &= \left\langle g(t,t'), 1_{I_j} C_{j,k} \right\rangle \\ &= \sqrt{\frac{2}{|I_j|}} \int_{\alpha'_j}^{\alpha'_{j+1}} g(t,t') \cos\left[\frac{\pi}{|I_j|}\left(k+\frac{1}{2}\right)\left(t'-\alpha'_j\right)\right] dt' \\ &\simeq \sqrt{\frac{|I_j|}{N}} \sqrt{\frac{2}{N}} \sum_{i=0}^{N-1} g(t,t'_i) \cos\left[\frac{\pi}{N}\left(k+\frac{1}{2}\right)\left(i+\frac{1}{2}\right)\right] \end{aligned} \tag{7.4.19}$$

with $t'_i = \alpha'_j + (i+0.5)(\alpha'_{j+1} - \alpha'_j)/N$. The summation in the last expression of (7.4.19) can be computed using a fast discrete cosine transform, algorithm DCT-IV, which can be found in [34]. The integration with respect to t can be done in the same manner. The fast DCT provides us with an opportunity to generate the impedance matrix of the MoM very rapidly and accurately.

The SLC algorithm was programmed in C++ for quite general 2D cases. In the remainder of this section we have selected three examples to compare the computational efficiency and precision. All examples were executed on an HP-B2000 work station, which is a 64-bit machine with 400 MHz clock, 512 Mbytes RAM and under an HP-UX Unix operating system.

Although one example of a radar cross section is given, emphasis in the examples is on the computation of current distributions because the evaluation of accurate surface currents is a much more sensitive and stringent test of numerical precision.

Example 1 L-Shaped 2D PEC Scatterer, TM and TE Cases. For the scatterer in Fig. 7.35*a* we choose $\ell = 25.6\lambda$ and $\varphi_i = 45°$. This problem was taken from [21]. For TM and TE cases we split interval $[0, 25.6\lambda]$ into six subintervals with $I_j = 6.4\lambda$ for $j = 0, 5$ and $I_j = 3.2\lambda$ for $j = 1, 2, 3, 4$. On intervals $j = 2, 3$, we use 30 frequency components and 50 components on the remaining intervals. Therefore, the total number of unknown coefficients is equal to 240.

To verify the SLC results, we use 768 pulses (30 pulses per λ) for the MoM with the Dirac δ-function as a testing function. The resultant currents J_s (TM case) for both algorithms are presented in Fig. 7.36*a*. Because of the symmetry, currents are plotted only for half of the L-shaped scatterers. The reader could also compare our results with the results published in [21]. The normalized scattering coefficient

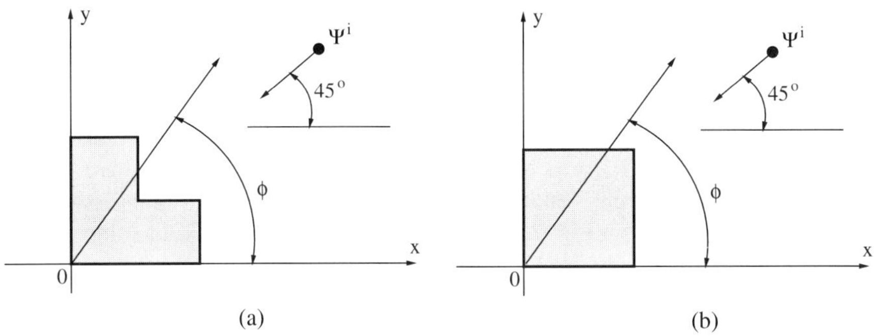

FIGURE 7.35 Geometry of the two-dimensional (*a*) L-shaped scatterer and (*b*) square-shaped scatterer.

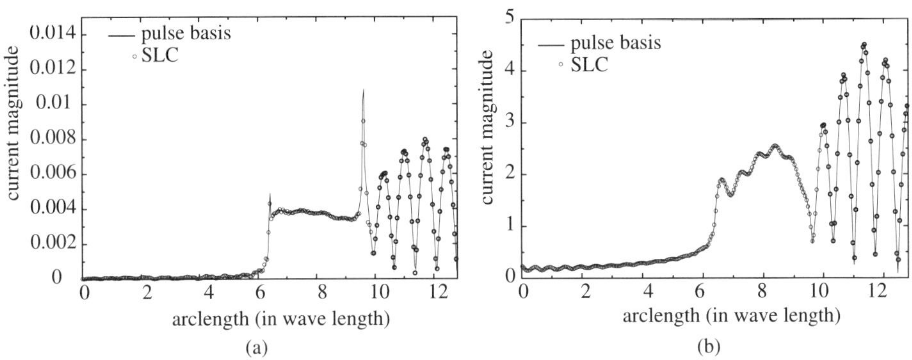

FIGURE 7.36 Current magnitude versus arclength (in λ) for the L-shaped scatterer: (*a*) TM case; (*b*) TE case.

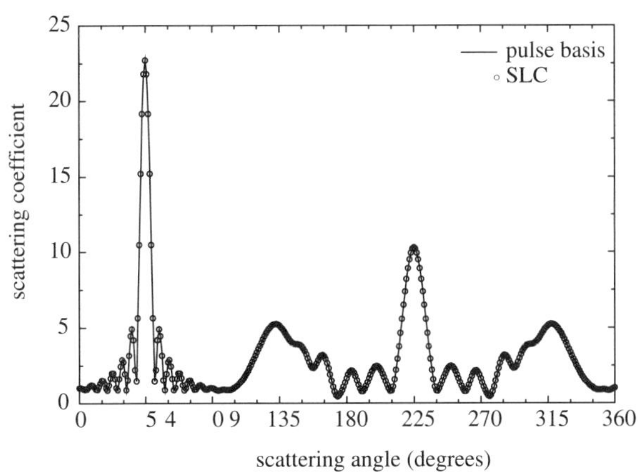

FIGURE 7.37 Normalized scattering coefficient $\sqrt{\sigma/\lambda}$ versus scattering angle ϕ (degrees) for the L-shaped scatterer, TM case.

$\sqrt{\sigma/\lambda}$ (TM case) is obtained using (7.4.12), and the results are shown in Fig. 7.37. As expected, excellent agreement between the SLC and pulse scheme is observed in the scattering coefficient. The far-zone field has been smoothed out and thus is less sensitive to the numerical errors as compared to the induced current. Therefore, for all other examples we omit scattering coefficient plots. The induced current J_s for the TE case is presented in Fig. 7.36*b*. Again, we used 240 SLCs and compared the results with, the pulse-collocation based MoM of 768 unknowns.

Example 2 Square-Shaped 2D PEC Scatterer, TE Case. A 2D square-shaped scatterer is shown in Fig. 7.35*b*. The TE plane wave excitation is considered. This example was taken from [35]. We specify the circumference $\ell = 32\lambda$ and incident angle $\varphi_i = 45°$. The pulse-collocation based MoM solution has been obtained using 2048 pulses. For the SLC technique we split interval $[0, 32\lambda]$ into four subintervals with $I_j = 8\lambda$, $j = 0, 1, 2, 3$. On each interval, only 64 frequency components are used. Hence the total number of unknown coefficients is 256. The results are shown in Fig. 7.38. The magnitude of the error $(J_s - \tilde{J}_s)$ is estimated and presented in Fig. 7.38, where $\tilde{J}_s$ denotes the benchmark (b.m.) solution is obtained using 2048 SLCs. Table 7.4 lists the error using the following expression:

$$\text{Error}(\%) = 100 \times \frac{\| J_s - \tilde{J}_s \|_2}{\| \tilde{J}_s \|_2}, \tag{7.4.20}$$

where the l^2 norm is defined by

$$\| f \|_2 = \sqrt{\frac{\sum_{i=1}^{N} | f_i |^2}{N}}. \tag{7.4.21}$$

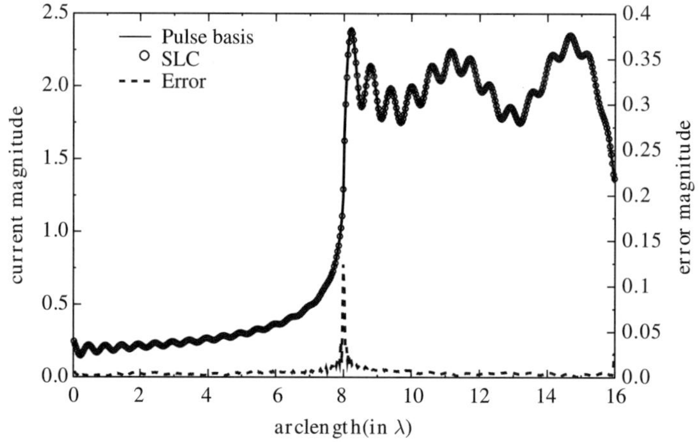

FIGURE 7.38 Current magnitude versus arclength (in λ) for the square-shaped scatterer, TE case.

TABLE 7.4. Error in the Induced Current for the Square-Shaped Scatterer

	Pulse Basis		SLC		Daubechies Wavelets	
Number of Unknowns	Error (%)	Time (s)	Error (%)	Time (s)	Error (%)	Time (s)
2048	0.53473	982.2	b.m.	414.4	N/A	N/A
1024	3.67309	247.9	0.55411	104.7	0.59112	1097.5
512	8.12372	62.5	0.68923	54.5	1.13619	277.3
256	16.67821	15.9	1.15151	39.4	3.86526	67.1
128	N/A	N/A	2.88080	10.1	N/A	N/A

The CPU time is also shown in Table 7.4. As the reference benchmark (b.m.) solution $\tilde{J}_s$ (7.4.20), we used the numerical solution obtained using 2048 SLCs.

To make a fair comparison among the performance of the different approaches, one, needs to specify an error and then compare computational time and the number of unknowns. For instance, let us select the case of an error of 2.8%. From Table 7.4 it follows that we may use only 128 SLCs to, achieve the numerical precision with respect to the benchmark solution of error 2.88080%. The corresponding CPU time is 10.1 seconds. At the same time we found (not shown in Table 7.4) that, we need 1800 pulses to achieve an error of 2.80753%, with the computation time in 757.4 s. Hence the improvements are by a factor $757.4/10.1 \approx 75$ in CPU time, and $1800/128 \approx 14$ in memory requirements.

In Table 7.4 we also present the results obtained from the MoM using Daubechies compactly supported wavelets of order $N = 2$. From this table the superiority of the SLC, over the traditional MoM and standard wavelet approach is clearly evident.

Example 3 Elliptic 2D PEC Scatterer, TE Case. Figure 7.39 illustrates the scattering of a TE plane wave from a 2D PEC elliptic cylinder. We will first discuss a medium size, scatterer of semi-major and semi-minor axis of $a = \lambda$ and $b = \lambda/4$.

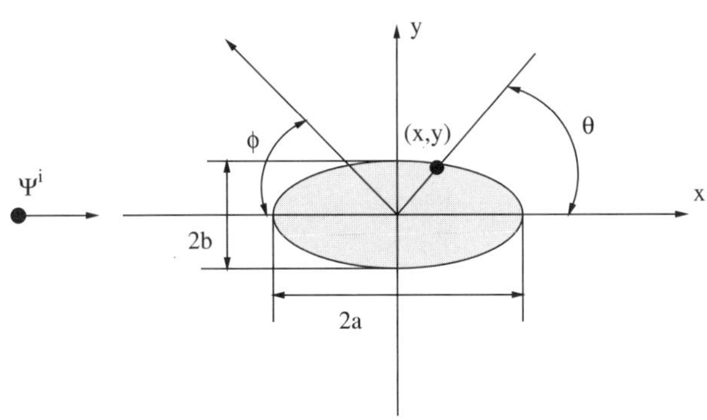

FIGURE 7.39 Geometry of the 2D elliptic scatterer.

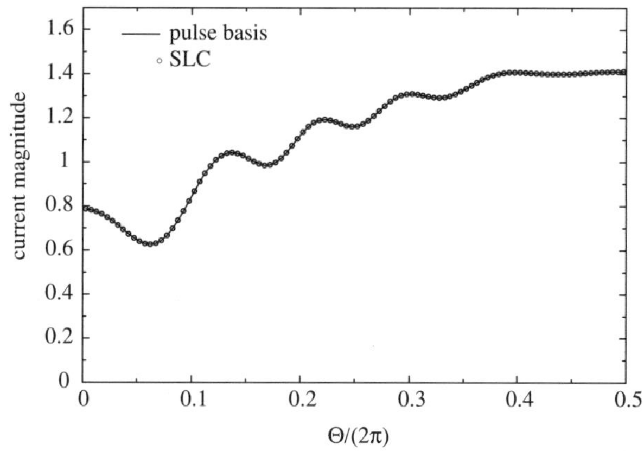

FIGURE 7.40 Current magnitude versus normalized angle $\theta/(2\pi)$ for the elliptic scatterer, TE case.

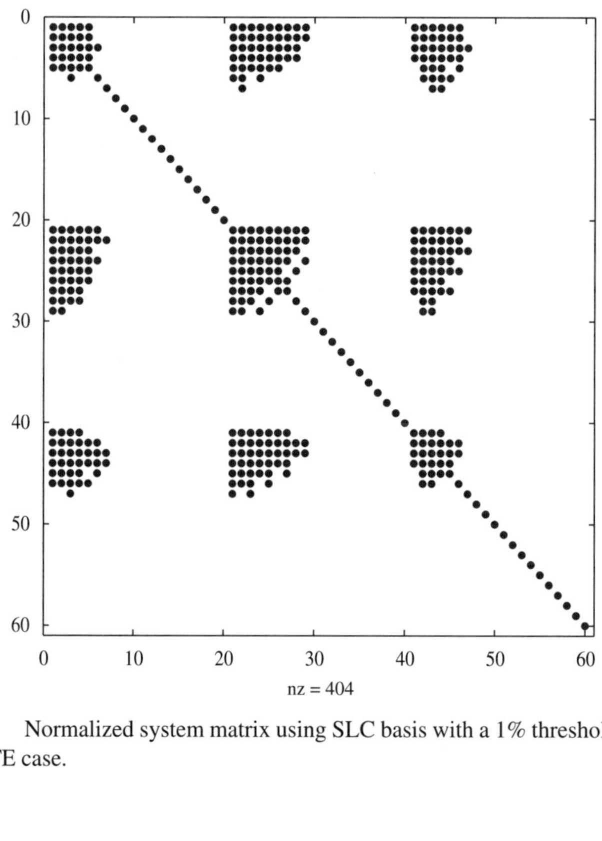

FIGURE 7.41 Normalized system matrix using SLC basis with a 1% threshold for the elliptic scatterer, TE case.

The unknown current J_s is defined on the interval of $t \in [0, 1]$, where $t = \theta/(2\pi)$ is the normalized angle. We employ 200 pulse basis functions to obtain a pulse-based MoM solution. For the SLC method we divided the interval [0, 1] into 3 subintervals and use 20 frequency components on each interval, totaling 60 unknowns. The resulting induced current J_s is presented in Fig. 7.40. Again, due to symmetry, we have plotted, current only for one-half of the elliptic scatterer.

Figure 7.41 demonstrates the normalized system matrix for the TE case above after a 1% thresholding. The sparsity of a system matrix is defined as the percentage of nonzero entries after thresholding. To reveal the connection between matrix sparsity and accuracy of computations, we change the threshold levels to vary the sparsity of the system matrix. We solve corresponding systems of liner equations. The system matrix before thresholding contains $60 \times 60 = 3600$ elements. The obtained numerical results are, summarized in Table 7.5. The pulse basis solution and the SLC result after 1% thresholding of the system matrix are depicted in Fig. 7.42.

Finally, let us reconsider Fig. 7.39 with a larger size given by $a = 4\lambda$ and $b = \lambda$. We use 2048 pulses to obtain an accurate solution as compared to the solution employing 256 SLCs. The results for the induced current are, shown in Fig. 7.43. The magnitude of the error $(J_s - \tilde{J}_s)$ ($\tilde{J}_s$ is the solution, which is obtained using 2048 pulses) is also presented in Fig. 7.43. Similar to Example 2, we compare the SLCs

TABLE 7.5. Matrix Sparsity versus Error in the Induced Current for 2D Elliptic Scatterer, TE Case

Threshold (%)	Sparsity (%)	Error (%)
0.1	20.7778	0.3029
0.5	14.0556	1.4694
1.0	11.2222	2.7921
1.5	9.7778	4.7809

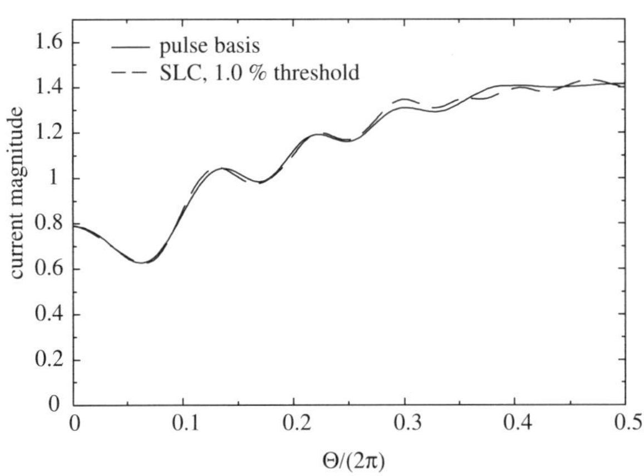

FIGURE 7.42 Comparison of current magnitude between pulse and 1% threshold SLC for the elliptic scatterer, TE case.

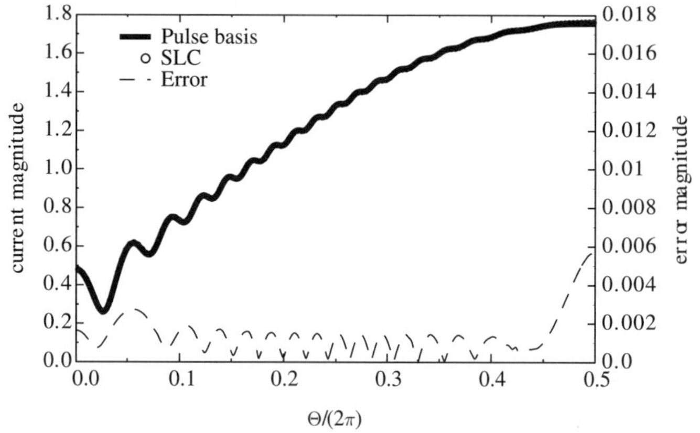

FIGURE 7.43 Current magnitude and error versus normalized angle $\theta/(2\pi)$ for the elliptic scatterer, TE case.

against the pulses and Daubechies wavelets in terms of numerical error, CPU time, and number of unknowns. The results are presented in Table 7.6. It can be seen clearly that the MoM with SLC-basis has the best overall performance. We adopt the result obtained using the SLC-based MoM with 2048 unknowns (16 segments and 128 bases per segment) as the accurate benchmark (b.m.) solution for all results presented in Table 7.6.

From Table 7.6 we find that the pulse-based MoM of 2048, unknowns runs 957.2 seconds to reach an error of 0.066% with respect to the above-mentioned benchmark result. A similar precision of 0.068% can be, achieved by only 128 SLC bases with a CPU time in 8.1 seconds. In other words, the SLC-based MoM is $957.2/29 \approx 118$ times faster than the pulse-based MoM with a factor of 16 in memory savings. Table 7.6 indicates that this method exhibits a much higher order of convergence than the pulse-based MoM, dramatic error reduction as the number of bases increases. This behavior seems to be the super-algebraic convergence, although we do not have a rigorous proof.

TABLE 7.6. Error in the Induced Current for the Elliptic Scatterer

	Pulse Basis		SLC			Wavelet Basis	
Number of Unknowns	Error (%)	Time (s)	Number of Segments × Bases/Segment	Error (%)	Time (s)	Error (%)	Time (s)
2048	0.06624	957.2	benchmark (b.m.)	b.m.	236.9	N/A	N/A
1024	2.78930	239.6	16 × 64	0.00125	138.7	0.18046	1629.6
512	6.23674	60.9	16 × 32	0.00805	111.0	0.59502	407.7
256	12.78981	15.5	8 × 32	0.06726	30.4	2.49634	102.6
128	N/A	N/A	4 × 32	0.20079	8.1	10.05213	26.4
64	N/A	N/A	4 × 16	3.21048	7.4	N/A	N/A

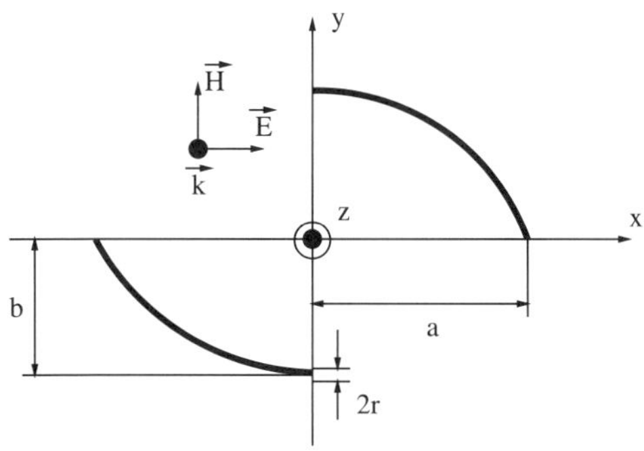

FIGURE 7.44 Geometry of thin-wire problem.

7.4.4 Application of the SLC to Thin-Wire Scatterers and Antennas

Let us re-examine the thin-wire problems presented in Section 7.3. We now solve the Pocklington's equation with the SLC and compare the results with the pulse based MoM.

Example 1 Thin-Wire Scatterer. Shown in Fig. 7.44 is a thin-wire scatterer or antenna, consisting of two elliptic arc wires of radius of $r = 0.01\lambda$, $a = 1.6\lambda$, and $b = 0.8\lambda$. The scatterer is excited by a plane wave, which is also shown in Fig. 7.44. This scattering problem has been solved using the standard pulse based MoM and intervallic Coiflets in Section 7.3.2.

In Fig. 7.45 we plotted the induced current, from the pulse and SLC-based MoM. The induced current for one wire is depicted versus the normalized arclength, starting at the major axis and ending at the minor axis. It has been found numerically that we need at least 128 pulses per wire to obtain a stable solution, which agrees well with

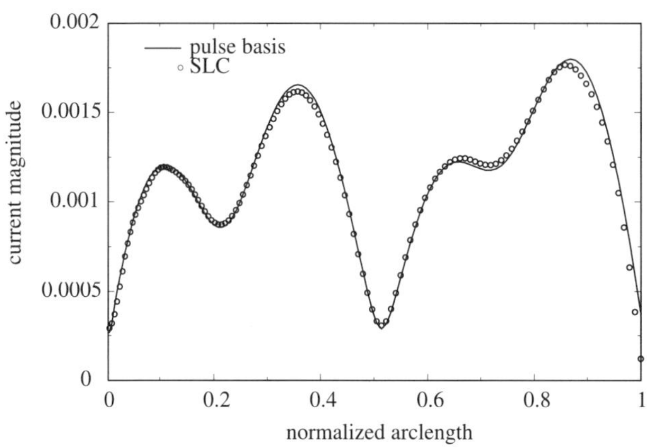

FIGURE 7.45 Normalized current magnitude $| J_s/E^i |$ versus normalized arclength.

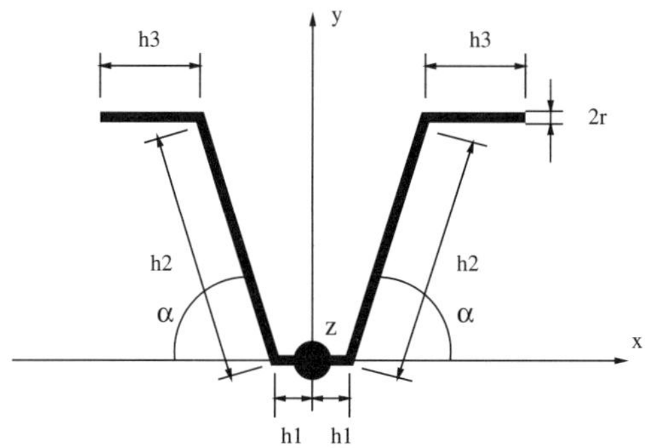

FIGURE 7.46 Geometry of the gull-shaped piecewise linear antenna.

the results published in [14]. The CPU time on an HP-B2000 work station is 147.9 seconds. Meanwhile we need only 24 SLCs per wire to reach an accurate numerical result showing in Fig. 7.45. The corresponding CPU time on the same platform is only 5.5 seconds. It demonstrates that the SLC-based MoM is $147.9/5.5 \approx 27$ times faster than the standard MoM, and uses approximately $128/24 \approx 5$ times less unknowns.

Example 2 Gull-Shaped Antenna. A more general gull-shaped linear antenna is shown in Fig. 7.46, which consists of several segments. The antenna has the following parameters: $h1 = 0.0714\lambda$, $h2 = 0.4286\lambda$, $h3 = 0.25\lambda$, $r = 0.005\lambda$, and $\alpha = 50^0$. This example has been reported in [37] and studied in Section 7.3.2.

Using the SCL with a window to force the current to zero at the ends of the wire, we obtained very good answers. Figure 7.47 shows the normalized current magnitude for this gull-shaped antenna. Due to symmetry, current is plotted only

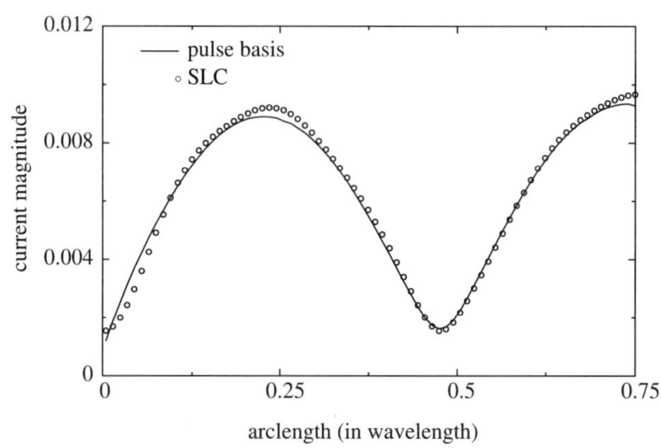

FIGURE 7.47 Normalized current magnitude versus arclength.

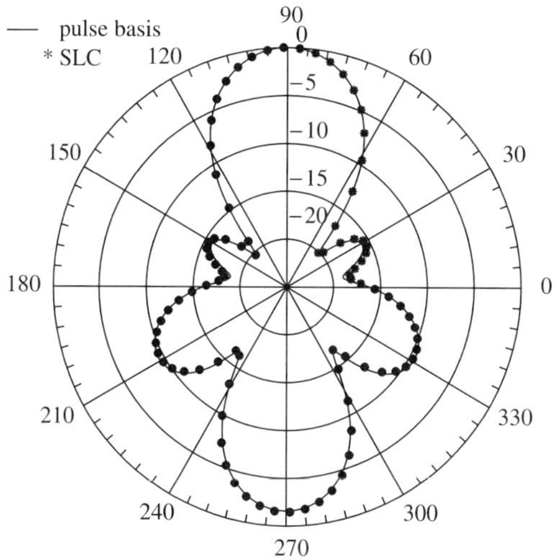

FIGURE 7.48 Radiation pattern of the gull-shaped antenna in the xy plane.

for one-half of the antenna. We used 151 pulses and only 20 SLCs to obtain the numerical results, which is presented in Fig. 7.47. The computation time for the pulse-based MoM is 123.0 seconds. In the mean time, SLC runs only 10.8 seconds. This gives a factor of $123.0/10.8 \approx 11$ and $151/20 \approx 7$ of the CPU time and memory savings. Finally, Fig. 7.48 presents the radiation pattern of the gull-shaped antenna in the xy plane.

7.5 MICROSTRIP ANTENNA ARRAYS

Microstrip antennas and arrays are used in mobile communications and phased array radars. The popular method in analyzing a patch antenna array is to approximate the structure by an infinite array. Thus the analysis of the array is reduced to the study of a single element with the Floquet condition. This approach cannot represent the edge effects due to the finite array elements and the effects of the feed and termination network. More advanced methods that deal with finite element arrays are spectral domain method [38, 39], mixed potential integral equation method (MPIE) in conjunction with the conjugate gradient (CG) and the fast Fourier transform (FFT) [40], among others. Figure 7.49 demonstrates general configuration of a microstrip antenna array.

In this section we present the time domain approach for antenna pattern prediction based on the FDTD and SBTD studied in Chapter 5. The field quantities are obtained from the SBTD or FDTD, and then they are transformed into the phasor form of complex values in the frequency domain by the Fourier transform.

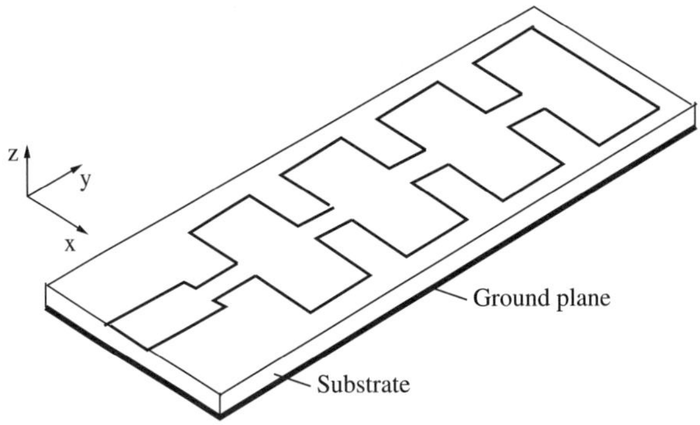

FIGURE 7.49 Microstrip antenna array configuration.

7.5.1 Impedance Matched Source

The impedance Z_0 of the feeding microstrip line at frequency f_0 is obtained by a semiempirical formula [41]

$$Z_0(f_0) = Z_0(0)\sqrt{\frac{\epsilon_{r,eff}(0)}{\epsilon_{r,eff}(f_0)}},$$

where $\epsilon_{r,eff}$ is the effective permittivity.

For $w_{eff}(0)/h > 1$ (h is the substrate thickness, $w_{eff}(0)$ is the effective width of the feeding microstrip)

$$Z_0(0) = \frac{120\pi/\sqrt{\epsilon_{r,eff}(0)},}{(w_{eff}(0)/h) + 1.393 + 0.667\ln[(w_{eff}(0)/h) + 1.444]},$$

$$\epsilon_{r,eff}(0) = \frac{\epsilon_r + 1}{2} + \frac{\epsilon_r - 1}{2}\left[1 + 12\frac{h}{w_{eff}(0)}\right]^{-1/2},$$

$$\epsilon_{r,eff}(f_0) = \epsilon_r - \left[\frac{\epsilon_r - \epsilon_{r,eff}(0)}{1 + (\epsilon_{r,eff}(0)/\epsilon_r)\,(f_0/f_r)^2}\right],$$

$$f_r = \frac{Z_0}{2h\mu_0}.$$

The matched source in the FDTD mesh consists of $N_d \times N_\parallel$ lumped elements, each of them has a source voltage of V_s and source resistance of R_s, as shown in Fig. 7.50. They are given by

$$R_s = Z_0\frac{N_\parallel}{N_d},$$

$$_{k+(1/2)}V_s = -\,_{k+(1/2)}E_s^z\,\Delta z$$

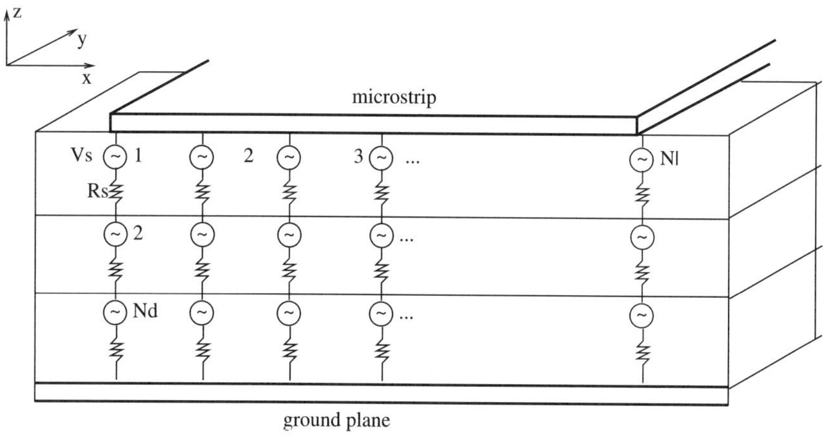

FIGURE 7.50 Impedance matched source.

where the excitation E_s^z can be a modulated Gaussian pulse of the form

$$E_s^z(t) = E_0 e^{-[(t-t_0)/T]^2} \sin(2\pi f_0(t-t_0)).$$

The feeding port update equation [43] is given by

$$\begin{aligned} {}_{k+1}E^z_{l,m_0,n+(1/2)} &= \frac{1-t_p}{1+t_p}\, {}_kE^z_{l,m_0,n+(1/2)} + \frac{\Delta t/\epsilon_r}{1+t_p}\, {}_k(\nabla \times \mathbf{H})^z \\ &\quad + \frac{\Delta t/R_s\epsilon_r\,\Delta x\,\Delta y}{1+t_p}\, {}_{k+(1/2)}V_s, \end{aligned}$$

where

$$\begin{aligned} t_p &= \frac{\Delta t\,\Delta z}{2R_s\epsilon_r\,\Delta x\,\Delta y}, \\ R_s &= Z_0\frac{N_\|}{N_d}, \\ {}_{k+(1/2)}V_s &= -\,{}_{k+(1/2)}E_s^z\,\Delta z, \end{aligned}$$

and for Yee's FDTD scheme

$$\begin{aligned} {}_k(\nabla \times \mathbf{H})^z = &\frac{{}_{k+(1/2)}H^y_{l+(1/2),m_0,n+(1/2)} - {}_{k+(1/2)}H^y_{l-(1/2),m_0,n+(1/2)}}{\Delta x} \\ &- \frac{{}_{k+(1/2)}H^x_{l,m_0+(1/2),n+(1/2)} - {}_{k+(1/2)}H^x_{l,m_0-(1/2),n+(1/2)}}{\Delta y}. \end{aligned}$$

In case a soft-excitation source is used [44], the update equation is

$${}_{k+1}E^z_{l,m_0,n+(1/2)} = {}_kE^z_{l,m_0,n+(1/2)} + E_s^z(t_k).$$

Fields at other spatial nodes are updated using the source-free time domain equations [44, 45]. The FDTD updating equation for E_z in a lossless medium is

$$_{k+1}E^z_{l,m,n+(1/2)} = {}_kE^z_{l,m,n+(1/2)} + \frac{\Delta t}{\epsilon_{l,m,n+(1/2)}} {}_k(\nabla \times \mathbf{H})^z.$$

In contrast, the SBTD updating equation for E_z in a lossless medium is

$$\begin{aligned} _{k+1}E^z_{l,m,n+(1/2)} = {}_kE^z_{l,m,n+(1/2)} &+ \frac{\Delta t}{\epsilon_{l,m,n+(1/2)}} \\ &\times \left[\frac{1}{\Delta x} \sum_{i=-3}^{2} c_i \cdot {}_{k+(1/2)}H^y_{l+(1/2)+i,m,n+(1/2)} \right. \\ &\left. - \frac{1}{\Delta y} \sum_{i=-3}^{2} c_i \cdot {}_{k+(1/2)}H^x_{l,m+(1/2)+i,n+(1/2)} \right], \end{aligned}$$

where coefficients c_i are

i	c_i
0	1.22916661
1	−0.09374998
2	0.01041667

and $c_{-1-i} = -c_i$. Other updating equation are similar, and they were given in Chapter 5.

7.5.2 Far-Zone Fields and Antenna Patterns

The frequency domain radiated fields in in the far zone are

$$\begin{aligned} E_r &= 0, \quad H_r = 0, \\ E_\theta &\simeq (\eta N_\theta + L_\phi)\frac{-j\beta e^{-j\beta r}}{4\pi r}, \\ E_\phi &\simeq (-\eta N_\phi + L_\theta)\frac{j\beta e^{-j\beta r}}{4\pi}, \end{aligned} \tag{7.5.1}$$

where $\eta = \sqrt{\mu/\epsilon}$ is the intrinsic impedance, and $\beta = \omega\sqrt{\mu\epsilon}$ is the propagation constant.

We applied the equivalent source to evaluate the radiation fields. To this end a fictitious plane is set in parallel and close to the substrate-air interface, that is, in parallel to the *XOY* plane according to Fig. 7.49. Thus the surface electric current $\mathbf{J}_s$ and magnetic current $\mathbf{M}_s$ may be imposed to compute the quantities in (7.5.1) as

$$N_\theta = \iint_s (J_{sx} \cos\theta \cos\phi + J_{sy} \cos\theta \sin\phi) e^{+j\beta r' \cos\psi}\, ds',$$

$$N_\phi = \iint_s (-J_{sx} \sin\phi + J_{sy} \cos\phi) e^{+j\beta r' \cos\psi}\, ds',$$

$$L_\theta = \iint_s (M_{sx} \cos\theta \cos\phi + M_{sy} \cos\theta \sin\phi) e^{+j\beta r' \cos\psi}\, ds',$$

$$L_\phi = \iint_s (-M_{sx} \sin\phi + M_{sy} \cos\phi) e^{+j\beta r' \cos\psi}\, ds',$$

where

$$\mathbf{J}_s = \hat{n} \times \mathbf{H}_s, \quad \mathbf{M}_s = -\hat{n} \times \mathbf{E}_s,$$

$$\cos\psi = \cos\theta \cos\theta' + \sin\theta \sin\theta' \cos(\phi - \phi').$$

In the previous equations, the equivalent surface currents are the frequency domain fields on a chosen surface. These surface currents are obtained by the Fourier transform (FT) [44] or by the discrete Fourier transform (DFT) of the time domain field quantities, which were carried out concurrently during the FDTD or SBTD procedure in the space positions [45–47].

The computation of antenna patterns may be accomplished using the tangential components of E field only on the basis of surface equivalence principle [42]. In this case, the previous for far-zone formulas (7.5.1) reduce to

$$E_r = 0, \quad H_r = 0,$$

$$E_\theta \simeq (L_\phi) \frac{-j\beta e^{-j\beta r}}{4\pi r},$$

$$E_\phi \simeq (L_\theta) \frac{j\beta e^{-j\beta r}}{4\pi},$$

with

$$N_\theta = 0, \quad N_\phi = 0,$$

$$L_\theta = \iint_s (M_{sx} \cos\theta \cos\phi + M_{sy} \cos\theta \sin\phi) e^{+j\beta r' \cos\psi}\, ds',$$

$$L_\phi = \iint_s (-M_{sx} \sin\phi + M_{sy} \cos\phi) e^{+j\beta r' \cos\psi}\, ds',$$

where

$$\mathbf{J}_s = 0, \quad \mathbf{M}_s = -2\hat{n} \times \mathbf{E}_s.$$

Valuable economy in computer memory and CPU time is achieved as a consequence of the reformulation above, because there is no need to use on each time step

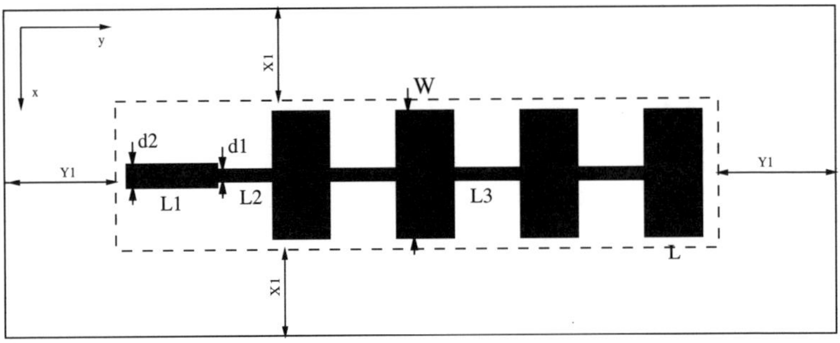

FIGURE 7.51 Geometry of antenna array.

in the FT or DFT procedure the $3D$ data arrays storing the time domain H fields. In addition the $2D$ complex data array for the frequency domain H fields are excluded as well.

The geometry and dimensions of a microstrip antenna array is shown in Fig. 7.51 with the following parameters: $\epsilon_r = 2.1$, $W = 11.79$ mm, $L_1 = 23.6$ mm, $L_2 = 13.4$ mm, $L_3 = 12.32$ mm, $d_1 = 1.3$ mm, $d_2 = 3.93$ mm, and the thickness of substrate $h = 1.57$ mm. The discretization numbers corresponding to the antenna array of Fig. 7.51 are $\Delta x = 0.433$ mm, $\Delta y = 1.12$ mm, and $\Delta z = 0.787$ mm. The size of the fictitious tangential plane is $175\,\Delta x \times 208\,\Delta y$; $X_1 = 70\,\Delta x$ and $Y_1 = 50\,\Delta y$ are the distances to the absorbing boundary or the perfectly matched layers (PMLs). The measured data of the antenna pattern are quoted from [48].

Figure 7.52 illustrates the antenna pattern obtained by using the impedance-matched source. It should be noted that the tangential plane may be set on the

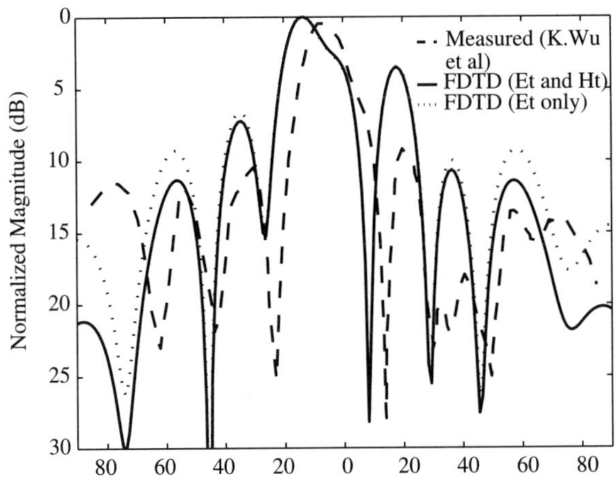

FIGURE 7.52 Pattern from the FDTD with impedance matched source.

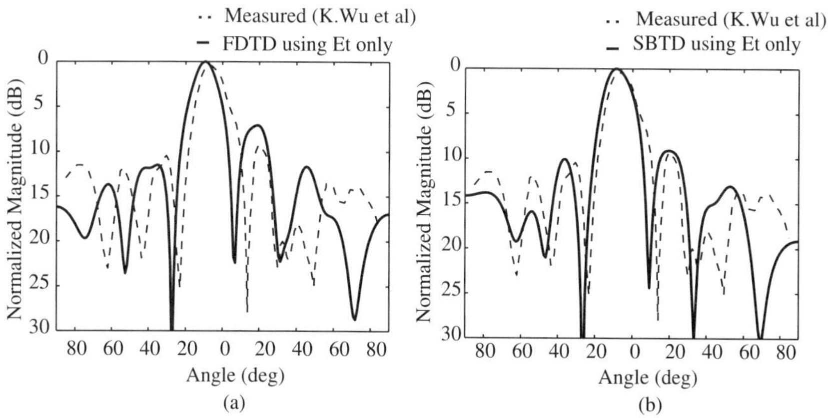

FIGURE 7.53 Antenna pattern from soft-excitation source for (*a*) FDTD and (*b*) SBTD.

interface or just one grid above it, if only $\hat{n} \times \mathbf{E}$ is used with the equivalence principle to compute the far-zone fields. When both $\hat{n} \times \mathbf{E}$ and $\hat{n} \times \mathbf{H}$ are in use, the fictitious plane must be set in a higher position. The results from tangential E as well as tangential E and tangential H are all ploted in the figures. Figure 7.53*a* shows the antenna pattern obtained from the FDTD with a soft-excitation source. Fig. 7.53*b* presents the antenna pattern computed by the SBTD under the same excitation as in Fig. 7.53*a*. It is very clear that the SBTD results are in better agreement with the measurement than the FDTD, and this is expected.

BIBLIOGRAPHY

[1] Y. Meyer, *Wavelets: Algorithms and Applications*, R. Ryan, Trans., SIAM, Philadelphia, 1993.

[2] Y. Shifman and Y. Leviatan, "A wavelet-based hybrid PO-MoM analysis of scattering by a groove in a conducting plane," *IEEE Trans. Ant. Propg.*, **49**(12), 1807–1811, 2001.

[3] C. Balanis, *Advanced Engineering Electromagnetics*, John Wiley, New York, 1989.

[4] L. Kantorovich and V. Krylov, *Approximate Methods of Higher Analysis*, 4th ed., C. D. Benster, Trans., John Wiley, New York, 1959, at ch. IV.

[5] W. Press, S. Teukolsky, W. Vetterling, and B. Flannery, *Numerical Recipes in C*, Cambridge University Press, Cambridge, 1992.

[6] H. Van der Vorst, "Bi-CGSTAB: A fast and smoothly converging variant of Bi-CG for the solution of nonsymmetric linear systems," *SIAM J. Sci. Statist. Comput.*, **13**(2), 631–644, Mar. 1992.

[7] J. Kong, *Electromagnetic Wave Theory,* 2nd ed., John Wiley, New York, 1990, p. 508.

[8] G. Pan, M. Toupikov, and B. Gilbert, "On the use of Coifman intervallic wavelets in the method of moments for fast construction of wavelet sparsified matrices," *IEEE Trans. Ant. Propg.*, **47**(7), 1189–1200, 1999.

[9] R. Harrington, *Field Computation by Moment Method*, Krieger Publishing, 1968.

[10] X. Min, W. Sun, W. Gesang, and K. Chen, "An efficient formulation to determine the scattering characteristics of a conducting body with thin magnetic coatings," *IEEE Trans. Ant. Propg.*, **39**, 448–454, Apr. 1991.

[11] A. Poggio and E. Miller, "Integral equation solutions of three-dimensional scattering problems," in *Computer Techniques for Electromagnetics*, R. Mittra, Ed., Pergamon Press, New York, 1973.

[12] D. Moffatt and E. Kennaugh, "The axial echo area of a perfectly conducting prolate spheroid," *IEEE Trans. Ant. Propg.*, **13**, 401–4099, 1965.

[13] J. Wang, *Generalized Moment Methods in Electromagnetics: Formulation and Computer Solution of Integral Equations*, John Wiley, New York, 1991.

[14] G. Wang, "Application of wavelets on the interval to the analysis of thin-wire antennas and scatters," *IEEE Trans. Ant. Propg.*, **45**(5), 885–893, May 1997.

[15] B. Steinberg and Y. Leviatan, " On the use of wavelet expansions in the method of moments," *IEEE Trans. Ant. and Propg.*, **41**, 610–619, May 1993.

[16] G. Wang and G. Pan, "Full wave analysis of microstrip floating line structures by wavelet expansion method," *IEEE Trans. Microw. Theory Tech.*, **43**, 131–142, Jan. 1995.

[17] G. Wang, G. Pan, and B. Gilbert, "A hybrid wavelet expansion and boundary element analysis for multiconductor transmission line in multilayered dielectric media," *IEEE Trans. Microw. Theory Tech.*, **43**, 664–675, Mar. 1995.

[18] J. Goswami, A. Chan, and C. Chui, "On solving first-kind integral equations using wavelets on a bounded interval," *IEEE Trans. Ant. Propg.*, **43**, 614–622, June 1995.

[19] G. Pan, "Orthogonal wavelets with applications in electromagnetics," *IEEE Trans. Magnet.*, **3**(3), 975–983, May 1996.

[20] G. Pan and J. Du, "The intervallic wavelets with applications in the surface integral equations," *11th Ann. Rev. Progr. ACES*, 993–999, Mar. 1995.

[21] R. Wagner and C. Chew, "A study of wavelets for the solution of electromagnetic integral equations," *IEEE Trans. Ant. Propg.*, **43**, 802–810, Aug. 1995.

[22] X. Zhu, G. Lei, and G. Pan, "On application of first and adaptive periodic Battle-Lemarie wavelets to modeling of multiple lossy transmission lines," *J. Comput. Phys.*, **132**, 299–311, Apr. 1997.

[23] Z. Xiang and Y. Lu, "An effective wavelet matrix transform approach for efficient solutions of electromagnetic integral equations," *IEEE Trans. Ant. Propg.*, **45**, 1205–1213, Aug. 1997.

[24] G. Pan, M. Toupikov, J. Du, and B. Gilbert, "Use of Coffman intervallic wavelets in 2D and 3D scattering problems," *IEE Proc. Microw. Ant. Propg.*, **145**(6), Dec. 1998.

[25] M. Toupikov, G. Pan, and B. Gilbert, "Weighted wavelet expansion in the method of moments," *IEEE Trans. Magn.*, **35**(3), 1550–1553, May 1999.

[26] D. Zahn, K. Sarabandi, K. Sabet, and J. Harvey, "Numerical simulation of scattering from rough surfaces: A wavelet-based approach," *IEEE Trans. Ant. Propg.*, **48**(2), 246–253, 2000.

[27] S. Barmada and M. Raugi, "New wavelet based approach for time domain simulations," *IEEE Trans. Ant. Propg.*, accepted July 2002.

[28] B. Alpert, G. Beylkin, R. Coifman, and V. Rokhlin, "Wavelet-like bases for the solution of second-kind integral equations," *SIAM J. Sci. Comput.*, **14**(1), 159–184, Jan. 1993.

[29] H. Malvar and D. Staelin, "Reduction of blocking effects in image coding with a lapped orthogonal transform," *Proc. ICASSP 88*, April 1988, pp. 781–784, New York.

[30] R. Coifman and Y. Meyer, "Remarques sur l'analyse de Fourier a fenetre," *C. R. Acad. Sci. Paris Ser. I. Math.*, **I**(312), 259–261, 1991.

[31] W. Sweldens, "Wavelets: What next?" *Proc. IEEE*, **84**(4), 680–685, 1996.

[32] H. Malvar. "Lapped transforms for efficient transform/subband coding," *IEEE Trans. Acoust. Speech Signal Process.*, **38**, 969–978, 1990.

[33] M. Wickerhauser, *Adapted Wavelet Analysis. From Theory to Software*, Wellesley, MA, A.K. Peters, 1994.

[34] K. Rao and P. Yip, *Discrete Cosine Transform: Algorithms, Advantages, and Applications*, Academic Press, Harcourt Brace Jovanovich, Boston, 1990.

[35] G. Pan, Y. Tretiakov, and B. Gilbert, "Smooth local cosine based Galerkin method for scattering problems," *IEEE Trans. Ant. Propg.*, to appear in Dec. 2002.

[36] B. Alpert,"Wavelets and other bases for fast numerical linear algebra," in *Wavelets: A Tutorial in Theory and Applications*, C. Chui, Ed., Academic Press, New York, 1992.

[37] M. Kominami and K. Rokushima, "On the integral equation of piecewise linear antennas," *IEEE Trans. Ant. Propg.*, **AP-29**, 787–792, Sept. 1981.

[38] D. Pozar, "Finite phased arrays of rectangular microstrip patches," *IEEE Trans. Ant. Propg.*, **AP-34**, pp. 658–665, May 1986.

[39] A. King and W. Bow, "Scattering from a finite array of microstrip patches," *IEEE Trans. Ant. Propg.*, **40**, 770–774, July 1992.

[40] C. Wang, F. Ling and J. Jin, "A fast full-wave analysis of scattering and radiation from large finite arrays of microstrip antennas," *IEEE Trans. Ant. Propg.*, **46**(10), 1467–1474, Oct. 1998.

[41] D. Pozar, *Microwave Engineering*, Addison-Wesley, Reading, MA, New York, 1990.

[42] C. Balanis, *Advanced Engineering Electromagnetics*, John Wiley, New York, 1989.

[43] M. Piket-May, A. Taflove, and J. Barron, "FD-TD modeling of digital signal propagation in 3-D circuits with passive and active loads," *IEEE Trans. Microw. Theory Tech.*, **42**(8), 1514–1523, Aug. 1994.

[44] D. M. Sullivan, *Electromagnetic Simulation Using the FDTD Method*, IEEE Press, New York, 2000.

[45] A. Taflove, *Computational Electromagnetics: The Finite-difference Time Domain Method,* Artech House, Norwood, MA, 1995.

[46] W. Sui, *Time-Domain Computer Analysis of Nonlinear Hybrid Systems,* CRC Press, New York, 2002.

[47] K.-L. Wu, M. Spenuk, J. Litva, and D.-G. Fang, "Theoretical and experimental study of feed network effects on the radiation pattern of series-fed microstrip antenna arrays," *IEE Proc.-H*, **138**(3), 238–242, June 1991.

[48] D. M. Pozar and D. H. Schaubert, Eds., *Microstrip Antennas: The Analysis and Design of Microstrip Antennas and Arrays,* IEEE Press, New York, 1995.

CHAPTER EIGHT

Wavelets in Rough Surface Scattering

In this chapter we will study scattering of electromagnetic waves from rough surfaces numerically, using the Coifman wavelets. Owing to the orthogonality, vanishing moments, and multiresolution analysis, a very sparse moment matrix is obtained. In addition the wavelet bases are continuous. Hence the sampling rate for wavelet bases is reduced to one-half the rate of the pulse cases, allowing the same computer resource to deal with quadruple the truncated surface area. More important, the Coiflets allow the development of one-point quadrature formula, which reduces the computational effort in filling matrix entries to $O(n)$. As a result the wavelet-Galerkin method with twofold integrals is faster than the traditional pulse-collocation approach with onefold integrals.

8.1 SCATTERING OF EM WAVES FROM RANDOMLY ROUGH SURFACES

Rough surface scattering has potential applications in remote sensing, semiconductor processing, radar, and sonar, among others. Figure 8.1 demonstrates a computer generated random surface, which will be discussed in Section 8.2.

Scattering of electromagnetic waves from rough surfaces has been studied by analytical [1, 2], numerical [3–6], and experimental means [7–9]. Analytic methods provide fast solutions and allow users to foresee the effects and trends of the solution due to individual parameters in the formulas. However, there are many geometric and physical limitations restricting the utility of analytical models in general applications. For instance, the tangential plane approximation, known as the Kirchhoff model, works only for undulating surfaces without shadowing, while the small perturbation method, known as the Rice model, is valid only for small roughness. Attempts were made to extend these analytical models, including the iterated Kirchhoff [10, 11] and Wiener–Hermite expansion [12], among others. Nevertheless, the modified analytical models still operate under certain assumptions and conditions. Experimental

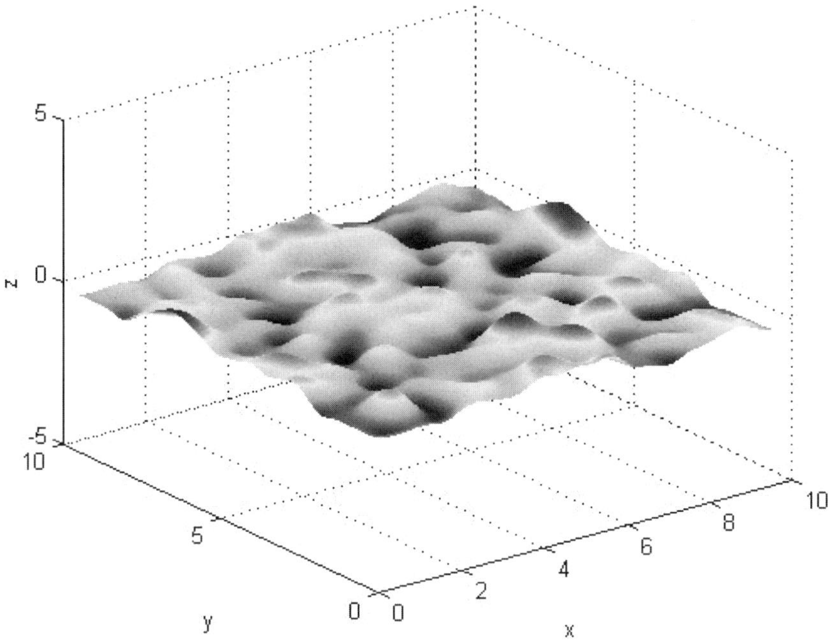

FIGURE 8.1 Computer generated random surface with Gaussian distribution $\sigma = 0.2\lambda$ and Gaussian correlation $\ell_x = \ell_y = 0.6\lambda$.

method requires fabrication of rough surfaces with specified statistical parameters, and it requires high-tech equipment that is costly and is not versatile. With advances in today's computers, it seems ideas to develop numerical methods that are accurate, versatile and relatively inexpensive. In the numerical approaches, the 1D Monte Carlo was developed several decades ago using the MoM [3]. In the Monte Carlo simulation, many sample surfaces with desired roughness statistics are generated and then the scattering solution for each sample surface, or realization is obtained using the MoM. These solutions are then averaged numerically to approximate the required statistical quantities. Clearly, from the nature of physics and statistics, rough surface scattering problems are electrically large problems. Traditional MoM in conjunction with the Galerkin procedure requires that the computation time be on the order of n^2 for matrix filling and n^3 for matrix inversion if Gaussian elimination is employed. Tsang et al. reported the band matrix iterative method (BMIA) [6] and applied the method to 3D scattering problems. Nevertheless, in the BMIA computation, humans must have interact with computers to set up the strong or weak terms in the system matrix.

Recently wavelets have appeared in applied mathematics [13] and have been successfully used to solve integral equations [14]. In electromagnetics, wavelets have been applied to guidedwave, radiation, object scattering, nonlinear device modeling, and target identification [15–17]. Wavelets have also been employed in rough surface scattering [18, 19]. In [18] the Daubechies wavelets were employed as a transformation matrix that converts the dense matrix generated from the MoM into a sparse matrix. This approach follows the idea in [16, 17, 20]. In [19] wavelets are directly used as the basis and testing functions to create a sparse impedance matrix, bypassing the MoM computation to fill the matrix. Despite the differences in the two approaches, both of them require massive computation to fill the entire entries of the impedance matrix on the order of $O(n^2)$. Here we apply wavelets to the 2D and 3D scattering of electromagnetic waves from perfectly conducting random surfaces. The integral equations for both the HH and VV polarizations are solved using the Galerkin procedure. More specifically, we choose the Coifman wavelets, which are orthogonal and compactly supported with zero moments of both the wavelets and scalets. As a consequence, a property similar in nature to the Dirac δ is evolved that allows fast computation of the most off-diagonal elements in the impedance matrix using the single-point quadrature formula. Hence only the "strong" elements around the diagonal of the matrix need to be evaluated via numerical quadrature; they are on the order of $O(n)$. The resultant impedance matrix is sparse and can be solved with iterative methods (e.g., conjugate gradient) or newly developed nonstandard LU factorization [21] on the order of $O(n)$. As a result, the wavelet-Galerkin method with twofold integrals is faster than the traditional pulse-collocation approach with one-fold integrals.

Numerical examples of the wavelet-Galerkin method are compared with those obtained from the standard MoM that employs pulse basis and a point match in scheme. Excellent agreement was observed between new approach and previously published results.

8.2 GENERATION OF RANDOM SURFACES

In order to perform numerical simulations, a realization has to be generated in a randomly rough surface with prescribed surface distribution and autocorrelation functions. The spectral method [22] for the generation of a random surface profile has been found more convenient than the autoregressive (AR) method used in [23], especially for surface derivatives. The description of the method for the case of the 1D random surface can be found in [24] and for the 2D case in [9]. A surface is called simple if its correlation function has only one correlation length parameter; it is called composite if more than one parameter is required to describe its correlation function.

In most research articles, the rough surface profile is described in terms of its deviation from a flat "reference plane." In general, the reference plane is assumed to be located at $z = 0$. The random fluctuations from this reference plane are described by the probability density function (p.d.f.).

For analytical convenience, one usually uses the Gaussian type p.d.f.

$$p(z) = \frac{1}{\sigma\sqrt{2\pi}} \exp\left(-\frac{z^2}{2\sigma^2}\right), \tag{8.2.1}$$

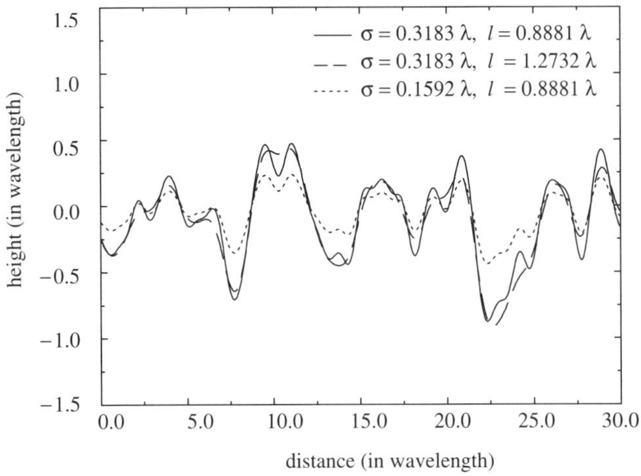

FIGURE 8.2 Random surfaces with different standard deviations and correlation lengths.

where we have assumed a zero mean $\langle z \rangle = 0$ and variance $\langle z^2 \rangle = \sigma^2$. In the previous case the rough surface is generated by a 1D stationary (in the wide sense), normal, random process with zero mean and standard deviation σ. The height coordinate z of the surface is a realization of the random process $z(x)$, which is a function of the x coordinate. The relations between surface points $z_1 = z(x_1)$ and $z_2 = z(x_2)$ are specified by the correlation function, which we consider also to be a Gaussian-type

$$R(\tau) = \langle z(x_1), z(x_2) \rangle = \sigma^2 \exp\left(-\frac{\tau^2}{l^2}\right), \tag{8.2.2}$$

where $\langle \cdot \rangle$ denotes the ensemble average, $\tau = x_1 - x_2$, and l is a correlation length in the x direction.

We will describe two methods of generating a random surface profile, the autocorrelation approach and spectral domain approach. In Fig. 8.2 we plotted the random surface profiles generated with a Gaussian probability density function p.d.f. and Gaussian correlation function. We used different parameters of standard deviation σ and correlation length l in the figure. Plotted in Fig. 8.3 is the p.d.f. of the height estimated from the actual profile. In order to compare the obtained numerical results we also plotted in Fig. 8.3 the p.d.f. calculated by using (8.2.1). In Fig. 8.4 two Gaussian correlation functions with different parameters l are shown. As for the case of the p.d.f., we estimated the correlation functions from a numerically generated random surface profile and plotted the corresponding correlation functions using (8.2.2). To create Fig. 8.5*a*, we used the Gaussian p.d.f. and two different correlation functions, namely the Gaussian and exponential functions. The small-scale roughness in Fig. 8.5*a* of the random surface profile with the exponential correlation function gives rise to the high-frequency tail of the exponential spectrum. Figure 8.5*b* depicts these correlation functions that are calculated from the actual random surface profiles by using theoretical expressions. All curves in Fig. 8.5*b* have been normalized

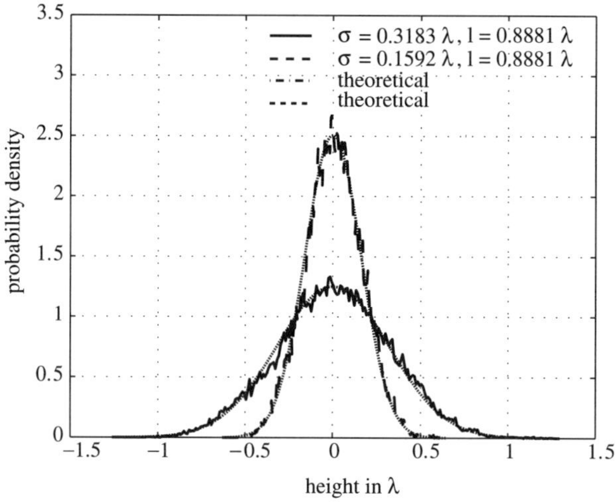

FIGURE 8.3 Probability density function of height for simple surface.

to the maximum value of unity. In Fig. 8.6 we also illustrate simple and composite random surface profiles. A composite surface is a superposition of two surfaces with clearly distinct vertical and horizontal scales.

8.2.1 Autocorrelation Method

This method was suggested in [23]. We begin with a numerically generated sequence of independent Gaussian variables $\{X_k\}$ with zero mean and a standard deviation

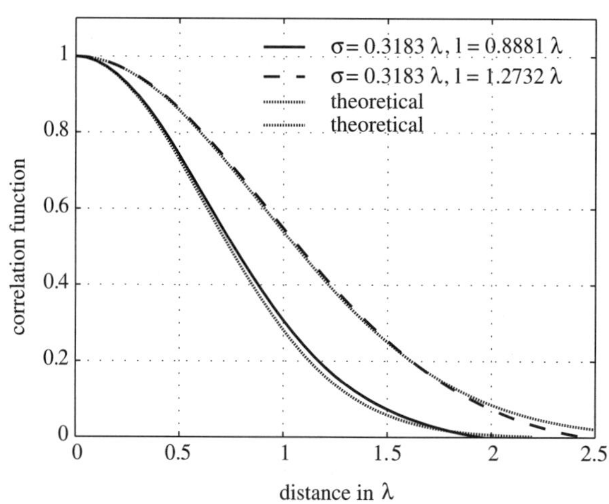

FIGURE 8.4 Normalized correlation function of simple surface.

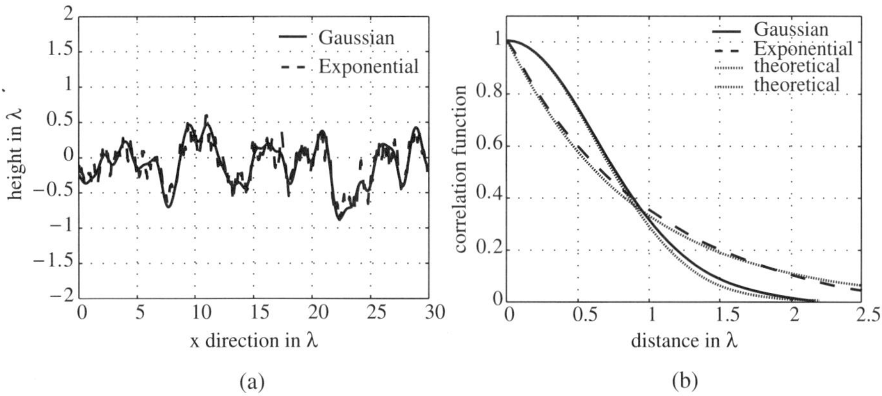

FIGURE 8.5 Random surfaces of Gaussian distribution with Gaussian and exponential correlation functions: (*a*) 1D rough surfaces, (*b*) corresponding correlation functions.

of unity. This sequence can be obtained utilizing a commercial software package such as the IMSL, Matlab, or NAG. From this uncorrelated sequence of normally distributed samples, a sequence of correlated normal samples $\{C_k\}$ can be obtained by digitally filtering in the manner

$$C_k = \sum_{j=-N}^{N} W_j X_{j+k}, \tag{8.2.3}$$

where W_j are the correlation weights yet to be determined. The expectation

$$E\{C_k C_{k+i}\} = \sum_j \sum_n W_j W_n E\{X_{j+k} X_{n+k+i}\}, \tag{8.2.4}$$

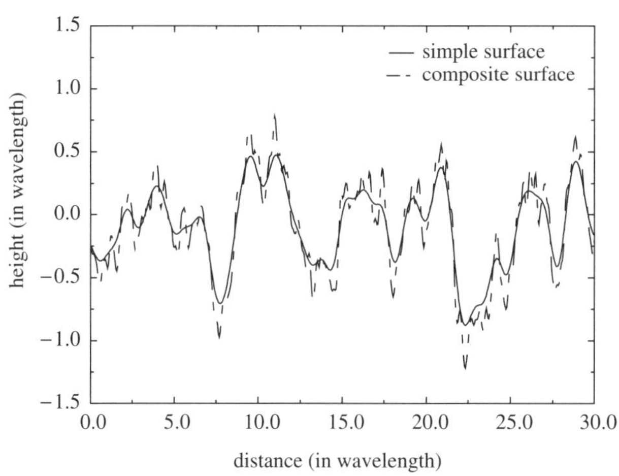

FIGURE 8.6 Simple and composite random surfaces.

and $\{X_k\}$ is an independent sequence, satisfying

$$E\{X_{j+k}X_{n+k+i}\} = \begin{cases} 0, & j \neq n+i \\ 1, & j = n+i. \end{cases} \tag{8.2.5}$$

Hence

$$E\{C_k C_{k+i}\} = \sum_j W_j W_{j-i}. \tag{8.2.6}$$

The previous equation states that the autocorrelation function of the correlated normal sample $\{C_k\}$ is identical to the convolution of the digital weights. It follows also that the Fourier transform of the correlation is equal to the product of the Fourier transforms of the digital filter weights. Thus the inverse transform of the square root of the prescribed spectrum is the filter weight. For instance, let the correlation function be Gaussian

$$\rho = \exp\left(-\frac{j^2}{l^2}\right). \tag{8.2.7}$$

Its spectrum is

$$\rho_s = (l\sqrt{\pi}) \exp\left(-\frac{l^2 f^2}{4}\right), \tag{8.2.8}$$

and the square root of ρ_s is

$$(\rho_s)^{1/2} = (l\sqrt{\pi})^{1/2} \exp\left(-\frac{l^2 f^2}{8}\right). \tag{8.2.9}$$

The inverse Fourier transform of (8.2.9) is the filter weight and can be written as

$$W_j = \left(\frac{2}{\sqrt{\pi} l}\right)^{1/2} \exp\left(-2\frac{j^2}{l^2}\right). \tag{8.2.10}$$

Notice that expression (8.2.3) with W_j as defined in (8.2.10) produces correlated samples of z with standard deviation of unity and with a sampling interval of unity in the x direction. For a general case where the correlated samples of z create a random surface with a standard deviation σ, correlation length l, and a sampling interval Δx units, we will have the following modified expression for the weight W_j:

$$W_j = \left(\frac{2\sigma^2 \Delta x}{\sqrt{\pi} l}\right)^{1/2} \exp\left(-2\frac{(j\,\Delta x)^2}{l^2}\right). \tag{8.2.11}$$

A realization of a random surface $\{C_k\}$ with the properties above will be generated at points $x_k = k\,\Delta x$ ($k = 0, \ldots, N$) with standard deviation σ, correlation length l, and root mean square (rms) slope $\rho_x = \sqrt{2}\sigma/l$. The first derivative of the surface at each sampling point can be approximated using the finite difference scheme

$$\left(\frac{dz}{dx}\right)_{x=x_k} \approx \frac{C_{k+1} - C_k}{\Delta x}. \tag{8.2.12}$$

The derivative will be stored for future numerical computations.

8.2.2 Spectral Domain Method

The second method, described in [24], imposes a roughness spectral density since the inverse Fourier transform can be done very quickly by the implementation of the standard fast Fourier transform (FFT) algorithm. For this method we use a corresponding roughness spectral density of the correlation function to generate a realization of a random surface profile. If we assume a Gaussian correlation function of (8.2.2), then the corresponding roughness spectral density is

$$W(k) = \frac{\sigma^2 l}{\sqrt{4\pi}} \exp\left(-\frac{k^2 l^2}{4}\right) = \frac{1}{2\pi} \int_{-\infty}^{+\infty} R(\tau) e^{ik\tau}\, d\tau. \tag{8.2.13}$$

An alternative correlation function, such as the exponential function, more precisely describes surfaces with very sharp peaks. This correlation has the form

$$R(\tau) = \sigma^2 \exp\left(-\frac{|\tau|}{l}\right) \tag{8.2.14}$$

and the corresponding roughness spectral density

$$W(k) = \frac{\sigma^2 l}{\sqrt{4\pi}} \left(\frac{1}{1 + k^2 l^2}\right). \tag{8.2.15}$$

In turbulence modeling, the power law spectrum is used to model the random fluctuation of the propagation characteristics for the medium. Its corresponding spectrum is given by

$$W(k) = \frac{\sigma^2 l}{\sqrt{4\pi}} \left\{1 + \pi \left[\frac{(2n-3)!!}{(2n-2)!!}\right]^2 \frac{k^2 l^2}{4}\right\}^{-n}, \tag{8.2.16}$$

where $(2n-2)!! = 2 \times 4 \times \cdots (2n-2)$, $(2n-3)!! = 1 \times 3 \times \cdots (2n-3)$, $(-1)!! = 1$ and n is the order of the power law spectrum. The power law spectrum converges to the Gaussian spectrum for large order n, and is almost equivalent to the Laurentzian spectrum for order $n = 1$. Moreover, for any given order, the power law spectrum reduces to k^{-2n} for large k. No closed-form expression is available for the autocorrelation of a surface with the power law spectrum.

Suppose that we have a roughness spectrum $W(k)$. For the scattering computation, surface realization (heights and first derivatives) are needed as a set of N points with spacing Δx over length $L = N\,\Delta x$. Realizations with the desired properties can be generated at points $x_k = (k+0.5)\,\Delta x$ $(k = 0, \ldots, N-1)$ using the discrete Fourier transform (DFT) method. The rough surface profile $z = f(x_k)$ is related to the 1D DFT of the surface spectrum by

$$f(x) = \frac{1}{L} \sum_{n=-N/2}^{N/2-1} F(K_n) \exp(i K_n x), \tag{8.2.17}$$

where

$$F(K_n) = \sqrt{2\pi L W(K_n)} \begin{cases} \dfrac{N(0,1) + i N(0,1)}{\sqrt{2}}, & n \neq 0, N/2 \\ N(0,1), & n = 0, N/2 \end{cases}$$

$$K_n = \frac{2\pi n}{L}, \quad i = \sqrt{-1},$$

and $N(0, 1)$ denotes an independent sample taken from a zero mean with unit standard variance Gaussian distribution.

For the Fourier coefficients of the first derivative of a random surface profile we have

$$F_{\partial x}(K_n) := F(K_n) \times i K_n. \tag{8.2.18}$$

The first derivative of a rough surface profile at each sampling point can be obtained by using the DFT in the same manner as in (8.2.17).

Equation (8.2.17) can be computed by means of a fast Fourier transform (FFT), as can the first derivative of $f(x)$. For a p.d.f. of height with another distribution, such as a gamma distribution, it suffices to replace $N(0, 1)$ by such an appropriate distribution. The two-point statistics are governed by the magnitude of the Fourier spectrum, which follows the surface spectrum $W(k)$. Since the surface must be represented by a sequence of real numbers, the phase of the Fourier coefficients must satisfy certain requirements. In order to generate a real sequence, the Fourier coefficients must be Hermitian, namely

$$F(K_n) = F^*(-K_n). \tag{8.2.19}$$

The requirement above is also important in the synthesis of 2D surfaces. The use of the DFT in rough surface generation requires that the surface lengths be at least five correlation lengths so that no spectral aliasing is present in the resulting surface. Furthermore the resulting rough surface is a periodic function in which the surface height and the slope are periodic in space. It is important to note that due to a finite surface length in the discrete synthesis process, the surface autocorrelation does not completely decay to zero and some oscillations are presented. In practice, the surface spectrum can be estimated from the actual surface profile by the expression

$$W(k) = \frac{1}{2\pi L} \left\langle \left| \int_{-L/2}^{L/2} g(x) f(x) e^{-ikx}\, dx \right|^2 \right\rangle. \tag{8.2.20}$$

The purpose of the window function $g(x)$ with an appropriate tapering is to minimize spectral sidelobes, also known as the "Gibbs phenomenon" in the Fourier analysis, due to the finite length involved.

Most of the statistics used to describe 1D rough surfaces can be extended in the 2D case. The 2D rough surface is described by $z = f(x, y)$, which is a random function of position (x, y). Various two-dimensional spectra and autocorrelations, which are basically extensions of the one-dimensional case, can be used to generate the 2D rough surface. For reasons of practicality in surface manufacturing, only surfaces with Gaussian roughness and Gaussian spectrum are considered. The correlation function $R(\tau_x, \tau_y)$ that describes the coherence between different points on the surface separated by the distance $d = \sqrt{\tau_x^2 + \tau_y^2}$ and is given by

$$R(\tau_x, \tau_y) = \sigma^2 \exp\left(-\frac{\tau_x^2}{2l_x^2} - \frac{\tau_y^2}{2l_y^2}\right), \tag{8.2.21}$$

where τ_x and τ_y describe the separation between any two points along the x and y directions. The coherence length of the surface profiles is given by l_x and l_y. The power spectral density function of the surface $W(k_x, k_y)$ is related to the correlation function via a 2D Fourier transform. For a Gaussian correlation function given by (8.2.21), we have

$$W(k_x, k_y) = \frac{l_x l_y \sigma^2}{4\pi} \exp\left(-\frac{k_x^2 l_x^2}{4} - \frac{k_y^2 l_y^2}{4}\right). \tag{8.2.22}$$

It is important to note that in (8.2.22), there are two distinct correlation lengths, l_x and l_y. The surface is isotropic when $l_x = l_y$, and anisotropic if $l_x \neq l_y$. In the other extreme, if one of the correlation lengths is much greater than the other, the 2D surface becomes essentially a 1D surface for the purpose of the experiments and numerical calculations. The corresponding rms slopes are defined respectively by $\rho_x = \sqrt{2}\sigma/l_x$ and $\rho_y = \sqrt{2}\sigma/l_y$.

Similarly to the 1D case, the rough surface profile $z = f(x, y)$ is related to the 2D DFT of the power spectrum as

$$f(x, y) = \frac{1}{L_x L_y} \sum_{m=-(N_x/2)}^{(N_x/2)-1} \sum_{n=-(N_y/2)}^{(N_y/2)-1} F(K_{xm}, K_{yn}) \exp(iK_{xm}x + iK_{yn}y), \tag{8.2.23}$$

where

$$F(K_{xm}, K_{yn}) = 2\pi\sqrt{L_x L_y W(K_{xm}, K_{yn})} \begin{cases} \dfrac{N(0,1) + iN(0,1)}{\sqrt{2}}, & m \neq 0, N_x/2,\ n \neq 0, N_y/2 \\ N(0,1), & m = 0, N_x/2 \text{ or } n = 0, N_y/2 \end{cases} \tag{8.2.24}$$

and

$$K_{xm} = \frac{2\pi m}{L_x}, \quad K_{yn} = \frac{2\pi n}{L_y}, \quad i = \sqrt{-1}. \tag{8.2.25}$$

In the expressions above, K_{xm} and K_{yn} are the discrete set of spatial frequencies; L_x and L_y are surface profile lengths in x and y directions, respectively. To generate a real sequence, the requirement for $F(K_{xm}, K_{yn})$ is as follows:

$$\begin{aligned} F(K_{xm}, K_{yn}) &= F^*(-K_{xm}, -K_{yn}), \\ F(K_{xm}, -K_{yn}) &= F^*(-K_{xm}, K_{yn}). \end{aligned} \tag{8.2.26}$$

Under these two conditions, the 2D sequence is "conjugate symmetrical" about the origin. This means that the reflection of any point about the origin is its complex conjugate. By using the Fourier coefficients (8.2.24), we can also find the corresponding Fourier coefficients for the surface derivatives in the x and y directions

$$\begin{aligned} F_{\partial x}(K_{xm}, K_{ny}) &:= F(K_{xm}, K_{ny}) \times iK_{xm}, \\ F_{\partial y}(K_{xm}, K_{ny}) &:= F(K_{xm}, K_{ny}) \times iK_{yn}. \end{aligned} \tag{8.2.27}$$

By taking the inverse 2D DFT with the Fourier coefficients given in (8.2.27), we can also obtain at each sampling point the first derivatives of a random surface profile in both the x and y directions. Figure 8.1 is generated from the 2D spectral method discussed above.

8.3 2D ROUGH SURFACE SCATTERING

2D scattering cases are simpler than 3D cases, but they address the main features, such as discretization rate, single-point quadrature, and singularity treatment. The experience one has gained from 2D scattering illuminates the more advanced study of 3D scattering problems. Figure 8.7 demonstrates both horizontal and vertical polarizations with physical and geometric parameters indicated.

8.3.1 Moment Method Formulation of 2D Scattering

The standard MoM [25] is employed to formulate the noncoherent backscattering coefficient of a random surface profile. The geometry of the scattering problem is shown in Fig. 8.7.

To compute the scattering coefficient from a computer-generated, random, perfectly conducting surface, it is necessary to find the surface current density $J(x)$ which is induced by a given incident plane wave over the entire illuminated area. The MoM is employed to solve for the induced current density from which the scattered fields and radar cross sections are computed. In practice, the Gaussian taper function in the form $\exp(-g^{-2}x^2\cos^2\theta)$ is applied to the incident field to suppress the artifacts of current at the edges of the illuminated area, so as to obtain stable esti-

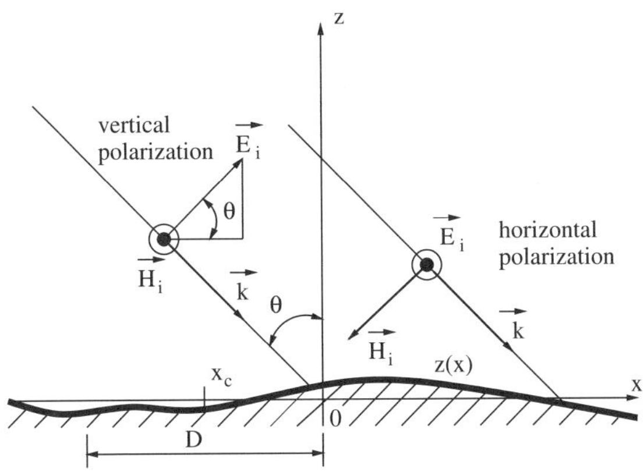

FIGURE 8.7 Geometry of 2D scattering problem.

mates of the scattering coefficients [26]. Due to finite computer storage and practical restrictions on the matrix size, the illuminated segment length D must be finite. We repeat calculations of M segments to obtain meaningful estimates of the backscattering coefficient. The choice of parameters g, D, and M is discussed in detail in [3].

Let us consider the case of the HH polarization, where the second H denotes the horizontal incident wave and the first H implies horizontal polarization of the scattered wave. The time convention $e^{j\omega t}$ is assumed and suppressed. The incident plane wave

$$\begin{aligned}\mathbf{E}^i(x) &= -\hat{y}\cdot\exp(jk_0[(x-x_c)\sin\theta + z(x)\cos\theta)]) \\ &= -\hat{y}\cdot E^i(x)\end{aligned} \tag{8.3.1}$$

is impinging upon a random surface $z(x)$. In (8.3.1), θ is the angle of incidence and x_c is the center point of the illuminated segment with the length D as shown in Fig. 8.7. The integral equation governing the surface current is

$$\begin{aligned}E^i(x, z(x)) = \frac{k_0\eta}{4}\int_{x_c-D/2}^{x_c+D/2} J_i(x')H_0^{(2)}\left(k_0\sqrt{(x-x')^2 + (z(x)-z(x'))^2}\right) \\ \cdot\sqrt{1+\left(\frac{dz(x')}{dx'}\right)^2}\,dx'\end{aligned} \tag{8.3.2}$$

where k_0 is the wavenumber in free space, η is the intrinsic impedance of free space, D is the width of the illuminated segment, $(x, z(x))$ is a point on the surface, and $H_0^{(2)}(x)$ is the zero-order Hankel function of the second kind. Upon breaking the segment into P subsegments with widths $\Delta x = D/P$, integral equation (8.3.2) is solved by the method of moments [25], which converts (8.3.2) into a matrix equation of the form

$$[Q][I] = [V], \tag{8.3.3}$$

where the mnth element of the impedance matrix $[Q]$ is given by

$$Q_{m,n} = \frac{k_0 \eta}{4} \int_{(n-1)\Delta x + x_c - D/2}^{n\,\Delta x + x_c - D/2} H_0^{(2)}\left(k_0 \sqrt{(x_m - x')^2 + (z_m - z')^2}\right) \cdot \sqrt{1 + \left(\frac{dz'}{dx'}\right)^2}\, dx' \tag{8.3.4}$$

with $x_m = (m - 1/2)\,\Delta x + x_c - D/2$, $z_m = z(x_m)$, $I_n = J_i(x_n)$, and $V_m = E^i(x_m)$. The matrix $[Q]$ may be viewed as a $P \times P$ generalized impedance matrix.

It should be noted that in (8.3.4) for the diagonal elements $Q_{n,n}$, the Hankel function has an integrable singularity. By using small-argument expansion of the Hankel function and approximating the subsegment by a straight line, we have

$$Q_{n,n} \approx \frac{k_0 \eta}{4}\,\Delta d \left\{1 - j\frac{2}{\pi}\left[\ln\left(\frac{k_0\,\Delta d}{4e}\right) + \gamma\right]\right\}, \tag{8.3.5}$$

where $\gamma = 0.5772156649$, $e = 2.718281828$, $\Delta d = [1 + (dz'/dx')_n^2]^{1/2}\,\Delta x$, and dz'/dx' is the slope at x_n' which is calculated numerically from the surface profile.

The numerical solution of (8.3.3) provides the estimate of the induced surface current density at each segment. With the surface current obtained over the ith segment, the far-zone backscattered field due to the segment is obtained by

$$E^s(\theta) = \eta k_0 \frac{e^{-j(k\rho_0 + 3\pi/4)}}{\sqrt{8\pi k\rho_0}} \cdot \int_{x_c - D/2}^{x_c + D/2} J_i(x') e^{(jk_0[(x' - x_c)\sin\theta + z(x')\cos\theta])} \cdot \sqrt{1 + \left(\frac{dz'}{dx'}\right)^2}\, dx' \tag{8.3.6}$$

where ρ_0 is the distance to the far-field point from the illuminated zone. If we approximate the rough surface between two sample points by a straight line with a constant slope, then the expression above can be evaluated numerically as

$$E^s(\theta) = \eta k_0 \frac{e^{-j(k\rho_0 + 3\pi/4)}}{\sqrt{8\pi k\rho_0}} \sum_n \sqrt{1 + \left(\frac{dz'}{dx'}\right)_n^2}\, I_n \cdot e^{(jk_0[(x_n' - x_c)\sin\theta + z(x_n')\cos\theta])}\,\Delta x, \tag{8.3.7}$$

where $x_n' = (n - 1/2)\,\Delta x + x_c - D/2$. As stated previously, a taper function of the form

$$G(x_m - x_c) = \exp[-g^{-2}(x_m - x_c)^2 \cos^2\theta] \tag{8.3.8}$$

was adopted and multiplied to the incident field. The effective (associated with the scattered power) illuminated width L_{eff} due to this illumination is

$$L_{\text{eff}} = \int_{-\infty}^{+\infty} \exp(-2g^{-2}x^2\cos^2\theta)\,dx = \frac{g\sqrt{\pi/2}}{\cos\theta}. \tag{8.3.9}$$

The average noncoherent backscattering coefficient from M independent segments can be written as

$$\sigma^0(\theta) = \frac{2\pi\rho_0}{ML_{\text{eff}}}\left[\sum_{j=1}^{M} |E_j^s|^2 - \frac{1}{M}\left|\sum_{j=1}^{M} E_j^s\right|^2\right]. \tag{8.3.10}$$

For a vertical polarization, the integral equation is cast in terms of the incident magnetic field $\mathbf{H}^i$, written as

$$\begin{aligned} -H^i(x, z(x)) = \frac{1}{2}J_i(x) + \frac{jk_0}{4}\int_{x_c-D/2}^{x_c+D/2} J_i(x')\sqrt{1+\left(\frac{dz'}{dx'}\right)^2} \\ \cdot\cos\phi\cdot H_1^{(2)}\left(k_0\sqrt{(x-x')^2+(z(x)-z(x'))^2}\right)dx', \end{aligned} \tag{8.3.11}$$

where

$$\mathbf{H}^i = -\hat{y}\cdot\exp(jk_0[(x-x_c)\sin\theta + z(x)\cos\theta)]). \tag{8.3.12}$$

$H_1^{(2)}(x)$ is the first-order Hankel function of the second kind, and

$$\cos\phi = \frac{(\boldsymbol{\rho}-\boldsymbol{\rho}')\cdot\hat{n}'}{|\boldsymbol{\rho}-\boldsymbol{\rho}'|}, \tag{8.3.13}$$

where $\hat{n}'$ is the unit vector normal to the surface at point $(x', z(x'))$.

Applying the MoM procedures, the integral equation (8.3.11) is again converted into a matrix equation of the form (8.3.3), with the mnth element

$$\begin{aligned} Q_{m,n} = \frac{1}{2}\delta_{m,n} + \frac{jk_0}{4}\int_{(n-1)\,\Delta x+x_c-D/2}^{n\,\Delta x+x_c-D/2} H_1^{(2)}\left(k_0\sqrt{(x_m-x')^2+(z_m-z')^2}\right) \\ \cdot\cos\phi_m\sqrt{1+\left(\frac{dz'}{dx'}\right)^2}\,dx', \end{aligned} \tag{8.3.14}$$

where

$$\begin{aligned} \delta_{m,n} &= \text{Kronecker delta}, \\ \cos\phi_m &= \frac{(\boldsymbol{\rho}_m-\boldsymbol{\rho}')\cdot\hat{n}'}{|\boldsymbol{\rho}_m-\boldsymbol{\rho}'|}, \\ I_n &= J_i(x_n), \\ V_m &= -H^i(x_m). \end{aligned} \tag{8.3.15}$$

The induced surface current is obtained by solving the matrix equation for I_n. A direct solver of Gaussian elimination or an iterative solver such as the conjugate gradient method may be applied. This current I_n is then employed to compute the far-zone backscattered field $H^s(\theta)$ via the formula

$$H^s(\theta) = k_0 \frac{e^{-j(k\rho_0+3\pi/4)}}{\sqrt{8\pi k\rho_0}} \sum_n \sqrt{1+\left(\frac{dz'}{dx'}\right)_n^2} I_n \cos\psi_n \cdot e^{(jk_0[(x_n'-x_c)\sin\theta+z(x_n')\cos\theta])} \Delta x, \tag{8.3.16}$$

where

$$\cos\psi_n = \hat{n}\cdot\hat{R}, \quad \hat{R} = \boldsymbol{\rho}_0/|\,\boldsymbol{\rho}_0\,| \tag{8.3.17}$$

and $\boldsymbol{\rho}_0$ is the radial vector from the center of the segment to the observation point.

Finally, the averaged noncoherent backscattering coefficient is calculated by

$$\sigma^0(\theta) = \frac{2\pi\rho_0}{ML_{\text{eff}}}\left[\sum_{j=1}^{M} |\,H_j^s\,|^2 - \frac{1}{M}\left|\sum_{j=1}^{M} H_j^s\right|^2\right]. \tag{8.3.18}$$

8.3.2 Wavelet-Based Galerkin Method for 2D Scattering

The Coifman scalets of order $\mathcal{L}=4$ and resolution level j_0 are employed to expand the unknown surface current $J_i(x')$ in (8.3.2) in the form

$$J_i(x') = \sum_n a_n^{j_0}\varphi_{j_0,n}(x'), \tag{8.3.19}$$

where $\varphi_{j_0,n}(x) = 2^{j_0/2}\varphi(2^{j_0}x-n)$. In the Galerkin procedure the testing functions are the same as the basis functions. After testing the integral equation (8.3.2) with the same Coifman scalets $\{\varphi_{j_0,n}\}$, we convert the integral equation into a matrix equation of the form (8.3.3) with the mnth entry

$$Q_{m,n} = \int_{S_m}\int_{S_n} \varphi_{j_0,m}(x)\varphi_{j_0,n}(x')K(x,x')\,dx'\,dx \tag{8.3.20}$$

and

$$V_m = \int_{S_m} \varphi_{j_0,m}(x)E^i(x)\,dx, \tag{8.3.21}$$

where $K(x,x')$ is the kernel of the integral equation under consideration, S_n and S_m are, respectively, the supports of the expansion and testing functions.

The previously discussed Dirac δ-like property of the Coiflets can be used for the construction of the one-point quadrature formula when the kernel $K(x,x')$ is free of singularities within the interval of integration. The detailed treatment and error estimate of the one-point quadrature are contained in Section 7.2.3. The kernel of the integral equation (8.3.2) has a singularity when $m = n$ in (8.3.20). In the impedance matrix Q, the diagonal elements and elements adjacent to the diagonal are computed using standard Gauss–Legendre quadrature. The Coifman scalet has a support of $[-4, 7]$. However, the scalet dies down quickly, and the support is truncated into $[-3, 3]$. We have used the square shape with 9 points per unit, and we have dropped the singular point at the square center. Another way of performing numerical integration is to divide the truncated support into three equal intervals of $[-3, -1]$, $[-1, 1]$ and $[1, 3]$. In each interval we employ Gaussian quadrature of 8 source points by 10 field points. Since no source and field points coincide, singularity is avoided. Both techniques perform roughly equivalently. For all other matrix elements we used the one-point quadrature formula of the form

$$Q_{m,n} \approx 2^{-j_0} K(2^{-j_0}m, 2^{-j_0}n). \tag{8.3.22}$$

The application of the one-point quadrature formula (8.3.22) has significantly accelerated the generation of the system matrix Q for each realization of the random surface profile in the Monte Carlo simulation. Savings in computation time will prove more profound when the impedance matrix is very large. Indeed, this technique is particularly powerful when 2D surfaces are considered, since the matrix size for 2D cases will be the square of that for cases of a 1D surface.

Suppose that the number of unknowns is large, say 10 thousand; then we prefer to use iterative techniques to solve matrix equation (8.3.3). The standard collocation technique with pulse basis and Dirac δ testing may lead to a dense system matrix. The approach used in [16, 17] is to apply the wavelet transform to sparsify the resultant dense system matrix, then use the conjugate gradient method to solve the transformed matrix. Despite the gain in solving the sparsified matrix, one has to pay an overhead in converting the MoM matrix to the sparse matrix. If the MoM matrix is too large to be generated, there will be no way to obtain the sparse matrix. In contrast, for our approach, the impedance matrix is generated directly from the wavelet basis without the original MoM matrix. Furthermore the operation count for the impedance matrix is $O(n)$ rather than $O(n^2)$. In fact we can use scalets at the highest resolution level j_0 to create a system matrix and then apply the fast wavelet transform to go down a few resolution levels [27]. By doing that, we introduce wavelets into the expansion for the unknown current $J_i(x)$. The combination of scalets and wavelets makes the system matrix extremely sparse. These sparse matrices can be solved with iterative methods, or newly developed nonstandard LU factorization [21] on the order of $O(n)$. This procedure is also helpful when we have to solve matrix equation (8.3.3) several times for different right-hand sides with the same matrix Q.

8.3.3 Numerical Results of 2D Scattering

The backscattering coefficients for simple rough surfaces with Gaussian *p.d.f.* and Gaussian correlation functions are shown in Fig. 8.8, where different parameters σ and l were used. In Fig. 8.9 we plotted the backscattering coefficients that were calculated for the simple surfaces with Gaussian and exponential correlation functions.

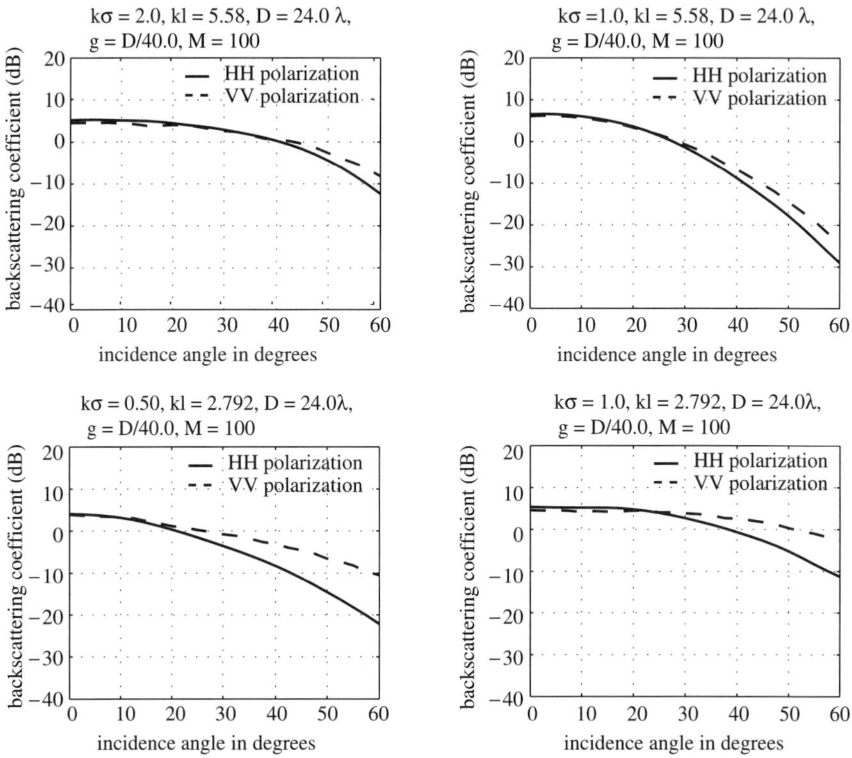

FIGURE 8.8 Backscattering coefficients of simple surfaces with different parameters.

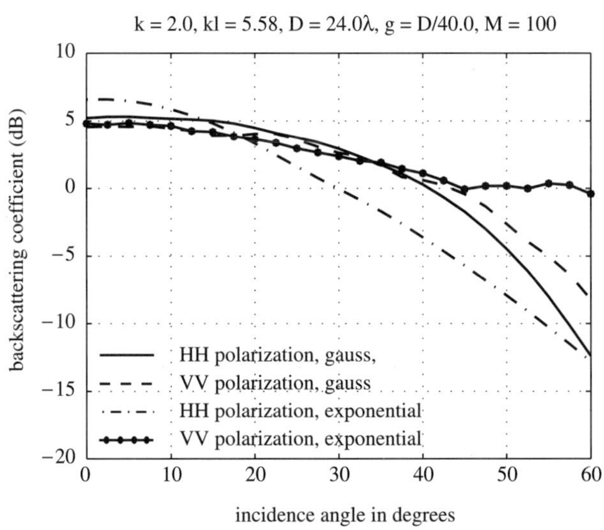

FIGURE 8.9 Backscattering coefficient of the simple surface.

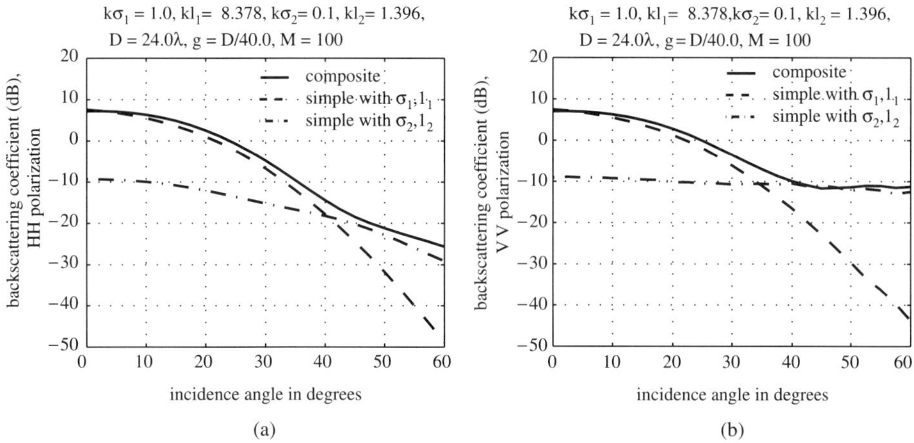

FIGURE 8.10 Backscattering coefficient of the simple and composite surfaces: (*a*) HH polarization and (*b*) VV polarization.

In Fig. 8.10 we depict the radar cross section from composite rough surfaces where the correlation function is

$$\rho(\tau) = ae^{-\tau^2/l_1^2} + (1-a)e^{-\tau^2/l_2^2}, \qquad a = 0.01746. \tag{8.3.23}$$

For all the basic cases presented thus far, we have used $D = 24.0\lambda$, $g = D/40.0$, $M = 100$ (number in average), and $\Delta x = 0.05\lambda$. In Figs. 8.8 to 8.10 the matrix size for the pulse basis is 480×480 in each case. The sampling rate used in the numerical calculations is 20 pulses per wavelength, or $\Delta x = 0.05\lambda$, as recommended in [28].

Figure 8.11 shows the backscattering coefficients of a simple random surface with HH and VV polarizations, respectively. The following nominal parameters are used: $k\sigma = 1.0$, $kl = 5.58$, $D = 34.5\lambda$, $g = D/40.0$, $M = 50$, and a mean height of zero. The sampling rate of 0.0625λ or 16 points per wavelength was adopted for the generation of all random surface samples. The numerically created random surface

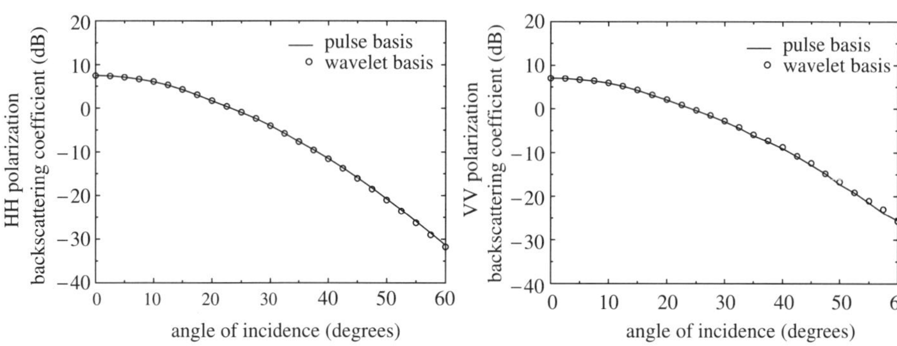

FIGURE 8.11 Backscattering coefficient of the simple surface in HH and VV polarization.

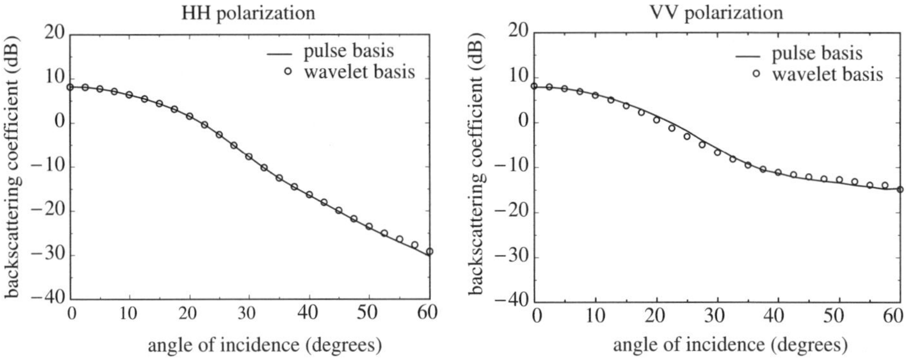

FIGURE 8.12 Backscattering coefficient of the composite surface in HH and VV polarization.

profile has the following actual parameters: $k\sigma \approx 0.9566$, $kl \approx 5.5916$, and 0.003λ mean height.

To obtain numerical data for Fig. 8.11, we imposed two different expansion schemes, namely the pulse collocation with 276 unknowns and wavelet Galerkin approach with 128 Coifman scalets. The resolution level was $j_0 = 2$, meaning 4 scalets per wavelength. From Fig. 8.11, good agreement is observed between the two methods. The results for the scattering from a composite random surface of composition (8.3.23), with HH and VV polarizations are presented in Fig. 8.12. The following parameters have been used to generate the random surface profile: $k\sigma_1 = 1.0$, $kl_1 = 8.45$, $k\sigma_2 = 0.1$, $kl_2 = 1.85$, $D = 34.5\lambda$, $g = D/40.0$, and $M = 50$. In Fig. 8.12 the Coiflets have achieved roughly a factor 6 in CPU acceleration and factor 2 in memory reduction as in Fig. 8.11. All numerical simulations presented here were executed on a Sun Blade-1000 workstation.

Table 8.1 summarizes the numerical results in terms of number of unknowns and corresponding computational time for the simple surface. The impedance matrix obtained from the Coifman scalets can be further sparsified by the introduction of wavelets. This fact is due to the vanishing moment property, localization, and multiresolution analysis of the wavelet basis. There are two kinds of matrix representation in the wavelet basis, namely the standard and nonstandard forms [14]. Here we select the standard matrix form that is obtained by using the fast wavelet transform

TABLE 8.1. Computational Time: Simple Surface

Number of	Pulse Basis		Wavelet Basis	
Unknowns	HH Time (s)	VV Time (s)	HH Time (s)	VV Time (s)
512	1350	1367	1121	1115
256	229	243	167	165
128	46	53	32	31

(FWT). The sparse matrix is then stored in the computer memory using a special algorithm [29]. Then the Bi-CGSTAB [30] iterative solver is employed to solve the system of linear equations.

Tables 8.2 to 8.4 summarize the numerical results in terms of the number of unknowns and the corresponding computational time required for electrically large simple surfaces. We use $M = 50$, and 25 incident angles to calculate backscattering coefficient. Note that fair comparison between the Coiflet and pulse in Table 8.1 to 8.4 should be in terms of numerical accuracy, that is, 512 pulses versus 256 wavelets, 2048 pulses versus 1024 wavelets, and so on.

The threshold level of 10^{-3} and 4 resolution levels are employed to get the sparse standard matrix form. We settle on a relative error of 10^{-2} as the stopping criterion in the Bi-CGSTAB solver. The results obtained by using the standard LU decomposition [29] to solve a system of linear equations in the MoM are also presented for comparison. Depicted in Fig. 8.13 is the standard form [21] of the impedance matrix. The initial impedance matrix was calculated using only Coifman scalets and then was further decomposed into 3 resolution levels using the FWT. It is clearly evident that such an impedance matrix is much sparser than the MoM matrix, which would be a totally dark square patch when plotted. In Fig. 8.13 the threshold was chosen as 1% of the maximum entry in terms of its absolute value.

In Fig. 8.14 we plotted the induced current of the HH polarization for both types of expansion functions, wavelet and pulse bases. Excellent agreement can be seen.

TABLE 8.2. Computational Time: Simple Surface, Pulse Basis

Number of Unknowns	Pulse Basis: HH Time (s)	Pulse Basis: VV Time (s)
2048	84269	84729
1024	9832	9963

Note: Results obtained using LU decomposition.

TABLE 8.3. Computational Time: Simple Surface, Wavelet Basis

Number of	Wavelet basis, HH Time (s)		Sparsity
Unknowns	LU Decomposition	Bi-CGSTAB	(%)
2048	80445	14264	14.4
1024	9150	3638	15.7

TABLE 8.4. Computational Time: Simple Surface, Wavelet Basis

Number of	Wavelet basis, VV Time (s)		Sparsity
Unknowns	LU Decomposition	Bi-CGSTAB	(%)
2048	80574	8259	10.4
1024	9190	2250	13.2

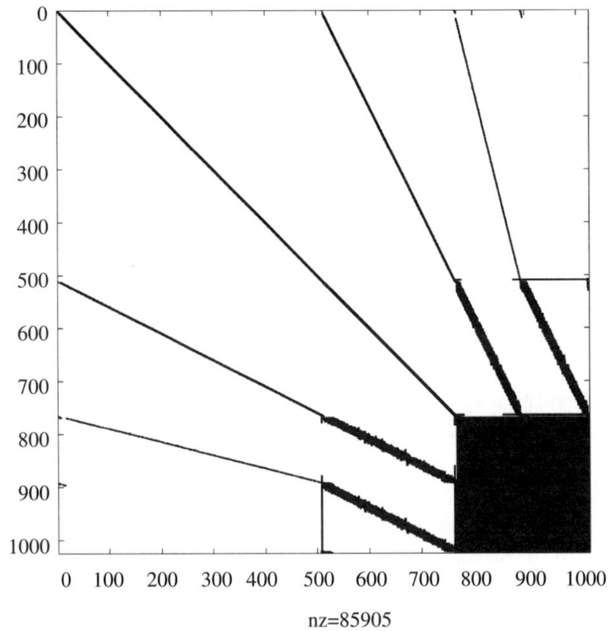

FIGURE 8.13 Standard form of the impedance matrix in HH polarization.

We should note here that the wavelet solution in Fig. 8.14 is obtained using five resolution levels and 0.1% relative threshold level for the standard matrix form.

It can be seen from Tables 8.1 to 8.4 that the improvements of the Coiflet over the pulse are threefold:

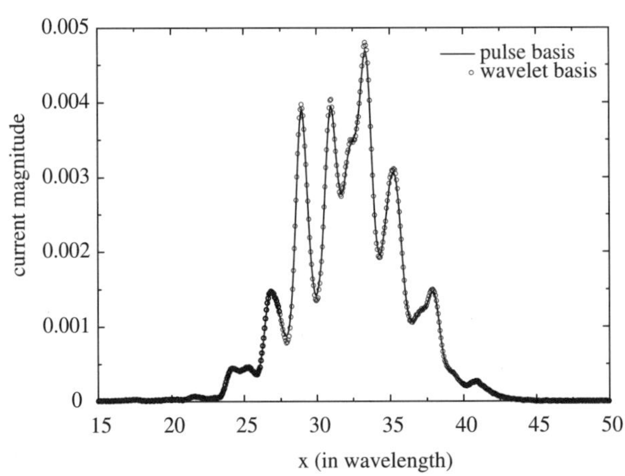

FIGURE 8.14 Induced current in HH polarization.

(1) Owing to single-point quadrature, the Coiflet method is about 5–70% faster than the pulse approach with the same number of unknowns.
(2) Because of pulse discontinuity, pulse basis requires approximately twice as many unknowns as Coiflets to reach the same precision.
(3) Coiflet matrix can be sparsified using the FWT, similar to the FFT. The sparse matrix can be solved using the Bi-CGSTAB, gaining an additional factor of 2–9 in CPU time.

In combination, the Coiflet approach can gain one order of magnitude in terms of the computational speed over the standard pulse-collocation based MoM.

8.4 3D ROUGH SURFACE SCATTERING

The spectral method was used to generate 1D as well as 2D random surfaces. The isotropic 2D rough surface with prespecified statistics is illustrated in Fig. 8.1. Figure 8.15 sketches a general configuration of 3D scattering, where the elevation angle θ_i, azimuthal angle ϕ_i, plane of incidence, and so on, are clearly marked for a horizontally incident case. For the numerical study of 3D rough surface scattering, a truncation of the surface is required because of the limitations on computational resources. The truncation may produce anomalous results owing to the artifacts of edge diffraction when plane waves are impinging upon the system. The tapered wave is introduced to provide an illumination that resembles the plane wave near the scattering center, and decays rapidly to a negligibly weak intensity before reaching the surface edge. A simple tapering multiplier to a plane wave in the form of $e^{-(x^2+y^2)}$ does not work because the resulting product does not satisfy Maxwell's equations. The Thorsos wave [24, 31] has provided good solutions to the tapering mainly for scalar cases. In this section we apply a more advanced formulation of the vector-tapered waves. For ease of reference, the main vector tapering formulation is briefly

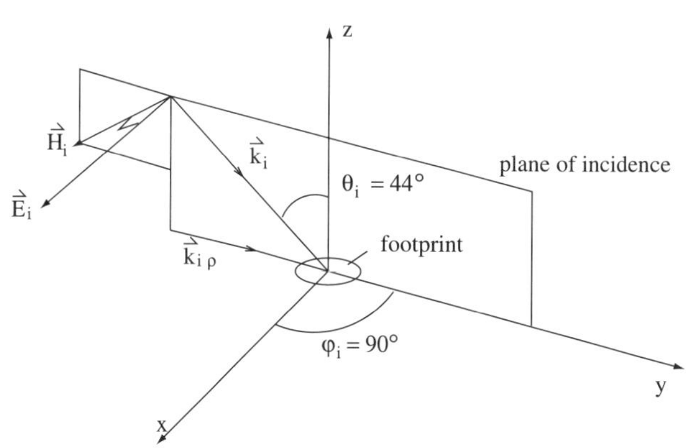

FIGURE 8.15 Configuration of 3D scattering.

summarized in the next subsection. For detailed derivations, discussions, and error analysis, the reader is referred to [32].

8.4.1 Tapered Wave of Incidence

An ideal tapered wave should be free of problems at an arbitrary angle of incidence and should provide clean footprints and clear polarization. Considering a homogeneous, isotropic medium with real wave number k and wave impedance η, we will use the superposition of a 2D spectrum of plane waves to obtain a wave incident upon the $x - y$ plane from $z > 0$, namely

$$\mathbf{E}_i(\mathbf{r}) = \int_{-\infty}^{\infty} d\boldsymbol{\kappa}_\rho e^{i(\boldsymbol{\kappa}_\rho \cdot \boldsymbol{\rho} - \kappa_z z)} \psi(\boldsymbol{\kappa}_\rho) \mathbf{e}(\boldsymbol{\kappa}_\rho), \tag{8.4.1}$$

$$\mathbf{H}_i(\mathbf{r}) = \int_{-\infty}^{\infty} d\boldsymbol{\kappa}_\rho e^{i(\boldsymbol{\kappa}_\rho \cdot \boldsymbol{\rho} - \kappa_z z)} \frac{\psi(\boldsymbol{\kappa}_\rho)}{\eta} \mathbf{h}(\boldsymbol{\kappa}_\rho). \tag{8.4.2}$$

The expressions (8.4.1) and (8.4.2) are exact solutions to the Maxwell equations, and the variables in the expressions are

$$\begin{aligned}
\mathbf{r} &= \boldsymbol{\rho} + \hat{z} z, \\
\boldsymbol{\kappa}_\rho &= \hat{x}\kappa_x + \hat{y}\kappa_y, \\
\kappa_z &= \kappa_z(\kappa_\rho) = \begin{cases} \sqrt{k^2 - \kappa_\rho^2}, & 0 \le \kappa_\rho \le k, \\ -i\sqrt{\kappa_\rho^2 - k^2}, & \kappa_\rho > k, \end{cases} \\
k^2 &= \omega^2 \mu \epsilon.
\end{aligned}$$

The spectrum $\psi(\boldsymbol{\kappa}_\rho)$ carries information about the shape of the footprint of the incident field and κ_ρ is assumed to be centered about the incident direction

$$\begin{aligned}
\mathbf{k}_{i\rho} &= \hat{x} k_{ix} + \hat{y} k_{iy} \\
&= k \sin\theta_i (\hat{x} \cos\phi_i + \hat{y} \sin\phi_i),
\end{aligned}$$

where θ_i and ϕ_i are the polar and azimuthal angles of the incident wave. A Gaussian-shaped footprint where the amplitude at $\rho = \tau$ has been reduced to $1/e$ of the magnitude at the center is implemented by choosing

$$\psi(\boldsymbol{\kappa}_\rho) = \frac{\tau^2}{4\pi} e^{-\tau^2 |\boldsymbol{\kappa}_\rho - \mathbf{k}_{i\rho}|^2/4}. \tag{8.4.3}$$

When $\tau \to \infty$, the tapered wave becomes a pure plane wave. The footprint of the tapered wave can be controlled at will by varying the parameters in the expression (8.4.3) or selecting different functional forms of ψ. In addition to the Gaussian shape, we may use exponential, transformed exponential, and two-parameter tapering, among other forms. The tapered wave in the spatial domain is obtained by means of (8.4.1) and (8.4.2) by integrating ψ in the $\kappa_x - \kappa_y$ plane about its center $\mathbf{k}_{i\rho}$. The

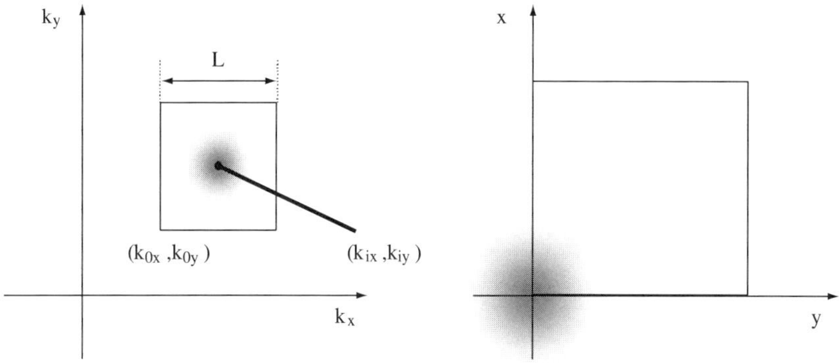

FIGURE 8.16 Construction of a beam from its spectrum.

prescribed footprint itself is fixed with respect to the angle of incidence. Figure 8.16 illustrates the integration of the plane waves ψ to obtain the tapered waves in the spatial domain.

The general form of the polarization vectors **e** and **h** can be written as

$$\mathbf{e}(\boldsymbol{\kappa}_\rho) = e_h(\boldsymbol{\kappa}_\rho)\hat{h}(\boldsymbol{\kappa}_\rho) + e_v(\boldsymbol{\kappa}_\rho)\hat{v}(\boldsymbol{\kappa}_\rho),$$
$$\mathbf{h}(\boldsymbol{\kappa}_\rho) = e_v(\boldsymbol{\kappa}_\rho)\hat{h}(\boldsymbol{\kappa}_\rho) - e_h(\boldsymbol{\kappa}_\rho)\hat{v}(\boldsymbol{\kappa}_\rho).$$

The unit vectors $\hat{h}$ and $\hat{v}$ are respectively perpendicular to and within the incident plane. Figure 8.15 sketches a horizontal incident of $\theta_i = 44°$, $\phi_i = 90°$. Notice that both vectors $\hat{h}$ and $\hat{v}$ are functions of $\boldsymbol{\kappa}_\rho$ such that

$$\hat{h}(\boldsymbol{\kappa}_\rho) = \begin{cases} \hat{x}\sin\phi_i - \hat{y}\cos\phi_i, & \kappa_\rho = 0 \\ \dfrac{1}{\kappa_\rho}(\hat{x}\kappa_y - \hat{y}\kappa_x), & \kappa_\rho > 0, \end{cases}$$

$$\hat{v}(\boldsymbol{\kappa}_\rho) = \begin{cases} \hat{x}\cos\phi_i + \hat{y}\sin\phi_i, & \kappa_\rho = 0 \\ \dfrac{\kappa_z}{k\kappa_\rho}(\hat{x}\kappa_x + \hat{y}\kappa_y) + \hat{z}\dfrac{\kappa_\rho}{k}, & \kappa_\rho > 0. \end{cases}$$

In these expressions $\kappa_\rho = 0$ corresponds to the individual plane wave that is normally incident on the xOy plane. In order to construct a wave with clear polarization, we employ

$$e_h(\boldsymbol{\kappa}_\rho) = \mathbf{e}_i \cdot \hat{h}(\boldsymbol{\kappa}_\rho),$$
$$e_v(\boldsymbol{\kappa}_\rho) = \mathbf{e}_i \cdot \hat{v}(\boldsymbol{\kappa}_\rho),$$

with the polarization vector of the central plane wave

$$\mathbf{e}_i = \mathbf{e}(\mathbf{k}_{i\rho}) = E_h\hat{h}(\mathbf{k}_{i\rho}) + E_v\hat{v}(\mathbf{k}_{i\rho}).$$

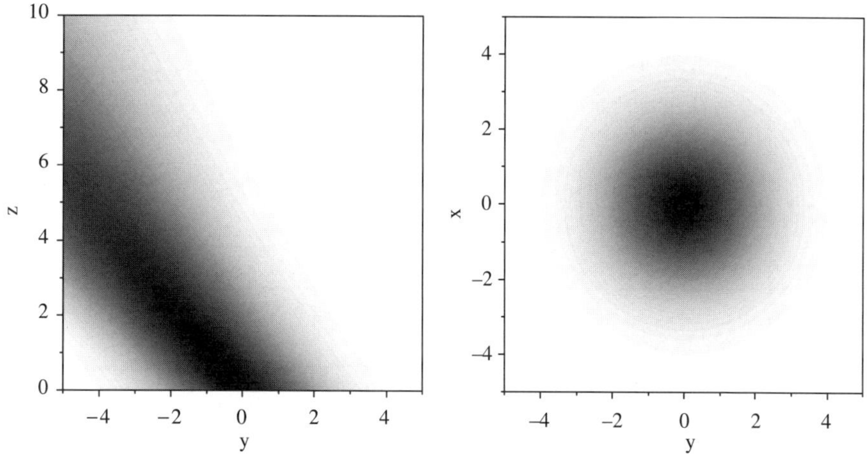

FIGURE 8.17 Beam side view and top view for $\theta_i = 40°$, $\phi_i = 90°$.

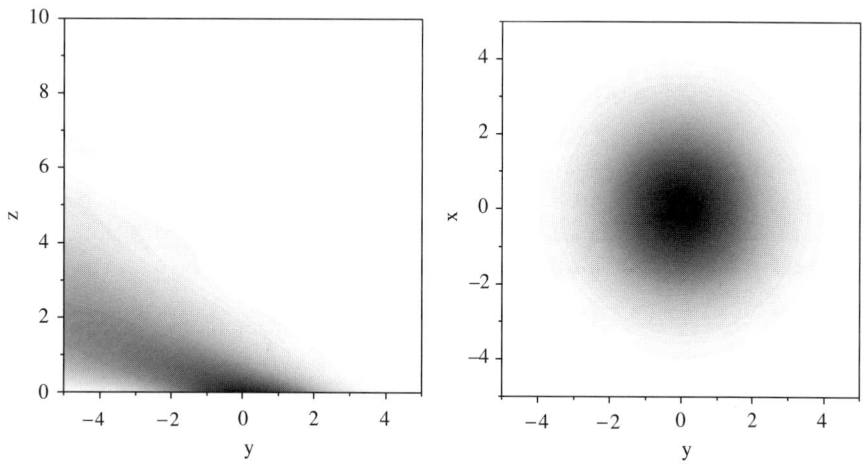

FIGURE 8.18 Side view and top view for grazing incident for $\theta_i = 90°$, $\phi_i = 90°$.

The dominant polarization state of the tapered wave is then determined by the choice of E_h and E_v, which describe the polarization of the central plane wave. Figures 8.17 and 8.18 describe the beamwidth of the tapered wave at oblique incidence and at grazing incidence. It is significant that the footprints of the synthesized tapered waves are always circles in the xOy plane, regardless of the angle of incidence.

The integration of (8.4.1) may be implemented by the fast Fourier transform (FFT) as derived below (see Fig. 8.16):

$$\mathbf{E}(\mathbf{r}) = \int_{-\infty}^{+\infty} d\boldsymbol{\kappa}_\rho e^{i(\boldsymbol{\kappa}_\rho \cdot \boldsymbol{\rho} - \kappa_z z)} \psi(\boldsymbol{\kappa}_\rho)\mathbf{e}(\boldsymbol{\kappa}_\rho)$$

$$\approx \int_{k_{ix}-L/2}^{k_{ix}+L/2} \int_{k_{iy}-L/2}^{k_{iy}+L/2} d\kappa_x\, d\kappa_y e^{i(\kappa_x x+\kappa_y y)} e^{-i\kappa_z z} \psi(\kappa_x, \kappa_y)\mathbf{e}(\kappa_x, \kappa_y)$$

$$= \int_0^L \int_0^L d\kappa_x\, d\kappa_y e^{i[(\kappa_x+k_{0x})x+(\kappa_y+k_{0y})y]} e^{-i\kappa_z z}$$
$$\psi(\kappa_x + k_{0x}, \kappa_y + k_{0y})\mathbf{e}(\kappa_x + k_{0x}, \kappa_y + k_{0y}). \tag{8.4.4}$$

When $z = 0$,

$$E_\alpha(\mathbf{r}) \approx e^{i(k_{0x}x+k_{0y}y)} \int_0^L \int_0^L d\kappa_x\, d\kappa_y e^{i(\kappa_x x+\kappa_y y)} P_{x,y,z}(\kappa_x + k_{0x}, \kappa_y + k_{0y})$$
$$\approx e^{i(k_{0x}x+k_{0y}y)} \sum_{\kappa_1=0}^{N-1} \sum_{\kappa_2=0}^{N-1} e^{i(\kappa_1 L/N)x} e^{i(\kappa_2 L/N)y}$$
$$P_{x,y,z}\left(\frac{\kappa_1 L}{N} + k_{0x}, \frac{\kappa_2 L}{N} + k_{0y}\right)\left(\frac{L}{N}\right)^2$$
$$= e^{i(k_{0x}x+k_{0y}y)} FFT\left\{P_\alpha\left(\frac{\kappa_1 L}{N} + k_{0x}, \frac{\kappa_2 L}{N} + k_{0y}\right)\cdot\left(\frac{L}{N}\right)^2\right\},$$

where $\alpha = x, y, z$. In this equation we have used

$$x, y = 0 \sim \frac{2\pi(N-1)}{L},$$
$$\Delta x = \Delta y = \frac{2\pi}{L}, \quad \Delta\kappa_x = \Delta\kappa_y = \frac{L}{N},$$
$$\mathbf{P}(\kappa_x + k_{0x}, \kappa_y + k_{0y}) := \psi(\kappa_x + k_{0x}, \kappa_y + k_{0y})\mathbf{e}(\kappa_x + k_{0x}, \kappa_y + k_{0y}),$$

where $k_{0x} = k_{ix} - (L/2)$, $k_{0y} = k_{iy} - (L/2)$. We recommend the direct integration of (8.4.4) instead of the FFT because of the oscillatory nature of the expressions (8.4.1) and (8.4.2). In fact the truncation error may produce large systematic bias [29]. A comparison of phase information in Fig. 8.19 reveals that while the direct integration produces smooth phase distribution, the FFT leads to a systematic phase distortion.

8.4.2 Formulation of 3D Rough Surface Scattering Using Wavelets

The method of moments (MoM) is employed for this numerical study. The basis and testing functions are the Coifman scalets, as in the 2D cases. The formulation here

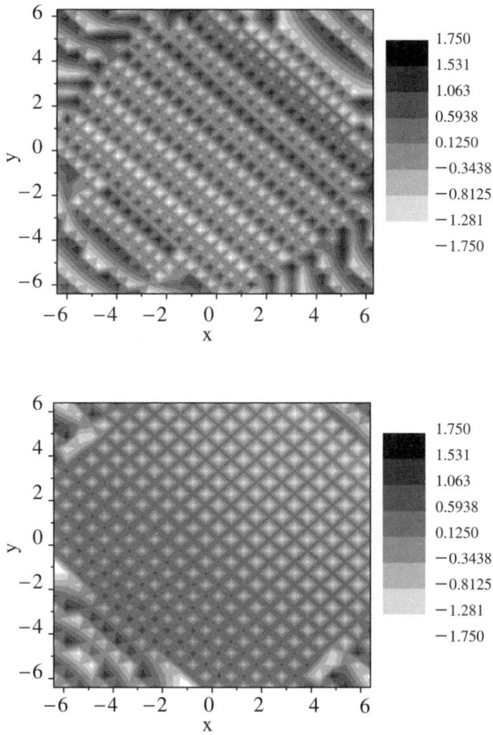

FIGURE 8.19 Phase comparison between FFT and direct integration.

in the 3D cases is based on the magnetic field integral equation (MFIE):

$$\begin{aligned}\frac{F_x(\mathbf{r})}{2} &+ \frac{\partial f(x,y)}{\partial y}\int dx'\,dy' G(R)[(x-x')F_y(\mathbf{r}') - (y-y')F_x(\mathbf{r}')]\\ &+ \int dx'\,dy' G(R)\left\{\left[-(x-x')\frac{\partial f(x',y')}{\partial x'} + [f(x,y) - f(x',y')]\right]F_x(\mathbf{r}')\right.\\ &\left. -(x-x')\frac{\partial f(x',y')}{\partial y'}F_y(\mathbf{r}')\right\} = \frac{\partial f(x,y)}{\partial y}H_z^i(\mathbf{r}) - H_y^i(\mathbf{r})\end{aligned} \tag{8.4.5}$$

and

$$\begin{aligned}\frac{F_y(\mathbf{r})}{2} &- \frac{\partial f(x,y)}{\partial x}\int dx'\,dy' G(R)[(x-x')F_y(\mathbf{r}') - (y-y')F_x(\mathbf{r}')]\\ &- \int dx'\,dy' G(R)\left\{\left[(y-y')\frac{\partial f(x',y')}{\partial y'} - [f(x,y) - f(x',y')]\right]F_y(\mathbf{r}')\right.\\ &\left. +(y-y')\frac{\partial f(x',y')}{\partial x'}F_x(\mathbf{r}')\right\} = \frac{\partial f(x,y)}{\partial x}H_z^i(\mathbf{r}) + H_y^i(\mathbf{r}).\end{aligned} \tag{8.4.6}$$

In Eqs. (8.4.5) and (8.4.6),

$$G(R) = \frac{(ikR-1)e^{ikR}}{4R^3},$$
$$R = \sqrt{(x-x')^2 + (y-y')^2 + [f(x,y) - f(x',y')]^2}.$$

Here f is the profile of the rough surface, and

$$F_x(\mathbf{r}) = \left\{1 + \left[\frac{\partial f(x,y)}{\partial x}\right]^2 + \left[\frac{\partial f(x,y)}{\partial y}\right]^2\right\}^{1/2} \hat{n} \times \mathbf{H}^s(\mathbf{r}) \cdot \hat{x},$$
$$F_y(\mathbf{r}) = \left\{1 + \left[\frac{\partial f(x,y)}{\partial x}\right]^2 + \left[\frac{\partial f(x,y)}{\partial y}\right]^2\right\}^{1/2} \hat{n} \times \mathbf{H}^s(\mathbf{r}) \cdot \hat{y},$$

are functions to be solved. H^i and H^s are, respectively, the incident and scattering magnetic fields. To solve the coupled integral equations (8.4.5) and (8.4.6), we applied the Galerkin procedure. This method expands the unknown functions $F_x(\mathbf{r})$ and $F_y(\mathbf{r})$ in terms of the Coffman wavelets and tests the discretized equations with weighting functions, the same as the expansion functions

$$F_x(\mathbf{r}) = \sum_i \sum_j \alpha_{ij} \varphi_{j_0,i}(x) \varphi_{j_0,j}(y), \tag{8.4.7}$$

$$F_y(\mathbf{r}) = \sum_i \sum_j \beta_{ij} \varphi_{j_0,i}(x) \varphi_{j_0,j}(y). \tag{8.4.8}$$

We substitute (8.4.7) and (8.4.8) into the integral function (8.4.5), multiply $\varphi_{j_0,m}(x)$ $\varphi_{j_0,n}(y)$, and integrate to arrive at

$$\begin{aligned}
&\sum_{ij} \frac{1}{2} \alpha_{ij} \int \varphi_{j_0,i}(x) \varphi_{j_0,j}(y) \varphi_{j_0,m}(x) \varphi_{j_0,n}(y)\, dx\, dy \\
&\quad + \sum_{ij} \int dx'\, dy'\, dx\, dy G(R) \varphi_{j_0,i}(x') \varphi_{j_0,j}(y') \varphi_{j_0,m}(x) \varphi_{j_0,n}(y) \\
&\quad \cdot \left\{ \alpha_{ij} \left[(f(x,y) - f(x',y')) - (y-y') \frac{\partial f(x,y)}{\partial y} - (x-x') \frac{\partial f(x',y')}{\partial x'} \right] \right. \\
&\quad \left. + \beta_{ij} \left[(x-x') \frac{\partial f(x,y)}{\partial y} - (x-x') \frac{\partial f(x',y')}{\partial y'} \right] \right\} \\
&= \int \varphi_{j_0,m}(x) \varphi_{j_0,n}(y) \left[-\frac{\partial f(x,y)}{\partial y} H_z^i(\mathbf{r}) - H_y^i(\mathbf{r}) \right] dx\, dy.
\end{aligned} \tag{8.4.9}$$

Using orthogonality and a one-point quadrature, we obtain from (8.4.9),

$$\frac{1}{2}\alpha_{mn} + \sum_{ij}\left(\frac{1}{2^{j_0/2}}\right)^4 G(x_i, y_j; x_m, y_n)\bigg\{\alpha_{ij}\bigg[f(x_m, y_n) - f(x_i, y_j)$$

$$- (y_n - y_j)\frac{\partial f(x_m, y_n)}{\partial y} - (x_m - x_i)\frac{\partial f(x_i, y_i)}{\partial x}\bigg]$$

$$+ \beta_{ij}\bigg[(x_m - x_i)\frac{\partial f(x_m, y_n)}{\partial y} - (x_m - x_i)\frac{\partial f(x_i, y_j)}{\partial y}\bigg]\bigg\}$$

$$= \left(\frac{1}{2^{j_0/2}}\right)^2 \left[-\frac{\partial f(x_m, y_n)}{\partial y} H_z^i(x_m, y_n) - H_y^i(x_m, y_n)\right]. \tag{8.4.10}$$

By the same token from (8.4.6), we obtain the other integral equation in terms of wavelet coefficients. After solving F_x, F_y, the normalized bistatic scattering coefficients for horizontal incident are evaluated by

$$\sigma_{\alpha h} = \gamma_{\alpha h}(\theta_s, \phi_s) = \frac{|E_\alpha^s|^2}{2\eta P_h^{\text{inc}}}, \tag{8.4.11}$$

where α can be h or v, and

$$P_h^{\text{inc}} = \text{incident power density},$$

$$E_h^s = \frac{\eta i k}{4\lambda}\int_{ds'} dx'\, dy' \exp(-ik\beta')$$

$$[F_x(x', y') \sin phi_s - F_{-y}(x', y') \cos\phi_s],$$

$$E_v^s = \frac{\eta i k}{4\lambda}\int_{ds'} dx'\, dy' \exp(-ik\beta')\, \{F_x(x', y')$$

$$\left[\frac{\partial f(x', y')}{\partial x'}\sin\theta_s - \cos\theta_s\cos\theta_s\right] + F_y(x', y')$$

$$\left[\frac{\partial f(x', y')}{\partial y'}\sin\theta_s - \cos\theta_s\sin\theta_s\right]\bigg\},$$

with

$$\beta' = x'\sin\theta_s\cos\phi_s + y'\sin\theta_s\sin\phi_s + f(x', y')\cos\theta_s.$$

8.4.3 Numerical Results of 3D Scattering

Case 1: Testing Example. The rough surface under investigation has a Gaussian p.d.f. and Gaussian correlation function with $\sigma = 0.2\lambda$, $l_x = l_y = 0.6\lambda$. Horizontal incident is specified as $\theta_i = -20°$, $\phi_i = 90°$. The rough surface is truncated into $8\lambda \times 8\lambda$, where we applied four Coifman scalets per λ. The number of unknowns for the surface current expansion is $n = (8 \times 4)^2 \times 2 = 2048$, where the factor 2 stands for both the x and y direction current. The system matrix of 2048×2048 is complex, and we used only single precision to save memory. Even so, the matrix consumes the RAM space $2048^2 \times 2 \times 4 = 34$ megabytes. If pulse bases are employed, the sampling rate needs to be doubled due to the discontinuity of the pulse function and related Gibb's phenomenon. The number of unknowns will be increased to 8192, and the RAM will grow by a factor of 16 to 537 megabytes. We have taken advantage

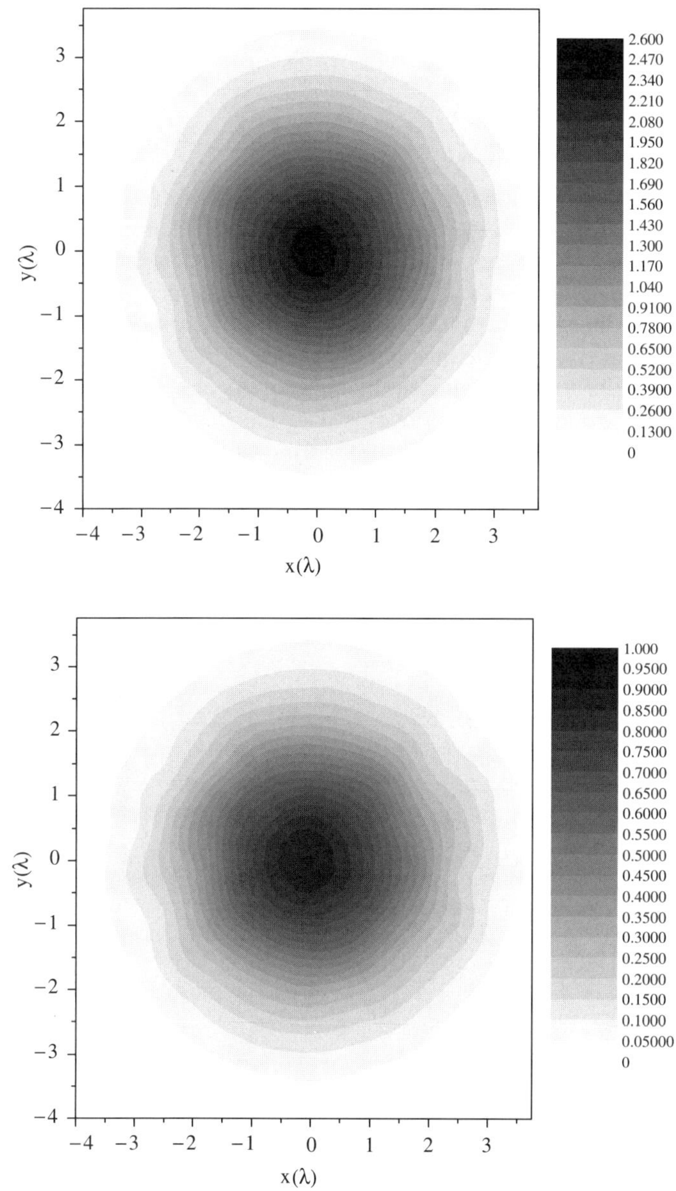

FIGURE 8.20 Incident magnetic field H_y and H_z on rough surface for $\theta_i = 40°$, $\phi_i = 90°$.

of the single-point quadrature to obtain all but tri-diagonal entries. As a result the matrix filling is in $O(n)$, instead of $O(n^2)$. The system matrix is then solved using the iterative technique, such as the Bi-CGSTAB solver.

The bottleneck for the entire computation process is the evaluation of the tri-diagonal elements of the matrix. We used a 2D quadrature formula for each source-field pair, that is,

$$\iint_S f(x, x')\, dx\, dx' = 4h^2 \sum_{i=1}^{n} w_i f(x_i, x_i') + \varepsilon$$

where the error $\varepsilon = O(h^4)$. To speed up the computation, we increased the order of the Coiflets to 10 but truncated the support in $[-3, 3]$. Numerically the external integral is carried out by one-point quadrature, while the interior integral is treated by generalised Gaussian quadrature (GGS) [33] using $x, x^p \ln(x), p = 0, 1, 2, \ldots, N$ as bases. To match the GGS style with the singularity at one end of the interval, we have folded the integrand about the singular point [34]. In each folded scalet five quadrature points were used to guarantee precision.

Figure 8.20 illustrates the incident footprint of H_y^i and H_z^i on the rough surface, where the idea circles have deformed because the surface is not flat. Figure 8.21 shows the power density of the incident footprint, where the intensity decays to negligible level before it reaches the edges of the surface.

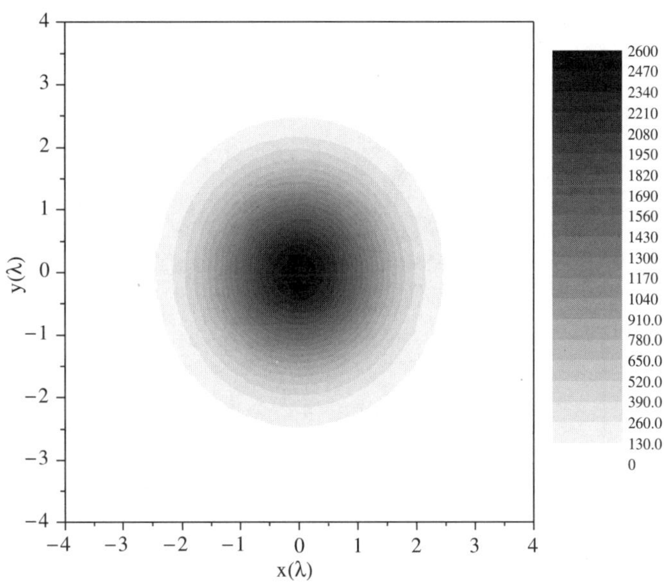

FIGURE 8.21 Power density of the footprint for $\theta_i = 40°, \phi_i = 90°$.

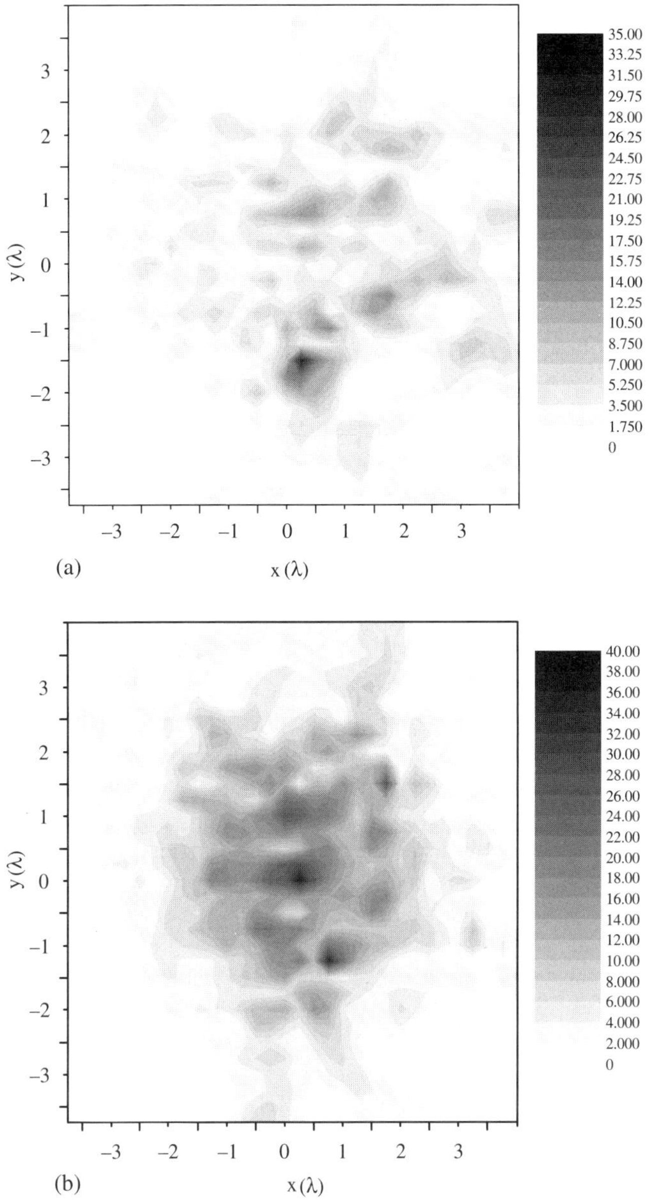

FIGURE 8.22 Induced surface current for $\theta_i = 40°$, $\phi_i = 90°$: (a) J_y, (b) J_x.

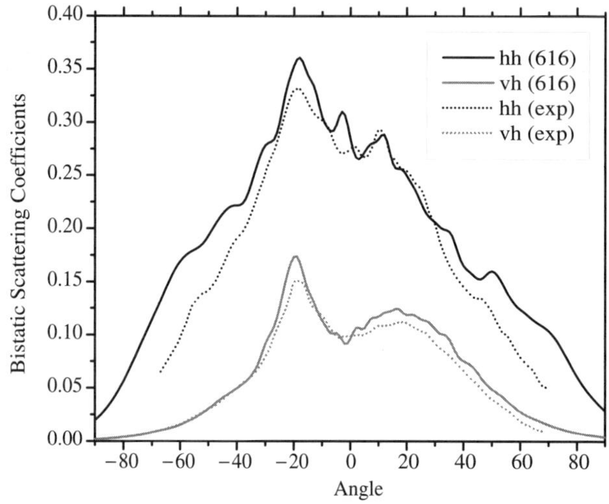

FIGURE 8.23 Radar cross-sections of benchmark structure: $\theta_i = 20°$, $\phi_i = 180°$.

The far-zone field is being smoothed out. Therefore the radar cross section computed from the far-zone fields is less sensitive to numerical errors as compared with the induced current. To demonstrate the characteristics of the induced current, we have also included Fig. 8.22, where currents J_x and J_y are plotted over the entire surface. It can be seen clearly that the magnitude of J_x is larger than that of J_y because the polarization is along the x axis. This result agrees with intuition.

Case 2: Benchmark Simulation. The benchmark structures of 2D randomly rough surfaces were produced by computer-aided manufacturing (CAM) at the University of Washington [9]. From the benchmark structures we selected a metallic surface with the following parameters: standard deviation $\sigma = 1\lambda$, correlation length $\ell_x = \ell_y = 2\lambda$, and truncated surface size $L_x = L_y = 16\lambda$. Horizontal incident is specified as $\theta_i = 20°$, $\phi_i = 180°$.

We applied four Coifman scalets per λ. As a result, the number of unknowns is 8,192 and the RAM space is 537 megabytes. The algorithm is programmed in C^{++} and executed on the DEC-Alpha 433 MHz workstation. For each realization the CPU time of one complete bistatic scattering computation is 10,481.5 seconds. Figure 8.23 demonstrates good agreement between our numerical results versus the experimental data. The numerical results are the average of 616 realizations. At large angles from nadir the computed copolarization scattering coefficient is higher than the experiment. This is due to the fact that the truncated surface is not large enough. In fact the induced surface current near the edges is about 10% in magniture of the peak value at the illumination center. Both numerical and experiment data exhibit backscattering enhancement, for the copolarization as well as crosspolarization. The numerical results presented here have closest agreement with the experiments and the CPU time is the most economical.

BIBLIOGRAPHY

[1] P. Beckmann and A. Spizzichino, *The Scattering of Electromagnetic Waves from Rough Surfaces*. Artech House, Norwood, MA, 1987.

[2] S. Rice, "Reflection of electromagnetic waves from slightly rough surfaces," *Comm. Pure Appl. Math.*, **4**(2/3), 361–378, 1951.

[3] R. Axline and A. Fung, "Numerical computation of scattering from a perfectly conducting random surface," *IEEE Trans. Ant Propg.*, **26**(3), 48–488, May 1978.

[4] M. Nieto-Vesperinas and J. Soto-Crespo, "Monte Carlo simulations for scattering of electromagnetic waves from perfectly conducting random surfaces," *Opt. Lett.*, **12**, 979–981, 1987.

[5] D. Kapp and G. Brown, "A new numerical model for rough surface scattering calculations," *IEEE Trans. Ant. Propg.*, **44**(5), 711–721, 1996.

[6] C. Chan, L. Tsang, and Q. Li, "Monte Carlo simulations of large-scale one-dimensional random rough-surface scattering at near-grazing incidence: Penetrable case," *IEEE Trans. Ant. Propg.*, **46**(1), 142–149, 1998.

[7] R. Axline and R. Mater, "Theoretical and experimental study of wave scattering from composite rough surfaces," CRES Technical Report 234-2, Remote Sensing Laboratory, University of Kansas Center for Research Inc. May 1974. U.S. Army Engineering Topographic Laboratories Fort Belvoir, VA 22060, Contract DAAK02-73-C-0106.

[8] Z. Gu, S. Dummer, A. Maradudin and A. McGurn, "Experimental study of the opposite effect in the scattering of light from a randomly rough metal surface," *Appl. Opt.*, **28** 537–543, 1989.

[9] P. Phu, A. Ishimaru, and Y. Kuga, "Copolarized and cross-polarized enhanced backscattering from 2D very rough surfaces at millimeter wave frequencies," *Radio Sci.*, **29**(5), 1275–1291, 1994.

[10] A. Fung and G. Pan, "A scattering model for perfectly conducting random surfaces I. Model development," *Int. J. Remote Sens.*, **8**(11), 1579–1593, 1987.

[11] G. Pan and A. Fung, "A scattering model for perfectly conducting random surfaces II. Range of validity," *Int. J. Remote Sens.*, **8**(11), 1595–1605, 1987.

[12] C. Eftimiu and G. Pan, "First-order Wiener–Hermite expansion in the electromagnetic scattering by dielectric rough interfaces," *Radio Sci.*, **25**(1), 1–8, 1990.

[13] I. Daubechies, *Ten Lectures on Wavelets*, SIAM, Philadelphia, 1992.

[14] G. Beylkin, R. Coifman, and V. Rokhlin, "Fast wavelet transforms and numerical algorithm I," *Comm. Pure Appl. Math.*, **44**, 141–183, 1991.

[15] B. Steinberg and Y. Leviatan, "On the use of wavelet expansions in the method of moments," *IEEE Trans. Ant. Propg.*, **41**(5), 610–119, 1993.

[16] R. Wagner and W. Chew, "A study of wavelets for the solution of electromagnetic integral equations," *IEEE Trans. Ant. Propg.*, **43**(8), 802–810, 1995.

[17] H. Deng and H. Ling, "Fast solution of electromagnetic integral equations using adaptive wavelet packet transform," *IEEE Trans. Ant. Propg.*, **47**(4), 674–682, 1999.

[18] C. Lin, C. Chan, and L. Tsang, "Monte Carlo simulations of scattering and emission from lossy dielectric random rough surfaces using the wavelet transform method," *IEEE Trans. Geo. Remote Sens.*, **37**(5), 2295–2304, 1999.

[19] D. Zahn, K. Sarabandi, K. Sabet and J. Harvey, "Numerical simulation of scattering from rough surfaces: A wavelet-based approach," *IEEE Trans. Ant. Propg.*, **48**(2), 246–253, 2000.

[20] B. Alpert, G. Beylkin, R. Coifman, and V. Rokhlin, "Wavelet-like bases for the fast solution of second-kind integral equations," *SIAM J. Sci. Comput.*, **14**(1), 159–184, 1993.

[21] D. Gines, G. Beylkin, and J. Dunn, "LU factorization of non-standard forms and direct multiresolution solvers," *Appl. Comput. Harmon. Anal.*, **5**, 156–201, 1998.

[22] E. Thorsos, "Exact numerical methods versus the Kirchoff approximation for rough surface scattering," *Computational Acoustics,* **2**, *Algorithms and Applications*, D. Lee, R. Sternberg, and M. Shultz, Eds., North-Holland, Amsterdam, 1988.

[23] J. Seltzer, "A modified specular point theory for radar backscatter," Ph.D. dissertation, Purdue University, 1971.

[24] E. Thorsos, "The validity of the Kirchoff approximation for rough surface scattering using a Gaussian roughness spectrum," *J. Acoust. Soc. Am.*, **83**(1), 78–92, Jan. 1988.

[25] R. Harrington, *Field Computation by Moment Methods*, IEEE Press, New York, 1993.

[26] A. Fung and M. Chen, "Numerical simulation of scattering from simple and composite random surfaces," *J. Opt. Soc. Am.*, **2**(12), 2274–2284, Dec. 1985.

[27] G. Pan, "Orthogonal wavelets with applications in electromagnetics," *IEEE Trans. Magn.*, **32**, 975–983, 1996.

[28] E. Thorsos, "Backscattering enhancement with the Dirichlet boundary condition," *Workshop on Enhanced Backscatter*, Boston University, July 28–29, 1989.

[29] W. Press, S. Teukolsky, W. Vetterling, and B. Flannery, *Numerical Recipes in C*, Cambridge University Press, Cambridge, 1992.

[30] H. Van der Vorst, "Bi-CGSTAB: A fast and smoothly converging variant of Bi-CG for the solution of nonsymmetric linear systems." *SIAM J. Sci. Statist. Comput.*, **13**(2), 631–644, Mar. 1992.

[31] E. Thorsor and D. Jackson, "Studies of scattering theory using numerical methods," *Waves Random Media*, **1**, 165–190, July 1991.

[32] H. Braunisch, Y. Zhang, C. Ao, S. Shih, Y. Yang, K. Ding, J. Kong, and L. Tsang, "Tapered wave with dominant polarization state for all angles of incidence," *IEEE Trans. Ant. Propg.*, **48**, 1086–1096, July 2000.

[33] J. Ma, V. Rokhlin, and S. Wandzura, "Generalized Gaussian quadrature rules for systems of arbitrary functions," *SIAM J. Numer. Anal.*, **33**(3), 971–996, Jan. 1996.

[34] J. Mosig and F. Gardiol, "A dynamical radiation model for microstrip structures," *Advances in Electronics and Electron Physics*, Academic Press, New York, 1982, 138–236.

CHAPTER NINE

Wavelets in Packaging, Interconnects, and EMC

In this chapter we will study multiconductor, multilayered transmission lines (MMTL) employing quasi-static, quasi-dynamic, and full-wave analyses. We extract from MMTL the distributed (parasitic) parameters in matrix form of the capacitance $[C]$, inductance $[L]$, resistance $[R]$ and conductance $[G]$, or the $[Z]$-parameters, $[Y]$-parameters, or more generally the scattering matrix $[S]$. MMTL systems are commonly found in high-speed, high-density digital electronics at the levels of individual chip carriers, printed circuit boards (PCBs), and more recently, multichip modules (MCMs). Previous methods for extraction of the distributed circuit parameters include the quasi-TEM solutions [1–5], and more rigorous techniques [6–9]. They also included full-wave analysis algorithms [10–15].

We begin with the quasi-static formulation (QSF) [1], which provides the parasitic capacitance $[C]$, inductance $[L]$, resistance $[R]$, and conductance $[G]$. Due to the limitation of its assumptions, the QSF results for L, C, R, and G are independent of frequency values. This characteristic is accurate only under special circumstances. The comparison of the QSF solution with the full-wave finite element method (FEM) data indicates that the capacitance $[C]$ values from the QSF are accurate to at least 50 GHz [16], while the $[L]$ and $[R]$ may have large errors. For most practical applications, conductance $[G]$ is negligibly small. Therefore, in the quasi-static formulations of Sections 9.1 and 9.2, we will focus mainly on capacitance extraction.

In Section 9.3 we will introduce an intermediate formulation between that of the quasi-static and full-wave, referred to as the quasi-dynamic formulation (QDF). The QDF provides us with frequency-dependent parameters of the skin effect resistance and total (internal plus external) inductance. The comparison of the QDF with the FEM [17] and laboratory tests [18] reveals that the $[L]$ and $[R]$ matrices from the QDF are accurate from 1 MHz to at least 10 GHz.

Following this we will present the full-wave analysis in Sections 9.4 and 9.5, from which we extract the scattering parameters $[S]$. The emphasis of this chapter will be given to packaging and interconnects of high-speed digital circuits and systems and the implementation of numerical algorithms using wavelets.

9.1 QUASI-STATIC SPATIAL FORMULATION

In this section the wavelet expansion method in conjunction with the boundary element method (BEM) is applied to the evaluation of the capacitance and inductance matrices of multiconductor transmission lines in multilayered dielectric media. The integral equations obtained by using a Green function above a grounded plane are solved by Galerkin's method, with the unknown total charge expanded in terms of orthogonal wavelets in $L^2([0, 1])$. The unknown functions defined in finite intervals are expanded in terms of wavelets in $L^2([0, 1])$, as discussed in Chapter 4. Adopting the geometric representation of the BEM converts the 2D problem into a 1D problem and provides a versatile and accurate treatment of curved conductor surfaces and dielectric interfaces. A sparse matrix equation is developed from the set of integral equations. This equation is extremely valuable for solving a large system of equations. We will compare the numerical QSF results with previously published data and demonstrate good agreement between the two sets of results.

Recently Nekhla reported in [19] that by modifying our wavelet-BEM approach [20], "The proposed algorithm has a major impact on the speed and accuracy of physical interconnect parameter extraction with speedup reaching 10^3 for even moderately sized problems."

9.1.1 What Is Quasi-static?

In digital and microwave circuits and systems, the electromagnetic (EM) modeling was based on the quasi-static method. The distributed circuit parameters obtained are inductance $L(H/m)$, capacitance $C(F/m)$, resistance $R(\Omega/m)$, and conductance $G(S/m)$, all expressed per unit length. These parameters are frequency-independent under the quasi-static assumption. The quasi-static method assumes:

(1) The wavelength of interest is much greater than the dimensions of the circuit/subsystems under consideration. Typically $f < 3 \sim 5\,\text{GHz}$.

(2) The longitudinal fields and transverse currents are negligible, which leads to $k^2 = k_x^2 + k_y^2 + k_z^2 \approx k_z^2$, where k_z is the wavenumber in the direction of propagation.

(3) Ohmic loss is low so that small perturbation is applicable.

(4) The linear dimension of the transmission line cross section is much greater than δ (skin depth). As a result current flows only on the conductor surface. Equivalently the microstrip thickness t and width w satisfy $w \gg t \gg \delta$, and thus internal inductance L_{int} can be neglected, and $L = L_{\text{ext}}$.

These assumptions no longer hold for high-speed electronic packaging applications. For instance, for typical multichip module (MCM) structures, the cross section of the transmission lines is $w \times t = 8 \times 6\,\mu$m. For such a structure the dc resistance $\approx 400\,\Omega$/m at 1 GHz with copper of conductivity $\sigma = 5.8 \times 10^7$ S/m and skin depth $\delta = 1/\sqrt{\pi f \mu \sigma} \approx 2\,\mu$m. The signal frequency bandwidth ranges from 10 MHz to 10 GHz, and the corresponding skin depths are from $\delta = 20$ to $\delta = 0.7\,\mu$m. Thus the surface resistance formula

$$R_s = \frac{1}{\sigma\,\delta} = \sqrt{\frac{\pi f \mu}{\sigma}}$$

is not applicable, since we do not have $w \gg t \gg \delta$. In addition the small perturbation approach does not apply due to relatively high ohmic losses. Nonetheless, the quasi-static approximation is still widely used, in particular, for capacitance computations.

The wave phenomena are governed by the Helmholtz equation

$$(\nabla^2 + k^2)\phi(x, y, z) = 0, \tag{9.1.1}$$

where $\phi(x, y, z)$ is the potential, $k = \omega\sqrt{\mu\epsilon} = \sqrt{k_x^2 + k_y^2 + k_z^2}$ is the wavenumber.

Let

$$\phi(x, y, z) = V(x, y)e^{\pm jk_z z}, \tag{9.1.2}$$

where $V(x, y)$ is the potential profile in the transverse plane. Substituting (9.1.2) into (9.1.1), we obtain

$$\left[\left(\frac{\partial^2}{\partial x^2} + \frac{\partial^2}{\partial y^2}\right) + (k^2 - k_z^2)\right] V(x, y) = 0. \tag{9.1.3}$$

Under quasi-static assumption (2), one has $k \approx k_z$. Hence (9.1.3) becomes

$$\left(\frac{\partial^2}{\partial x^2} + \frac{\partial^2}{\partial y^2}\right) V(x, y) = 0. \tag{9.1.4}$$

Equtation (9.1.4) is a 2D Laplace equation, which is much simpler then the Helmholtz equation (9.1.1). The static nature of (9.1.4) gives the name of this approach as quasi-static. The prefix "quasi-" is necessary because the wave does propagate along the $\mp\hat{z}$ direction. The quasi-static (quasi-TEM) method is very popular because of its simplicity in mathematics.

9.1.2 Formulation

Figure 9.1 shows the transmission line system under consideration. An arbitrary number of conductors N_c is embedded in a dielectric slab consisting of an arbitrary number of individual layers N_d. A perfectly conducting ground plane extends from $x = -\infty$ to $x = \infty$. The system is uniform in the y direction. The conductors are

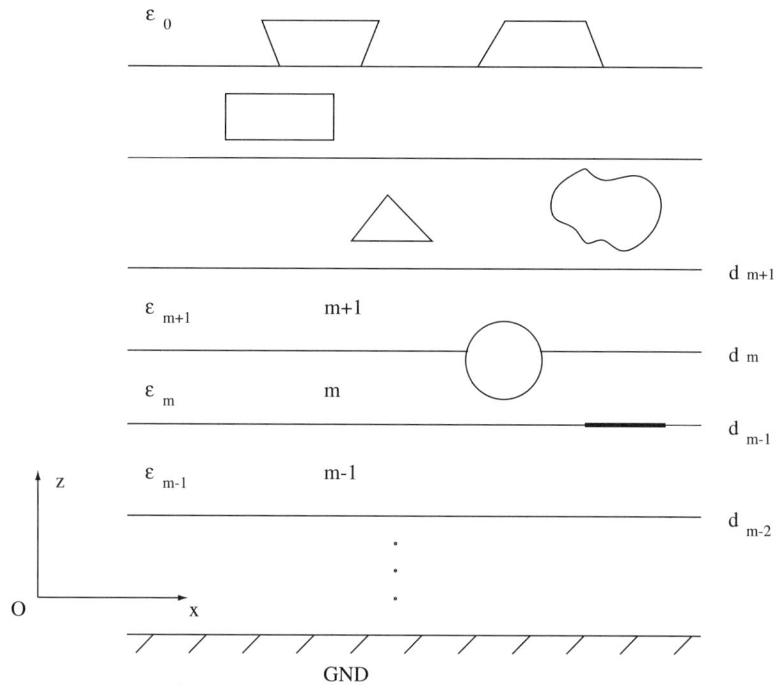

FIGURE 9.1 Geometry of multiconductor multilayer transmission lines (MMTL).

perfectly lossless and can possess either a finite cross section or be infinitesimally thin.

The integral equation formulation for this system is derived in [1]. For ease of reference, we briefly repeat the basic formulation here. The integral equations solved for the unknown total charge distribution $\sigma_T(\boldsymbol{\rho})$ can be obtained as follows:

$$\frac{1}{2\pi\epsilon_0}\sum_{j=1}^{J}\int_{l_j}\sigma_T(\boldsymbol{\rho}')\ln\frac{|\boldsymbol{\rho}-\boldsymbol{\rho}''|}{|\boldsymbol{\rho}-\boldsymbol{\rho}'|}\,dl' = V_c(\boldsymbol{\rho}) = \text{const.} \tag{9.1.5}$$

on the conductor surfaces, and

$$\frac{\epsilon^+(\boldsymbol{\rho})+\epsilon^-(\boldsymbol{\rho})}{2\epsilon_0\left[\epsilon^+(\boldsymbol{\rho})-\epsilon^-(\boldsymbol{\rho})\right]}\sigma_T(\boldsymbol{\rho}) + \frac{1}{2\pi\epsilon_0}\sum_{j=1}^{J}\fint_{l_j}\sigma_T(\boldsymbol{\rho}')\cdot\left(\frac{\boldsymbol{\rho}-\boldsymbol{\rho}'}{|\boldsymbol{\rho}-\boldsymbol{\rho}'|^2}-\frac{\boldsymbol{\rho}-\boldsymbol{\rho}''}{|\boldsymbol{\rho}-\boldsymbol{\rho}''|^2}\right)$$
$$\cdot\,\hat{n}(\boldsymbol{\rho})\,dl' = 0 \tag{9.1.6}$$

on the dielectric-to-dielectric interface. Here $\rho = \sqrt{x^2+z^2}$, l_j is the contour of the jth interface above the ground plane, $\boldsymbol{\rho}''$ is the image point of $\boldsymbol{\rho}'$ about the ground plane, and J is the total number of the interfaces (including conductor-to-dielectric interfaces and dielectric-to-dielectric interfaces); $\fint$ denotes the Cauchy principal value of the integral, and $\hat{n}(\boldsymbol{\rho})$ is the unit normal vector at $\boldsymbol{\rho}$. The side of the curve l_j is referred to as the "positive" side if $\hat{n}(\boldsymbol{\rho})$ points away from the curve, while the other side is called its "negative" side; $\epsilon^+(\boldsymbol{\rho})$ and $\epsilon^-(\boldsymbol{\rho})$ denote the permittivity on the positive and negative sides, respectively, of the interface that $\boldsymbol{\rho}$ approaches.

In order to obtain the capacitance matrix $[C]$, the integral equations (9.1.5) and (9.1.6) must first be solved for the total charge distribution $\sigma_T(\boldsymbol{\rho})$, with V_c assigned as a unity voltage on each particular conductor surface l_j as zero voltage on the other conductors. After obtaining the total charge distribution $\sigma_T(\boldsymbol{\rho})$, the free charge distribution $\sigma_F(\boldsymbol{\rho})$ on the conductors can be evaluated by

$$\sigma_F(\boldsymbol{\rho}) = \frac{\epsilon(\boldsymbol{\rho})}{\epsilon_0}\sigma_T(\boldsymbol{\rho})$$

for the conductors of finite cross section, and

$$\sigma_F(\boldsymbol{\rho}) = \frac{\epsilon^+(\boldsymbol{\rho}) + \epsilon^-(\boldsymbol{\rho})}{2\epsilon_0}\sigma_T(\boldsymbol{\rho}) + \frac{\epsilon^+(\boldsymbol{\rho}) - \epsilon^-(\boldsymbol{\rho})}{2\pi\epsilon_0}\sum_{j=1}^{J}\fint_{l_j}\sigma_T(\boldsymbol{\rho}') \cdot \left(\frac{\boldsymbol{\rho}-\boldsymbol{\rho}'}{|\boldsymbol{\rho}-\boldsymbol{\rho}'|^2} - \frac{\boldsymbol{\rho}-\boldsymbol{\rho}''}{|\boldsymbol{\rho}-\boldsymbol{\rho}''|^2}\right)\cdot\hat{n}(\boldsymbol{\rho})\,dl' \qquad (9.1.7)$$

for infinitesimally thin strips. The total free charge Q_i (per unit length in the z direction) on conductor l_i corresponding to this potential distribution yields the element C_{ij} $(i, j = 1, 2, \ldots, N_c)$ of the capacitance matrix. The external inductance matrix $[L]$ is related to the vacuum capacitance matrix $[C_v]$ by the simple formula $[L] = \epsilon_0\mu_0[C_v]^{-1}$. The vacuum capacitance matrix $[C_v]$ itself is the capacitance matrix of the same conductor system where all dielectrics have been replaced by a vacuum.

The previous integral equations, (9.1.5) and (9.1.6), need to be solved numerically for the unknown charge distribution $\sigma_T(\boldsymbol{\rho})$. This distribution on each interface is expanded in terms of basis functions

$$\sigma_T(\boldsymbol{\rho}) \simeq \sum_{m=1}^{M} g_{m-1}(\boldsymbol{\rho})\sigma_{Tm}, \qquad (9.1.8)$$

where $g_{m-1}(\boldsymbol{\rho})$ $(m = 1, 2, \ldots, M)$ are the basis functions, σ_{Tm} $(m = 1, 2, \ldots, M)$ are the unknown coefficients to be determined, and M is the total number of the bases.

We use Galerkin's method for the testing procedure. Using (9.1.8), a set of linear algebraic equations in matrix form can be derived from integral equations (9.1.5) and (9.1.6) [1] as

$$[A_{nm}][\sigma_{Tm}] = [B_n], \qquad (9.1.9)$$

where the elements of the matrices are

$$A_{nm} = \sum_{j1=1}^{J} \int_{l_{j1}} g_{n-1}(\boldsymbol{\rho}) \cdot \left[\frac{1}{2\pi\epsilon_0} \sum_{j2=1}^{J} \int_{l_{j2}} g_{m-1}(\boldsymbol{\rho}') \cdot \ln\left(\frac{|\boldsymbol{\rho} - \boldsymbol{\rho}''|}{|\boldsymbol{\rho} - \boldsymbol{\rho}'|} \right) dl' \right] dl \tag{9.1.10}$$

$$B_n = \sum_{j1=1}^{J} \int_{l_{j1}} g_{n-1}(\boldsymbol{\rho}) V_c(\boldsymbol{\rho})\, dl, \tag{9.1.11}$$

for those $g_{n-1}(\boldsymbol{\rho})$ defined on the conductor-to-dielectric interfaces, and

$$\begin{aligned} A_{nm} = \sum_{j1=1}^{J} \int_{l_{j1}} g_{n-1}(\boldsymbol{\rho}) \cdot & \left[\frac{\epsilon^+(\boldsymbol{\rho}) + \epsilon^-(\boldsymbol{\rho})}{2\epsilon_0 \left[\epsilon^+(\boldsymbol{\rho}) - \epsilon^-(\boldsymbol{\rho})\right]} g_{m-1}(\boldsymbol{\rho}) \right. \\ & \left. + \frac{1}{2\pi\epsilon_0} \sum_{j2=1}^{J} \int_{l_{j2}} g_{m-1}(\boldsymbol{\rho}') \cdot \left(\frac{\boldsymbol{\rho} - \boldsymbol{\rho}'}{|\boldsymbol{\rho} - \boldsymbol{\rho}'|^2} - \frac{\boldsymbol{\rho} - \boldsymbol{\rho}''}{|\boldsymbol{\rho} - \boldsymbol{\rho}''|^2} \right) \cdot \hat{n}(\boldsymbol{\rho})\, dl' \right] dl \end{aligned} \tag{9.1.12}$$

$$B_n = 0, \tag{9.1.13}$$

for those $g_{n-1}(\boldsymbol{\rho})$ defined on the dielectric-to-dielectric interfaces.

After A_{nm} $(n = 1, 2, \ldots, M; m = 1, 2, \ldots, M)$ and B_n $(n = 1, 2, \ldots, M)$ have been calculated, (9.1.9) produces M simultaneous equations in M unknowns, σ_{Tm} $(m = 1, 2, \ldots, M)$. These simultaneous equations can then be solved for σ_{Tm} $(m = 1, 2, \ldots, M)$ in terms of the potential $V_c(\boldsymbol{\rho})$ on the conductors.

9.1.3 Orthogonal Wavelets in $L^2([0, 1])$

Orthogonal periodic wavelets in $L^2([0, 1])$ were studied in great detail in Chapter 4. We will review the relevant material briefly here.

Given a multiresolution analysis with scalet $\varphi(x)$ and wavelet $\psi(x)$ in $L^2(R)$, the wavelets in $L^2([0, 1])$ are

$$\varphi_{m,n}^{\text{per}}(x) = \sum_{k \in Z} \varphi_{m,n}(x + k), \tag{9.1.14}$$

$$\psi_{m,n}^{\text{per}}(x) = \sum_{k \in Z} \psi_{m,n}(x + k), \tag{9.1.15}$$

and $V_m^{\text{per}} = \text{clos}_{L^2([0,1])}\{\varphi_{m,n}^{\text{per}}(x) : n \in \mathcal{Z}\}$, $W_m^{\text{per}} = \text{clos}_{L^2([0,1])}\{\psi_{m,n}^{\text{per}}(x) : n \in \mathcal{Z}\}$. It can be shown that V_m^{per} are all identical one-dimensional spaces containing only the constant functions for $m \le 0$, and $W_m^{\text{per}} = \{\emptyset\}$ for $m \le -1$. Thus we only need to study V_m^{per} and W_m^{per} for $m \ge 0$. Moreover it can easily be verified that

$$V_{m+1}^{\text{per}} = V_m^{\text{per}} \oplus W_m^{\text{per}}$$

and

$$\text{clos}_{L^2}\left(\bigcup_{m\in\mathcal{N}} V_m^{\text{per}}\right) = L^2([0, 1]),$$

where $\mathcal{N}$ is the set of nonnegative integers. Hence there is a ladder of multiresolution spaces

$$V_0^{\text{per}} \subset V_1^{\text{per}} \subset V_2^{\text{per}} \subset \cdots$$

with successive orthogonal complement $W_0^{\text{per}}, W_1^{\text{per}}, W_2^{\text{per}}, \ldots$, and orthonormal bases $\{\varphi_{m,n}^{\text{per}}(x)\}_{n=0,\ldots,2^m-1}$ in V_m^{per}, $\{\psi_{m,n}^{\text{per}}(x)\}_{n=0,\ldots,2^m-1}$ in W_m^{per} for $m \in \mathcal{N}$. In particular, that

$$\{\varphi_{0,0}^{\text{per}}\}\bigcup\{\psi_{m,n}^{\text{per}} : m \in \mathcal{N}, n = 0, \ldots, 2^m - 1\}$$

constitute an orthonormal basis in $L^2([0, 1])$. For simplicity, we relabel this basis as follows:

$$\begin{aligned}
g_0(x) &= \varphi_{0,0}^{\text{per}}(x) = 1\\
g_1(x) &= \psi_{0,0}^{\text{per}}(x)\\
g_2(x) &= \psi_{1,0}^{\text{per}}(x)\\
g_3(x) &= \psi_{1,1}^{\text{per}}(x) = g_2\left(x - \tfrac{1}{2}\right)\\
&\vdots\\
g_{2^m}(x) &= \psi_{m,0}^{\text{per}}(x)\\
&\vdots\\
g_{2^m+n}(x) &= \psi_{m,n}^{\text{per}}(x)\\
&= g_{2^m}(x - n2^{-m}), \qquad 0 \le n \le 2^m - 1\\
&\vdots
\end{aligned}$$

These Daubechies periodic scalets were illustrated in Fig. 4.7. For any $f(x) \in L^2([0, 1])$, the approximation at the resolution 2^m can be defined as the projection in V_m^{per},

$$f(x) \simeq P_m f(x) = \sum_{k=0}^{2^m-1} f_k g_k(x)$$

where P_m is the orthogonal projection operator onto V_m^{per} and f_k is the inner product of $f(x)$ and $g_k(x)$.

9.1.4 Boundary Element Method and Wavelet Expansion

Geometrical Representation Before considering the details of this problem, we will assume that most curves $\{l_j\}$ are closed for the purpose of expressing the unknown charge distribution. Roughly speaking, there are four types of contours: (1) the contour of the conductor with finite cross section, (2) the contour along the infinitesimally thin metal strip, (3) the contour along the dielectric-to-dielectric interface from $-\infty$ to $+\infty$, and (4) the contour along the dielectric-to-dielectric interface from $-\infty$ to $+\infty$, with some spaces of discontinuity wherever there is a conductor along the interface. We will examine the four types of contours one by one. In the first place, all the contours except type (4) are geometrically continuous.

Moreover the contour of type (1) is closed geometrically. The contour of type (2) can be considered to be closed, since the charge distribution has the same behavior (singularity) at its two edge points. Similarly the contour of type (3) can also be viewed as closed since no charge exists at infinity, and thus the charge distribution gives the same value of zero at the two ends ($-\infty$ and $+\infty$) of the contour.

In the case of of type (4), the contour intersects the conductor at two points if the conductor is lying along the contour and creates a discontinuity space for that contour. We must employ intervallic wavelets, instead of periodic wavelets.

Since the periodized wavelets are defined in $L^2([0, 1])$, one must map each of the contours $\{l_j\}$ onto the interval $[0, 1]$. For an arbitrary contour l_j, we take two steps:

(1) Use the conventional boundary element method to discretize the contour into a series of boundary elements, and then map each of the boundary elements onto 1D standard elements through the shape functions or interpolation functions [3, 22].

(2) Map the standard elements into corresponding portions of interval $[0, 1]$. A linear map is sufficient for this step.

This procedure can be precisely formulated in mathematical language as well. In step (1), the global coordinates $\boldsymbol{\rho}$ are expressed in terms of the local coordinate ξ of a standard element [3]:

$$\boldsymbol{\rho} = \sum_{i=1}^{M_e} N_i(\xi)\boldsymbol{\rho}_i = \Omega_1(\xi), \tag{9.1.16}$$

where M_e is the number of the interpolation nodes in the local standard element, $N_i(\xi)$ is the shape function referred to node i of the local standard element, and $\boldsymbol{\rho}_i$ are the global coordinates of node i of the actual element. The shape functions $\{N_i(\xi)\}$ are given in standard finite element or boundary element books and literature (e.g., [3, 22]).

Upon inspecting (9.1.16), we can conclude that (9.1.16) maps the standard element in local coordinates onto the actual element, which may have a quite arbitrary or distorted shape, in global coordinates. The node $\boldsymbol{\rho}_i$ in the actual element corresponds to the node i in the standard element (by definition, $N_i(\xi)$ is assumed to have a unity value at node i and zero at all other nodes of the element).

In step (2), the standard elements corresponding to the actual elements from contour l_j are mapped into the subintervals $[\zeta_0, \zeta_1], [\zeta_1, \zeta_2], \ldots, [\zeta_{K_j-1}, \zeta_{K_j}]$ of interval [0, 1], where K_j is the number of the elements from contour l_j and $0 = \zeta_0 < \zeta_1 < \zeta_2 < \cdots < \zeta_{K_j} = 1$ (e.g., one can simply assume that $\zeta_k = k/K_j$, $k = 1, \ldots, K_j - 1$). The map between the local coordinate ζ in interval [0, 1] and the local coordinate ξ in the kth standard element of contour l_j can be written as

$$\zeta = \zeta_{k-1} + (\zeta_k - \zeta_{k-1}) \cdot \xi, \tag{9.1.17}$$

or

$$\xi = \frac{\zeta - \zeta_{k-1}}{\zeta_k - \zeta_{k-1}}, \tag{9.1.18}$$

where $k = 1, 2, \ldots, K_j$. Combining (9.1.16) and (9.1.18), we obtain a map between the global coordinates $\boldsymbol{\rho}$ and the local coordinate ζ in interval [0, 1]:

$$\boldsymbol{\rho} = \Omega_1\left(\frac{\zeta - \zeta_{k-1}}{\zeta_k - \zeta_{k-1}}\right) = \Omega_2(\zeta). \tag{9.1.19}$$

The maps (9.1.16) through (9.1.19) establish the conversions among the local coordinate ξ, the local coordinate ζ and the global coordinates $\boldsymbol{\rho}$.

Source Representation Now we may define the basis functions $\{g_{m-1}(\boldsymbol{\rho})\}$. For simplicity and generality, the basis functions will not be directly defined over all the contours in terms of a set of global coordinates, but rather over interval [0, 1] since each of the contours can be related to interval [0, 1] through the map described by (9.1.19). By using the conversion between the global coordinates $\boldsymbol{\rho}$ and the local coordinate ζ for each individual contour, we can easily obtain the basis functions of the individual contour in the set of global coordinates. For the unknown charge distribution along contour l_j, expansion (9.1.8) can now accurately be written as the projection in $V_{m_h}^{\text{per}}$ (about ζ):

$$\sigma_T(\boldsymbol{\rho}) \simeq P_{m_h}\sigma_T(\boldsymbol{\rho}) = \sum_{m=1}^{M_j} g_{m-1}\left[\Omega_2^{-1}(\boldsymbol{\rho})\right]\sigma_{Tm}, \tag{9.1.20}$$

where Ω_2^{-1} denotes the inverse map of Ω_2, $g_{m-1}(\zeta)$ represents the orthogonal wavelets in $L^2([0, 1])$, and $M_j = 2^{m_h}$ is the number of the wavelet bases used for expressing the unknown charge distribution on contour l_j. Because Ω_2^{-1} maps contour l_j into interval [0, 1], the basis functions $\{g_{m-1}[\Omega_2^{-1}(\boldsymbol{\rho})]\}$ are well defined.

It has been shown [24] that if σ_T is smooth with a finite number of discontinuities, the error between $\sigma_T(\zeta)$ and $P_{m_h}\sigma_T(\zeta)$ is bounded:

$$\|\, \sigma_T(\zeta) - P_{m_h}\sigma_T(\zeta) \,\| \le C 2^{-m_h s}, \tag{9.1.21}$$

where C and s are some positive constants, respectively, relating to $\|\sigma_T(\zeta)\|$ and the smoothness of $\sigma_T(\zeta)$. The function $\sigma_T(\zeta)$ with higher-order (piecewise) continuity has larger s value and thus faster error decay. Moreover the approximation error of expansion (9.1.20) can be estimated as

$$\begin{aligned} \|\sigma_T(\boldsymbol{\rho}) - P_{m_h}\sigma_T(\boldsymbol{\rho})\| &\le C_d \|\sigma_T(\zeta) - P_{m_h}\sigma_T(\zeta)\| \\ &\le C C_d 2^{-m_h s}, \end{aligned}$$

where C_d is the tight upper bound of the Jacobian of the transformation $\Omega_2(\zeta)$. That is, the approximation error of (9.1.20) has exponential decay with respect to the resolution level m_h.

Matrix Equation Based on the preceding source expansion, a set of linear algebraic equations is obtained from integral equations (9.1.5) and (9.1.6) by using Galerkin's method. This set is matrix form described by (9.1.9) if the elements of the matrices are computed by replacing $\{g_{m-1}(\boldsymbol{\rho})\}$ with $\{g_{m-1}[\Omega_2^{-1}(\boldsymbol{\rho})]\}$ in equations (9.1.10) through (9.1.13), namely

$$\begin{aligned} A_{nm} &= \sum_{j1=1}^{J} \int_{l_{j1}} g_{n-1}\left[\Omega_2^{-1}(\boldsymbol{\rho})\right] \\ &\quad \cdot \left[\frac{1}{2\pi\epsilon_0} \sum_{j2=1}^{J} \int_{l_{j2}} g_{m-1}\left[\Omega_2^{-1}(\boldsymbol{\rho}')\right] \cdot \ln\left(\frac{|\boldsymbol{\rho}-\boldsymbol{\rho}''|}{|\boldsymbol{\rho}-\boldsymbol{\rho}'|}\right) dl'\right] dl, \end{aligned} \tag{9.1.22}$$

$$B_n = \sum_{j1=1}^{J} \int_{l_{j1}} g_{n-1}\left[\Omega_2^{-1}(\boldsymbol{\rho})\right] V_c(\boldsymbol{\rho})\, dl, \tag{9.1.23}$$

for those $g_{n-1}[\Omega_2^{-1}(\boldsymbol{\rho})]$ defined on the conductor-to-dielectric interfaces, and

$$\begin{aligned} A_{nm} &= \sum_{j1=1}^{J} \int_{l_{j1}} g_{n-1}\left[\Omega_2^{-1}(\boldsymbol{\rho})\right] \cdot \left[\frac{\epsilon^+(\boldsymbol{\rho}) + \epsilon^-(\boldsymbol{\rho})}{2\epsilon_0\left[\epsilon^+(\boldsymbol{\rho}) - \epsilon^-(\boldsymbol{\rho})\right]} g_{m-1}\left[\Omega_2^{-1}(\boldsymbol{\rho})\right] \right. \\ &\quad + \frac{1}{2\pi\epsilon_0} \sum_{j2=1}^{J} \int_{l_{j2}} -g_{m-1}\left[\Omega_2^{-1}(\boldsymbol{\rho}')\right] \\ &\quad \left. \cdot \left(\frac{\boldsymbol{\rho}-\boldsymbol{\rho}'}{|\boldsymbol{\rho}-\boldsymbol{\rho}'|^2} - \frac{\boldsymbol{\rho}-\boldsymbol{\rho}''}{|\boldsymbol{\rho}-\boldsymbol{\rho}''|^2}\right) \cdot \hat{n}(\boldsymbol{\rho})\, dl' \right] dl, \\ B_n &= 0, \end{aligned} \tag{9.1.24}$$

for those $g_{n-1}[\Omega_2^{-1}(\boldsymbol{\rho})]$ defined on the dielectric-to-dielectric interfaces.

Evaluation of Integrals In practice, integrals in (9.1.22) through (9.1.24) can be evaluated numerically in either the ζ domain or the ξ domain. We choose the ξ domain for our numerical computations in accordance with the conventional boundary element analysis. Without loss of generality, let us consider the following integral

$$T_{l_j}(\boldsymbol{\rho}_0) = \int_{l_j} g_{m-1}\left[\Omega_2^{-1}(\boldsymbol{\rho})\right] R(\boldsymbol{\rho}_0, \boldsymbol{\rho})\, dl.$$

Note that the integrals in (9.1.22) through (9.1.24) are equivalent to this 1D integral with a particular form of the kernel function $R(\boldsymbol{\rho}_0, \boldsymbol{\rho})$. Using the maps (9.1.16), (9.1.17), and (9.1.19), we have

$$T_{l_j}(\boldsymbol{\rho}_0) = \sum_{k=1}^{k=K_j} \int_0^1 g_{m-1}\left[\zeta_{k-1} + (\zeta_k - \zeta_{k-1}) \cdot \xi\right] \cdot R\left[\boldsymbol{\rho}_0, \Omega_1(\xi)\right] |\, D\,|\, d\xi, \tag{9.1.25}$$

where $|\,D\,|$ is the Jacobian of the transformation between the global coordinates $\boldsymbol{\rho}$ and the local coordinate ξ of the kth standard element of contour l_j.

The Jacobian that defines the map of (9.1.16) can be obtained from the expression for the differential length

$$dl = \sqrt{(dx)^2 + (dz)^2} = \left[\sqrt{\left(\frac{dx}{d\xi}\right)^2 + \left(\frac{dz}{d\xi}\right)^2}\,\right] d\xi.$$

The Jacobian is then calculated from the following equation:

$$|\,D\,| = \sqrt{(D_x)^2 + (D_z)^2},$$

where

$$D_x = \frac{dx}{d\xi} = \sum_{i=1}^{M_e} \frac{dN_i(\xi)}{d\xi} x_i,$$

$$D_z = \frac{dz}{d\xi} = \sum_{i=1}^{M_e} \frac{dN_i(\xi)}{d\xi} z_i,$$

and where x_i and z_i are, respectively, the x and z components of $\boldsymbol{\rho}_i$.

From the case of the orthogonal wavelet on the real line, we can use definitions (9.1.14), (9.1.15) and (9.1.16) to obtain the periodic orthogonal wavelet $\{g_{m-1}(\zeta)\}$. Integration (9.1.25) can be readily performed by standard numerical algorithms such as Gaussian quadrature [25].

9.1.5 Numerical Examples

Based on the technique presented in the preceding subsections, a program has been designed to compute the capacitance and external inductance matrices of multiconductor transmission lines in multilayered dielectrics. Two numerical examples are presented in this subsection. When using wavelets on the real line to solve problems with finite intervals, improper selection of the wavelets can result in nonphysical solutions. In contrast, any type of wavelets on the real line can be used for the construction of the wavelets in $L^2([0, 1])$, although there may be some discrepancy in their smoothness, as seen in Chapter 4. However, since the derivatives of the unknown function $\sigma_T(\boldsymbol{\rho})$ are of order zero in the integral equations under consideration, a set of basis functions with C^0 continuity is sufficient to yield a convergent solution. In the following computations the Daubechies wavelets are employed to construct the orthogonal wavelets in $L^2([0, 1])$.

Example 1 Thin microstrip line of width W above a dielectric substrate of thickness H and $\epsilon_r = 6$. We have studied this example of an infinitesimally thin microstrip line. The characteristic impedances obtained by this technique were compared with those from the conventional boundary element method (BEM) [3], the method of moments (MoM) [1], and the more accurate formulas from [29] in Table 9.1. The results of the conventional BEM were obtained using 16 subsections (33 bases) on the strip and 30 subsections (62 bases) at the dielectric interface; those of the MoM were obtained by using 12 subsections on the strip and 30 subsections at the dielectric interface. Two sets of the results from this technique are presented in columns A and B, with $M_1 = M_2 = 32$ and $M_1 = M_2 = 16$ respectively, where M_1 is the number of the wavelet bases used on the strip while M_2 is the number at the dielectric interface.

Table 9.1 provides an interesting insight into the wavelet expansions. Taking the column labeled "Hammerstad" as a set of "ground truth" or standard references, we see that the results from this technique with 64 bases (column A) give approximately the same accuracy as the conventional BEM, although the BEM results are obtained by using about 50% more (total 95) bases. The results from this technique with 32 bases (column B) exhibit a higher degree of accuracy than the MoM despite the fact that the MoM uses approximately one-third more (total 42) bases for its calculations.

TABLE 9.1. Characteristic Impedances for the Thin Microstrip Line (in Ohms)

W/H	A	B	BEM	MoM	Hammerstad
0.4	90.5779	91.3783	90.7758	92.2785	90.3339
0.7	72.9504	73.2748	73.0898	73.9626	72.7516
1.0	62.0383	62.3342	62.1102	62.8109	61.8397
2.0	42.4233	42.5918	42.4118	42.9980	42.2600
4.0	26.5482	26.6498	26.5236	26.9709	26.4593
10.	12.7707	12.8134	12.7351	12.9961	12.7198

Source: G. Wang, G. Pan, and B. Gilbert. *IEEE Trans. Microw. Theory Tech.*, **43**(3), 664–675, March 1995; © 1995 IEEE.

Finally, comparison between the results of column A and column B shows that this technique gives better accuracy with higher resolution approximation.

Theoretically, it is not a surprise that the wavelet expansions converge more quickly; that is, fewer coefficients are required by wavelets to represent a given function than by other expansions, since this is a well-known result from wavelet theory and has been extensively studied in Chapter 2. One of the most attractive features of wavelets is that they give completely local information on the functions analyzed. It can be shown that if a function does not have uniform smoothness, for instance, if a smooth function possesses discontinuities, there is an optimal way to approximate the function using low resolution wavelets everywhere and adding high resolution wavelets near the singularities [24].

Example 2 Multiconductor Transmission Lines above a Thick Substrate. Shown in Fig. 9.2 is a 10-conductor transmission line system. This problem arises during the modeling of CMOS chips, where the transmission lines are far above the ground plane in comparison to the cross-sectional dimensions or the separations of the individual conductors. For such structures the MoM approach frequently yields either singular matrices or nonphysical solutions [3]. In order to test the stability of this technique, we applied it to a ten conductor transmission line with a thick substrate. Tables 9.2 to 9.5 list the resulting capacitance and inductance matrices computed with this technique and the BEM with special edge treatment [3]. The BEM solutions were computed by using 160 subsections (360 bases) on the conductor surfaces and 190 subsections (392 bases) at the dielectric interfaces. These solutions are taken from [3]. The results from this technique were obtained by using 160 bases on the conductor surfaces and 256 bases at the dielectric interfaces. The self-capacitance of the ith conductor can be obtained by summing up all the elements at the ith row of the capacitance matrix $[C]$. Each of the self-capacitance values must be positive; otherwise, the results will be nonphysical solutions.

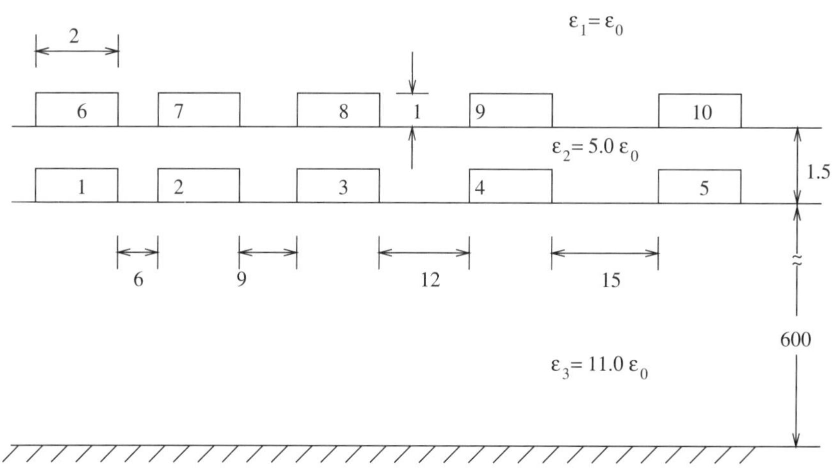

FIGURE 9.2 Ten conductors in a layered medium (in μm).

TABLE 9.2. Wavelet Technique: Capacitance Matrix [C] (in pF/m)

307.4	−41.10	−11.35	−6.330	−5.452	−219.6	−4.932	−1.389	−0.8246	−0.7600
−41.12	319.7	−27.96	−7.812	−5.043	−4.999	−217.5	−3.485	−0.9775	−0.6821
−11.35	−27.96	309.9	−24.24	−8.632	−1.377	−3.474	−218.4	−3.103	−1.154
−6.316	−7.794	−24.23	302.3	−24.70	−0.8192	−0.9580	−3.117	−218.9	−3.304
−5.440	−5.029	−8.624	−24.74	290.2	−0.7487	−0.6584	−1.136	−3.259	−221.5
−218.8	−5.019	−1.373	−0.8105	−0.7303	231.7	−2.063	−0.3899	−0.1799	−0.1349
−4.967	−216.6	−3.492	−0.9523	−0.6406	−2.064	231.6	−1.176	−0.2495	−0.1332
−1.386	−3.526	−217.3	−3.150	−1.127	−0.3896	−1.178	230.6	−0.8550	−0.2383
−0.8200	−0.9843	−3.162	−217.6	−3.306	−0.1803	−0.2511	−0.8580	229.6	−0.7478
−0.7467	−0.6755	−1.154	−3.343	−220.5	−0.1358	−0.1351	−0.2399	−0.7465	230.4

TABLE 9.3. Wavelet Technique: Inductance Matrix [L] (in nH/m)

1407.0	999.8	831.9	721.3	638.0	1306.0	998.7	831.8	721.4	638.1
999.8	1405.0	935.1	774.8	671.7	998.7	1304.0	934.7	774.8	671.8
831.9	935.1	1407.0	888.0	731.7	831.8	934.7	1307.0	887.7	731.8
721.3	774.8	888.0	1409.0	850.2	721.4	774.8	887.7	1309.0	850.1
638.0	671.7	731.7	850.2	1411.0	638.1	671.8	731.8	850.1	1310.0
1306.	998.7	831.8	721.4	638.1	1407.0	1000.0	832.1	721.6	638.3
998.7	1304.0	934.7	774.8	671.7	1000.0	1405.0	935.4	775.1	671.9
831.8	934.7	1307.0	887.7	731.8	832.1	935.4	1408.0	888.2	732.0
721.4	774.8	887.7	1309.0	850.1	721.6	775.1	888.2	1410.0	850.5
638.1	671.8	731.8	850.1	1310.0	638.3	671.9	732.0	850.5	1411.0

The sizes for matrix [A] are, respectively, 752×752 and 416×416 for the BEM and the wavelet technique. For such a relatively large matrix [A], the sparsity is more significant. As mentioned in Example 1 (9.1.6), is likely to produce sparse linear algebraic equations for both the wavelet-base approach and the BEM. Hence, we will only examine the sparsity for the upper part of matrix [A], which comes from (9.1.5). The upper part of matrix [A] is obtained by using this technique under a threshold of 10^{-3} and is a 160×416 sparse matrix. In sharp contrast, a 360×752 full dense matrix is generated by the BEM under the same threshold.

9.2 SPATIAL DOMAIN LAYERED GREEN'S FUNCTIONS

In this section we present a new approach to capacitance computation, which is more efficient than the method presented in the previous section. The major improvements are as follows:

(1) Under the formulation of the free-space Green function, polarization charges at the dielectric–dielectric interfaces have to be computed as unknowns in addition to the free charges on conductor surfaces. In contrast, we now use the layered Green's function that was proposed by DeZutter in [13] and approximated in closed forms in [32]. Under the layered Green function, only the free surface charges are unknown in the problem, resulting in much smaller impedance matrix.

(2) Only standard wavelets are employed to expand the free surface charges on closed contours of the conductor surfaces. No periodic or intervallic wavelets are necessary, and so a much simpler treatment is possible.

(3) Replacing the Daubechies wavelets with Coifman wavelets allows single-point quadrature and leads to fast matrix filling.

TABLE 9.4. BEM Solution: Capacitance Matrix [C] (in pF/m)

308.5	−41.50	−11.42	−6.335	−5.417	−219.6	−5.019	−1.402	−0.8288	−0.7474
−41.51	321.2	−28.25	−7.853	−5.038	−5.081	−217.8	−3.577	−0.9985	−0.6799
−11.43	−28.25	312.0	−24.48	−8.665	−1.384	−3.539	−219.2	−3.214	−1.164
−6.339	−7.854	−24.47	304.9	−24.93	−0.8126	−0.9598	−3.198	−220.3	−3.382
−5.417	−5.036	−8.660	−24.92	291.8	−0.7275	−0.6423	−1.137	−3.360	−222.1
−220.3	−5.073	−1.380	−0.8094	−.7240	233.4	−2.090	−0.3937	−0.1811	−0.1332
−5.019	−218.7	−3.542	−0.9590	−0.6409	−2.091	233.9	−1.201	−0.2544	−0.1336
−1.403	−3.580	−220.2	−3.200	−1.137	−0.3943	−1.201	233.7	−0.8819	−0.2420
−0.8282	−0.9984	−3.216	−221.3	−3.363	−.1814	−0.2545	−0.8820	233.5	−0.7688
−0.7448	−0.6777	−1.162	−3.377	−222.9	−0.1333	−0.1335	−0.2417	−0.7683	233.0

TABLE 9.5. BEM Solution: Inductance Matrix [*L*] (in nH/m)

1398.0	993.1	826.2	716.5	633.7	1297.0	992.1	826.2	716.6	633.8
993.0	1396.0	928.8	769.5	667.1	992.1	1295.0	928.4	769.6	667.2
826.2	928.9	1398.0	881.9	726.7	826.2	928.4	1298.0	881.8	726.8
716.5	769.6	882.0	1400.0	844.5	716.6	769.7	881.8	1300.0	844.4
633.7	667.1	726.8	844.5	1402.0	633.9	667.3	726.9	844.4	1301.0
1297.0	992.0	826.2	716.5	633.8	1398.0	993.4	826.6	716.8	634.0
992.0	1295.0	928.4	769.6	667.2	993.4	1396.0	929.1	769.9	667.4
826.2	928.4	1298.0	881.7	726.8	826.6	929.2	1399.0	882.3	727.1
716.6	769.6	881.7	1299.0	844.3	716.8	769.9	882.3	1401.0	844.8
633.8	667.2	726.8	844.4	1301.0	634.0	667.5	727.1	844.8	1402.0

As discussed in the previous section, the adoption of the geometric representation of the BEM converts a 2D problem into a 1D problem and provides a versatile and accurate treatment of curved conductor surfaces. The conductor cross sections of 2D problems are closed contours. The BEM representation of a contour utilizes the arc length ζ, varying from 0 to ℓ in circumference; it is in [0, 1] after normalization. In principle, one needs to utilize periodic wavelets in $\mathbf{L}^2([0, 1])$ when the domain of the problem is over a finite interval. Nevertheless, we find that the standard wavelets are sufficient to represent the contours. In fact, we now deploy the wavelet bases one by one on the contour that has been mapped by the BEM onto the interval [0, 1]. The portion of a wavelet basis that is beyond the interval will be lobbed off and relocated at the opposite end. This procedure is quite similar to the circular convolution in digital signal processing [33].

9.2.1 Formulation

Suppose that N_c perfect conductors are placed throughout N_d nonmagnetic dielectric layers and the geometry of the dielectric layers are assumed to be uniform in the x and y directions. The integral equation relating the electrostatic potential $V(\mathbf{r})$ to the charge density $\sigma(\mathbf{r})$ is

$$V(\mathbf{r}) = \int_{\Omega} G(\mathbf{r}, \mathbf{r}')\sigma(\mathbf{r}')d\mathbf{r}'. \tag{9.2.1}$$

Considering the case that a unit source is in layer m (see Fig. 9.1). The 3D Green's function satisfies Poisson's equation

$$\nabla^2 G^{3D}(x, y, z \mid x_0, y_0, z_0) = \frac{1}{\epsilon}\,\delta(x - x_0)\,\delta(y - y_0)\,\delta(z - z_0). \tag{9.2.2}$$

Spatial domain and spectral domain Green's functions are related by the 2D Fourier transform pair as

$$G^{3D}(x, y, z \mid x_0, y_0, z_0) = \frac{1}{(2\pi)^2} \int_{-\infty}^{\infty} \int_{-\infty}^{\infty} d\alpha\, d\beta e^{-j\alpha(x-x_0)-j\beta(y-y_0)} \times \tilde{G}^{3D}(\alpha, \beta, z \mid x_0, y_0, z_0)$$

and

$$\tilde{G}^{3D}(\alpha, \beta, z \mid x_0, y_0, z_0) = \int_{-\infty}^{\infty} \int_{-\infty}^{\infty} dx\, dy e^{j\alpha(x-x_0)-j\beta(y-y_0)} \times G^{3D}(x, y, z \mid x_0, y_0, z_0),$$

where $\tilde{G}^{3D}(\alpha, \beta, z \mid x_0, y_0, z_0)$ is the spectral domain Green function. By taking the 2D Fourier transform with respect to x and y, (9.2.2) becomes

$$\left(\frac{\partial^2}{\partial z^2} - \alpha^2 - \beta^2 \right) \tilde{G}^{3D}(\alpha, \beta, z \mid x_0, y_0, z_0) = \frac{1}{\epsilon} \delta(z - z_0).$$

Denoting $\gamma = \sqrt{\alpha^2 + \beta^2}$, we can write the z variation of the solution in region m as

$$\tilde{G}(z \mid z_0) = \frac{e^{-\gamma |z-z_0|} + B_m e^{\gamma z} + D_m e^{-\gamma z}}{2\epsilon_m \gamma}. \tag{9.2.3}$$

To find B_m and D_m, we need to use the constraint conditions at $z = d_{m-1}$ and $z = d_m$ (see Fig. 9.1). The descending wave for $z > z_0$ is a consequence of the reflection of the ascending wave for $z > z_0$ at $z = d_m$, namely

$$B_m e^{\gamma d_m} = \tilde{\Gamma}_{m,m+1}[e^{-\gamma(d_m - z_0)} + D_m e^{-\gamma d_m}]. \tag{9.2.4}$$

Similarly

$$D_m e^{-\gamma d_{m-1}} = \tilde{\Gamma}_{m,m-1}[e^{\gamma(d_{m-1} - z_0)} + B_m e^{\gamma d_{m-1}}]. \tag{9.2.5}$$

Rewriting (9.2.5) as

$$D_m = e^{\gamma d_{m-1}} \tilde{\Gamma}_{m,m-1}[e^{\gamma(d_{m-1} - z_0)} + B_m e^{\gamma d_{m-1}}] \tag{9.2.6}$$

and substituting D_m into (9.2.4), we have

$$B_m = \frac{\tilde{\Gamma}_{m,m+1}[e^{\gamma(-2d_m + z_0)} + \tilde{\Gamma}_{m,m-1} e^{\gamma(2d_{m-1} - 2d_m - z_0)}]}{1 - \tilde{\Gamma}_{m,m+1} \tilde{\Gamma}_{m,m+1} e^{2\gamma(d_{m-1} - d_m)}}. \tag{9.2.7}$$

Substituting (9.2.7) into (9.2.6), we arrive at

$$D_m = \frac{\tilde{\Gamma}_{m,m-1}[e^{\gamma(2d_{m-1}-z_0)} + \tilde{\Gamma}_{m,m+1}e^{\gamma(2d_{m-1}-2d_m+z_0)}]}{1 - \tilde{\Gamma}_{m,m-1}\tilde{\Gamma}_{m,m+1}e^{2\gamma(d_{m-1}-d_m)}}. \quad (9.2.8)$$

CASE 1 $z > z_0$. When $z > z_0$, we have

$$|z - z_0| = z - z_0.$$

Substituting (9.2.7) and (9.2.8) into (9.2.3) and letting

$$M_m = [1 - \tilde{\Gamma}_{m,m-1}\tilde{\Gamma}_{m,m+1}e^{2\Gamma(d_{m-1}-d_m)}]^{-1},$$

we arrive at

$$\tilde{G}(z \mid z_0) = \frac{M_m}{2\epsilon_m\gamma}[e^{-\gamma z} + \tilde{\Gamma}_{m,m+1}e^{-2\gamma d_m+\gamma z}][e^{\gamma z_0} + \tilde{\Gamma}_{m,m-1}e^{2\gamma d_{m-1}-\gamma z_0}].$$

CASE 2 $z < z_0$. In a similar way, for $z < z_0$, we have

$$\tilde{G}(z \mid z_0) = \frac{M_m}{2\epsilon_m\gamma}[e^{\gamma z} + \tilde{\Gamma}_{m,m-1}e^{2\gamma d_{m-1}-\gamma z}][e^{-\gamma z_0} + \tilde{\Gamma}_{m,m+1}e^{-2\gamma d_m+\gamma z_0}].$$

Furthermore, if we are looking for the field in region $n > m$, it can be found by using the recursive method. For $n > m$, $z > z_0$,

$$\tilde{G}(z \mid z_0) = \frac{A^+_{m,n}}{2\epsilon_m\gamma}(e^{-\gamma z} + \tilde{\Gamma}_{n,n+1}e^{-2\gamma d_n+\gamma z}),$$

$$A^+_{i,i+1} = A^+_{i,i}S^+_{i,i+1},$$

$$A^+_{m,n} = A^+_{m,m}\prod_{i=m}^{n-1} S^+_{i,i+1},$$

where $A^+_{m,m} = M_m[e^{\gamma z_0} + \tilde{\Gamma}_{m,m-1}e^{2\gamma d_{m-1}-\gamma z_0}]$.

For $n < m$, $z < z_0$,

$$\tilde{G}(z \mid z_0) = \frac{A^-_{m,n}}{2\epsilon_m\gamma}[e^{\gamma z} + \tilde{\Gamma}_{n,n-1}e^{2\gamma d_{n-1}-\gamma z}],$$

$$A^-_{i,i-1} = A^-_{i,i}S^-_{i,i-1},$$

$$A^-_{m,n} = A^-_{m,m}\prod_{n+1}^{m} S^-_{i,i-1},$$

where

$$A^-_{m,m} = M_m[e^{-\gamma z_0} + \tilde{\Gamma}_{m,m+1}e^{-2\gamma d_m+\gamma z_0}].$$

In the previous formulas

$$\tilde{\Gamma}_{i,i+1} = \frac{\Gamma_{i,i+1} + \tilde{\Gamma}_{i+1,i+2} e^{2\gamma(d_i - d_{i+1})}}{1 + \Gamma_{i,i+1}\tilde{\Gamma}_{i+1,i+2} e^{2\gamma(d_i - d_{i+1})}},$$

$$S^{+}_{i,i+1} = \frac{T_{i,i+1}}{1 - \Gamma_{i+1,i}\tilde{\Gamma}_{i+1,i+2} e^{2\gamma(d_i - d_{i+1})}},$$

$$\tilde{\Gamma}_{i,i-1} = \frac{\Gamma_{i,i-1} + \tilde{\Gamma}_{i-1,i-2} e^{2\gamma(d_{i-2} - d_{i-1})}}{1 + \Gamma_{i,i-1}\tilde{\Gamma}_{i-1,i-2} e^{2\gamma(d_{i-2} - d_{i-1})}},$$

$$S^{-}_{i,i-1} = \frac{T_{i,i-1}}{1 - \Gamma_{i-1,i}\tilde{\Gamma}_{i-1,i-2} e^{2\gamma(d_{i-2} - d_{i-1})}},$$

and

$$\Gamma_{i,j} = \frac{\epsilon_i - \epsilon_j}{\epsilon_i + \epsilon_j}, \qquad T_{i,j} = \frac{2\epsilon_i}{\epsilon_i + \epsilon_j}.$$

The parameters B_m, D_m, M_m, $A^{\pm}_{i,i+1}$, $S^{\pm}_{i,i+1}$, $\tilde{\Gamma}_{i,\pm 1}$, $\Gamma_{i,j}$, $T_{i,j}$, etc., are the static versions of their counterparts in [23]. The generalized reflection coefficient $\tilde{\Gamma}_{j,j+1}$ takes the value of 0 or -1 if the jth layer is a half-space or $(j+1)$th layer is a ground plane, respectively.

Rearranging these expressions by factoring out all z and z_0 dependencies, we obtain

$$\begin{aligned}\tilde{G}(z \mid z_0) = {} & \frac{1}{2\epsilon_m \gamma}[K^{+}_{m,n,1} e^{\gamma(z+z_0-2d_n)} + K^{+}_{m,n,2} e^{\gamma(z-z_0+2(d_{m-1}-d_n))} \\ & + K^{+}_{m,n,3} e^{\gamma(-z+z_0)} + K^{+}_{m,n,4} e^{\gamma(-z-z_0+2d_{m-1})}], \qquad z \geq z_0, \end{aligned} \tag{9.2.9}$$

$$\begin{aligned}\tilde{G}(z \mid z_0) = {} & \frac{1}{2\epsilon_m \gamma}[K^{-}_{m,n,1} e^{\gamma(z+z_0-2d_m)} + K^{-}_{m,n,2} e^{\gamma(z-z_0)} \\ & + K^{-}_{m,n,3} e^{\gamma(-z+z_0+2(d_{n-1}-d_m))} \\ & + K^{-}_{m,n,4} e^{\gamma(-z-z_0+2d_n-1)}], \qquad z \leq z_0, \end{aligned} \tag{9.2.10}$$

where

$$K^{+}_{m,n,1} = M_m \tilde{\Gamma}_{n,n+1} \prod_{j=m}^{n-1} S^{+}_{j,j+1}$$

$$K^{+}_{m,n,2} = M_m \tilde{\Gamma}_{n,n+1} \tilde{\Gamma}_{m,m-1} \prod_{j=m}^{n-1} S^{+}_{j,j+1}$$

$$K^{+}_{m,n,3} = M_m \prod_{j=m}^{n-1} S^{+}_{j,j+1}$$

$$K_{m,n,4}^{+} = M_m \tilde{\Gamma}_{m,m-1} \prod_{j=m}^{n-1} S_{j,j+1}^{+},$$

and

$$K_{m,n,1}^{-} = M_m \tilde{\Gamma}_{m,m+1} \prod_{j=n+1}^{m} S_{j,j-1}^{-}$$

$$K_{m,n,2}^{-} = M_m \prod_{j=n+1}^{m} S_{j,j-1}^{-}$$

$$K_{m,n,3}^{-} = M_m \tilde{\Gamma}_{m,m+1} \tilde{\Gamma}_{n,n-1} \prod_{j=n+1}^{m} S_{j,j-1}^{-}$$

$$K_{m,n,4}^{-} = M_m \tilde{\Gamma}_{n,n-1} \prod_{j=n+1}^{m} S_{j,j-1}^{-}.$$

Before we determine the closed-form spatial domain Green's function, we will approximate the coefficient functions $K_{m,n,i}^{\pm}$ of the exponentials in terms

$$K_{m,n,j}^{\pm}(\gamma) = K_{m,n,j}^{\pm\infty} + \sum_{i=1}^{N_{m,n,j}^{\pm}} C_{m,n,j}^{\pm,i} e^{a_{m,n,j}^{\pm,i}\gamma}, \qquad j = 1, 2, 3, 4, \tag{9.2.11}$$

where $K_{m,n,j}^{\pm\infty}$ denotes the asymptotic value of $K_{m,n,j}^{\pm}$, summation index $N_{m,n,j}^{\pm}$ is the number of exponential functions, $C_{m,n,j}^{\pm,i}$ and $a_{m,n,j}^{\pm,i}$ are Prony's coefficients given in Section 9.2.2.

By using the Fourier transform,

$$3D: \quad \frac{1}{2\pi} \int_{-\infty}^{+\infty} \int_{-\infty}^{+\infty} d\alpha \, d\beta e^{-j(\alpha x + \beta y)} \frac{e^{-\gamma|z|}}{\gamma} = \frac{1}{\sqrt{x^2 + y^2 + z^2}},$$

$$2D: \quad \frac{1}{2\pi} \int_{-\infty}^{+\infty} d\gamma e^{-j\gamma x} \frac{e^{-|\gamma z|}}{|\gamma|} = -\ln\left(\sqrt{x^2 + z^2}\right),$$

we can write the approximated Green's function for 2D and 3D cases as

$$G^{3D}(\mathbf{r}|\,\mathbf{r}_0) = \frac{1}{4\pi\epsilon_m} \sum_{j=1}^{4} f_j^{3D,\pm}(\mathbf{r}\,|\mathbf{r}_0),$$

$$G^{2D}(\boldsymbol{\rho} \mid \boldsymbol{\rho}_0) = -\frac{1}{2\pi\epsilon_m} \sum_{j=1}^{4} f_j^{2D,\pm}(\boldsymbol{\rho} \mid \boldsymbol{\rho}_0),$$

For the 2D case,

$$f_j^{2D,+}(\boldsymbol{\rho} \mid \boldsymbol{\rho}_0) = K_{m,n,j}^{+,\infty} \ln\left(\sqrt{(x-x_0)^2 + Z_j^{+\,2}}\right)$$

$$+ \sum_{i=1}^{N_{m,n,j}^{+}} C_{m,n,j}^{+,i} \ln\left(\sqrt{(x-x_0)^2 + (Z_j^{+} + a_{m,n,j}^{+,i})^2}\right), \tag{9.2.12}$$

$$f_j^{2D,-}(\boldsymbol{\rho} \mid \boldsymbol{\rho}_0) = K_{m,n,j}^{-\infty} \ln\left(\sqrt{(x-x_0)^2 + Z_j^{-\,2}}\right)$$

$$+ \sum_{i=1}^{N_{m,n,j}^{-}} C_{m,n,j}^{-,i} \ln\left(\sqrt{(x-x_0)^2 + (Z_j^{-} + a_{m,n,j}^{-,i})^2}\right), \tag{9.2.13}$$

where $j = 1, \ldots, 4$. For the 3D case, the formulas can be written in a similar way:

$$f_j^{3D,+}(\mathbf{r} \mid \mathbf{r}_0) = K_{m,n,j}^{+,\infty} \frac{1}{\sqrt{(x-x_0)^2 + Z_j^{+\,2} + (y-y_0)^2}}$$

$$+ \sum_{i=1}^{N_{m,n,j}^{+}} C_{m,n,j}^{+,i} \frac{1}{\sqrt{(x-x_0)^2 + (Z_j^{+} + a_{m,n,j}^{+,i})^2 + (y-y_0)^2}}, \tag{9.2.14}$$

$$f_j^{3D,-}(\mathbf{r} \mid \mathbf{r}_0) = K_{m,n,j}^{-,\infty} \frac{1}{\sqrt{(x-x_0)^2 + Z_j^{-\,2} + (y-y_0)^2}}$$

$$+ \sum_{i=1}^{N_{m,n,j}^{-}} C_{m,n,j}^{-,i} \frac{1}{\sqrt{(x-x_0)^2 + (Z_j^{-} + a_{m,n,j}^{-,i})^2 + (y-y_0)^2}}, \tag{9.2.15}$$

where $j = 1, \ldots, 4$ and

$$\begin{aligned}
Z_1^{+} &= z + z_0 - 2d_n, \\
Z_2^{+} &= z - z_0 + 2(d_{m-1} - d_n) \\
Z_3^{+} &= -z + z_0 \\
Z_4^{+} &= -z - z_0 + 2d_{m-1}, \\
Z_1^{-} &= z + z_0 - 2d_m \\
Z_2^{-} &= z - z_0 \\
Z_3^{-} &= -z + z_0 + 2(d_{n-1} - d_m) \\
Z_4^{-} &= -z - z_0 + 2d_{n-2}.
\end{aligned}$$

9.2.2 Prony's Method

The coefficients $C^{\pm,i}_{m,n,j}$ and $a^{\pm,i}_{m,n,j}$ in (9.2.12) to (9.2.15) may be computed by Prony's method [34]. For ease of reference, we briefly present the major steps of the Prony method below.

To determine an approximation of the form

$$f(x) \simeq C_1 e^{a_1 x} + C_2 e^{a_2 x} + \cdots + C_n e^{a_n x},$$

we assume that values of $f(x)$ are specified on a set of N equally spaced points. By using a linear change of variables, the data points become $x' = 0, 1, 2, \ldots, N-1$.

Say that the interval between x is Δ, and we have $x_k = k\,\Delta$, and $x'_k = (x_k/\Delta) - 1$. Now we have

$$\begin{aligned} f(x) &= C_1 e^{a_1 x} + C_2 e^{a_2 x} + \cdots + C_n e^{a_n x} \\ &= f(x') \\ &= C_1 e^{a_1 (x'+1)\,\Delta} + C_2 e^{a_2 (x'+1)\,\Delta} + \cdots + C_n e^{a_n (x'+1)\,\Delta} \\ &= e^{a_1 x'\,\Delta}(C_1 e^{a_1\,\Delta}) + e^{a_2 x'\,\Delta}(C_2 e^{a_2\,\Delta}) + \cdots + e^{a_n x'\,\Delta}(C_n e^{a_n\,\Delta}) \\ &= C'_1 e^{a'_1 x'} + C'_2 e^{a'_2 x'} + \cdots + C'_n e^{a'_n x'}, \end{aligned} \tag{9.2.16}$$

where

$$\begin{cases} C'_n = C_n e^{a_n\,\Delta} \\ a'_n = a_n\,\Delta. \end{cases}$$

Letting $\mu_n = e^{a'_n}$, we may rewrite (9.2.16) as

$$f(x') = C'_1 \mu_1^{x'} + C'_2 \mu_x^{x'} + \cdots + C'_n \mu_n^{x'}.$$

For $x' = 0, 1, \ldots, N-1$, the following equations are satisfied:

$$\begin{cases} C'_1 + C'_2 + \cdots \qquad + C'_n = f_0 \\ C'_1\mu_1 + C'_2\mu_2 + \cdots \qquad + C'_n\mu_n = f_1 \\ C'_1\mu_1^2 + C'_2\mu_2^2 + \cdots \qquad + C'_n\mu_n^2 = f_2 \\ \quad\vdots \\ C'_1\mu_1^{N-1} + C'_2\mu_2^{N-1} + \cdots + C'_n\mu_n^{N-1} = f_{N-1}. \end{cases} \tag{9.2.17}$$

When μ's are unknown, at least $2n$ equations are needed. Let $\mu_1,\ \mu_2, \ldots, \mu_n$ be the roots of the algebraic equation

$$\mu^n + \alpha_1 \mu^{n-1} + \alpha_2 \mu^{n-2} + \cdots + \alpha_{n-1}\mu + \alpha_n = 0. \tag{9.2.18}$$

In order to determine the coefficients $\alpha_1, \alpha_2, \ldots, \alpha_n$, let us take the first $(n+1)$ equations from (9.2.17). We multiply the first equation in (9.2.17) by α_n, the second by α_{n-1} and the nth equation by α_1, the $(n+1)$th equation by 1, and add up the results

$$\begin{cases} C_1'\alpha_n + C_2'\alpha_n + \cdots & + C_n'\alpha_n = f_0\alpha_n \\ C_1'\mu_1\alpha_{n-1} + C_2'\mu_2\alpha_{n-1} + \cdots & + C_n'\mu_n\alpha_{n-1} = f_1\alpha_{n-1} \\ \vdots & \\ C_1'\mu_1^{n-1}\alpha_1 + C_2'\mu_2^{n-1}\alpha_1 + \cdots & + C_n'\mu_n^{n-1}\alpha_1 = f_{n-1}\alpha_1 \\ C_1'\mu_1^n + C_2'\mu_2^n + \cdots & + C_n'\mu_n^n = f_n. \end{cases}$$

Hence

$$\begin{aligned} \text{LHS} &= C_1'(\alpha_n + \mu_1\alpha_{n-1} + \cdots + \mu_1^{n-1}\alpha_1 + \mu_1^n) \\ &\quad + C_2'(\alpha_n + \mu_2\alpha_{n-1} + \cdots + \mu_2^{n-1}\alpha_1 + \mu_2^n) \\ &\quad \vdots \\ &\quad + C_2'(\alpha_n + \mu_n\alpha_{n-1} + \cdots + \mu_n^{n-1}\alpha_1 + \mu_n^n) \\ &= 0 \\ \text{RHS} &= f_0\alpha_n + f_1\alpha_{n-1} + \cdots + f_{n-1}\alpha_1 + f_n. \end{aligned}$$

In a similar way a set of $N - n - 1$ additional equations are obtained

$$\begin{cases} f_n + f_{n-1}\alpha_1 + \cdots & + f_0\alpha_n = 0 \\ f_{n+1} + f_n\alpha_1 + \cdots & + f_1\alpha_n = 0 \\ \vdots & \\ f_{N-1} + f_{N-2}\alpha_1 + \cdots & + f_{N-n-1}\alpha_n = 0. \end{cases} \tag{9.2.19}$$

For $N = 2n$, the following procedures are used:

(1) For given $f_0, f_1, \ldots, f_{n-1}$, solve (9.2.19) for $\alpha_1, \alpha_2, \ldots, \alpha_n$.
(2) Using $\alpha_1, \alpha_2, \ldots, \alpha_n$, find roots of (9.2.18) to obtain $\mu_1, \mu_2, \ldots, \mu_n$.
(3) Using $\mu_1, \mu_2, \ldots, \mu_n$, solve (9.2.17) to find $C_1', C_2', \ldots, C_n'$.

Upon approximation of the coefficients $C_{m,n,j}^{\pm i}$ and $a_{m,n,j}^{\pm i}$ by using Prony's method, the Green's functions are expressed in an explicit formula with complex numbers.

9.2.3 Implementation of the Coifman Wavelets

The Coifman scalets are employed to solve the integral equation for the charge density. First we map the circumferences of the conductor contours onto the interval

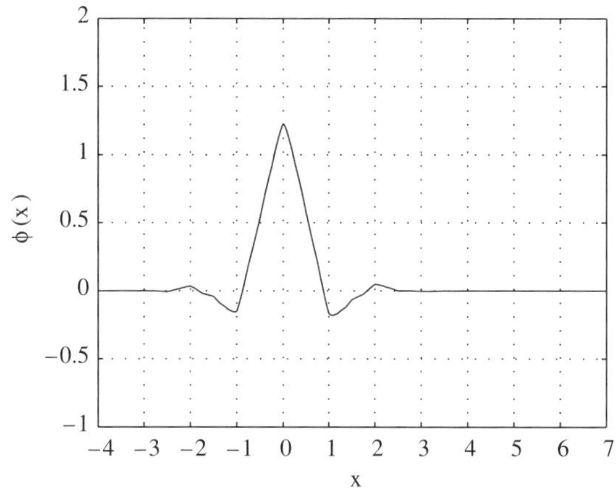

FIGURE 9.3 Coifman scalet of order $L = 4$.

[0, 1]. We then choose the scalets at a certain level ℓ and put them on the interval as a basis. In doing so, we convert the contour of each conductor with finite cross section into a finite 1D interval. Thus we have mapped a 2D problem into a 1D problem via a versatile and accurate treatment of curved conductor surfaces with arbitrary cross sections. As a result the global coordinates $\boldsymbol{\rho}$ are expressed in terms of the local coordinate ξ. The unknown charge density $\sigma(\boldsymbol{\rho}')$ has been expressed as $\sigma(\xi')$, which is then expanded in terms of Coifman scalet $\varphi_{l,m}(\xi)$ as shown in Fig. 9.3. Using the expansion

$$\sigma(\xi') = \sum \alpha_{l,m}\varphi_{l,m}(\xi'),$$

we write the integral equation (9.2.1) as

$$1 = \int_{\Omega} G(\xi, \xi') \sum \alpha_m \varphi_{l,m}(\xi)' \, d\xi'.$$

Applying Galerkin's testing procedure, we obtain

$$\int_{\triangle S_n} \varphi_{l,n}(\xi)\, d\xi = \sum \alpha_m \int_{\triangle S_n} \int_{\triangle S_m} G(\xi, \xi')\varphi_{l,m}(\xi')\varphi_{l,n}(\xi)\, d\xi'\, d\xi.$$

In matrix form we arrive at

$$[Z_{m,n}][\alpha_m] = [g_n],$$

where

$$Z_{m,n} = \int_{\triangle Sn} \int_{\triangle Sm} G(\xi, \xi')\varphi_{l,m}(\xi')\varphi_{l,n}(\xi)\, d\xi'\, d\xi,$$

$$g_n = \int_{\triangle Sn} \varphi_{l,n}(\xi)\, d\xi.$$

For the matrix entries, the diagonal elements are calculated using standard Gaussian quadrature, and the off-diagonal elements are calculated using the one-point quadrature technique

$$\int 2^{l/2}\varphi(2^l\xi - n) f(\xi)\, d\xi = f\left(\frac{n}{2^l}\right) 2^{-l/2}.$$

For the twofold integration the one-point quadrature is

$$\iint 2^{l_1/2}\varphi(2^{l_1}\xi_1 - n_1)2^{l_2/2}\varphi(2^{l_2}\xi_2 - n_2) f(\xi_1, \xi_2)\, d\xi_1\, d\xi_2 = 2^{-(l_1+l_2)/2} f\left(\frac{n_1}{2^{l_1}}, \frac{n_2}{2^{l_2}}\right).$$

From the shape of Coifman scalet, it can be seen in Fig. 9.3 that most of the nonzero values are between -3 and 3. Therefore we truncate the original Coifman scalet support from $[-7, 4]$ into $[-3, 3]$ and perform the integration with fewer intervals while maintaining almost the same precision. Actually, after shifting only one step, the correlation between adjacent scalets becomes very weak. The technique above is used to compute any off-diagonal element with a high accuracy.

After solving the coefficients $[\alpha_m]$, the charge density is obtained, and the entries of the capacitance matrix are calculated using

$$C_{mn} = \frac{\sigma_m}{V_n}.$$

9.2.4 Numerical Examples

The following five examples were executed on a DEC-Alpha workstation 600-5/333 MHz.

Example 1 Illustrated in Fig. 9.4 is the configuration for the 3D layered Green's function, where $h_1 = h_2 = 1$ mm, $\epsilon_1 = 9.8\epsilon_0$, and $\epsilon_2 = 2.55\epsilon_0$. The source is located at $z_0 = 3$ mm, and field point at $z = 0.5$ mm. We denote $\rho = \sqrt{(x - x_0)^2 + (y - y_0)^2}$. Table 9.6 shows the normalized (with ϵ_0) potential values from the two different algorithms, and they agree with at least four digits.

Example 2 We compared our Green's function with [35], again for the configuration of Fig. 9.4, with $z_0 = z = h_2$, $h_1 = h_2 = 1$ mm, and

$$\rho = \sqrt{(x - x_0)^2 + (y - y_0)^2}.$$

The results are listed in Table 9.7, where the high precision can be seen clearly.

Example 3 To demonstrate the capability in handling curved contour of conductors, we choose the geometry in Fig. 9.5 which is from [1] and [26]. The capacitance

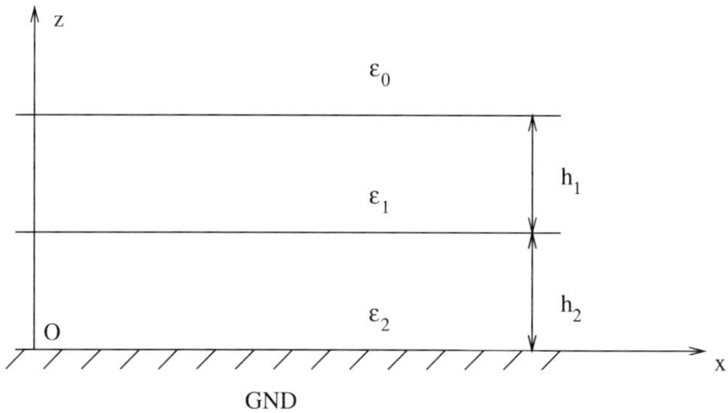

FIGURE 9.4 Point source in a substrate structure.

TABLE 9.6. Comparison of Normalized Potential Values between Two Algorithms

ρ (mm)	Our Results	UIUC [36]
0.1	44.324286	44.3243
0.6	42.079252	42.0793
1.1	37.418786	37.4188
1.6	31.706530	31.7066
2.1	26.054120	26.0542
3.1	16.866728	16.8637

TABLE 9.7. Normalized Potential Values for Fig. 9.4 with $z_0 = z = h_2$

	$\epsilon_{r1} = 9.80, h_1 = 1.0$ mm $\epsilon_{r2} = 2.55, h_2 = 1.0$ mm			$\epsilon_{r1} = 2.55, h_1 = 1.0$ mm $\epsilon_{r2} = 9.80, h_2 = 1.0$ mm		
ρ (mm)	This Method	Numerical Integration	Complex Image	This Method	Numerical Integration	Complex Image
0.1	1622.125705	1623.00	1622.12	1522.022279	1521.93	1522.00
0.6	270.200845	270.69	270.20	176.992714	176.94	176.99
1.1	142.910840	142.92	142.91	63.226447	63.23	63.22
1.6	91.897578	91.68	91.90	27.530837	27.55	27.53
2.1	63.520950	63.30	63.52	13.153255	13.17	13.15
3.1	33.313055	33.22	33.31	3.583560	3.58	3.58

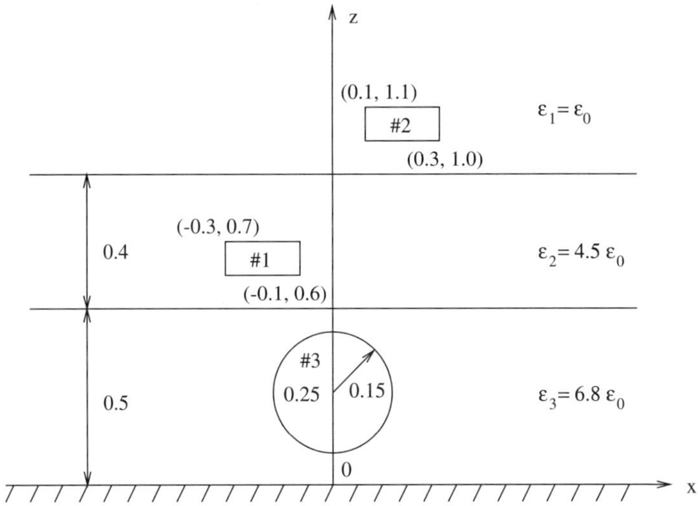

FIGURE 9.5 Three conductors in three different dielectric layers (in mm).

(more precisely, Maxwell static induction) matrix is shown in Table 9.8, which agrees well with that in the references. The CPU time is 1.6 seconds.

Example 4 A three-conductor microstrip and stripline system is demonstrated in Fig. 9.6, which has been reported in [26] and [32]. The resultant capacitance matrix is listed in Table 9.9. The CPU time is 1.5 seconds.

Example 5 Shown in Fig. 9.2 is an industry standard from our 1992 paper [3] to test the speed and accuracy of the CAD tools. As reported in [32] that the boundary element method requires 458.67 seconds on an IBM RS-6000 workstation. The new method in this section recorded the CPU time of 15.2 seconds to obtain the results shown in Table 9.10. The huge disparity in the vertical direction (e.g., 600 μm versus 1.5 μm) could lead to either singular matrices or nonphysical solutions. It is important to check if the sum of any row or column of the capacitance matrix is positive. This sum represents the capacitance value of the corresponding conductor while other conductors are grounded. If the sum is negative, then the stored electric energy

$$W = \tfrac{1}{2}CV^2$$

TABLE 9.8. Capacitance Matrix (pF/m) for Example 3

127.51	−13.13	−72.32
−13.13	34.23	−7.43
−72.32	−7.43	378.58

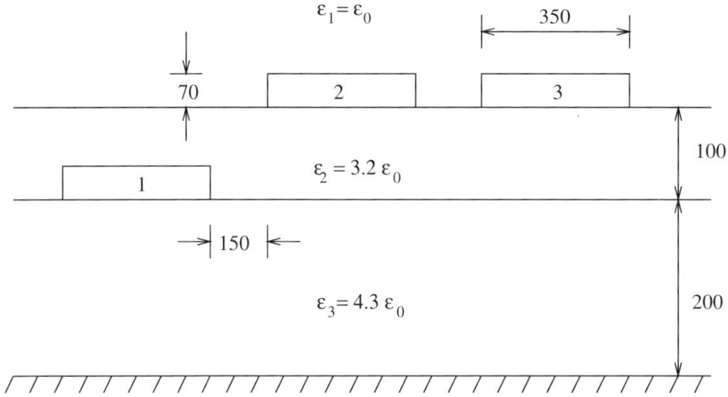

FIGURE 9.6 Three conductors in a layered medium (in μm).

TABLE 9.9. Capacitance Matrix (pF/m) for Example 4

142.66	−22.13	−0.93
−22.13	94.14	−18.39
−0.93	−18.39	88.37

would be negative, a violation of physics principle. It can be verified that values in Table 9.10 are not nonphysical.

9.3 SKIN-EFFECT RESISTANCE AND TOTAL INDUCTANCE

Multiconductor transmission lines (MTL) have been modeled by the distributed parameters R, L, C, and G in many commercial computer-aided design (CAD) packages. In this section we present a fast technique based on the integral equation method (IEM) for evaluating frequency dependences accurately while dramatically reducing the computation time by using wavelets.

As the amount of time required for processor cycles continually decreases and the density of integrated circuits within devices steadily increases, it becomes even more critical to have accurate frequency dependence estimates for MTL parameters. Many commercial CAD packages treat MTLs in a quasi-static way; that is, the distributed L, C, and G are assumed to be independent of frequency values while the resistance is assumed to be $\propto \sqrt{f}$. While the quasi-static models produce fairly accurate results for low frequencies and relatively large conductor cross sections, they become inaccurate as signal rise times become shorter and system clock rates increase, particularly for conductor dimensions in the micron range. As will be seen in this section, at 200 MHz the discrepancy between measured and quasi-statically computed values can exceed 50% for resistance and underestimate 30% for inductance,

TABLE 9.10. Capacitance Matrix (pF/m) for the 10-Conductor System

311.278	−41.6298	−11.5414	−6.50477	−5.70368	−222.949	−5.04494	−1.40393	−0.836176	−0.769878
−41.6298	323.965	−28.3225	−7.9522	−5.21854	−5.0205	−221.017	−3.56294	−0.971401	−0.661596
−11.5413	−28.3224	314.726	−24.5635	−8.83184	−1.35811	−3.48671	−222.487	−3.19803	−1.14693
−6.50484	−7.95221	−24.5638	307.621	−25.115	−0.799865	−0.933902	−3.1563	−223.559	−3.37668
−5.70372	−5.21858	−8.83204	−25.115	294.361	−0.742033	−0.639472	−1.13162	−3.34735	−225.512
−222.951	−5.02031	−1.35818	−0.80014	−0.742245	235.851	2.1539	−0.453029	−0.242798	−0.199872
−5.04527	−221.019	−3.48634	−0.933997	−0.639559	−2.1539	236.16	−1.2557	−0.308887	−0.191921
−1.40391	−3.56338	−222.489	−3.15623	−1.13174	−0.453028	−1.25571	235.88	−0.940147	−0.303317
−0.836231	−0.971658	−3.19824	−223.561	−3.34727	−0.242793	−0.308848	−0.940151	235.704	−0.833364
−0.770079	−0.661905	−1.14703	−3.37702	−225.514	−0.199812	−0.191968	−0.303316	−0.833386	235.409

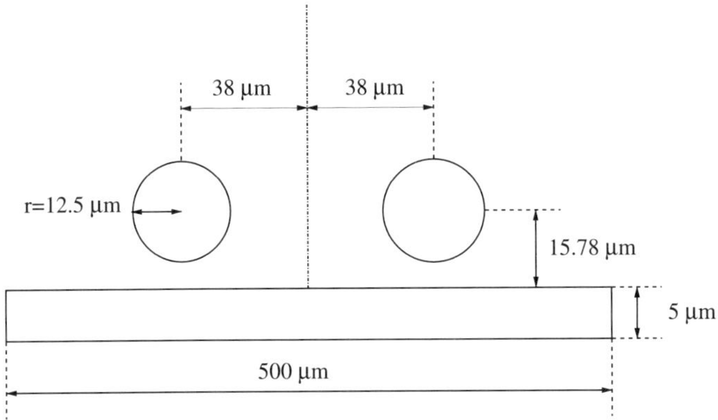

FIGURE 9.7 Cross section of the test structure.

respectively. For the test structure of Fig. 9.7, the error exceeds 300% for inductance values. These errors are due to oversimplified treatments of the skin-effect resistance and internal reactance of these conductors. Full-wave formulations may provide accurate solutions [26, 38, 37], but the computational cost of such methods increases dramatically. Here we have adopted an intermediate approach between a quasi-static and a full-wave solution that is based upon the integral equation formulation (IEF). This approach provides satisfactory precision at speeds up to 10 GHz for typical printed circuit board (PCB) and multichip module (MCM) configurations while retaining a level of computational complexity comparable to the quasi-static method. Similar work utilizing electromagnetically trained artificial neural networks for the CAD and CAE was reported [18, 39].

9.3.1 Formulation

To construct this technique, we adopted the surface integral equation proposed by Wu [40] and applied by Kong [8]. In this approach, inside the conductors the fields are modeled by the diffusion equation, meaning the full-wave formulation [13]

$$(\nabla^2 - j\omega\mu\sigma)J_z = 0. \tag{9.3.1}$$

Outside the conductors the fields are described by the quasi-static approximation

$$\nabla^2 A_z = 0.$$

The two unknown quantities J_z and A_z are related, forming coupled integral equations in terms of J_z. Here, for ease of reference, we present the major steps of the integral equation formulation. From Maxwell's equation

$$\hat{n} \times (\nabla \times \mathbf{E}) = -j\omega\mu(\hat{n} \times \mathbf{H}) = -j\omega\hat{n} \times (\nabla \times \mathbf{A}).$$

Substituting $\mathbf{J} = \sigma\mathbf{E}$ into the previous equation, we have

$$\hat{n} \times (\nabla \times \mathbf{J}) = -j\omega\sigma\hat{n} \times (\nabla \times \mathbf{A}).$$

Note that by definition,

$$\nabla \times \mathbf{J} = \hat{l}\frac{\partial J_z}{\partial n} - \hat{n}\frac{\partial J_z}{\partial l}.$$

Thus

$$\hat{n} \times \left(\hat{l}\frac{\partial J_z}{\partial n} - \hat{n}\frac{\partial J_z}{\partial l}\right) = -j\omega\sigma\hat{n} \times \left(\hat{l}\frac{\partial A_z}{\partial n} - \hat{n}\frac{\partial A_z}{\partial l}\right).$$

Hence we obtain

$$\frac{\partial J_z}{\partial n} = -j\omega\sigma\frac{\partial A_z}{\partial n}. \tag{9.3.2}$$

In a similar way, we arrive at

$$\frac{\partial J_z}{\partial l} = -j\omega\sigma\frac{\partial A_z}{\partial l}. \tag{9.3.3}$$

If the derivatives of two quantities along a line are equal, then those quantities must be equal to within a constant, namely

$$J_z = -j\omega\sigma[A_z - A_q],$$

where A_q is a constant depending on conductor q.

Upon numerical solution of J_z and $\partial J_z/\partial n$, we could represent the field quantities in the volume in terms of the surface values and their normal derivatives. Green's first identity is

$$\iint ds(\varphi\nabla^2\psi + \nabla\varphi\cdot\nabla\psi) = \oint dl\varphi\frac{\partial\psi}{\partial n}, \tag{9.3.4}$$

where the left-hand side of the equation is a 2D integral over a cross-sectional area and the right-hand side is a 1D integral along the closed contour bounding that area. As usual, the normals are defined as pointing outward from the region of interest. From (9.3.1) the total current flowing in a wire is

$$I = \iint dSJ_z = \frac{j}{\omega\mu\sigma}\iint dS(\nabla^2 J_z).$$

Using Green's first identity as described by (9.3.4) with $\psi = J_z$ and $\varphi = 1$, we get

$$I = \frac{j}{\omega\mu\sigma}\oint\frac{\partial J_z}{\partial n}dl, \tag{9.3.5}$$

which is an expression for the total current flowing in a wire in the $\hat{z}$ direction in terms of surface quantities. Integral equations for both A_z and J_z can now be obtained. For the outer equation

$$\oint_{\text{all wires}} dl' G_0(l,l') \frac{\partial A_z(l')}{\partial n'} = \oint_{\text{all wires}} dl' A_z(l') \left[\frac{\partial G_0(l,l')}{\partial n'} + \frac{1}{2}\,\delta(l-l') \right],$$

where Green's function outside the conductor is

$$G_0(\boldsymbol{\rho}, \boldsymbol{\rho}') = -\frac{1}{2\pi} \ln \left[\sqrt{(x-x')^2 + (y-y')^2} \right].$$

The range of integration is over the surface of every wire. Similarly, for the interior region of each wire, we have

$$\oint_{\text{wire } q} dl' G_d \frac{\partial J_z(l')}{\partial n'} = \oint_{\text{wire } q} dl' J_z(l') \left[\frac{\partial G_d(l,l')}{\partial n'} + \frac{1}{2}\,\delta(l-l') \right],$$

where

$$\begin{aligned} G_d(\boldsymbol{\rho}, \boldsymbol{\rho}') &= \frac{-j}{4} H_0^{(2)} \left(e^{-j(\pi/4)} \sqrt{\omega\mu\sigma} \sqrt{(x-x')^2 + (y-y')^2} \right) \\ &= \frac{1}{2\pi} \left[ker \sqrt{\omega\mu\sigma} \sqrt{(x-x')^2 + (y-y')^2} \right. \\ &\qquad \left. + \, j kei \sqrt{\omega\mu\sigma} \sqrt{(x-x')^2 + (y-y')^2} \right]. \end{aligned}$$

Using the boundary conditions (9.3.2) and (9.3.3) to eliminate A_z from the integral equation for the outside fields and adding the condition imposed on the total currents in (9.3.5), we arrive at a set of integral equations

$$\begin{aligned} &\oint_{\text{all wires}} dl' G_0(l,l') \frac{\partial J_z(l')}{\partial n'} = \oint_{\text{all wires}} dl' \left[J_z(l') - j\omega\sigma A_q \right] \\ &\times \left[\frac{\partial G_0(l,l')}{\partial n'} - \frac{1}{2}\,\delta(l-l') \right], \\ &\oint_{\text{wire } q} dl' G_d(l,l') \frac{\partial J_z(l')}{\partial n'} = \oint_{\text{wire } q} dl' J_z(l') \left[\frac{\partial G_d(l,l')}{\partial n'} + \frac{1}{2}\,\delta(l-l')) \right], \\ &\oint_{\text{wire } q} dl \frac{\partial J_z}{\partial n} = -\omega\mu\sigma I_q. \end{aligned} \tag{9.3.6}$$

9.3.2 Moment Method Solution of Coupled Integral Equations

The previous set of coupled integral equations, (9.3.6), is solved numerically by employing the method of moments with subdomain basis functions. Expanding the unknown functions J_z and $\partial J_z/\partial n$ in terms of the basis functions with unknown coefficients, we have

$$J_z = \sum_{\ell} j_\ell B_\ell(l),$$

$$\frac{\partial J_z}{\partial n} = \sum_{\ell} k_\ell B_\ell(l),$$

where the bases may be chosen as the Coifman scalets

$$B_m(l) = \varphi_{m,\ell}(l).$$

This discretization gives an approximation to the surface quantities. We implemented Galerkin's method, and have thus discretized the coupled integral equations (9.3.6) into a matrix equation

$$\begin{bmatrix} \bar{\bar{V}}_0 & \bar{\bar{W}}_0 & \bar{\bar{U}}_0 \\ \bar{\bar{S}} & 0 & 0 \\ \bar{\bar{V}}_d & 0 & \bar{\bar{U}}_d \end{bmatrix} \begin{bmatrix} K \\ -j\omega\sigma A_0 \\ J \end{bmatrix} = \begin{bmatrix} 0 \\ j\omega\mu\sigma I \\ 0 \end{bmatrix}.$$

The corresponding impedance matrix is plotted in Fig. 9.8 using the standard form that was presented in Chapter 4. The magnitude of the matrix is normalized accord-

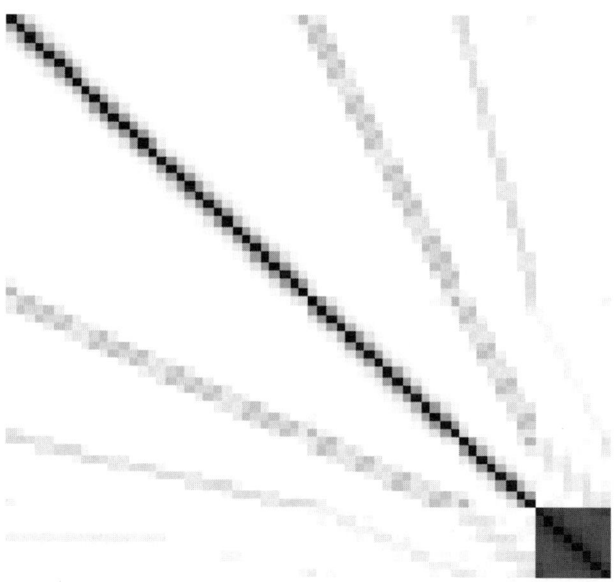

FIGURE 9.8 Impedance matrix of standard form from fast wavelet transform.

ing to the maximum entry and then digitized into 256 gray levels; the dark levels represent higher values. No threshold is applied.

9.3.3 Circuit Parameter Extraction

After the currents and their normal derivatives have been obtained, we can express the resistance and inductance per unit length in terms of the current and its derivatives normal to the surface (see Section 4.9). They are

$$R = \omega\mu \frac{\oint_{\text{all wires}} dl\, \text{Im}\{J_z(\partial J_z^\star/\partial n)\}}{|\oint_{\text{signal wire}} dl(\partial J_z/\partial n)\,|^2} \tag{9.3.7}$$

and

$$L = \omega\mu \frac{\oint_{\text{all wires}} dl \text{Re}\left\{-jA_q(\partial J_z^\star/\partial n)\right\}}{|\oint_{\text{signal wire}} dl(\partial J_z/\partial n)\,|^2}. \tag{9.3.8}$$

The mutual resistance and inductance are calculated from energy considerations and from the self-terms calculated above. If we specify that a current I_x flows on the ith wire and a current $-I_x$ flows on the jth wire, we can calculate the dissipated power, P_d,

$$2P_d = [I_x \;\; -I_x] \begin{bmatrix} R_{ii} & R_{ij} \\ R_{ji} & R_{jj} \end{bmatrix} \begin{bmatrix} I_x \\ -I_x \end{bmatrix}.$$

Hence

$$R_{ij} = \frac{1}{2}\left(R_{ii} + R_{jj} - 2\frac{P_d}{I_x^2}\right). \tag{9.3.9}$$

Similarly

$$L_{ij} = \frac{1}{2}\left(L_{ii} + L_{jj} - 4\frac{W_m}{I_x^2}\right), \tag{9.3.10}$$

where W_m is the stored magnetic energy. The computation of the surface normal derivative of the current gives inaccurate results at extremely low frequencies because the current density over the cross section approaches a constant. In such cases the "filament technique" in [41] may be employed to obtain the results, which are displayed in Figs. 9.13 and 9.14 for comparison. For the configuration of Fig. 9.7 we present the current distributions in the ground plane and in the two circular wires at 1 GHz in Figs. 9.9 and 9.10. As expected, the current distribution in the ground plane exhibits two peaks under the two conductors. As one penetrates inside the conductor, the current decays exponentially, in consistence with the well-known skin-effect phenomenon. As the cross section of the microstrip shrinks to the dimension of the skin depth, δ, the current no longer flows only on the conductor surface. As a result the

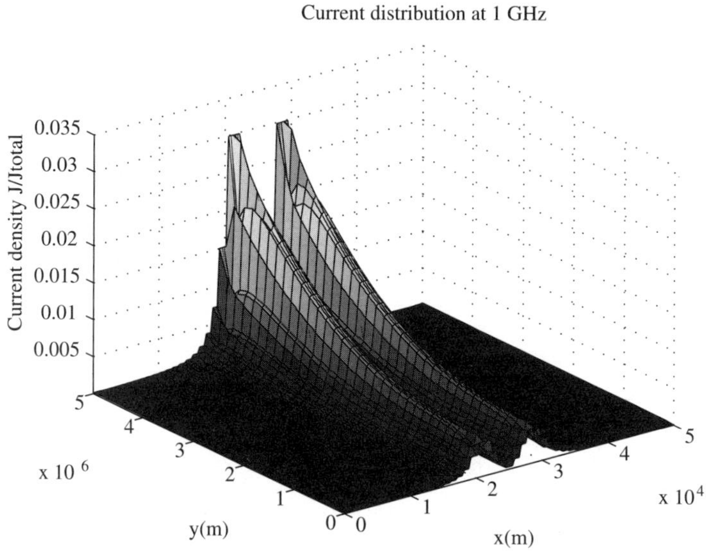

FIGURE 9.9 Current distribution in the ground plane.

inductance value computed from the quasi-static analysis can produce errors exceeding 300%. In a like manner the simple skin-effect resistance formula $R_s = \sqrt{\pi f \mu / \sigma}$ and internal inductance formula $L_i = \sqrt{\mu / \pi f \sigma}/2$ will both be inaccurate. Those formulas were, in fact, derived under the assumption that the metal thickness $t \gg \delta$.

The current distribution throughout the cross section of each conducting wire exhibits a conformity with a higher current density just above the ground plane; this is to be expected from the proximity effect.

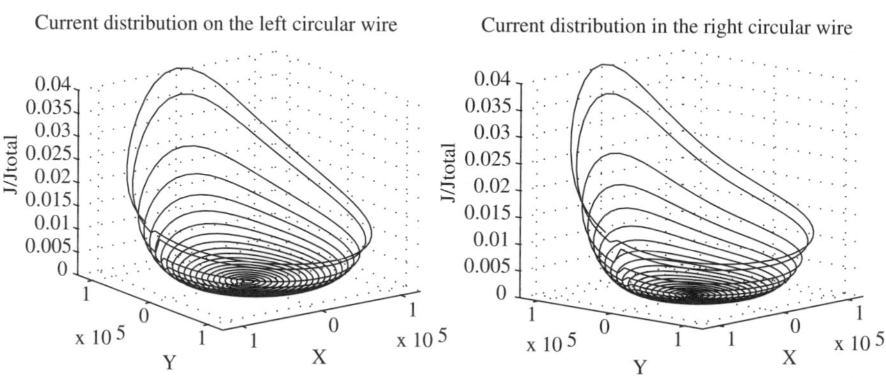

FIGURE 9.10 Current distribution in the left and right circular wire.

9.3.4 Wavelet Implementation

To improve the numerical accuracy and computational speed, Battle–Lemarie wavelets were employed to replace the pulse basis functions [9]. Here we use the Coifman scalets of order $L = 4$ for the Galerkin procedure. The Coiflet has a compact support of $[-4, 7]$. The dilation equation for the scalet can be written as

$$\varphi(x) = \sqrt{2}\sum_{-4}^{7} h_k \varphi(2x - k). \tag{9.3.11}$$

The corresponding low-pass filters can be found in Chapter 3 and used directly for construction of Coifman wavelets. One of the most important and useful properties for Coifman wavelets is its zeromoments

$$\int x^r \varphi(x)\, dx = 0, \qquad r = 1, \ldots, L-1. \tag{9.3.12}$$

By using this property, we approximate the integration by

$$\int f(x)\varphi_{m,n}(x)\, dx = \frac{1}{2^{m/2}} f\left(\frac{n}{2^m}\right). \tag{9.3.13}$$

When computing the impedance matrix, the off-diagonal elements are obtained from the one-point quadrature formula directly, which reduces the computational time dramatically. For the diagonal elements we still employ Gaussian quadrature.

Employing the wavelet-sparsified impedance matrix, we studied a microstrip line problem previously reported in [8]. This test structure is a lossy transmission line system consisting of three identical rectangular conductors of equal height above a ground plane. The dimensions of the conductors are provided in Fig. 9.11, where the ground plane is approximated by a conductor of cross section $1000 \times 300\,\mu$m. The self- and mutual resistances and inductances of the three transmission lines are plotted against frequency in Fig. 9.12. As a result of numerical errors, the standard MoM leads to unstable values of R_{12}, R_{13}, and L_{13}. Therefore no MoM curves of the mutual R or L_{13} were plotted. Moreover, by employing wavelet basis functions, we

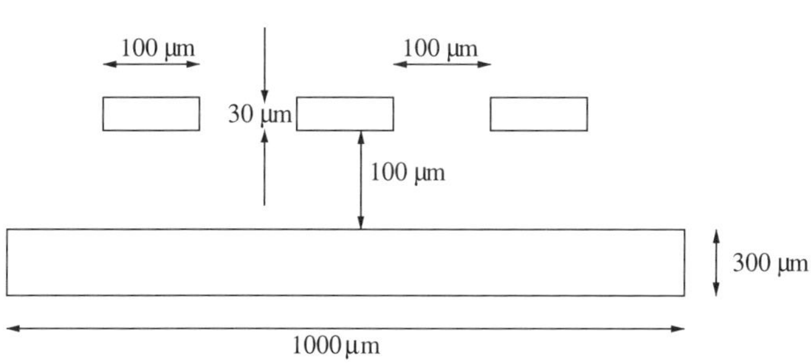

FIGURE 9.11 Three rectangular conductors over a ground plane.

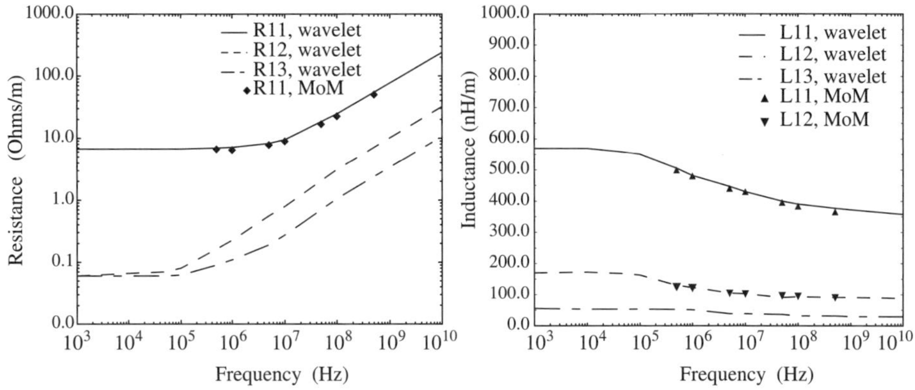

FIGURE 9.12 Self- and mutual resistances and inductances for three rectangular wires over a ground plane.

have extended the frequency range by three entire decades toward the low-frequency end of the spectrum.

9.3.5 Measurement and Simulation Results

Laboratory measurements of coupled wires operating from 100 KHz to 1 GHz were conducted using microwave testing equipment. In the test coupons the ground planes were fabricated as thin as 0.1 μm.

Figures 9.13 and 9.14 present a comparison among the neural network results, the mixed IEM-filament solver (an internal program), a commercial finite element code, and laboratory measurements. The results obtained by measurement or computation were for the geometry of Fig. 9.7. There was excellent agreement among the results of the different methods; this agreement provides strong motivation for

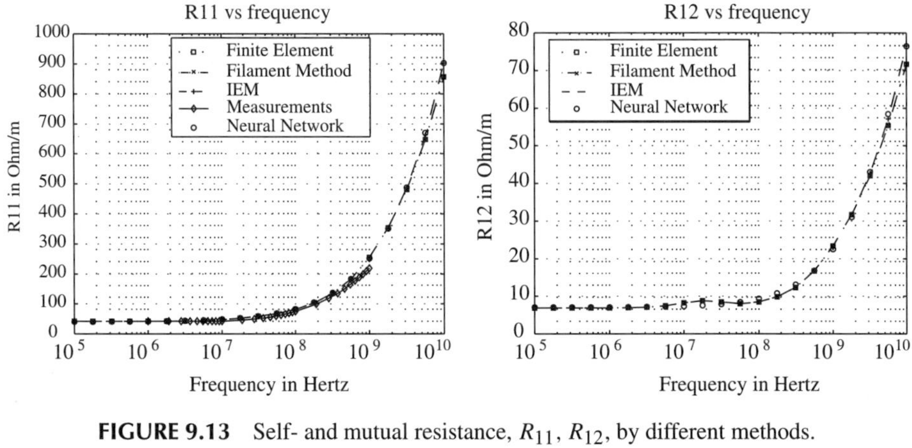

FIGURE 9.13 Self- and mutual resistance, R_{11}, R_{12}, by different methods.

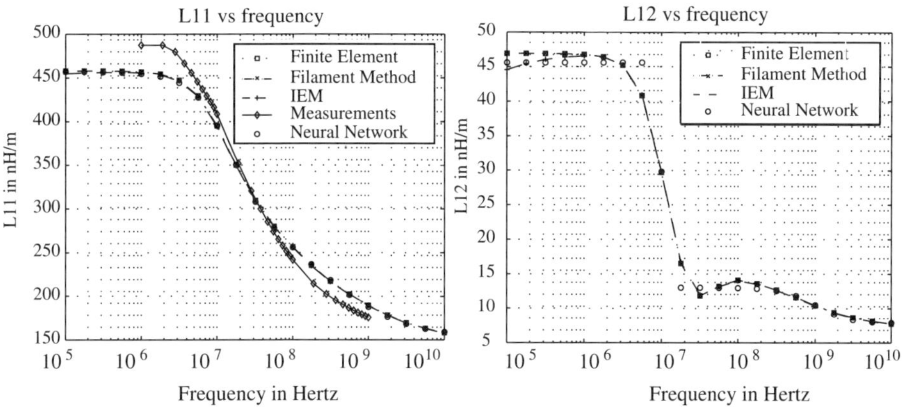

FIGURE 9.14 Self- and mutual inductance, L_{11}, L_{12}, by different methods.

further development of the technique described here. For further verification of our algorithm, a set of laboratory measurements is compared with our numerical solution in Table 9.11. We used an aluminum/polyimide MCM test coupon fabricated by Hughes, Inc.

The test coupon consisted of two groups of individual buried striplines with dimensions $w \times t = 25 \times 5\,\mu$m for the strips farthest from the ground plane and $w \times t = 125 \times 5\,\mu$m for the microstrips closest to the ground plane. The conduc-

TABLE 9.11. Comparison of Measurements against Computations

Line Parameters	Group 1	Group 2
H (μm)	10	20
Inductance L (nH/m)		
Hughes measured	8.4	10.5
Quasi-static	6.0	8.7
IEM	8.0	10.57
Resistance R (Ohms/m)		
Hughes measured	9.34	8.14
Quasi-static	14.07	12.37
IEM	8.85	8.53
Impedance Z_c (Ohms/m)		
Hughes measured		
LC meter	48.1	64.8
TDR	50.3	64.7
Quasi-static (real part)	39.0	59.0
IEM (real part)	45.0	65.03

tivity of the aluminum lines was $\sigma = 3.0 \times 10^7$s/m. The height of the stripline to the ground plane H was 10 μm for group 1 and 20 μm for group 2. The operating frequency was $f = 200$ MHz. The traditional quasi-static model [42] neglects the internal inductance and assumes that current is only flowing on the surface region of the cross section. This model may well underestimate the inductance value by as much as 30% and overestimate the resistance value by as much as 50%.

9.4 SPECTRAL DOMAIN GREEN'S FUNCTION-BASED FULL-WAVE ANALYSIS

So far most previous electromagnetic modeling work on high speed digital electronics has been based on quasi-static assumptions [1–4, 26–28, 42]. As a result dispersion and losses due to radiation and surface waves generated by discontinuities are not properly addressed. Advanced modeling of microstrip structures by the spectral domain method [45, 46] and by finite difference time domain (FDTD) [47, 48] have been reported. The surface integral equation method (SIE) with Green's functions serving as the integral kernels has been employed to study microstrip structures [11, 50]. To investigate microstrip discontinuities (open-end, gap, step change, T junction, etc.), a number of subsectional modes, referred to as the piecewise sinusoidal (PWS) basis functions, are used in the vicinity of the discontinuities to model the nonuniformity of the current in those regions [10, 11]. The SIE approach uses the exact Green function, taking into account the space wave and surface wave. Therefore it is an effective full-wave analysis for microstrip structures [50].

In this section the full-wave analysis of microstrip floating line structures is implemented by the wavelet expansion method, where a system of linear algebraic equations is obtained from the integral equation. The subsectional bases (a number of piecewise sinusoidal modes) employed in [10, 11] are replaced by a set of orthogonal wavelets. In the numerical example we demonstrate that while the PWS basis yields a full matrix, the wavelet expansion results in a nearly diagonal or block-diagonal matrix; both approaches effect very similar answers. However, as the geometry of the problem becomes more complicated, and consequently the resulting matrix size greatly increases, the advantage of having a sparse matrix over a full matrix will prove to be more profound.

9.4.1 Basic Formulation

Figure 9.15 shows the configuration of a buried microstrip floating line isolated by two gaps from a uniform transmission line, where the substrate is assumed to extend to infinity in the transverse directions and is made of a nonmagnetic, homogeneous, isotropic material of thickness d and relative permittivity ϵ_r. Both the bottom ground plane and conductor strip are considered to be infinitesimally thin perfect electric conductors in the following discussions. Furthermore, for simplicity, only $\hat{x}$-directed electric surface currents are assumed to flow on the lines; this has been found in

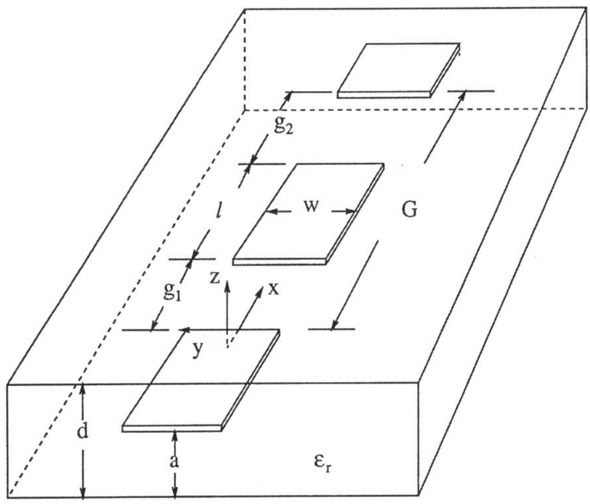

FIGURE 9.15 Configuration of embedded floating line.

many previous works [10, 45], etc. to be a good approximation as long as lines are narrow with respect to the wavelength of interest.

Green's Function for a Grounded Dielectric Slab and the Integral Equation

The dyadic Green's function for a grounded dielectric slab and the formulation of microstrip discontinuity were derived using magnetic vector potential **A** [10, 50] or using the normal components of **E** and **H** [23, 51]. Here we quote only the relevant equations for a dielectric slab backed by a perfectly electrically conducting (PEC) ground plane [12, 14]. The dyadic Green function for the grounded dielectric slab at $z = z' = a$ is

$$G_{\alpha\beta}(x \mid x'; y \mid y') = \iint Q_{\alpha\beta}\, dk_x\, dk_y \cdot e^{jk_x(x-x')}e^{jk_y(y-y')}, \tag{9.4.1}$$

where $\alpha, \beta = x, y, z$, and

$$Q_{xx} = -j\frac{Z_0}{r\pi^2\epsilon_r k_0} \cdot \left\{ \frac{(\epsilon_r k_0^2 - k_x^2)[k_2\cos(k_2(d-a))] + jk_1 \sin k_2(d-a)}{T_e} \right.$$

$$\left. \frac{\sin k_2 a}{k_2} + j\frac{(1-\epsilon_r)k_2 k_x^2 \sin^2 k_2 a}{T_e T_m} \right\}$$

$$Q_{yx} = j\frac{Z_0}{4\pi^2 k_0} \cdot \left\{ \frac{k_2 \cos k_2(d-a) + jk_1 \sin k_2(d-a)}{T_e k_2} \right.$$

$$\left. - \frac{j(1-\epsilon_r)k_2 \sin k_2 a}{T_e T_m} \right\} \cdot k_x k_y \sin(k_2 a),$$

$$Q_{xy} = Q_{yx}$$
$$Q_{yy} = Q_{xx} \mid k_x \leftrightarrow k_y,$$

that is, interchanging k_x with k_y in Q_{xx} we arrive at Q_{yy}. Another component Q_{zx} is obtained in an expression similar to that of Q_{yx}. Since the conductor thickness in the z dimension is much smaller than the width in y, we ignore the Q_{zx}. In the equations above

$$k_0 = \frac{\omega}{c} = \omega\sqrt{\mu_0\epsilon_0},$$
$$Z_0 = \sqrt{\mu_0/\epsilon_0},$$
$$k_1^2 = k_0^2 - k_x^2 - k_y^2, \qquad \mathrm{Im}(k_1) \le 0,$$
$$k_2^2 = \epsilon_r k_0^2 - k_x^2 - k_y^2, \qquad \mathrm{Im}(k_1) \le 0,$$
$$T_e = k_2\cos(k_2 d) + jk_1\sin(k_2 d),$$
$$T_m = \epsilon_r k_1\cos(k_2 d) + jk_2\sin(k_2 d),$$

and the time dependence $e^{j\omega t}$ is assumed and suppressed. When $a = d$, the previous equations are simplified [11].

As discussed in [50, 52], the zeros of T_e and T_m represent the TE and TM surface wave modes, respectively. T_m always has at least one zero in the whole frequency range and thus the first TM surface wave mode has no cutoff frequency [50].

The $\hat{x}$ component of the electric field at $z = a$ can be formulated from the dyadic Green's function as

$$E_x(x, y) = \iint G_{xx}(x, y \mid x', y')I_{sx}(x', y')\,dx'\,dy',$$

where I_{sx} is the longitudinal electric surface current density, which only exists over all metal regions. Since the lines are assumed to be perfect conductors, an integral equation for the surface current density can be obtained by requiring the $\hat{x}$ component of the electric field on the lines to be zero

$$\iint G_{xx}(x, y \mid x', y')I_{sx}(x', y')\,dx'\,dy' = 0 \tag{9.4.2}$$

for $(x, y) \in S$, where S is for all the lines.

Usually $I_{sx}(x, y)$ is written in the form of separated variables:

$$I_{sx}(x, y) = I_1(x) \cdot I_2(y), \tag{9.4.3}$$

where the y-dependent factor $I_2(y)$ can be assumed to be some known real function. For example, $I_2(y)$ was chosen as a function $1 + |\,2y/w\,|^3$ to model the edge effect of the $\hat{x}$- direction current distribution along the $\hat{y}$-dimension [3, 11]. Substituting the expressions (9.4.1) and (9.4.3) of G_{xx} and I_{sx} into integral equation (9.4.2), multiplying the equation by $I_2(y)$, and integrating the result with respect to y yields an integral equation about $I_1(x)$ as

$$\int P_{xx}(x,x')I_1(x')\,dx' = 0, \tag{9.4.4}$$

where the kernel is

$$P_{xx}(x,x') = \int_{-\infty}^{\infty}\int_{-\infty}^{\infty} Q_{xx}(k_x,k_y) \mid F_y(k_y) \mid^2 e^{jk_x(x-x')}\,dk_x\,dk_y. \tag{9.4.5}$$

The Fourier transform $F_y(k_y)$ of $I_2(y)$ is given by

$$F_y(k_y) = \int I_2(y)e^{-jk_y y}\,dy = \int_{-w/2}^{w/2} I_2(y)e^{-jk_y y}\,dy.$$

Current Distributions The total region under consideration consists of three subregions: the incident region, transient region, and transmitting region. In the incident region, the current density is approximated by the sum of incident and reflected waves, since the discontinuities are far away and the effects of discontinuities are negligible. Similarly, in the transmitting region, the current density is expressed by transmitted waves. In the transient region, the current density is nonuniformly distributed along the line under the influence of the discontinuities. Correspondingly the x-dependent factor of the $\hat{x}$-directed electric surface current densities consist of four different terms: the incident, reflected, and transmitted traveling waves $I^{\text{inc}}(x)$, $I^{\text{ref}}(x)$, and $I^{\text{tr}}(x)$ in addition to a term $I^{\text{loc}}(x)$ that is defined in the transient region, the vicinity of the discontinuities, and is used to model the nonuniform current there. Mathematically

$$I_1(x) = \begin{cases} I^{\text{inc}}(x) + I^{\text{ref}}(x), & -\infty < x < -L \\ I^{\text{loc}}(x), & -L \le x \le G+L \\ I^{\text{tr}}(x), & G+L < x < \infty, \end{cases} \tag{9.4.6}$$

where $G = g_1 + l + g_2$; g_1 and g_2 are respectively the width of gap 1 and gap 2; and l is the length of the floating line. L is a large enough real number that the effect of discontinuities is negligible beyond $x < -L$ or $x > G + L$.

Suppose that the incident wave is propagating along $\hat{x}$ direction. We can write the incident electric current as

$$I^{\text{inc}}(x) = e^{-jk_e x},$$

the reflected electric current as

$$I^{\text{ref}}(x) = -Re^{jk_e x},$$

and the transmitted electric current as

$$I^{\text{tr}}(x) = Te^{-jk_e(x-G)},$$

where R and T are the reflection and transmission coefficients, respectively; k_e is the effective propagation constant of the uniform infinite microstrip line, which can easily be evaluated (e.g., see [10, 11]). Moreover I^{loc} can be written as

$$I^{\text{loc}}(x) = I_0(x) + I(x), \tag{9.4.7}$$

where

$$I_0(x) = \begin{cases} \left[(1-R) f_s(k_e x + \frac{\pi}{2}) - j(1+R) f_s(k_e x)\right], & -L \le x \le 0 \\ T\left[f_s(k_e[G-x] + \frac{\pi}{2}) + j f_s(k_e[G-x])\right], & G \le x \le G+L \\ 0, & \text{elsewhere,} \end{cases}$$

$$f_s(u) = \begin{cases} \sin u, & u < 0 \\ 0, & \text{otherwise.} \end{cases} \tag{9.4.8}$$

Since the continuity condition must be satisfied by electric surface currents at the interfaces of the uniform current regions and transient region and the electric surface currents must be zero outside the lines, $I(x)$ is required to meet the homogeneous conditions

$$I(-L) = I(G+L) = I(x)|_{0 \le x \le g_1} = I(x)|_{g_1 + l \le x \le G} = 0. \tag{9.4.9}$$

By solving integral equation (9.4.4) with condition (9.4.9), we can obtain reflection coefficient R, transmission coefficient T, and the surface current distributions $I(x)$, $I^{\text{loc}}(x)$, and $I_1(x)$. In the next subsection the wavelet bases satisfying (9.4.9) will be introduced to expand the current density $I(x)$ in the transient region.

9.4.2 Wavelet Expansion and Matrix Equation

Now integral equation (9.4.4) is converted into a matrix equation by using the wavelet expansion technique.

Wavelet Expansions of an Integral Kernel Based on the wavelet theory in Chapter 4, the projection $A_m I(x)$ of the unknown function $I(x)$ on the subspace V_m of the real line R provides an approximation at resolution 2^m and the function $I(x)$ can be approximated as closely as desired by its projection $A_m I(x)$ as m increases. Let 2^{m_h} be the resolution at which the projection $A_{m_h} I(x)$ gives a sufficiently accurate approximation to $I(x)$. In the subspace V_{m_h} of R, a unique expansion (approximation) can be obtained as

$$I(x) \cong A_{m_h} I(x) = \sum_n \bar{I}_{m_h,n} \varphi_{m_h,n}(x),$$

where $\varphi_{m_h,n}(x)$ are the scalets in V_{m_h}. Since $I(x)$ is only defined for the conductors in the transient region, that is, the intervals $[-L, 0]$, $[g_1, g_1 + l]$ and $[G, G + L]$, the scalets beyond these three intervals should be deleted at the boundaries. However, this deletion may lead to a solution for which it is difficult to satisfy the condition (9.4.9). By using compactly supported wavelets [30], we can easily delete the scalets that are beyond the regions of interest. As a consequence condition (9.4.9) will be automatically satisfied.

To exercise the cancellation property of a wavelet basis, the preceding expansion about the scalet is further converted to a wavelet expansion through a multiresolution analysis

$$I(x) \cong A_{m_h} I(x) \tag{9.4.10}$$

$$= \sum_{m=m_l}^{m_h-1} \sum_{n} \tilde{I}_{m,n} \psi_{m,n}(x) + \sum_{n} \bar{I}_{m_l,n} \varphi_{m_l,n}(x), \tag{9.4.11}$$

where $\psi_{m,n}(x)$ is the wavelet function in W_m and $m_l \le m_h - 1$.

Next we expand the kernel in integral equation (9.4.4) as a two-variable function in the two-dimensional wavelet series

$$\begin{aligned} P_{xx}(x, x') = \sum_{m=m_l} \sum_{n,k} \big[&\alpha^m_{n,k} \psi_{m,k}(x') \psi_{m,n}(x) + \beta^m_{n,k} \psi_{m,k}(x') \varphi_{m,n}(x) \\ &+ \gamma^m_{n,k} \varphi_{m,k}(x') \psi_{m,n}(x) \big] \\ &+ \sum_{n,k} s^{m_l}_{n,k} \varphi_{m_l,k}(x') \varphi_{m_l,n}(x) \end{aligned} \tag{9.4.12}$$

where $\alpha^m_{n,k}$, $\beta^m_{n,k}$, $\gamma^m_{n,k}$, and $s^m_{n,k}$ are the 2D wavelet coefficients defined by the inner product of $P_{xx}(x, x')$ with $\psi_{m,k}(x')\psi_{m,n}(x)$, $\psi_{m,k}(x')\varphi_{m,n}(x)$, $\varphi_{m,k}(x')\psi_{m,n}(x)$, and $\varphi_{m,k}(x')\varphi_{m,n}(x)$, respectively.

Since $V_m = W_{m-1} \oplus \cdots \oplus W_{m_l} \oplus V_{m_l}$ for any $m \ge m_l + 1$, the scalet $\varphi_{m,n}(x) \in V_m$ can be expanded in terms of the wavelet functions $\{\psi_{m',n'}(x)\}_{m'=m-1,\ldots,m_l;n' \in Z}$ and the scalets $\{\varphi_{m_l,n'}(x)\}_{n' \in Z}$. Hence, the 2D wavelet expansion above can also be written in the following form:

$$P_{xx}(x, x') = \sum_{m,i=m_l-1} \sum_{n,k} P^{(m,i)}_{n,k} \tilde{\psi}_{i,k}(x') \tilde{\psi}_{m,n}(x), \tag{9.4.13}$$

where $\tilde{\psi}_{m,n}(x)$ is defined as

$$\tilde{\psi}_{m,n}(x) = \begin{cases} \psi_{m,n}(x) & \text{for } m \ge m_l \\ \varphi_{m_l,n}(x) & \text{for } m = m_l - 1, \end{cases}$$

$$\begin{aligned} P^{(m,i)}_{n,k} &= \langle P_{xx}(x, x'), \tilde{\psi}_{i,k}(x') \tilde{\psi}_{m,n}(x) \rangle \\ &\equiv \int_{-\infty}^{\infty} \int_{-\infty}^{\infty} P_{xx}(x, x') \tilde{\psi}_{i,k}(x') \tilde{\psi}_{m,n}(x)\, dx'\, dx. \end{aligned}$$

Usually (9.4.12) and (9.4.13) are referred to as the nonstandard form and the standard form, respectively. There exists a relationship between the coefficients $P^{(m,i)}_{n,k}$

of the standard form and the coefficients $\alpha^m_{n,k}$, $\beta^m_{n,k}$, $\gamma^m_{n,k}$, and $s^m_{n,k}$ of the nonstandard form [20]. Using the notation of $\tilde{\psi}_{m,n}(x)$ and setting $I_{m,n} = \tilde{I}_{m,n}$ if $m \geq m_l$ and $I_{m_l-1,n} = \bar{I}_{m_l,n}$, we can then rewrite (9.4.11) as

$$I(x) \cong A_{m_h} I(x) = \sum_{m=m_l-1}^{m_h-1} \sum_{n} I_{m,n} \tilde{\psi}_{m,n}(x). \tag{9.4.14}$$

For ease of notation, we order and count the wavelet bases in (9.4.13) and (9.4.14), and we replace the double subscripts (i, k) and (m, n) by their counting numbers l and q. We can then write (9.4.13) and (9.4.14) as

$$P_{xx}(x, x') = \sum_{q,l} P_{q,l} \tilde{\psi}_l(x') \tilde{\psi}_q(x)$$

and

$$I(x) \cong A_{m_h} I(x) = \sum_{q=1}^{M} I_q \tilde{\psi}_q(x), \tag{9.4.15}$$

where M is the total number of basis functions in (9.4.14). Equation (9.4.15) gives an approximation of $I(x)$ in the subspace V_{m_h}. Notice that the Daubechies scalet of support width $2N - 1$ gives rise to a wavelet whose expansions are Nth-order convergent [53]. Thus the truncation error $|| I(x) - A_{m_h} I(x) ||$ of the approximation $A_{m_h} I(x)$ to $I(x)$ is bounded as follows:

$$|| I(x) - A_{m_h} I(x) || \leq C 2^{-m_h N},$$

where C is some positive constant.

Matrix Equation Substitution of (9.4.6) and (9.4.7) into (9.4.4) leads to

$$\begin{aligned} \int_{-L}^{G+L} & P_{xx}(x, x') I(x')\, dx' + R\left[-F^{(\mathrm{irc})}(x) - jF^{(\mathrm{irs})}(x)\right] \\ & + T\left[F^{(\mathrm{trc})}(x) + jF^{(\mathrm{trs})}(x)\right] \\ & = \left[-F^{(\mathrm{irc})}(x) + jF^{(\mathrm{irs})}(x)\right], \end{aligned} \tag{9.4.16}$$

where

$$F^{(\mathrm{irc})}(x) = \int_{-\infty}^{\infty} P_{xx}(x, x') f_s\left(k_e x' + \frac{\pi}{2}\right) dx',$$

$$F^{(\mathrm{irs})}(x) = \int_{-\infty}^{\infty} P_{xx}(x, x') f_s(k_e x')\, dx',$$

$$F^{(\text{trc})}(x) = \int_{-\infty}^{\infty} P_{xx}(x, x') f_s\left(k_e[G - x'] + \frac{\pi}{2}\right) dx',$$

$$F^{(\text{trs})}(x) = \int_{-\infty}^{\infty} P_{xx}(x, x') f_s(k_e[G - x']) \, dx'.$$

Replacing $P_{xx}(x, x')$ and $I(x)$ in equation (9.4.16) with their wavelet expansions and multiplying $\tilde{\psi}_q(x)$ both sides and integrating with respect to x, we obtain

$$\sum_{l=1}^{M} P_{q,l} I_l + R P_{q,l+1} + T P_{q,l+2} = B_q \tag{9.4.17}$$

for $q = 1, 2, \ldots, M + 2$, where the orthogonality $\langle \tilde{\psi}_q(x), \tilde{\psi}_l(x) \rangle = \delta_{ql}$ has been used, and

$$P_{q,l} = \int_{-\infty}^{\infty} \int_{-\infty}^{\infty} P_{xx}(x, x') \tilde{\psi}_l(x') \tilde{\psi}_q(x) \, dx' \, dx,$$

$$P_{q,l+1} = -F_q^{(\text{irc})} - j F_q^{(\text{irs})},$$

$$P_{q,l+2} = F_q^{(\text{trc})} + j F_q^{(\text{trs})},$$

$$B_q = -F_q^{(\text{irc})} + j F_q^{(\text{irs})},$$

$$F_q^{(\text{let})} = \int_{-\infty}^{\infty} F^{(\text{let})}(x) \tilde{\psi}_q(x) \, dx \qquad (\text{let} = \text{irc, irs, trc, trs}).$$

Parameters $P_{q,l}$, $F_q^{(\text{irc})}$, $F_q^{(\text{irs})}$, $F_q^{(\text{trc})}$ and $F_q^{(\text{trs})}$ can be evaluated numerically. Equation (9.4.17) is the matrix equation for the unknown coefficients $R, T, I_1, I_2, \cdots, I_M$. The evaluation of the matrix elements involves the rigorous dyadic Green's function as a kernel. The intractable behavior of this Green's function, which includes singularities and strong oscillations, makes the computation of the expansion of the kernel in terms of wavelets very sensitive to the numerical treatment. The numerical aspects of expanding the kernel are described next.

9.4.3 Evaluation of Sommerfeld-Type Integrals

To evaluate the elements $P_{q,l}$, $F_q^{(\text{irc})}$, $F_q^{(\text{irs})}$, $F_q^{(\text{trc})}$, and $F_q^{(\text{trs})}$ is essentially to compute the Sommerfeld-type integral

$$P = \int_{-\infty}^{\infty} \int_{-\infty}^{\infty} P_{xx}(x, x') f_2(x') f_1(x) \, dx' \, dx, \tag{9.4.18}$$

where $f_1(x)$ is a wavelet basis, while $f_2(x')$ can be either a wavelet basis or a function related to $f_s(\cdot)$ as defined in (9.4.8). Substituting expression (9.4.5) of P_{xx} into (9.4.18) leads to

$$P = 4\int_0^\infty \int_0^\infty Q_{xx}(k_x, k_y)| F_y(k_y) |^2 \Re\{F_2(k_x)F_1^*(k_x)\}\, dk_x\, dk_y, \tag{9.4.19}$$

where superscript $*$ and the symbol $\Re$ indicate, respectively, the complex conjugate and the real part of a complex quantity, and

$$F_q(k_x) = \int_{-\infty}^{\infty} f_q(x)e^{-jk_x x}\, dx, \tag{9.4.20}$$

where $q = 1$ or 2. Equation (9.4.19) is a spectral domain formulation.

The poles of Q_{xx} come from zeros of the T_e and T_m functions, and represent the TE and TM surface waves. Moreover T_m always has at least one zero in the whole frequency range, indicating that the first TM surface wave mode has no cutoff frequency [49]. There are many techniques that can be used for treating the singularities caused by those TE and TM poles, including the contour deformation approach, the folding technique, the pole extraction method, and so on [50]. Here a pole extraction technique in conjunction with the conventional folding method was used [11].

Although the singularities relating to the zeros of T_e and T_m were readily treated, the integral in (9.4.19) has two other difficulties: (1) very slow convergence and (2) rapid oscillation of the integrand for large $\beta = \sqrt{k_x^2 + k_y^2}$. These two difficulties are the consequences of the following facts: Green's function (9.4.1) does not contain an explicit $1/R$ dependence for the decay of the fields away from the source and its image; this range dependence, representing the source and image singularities, must be synthesized by the continuous spectrum of plane waves. This is the nature of the spectrum representation. Fortunately the source and image singularities can be shown to be identical to the singularities arising from the same source in a grounded homogeneous medium of relative permittivity

$$\epsilon_e = \begin{cases} \frac{\epsilon_r+1}{2}, & a = d \\ \epsilon_r, & a < d \end{cases}$$

(see [11, 52]). The fields from this source and its image in the homogeneous medium can be evaluated in closed form. Thus it is possible to separate off the source and image singularities in closed form from Green's function for a grounded dielectric slab, yielding a remaining integral that is relatively well-behaved. The component of the Green's function for a grounded homogeneous medium of relative permittivity ϵ_e representing the $\hat{x}$ component of the electric field at (x, y, a) produced by a unit $\hat{x}$-directed infinitesimal dipole located at (x', y', a) is denoted by G_{xx}^h and has a simple closed-form expression as follows [52]:

$$G_{xx}^h(x, y \mid x', y') = -\frac{jZ_0}{4\pi k_0 \epsilon_e}\left(k_e^2 + \frac{\partial^2}{\partial x^2}\right)\left[\frac{e^{-jk_e R_{s0}}}{R_{s0}} - \frac{e^{-jk_e R_{i0}}}{R_{i0}}\right], \tag{9.4.21}$$

where

$$k_e = k_0\sqrt{\epsilon_e},\ R_{s0} = \sqrt{(x-x')^2 + (y-y')^2}$$

and

$$R_{i0} = \sqrt{(x-x')^2 + (y-y')^2 + (2a)^2}.$$

Moreover its spectral representation can be obtained [52]:

$$G_{xx}^h(x, y \mid x', y') = \int_{-\infty}^{\infty}\int_{-\infty}^{\infty} Q_{xx}^h(k_x, k_y)e^{jk_x(x-x')}e^{jk_y(y-y')}\, dk_x\, dk_y \quad (9.4.22)$$

with

$$Q_{xx}^h(k_x, k_y) = -\frac{jZ_0}{4\pi^2 k_0}\cdot\frac{k_e^2 - k_x^2}{2j\epsilon_e k_{e1}}\left[1 - e^{-2jk_{e1}a}\right],$$

where $k_{e1} = \sqrt{k_e^2 - k_x^2 - k_y^2}$. We rewrite the dielectric slab Green's function (9.4.1) as

$$G_{xx}(x, y \mid x', y') = G_{xx}^h(x, y \mid x', y') + \left[G_{xx}(x, y \mid x', y') - G_{xx}^h(x, y \mid x', y')\right].$$

Then (9.4.19) becomes

$$P = P^h + 4\int_0^{\infty}\int_0^{\infty}\left[Q_{xx}(k_x, k_y) - Q_{xx}^h(k_x, k_y)\right] \cdot \mid F_y(k_y)\mid^2 \Re\{F_2(k_x)F_1^*(k_x)\}\, dk_x\, dk_y, \quad (9.4.23)$$

where

$$P^h = \int_{-\infty}^{\infty}\int_{-\infty}^{\infty} P_{xx}^h(x, x')f_2(x')f_1(x)\, dx'\, dx \quad (9.4.24)$$

$$P_{xx}^h(x, x') = \int_{-\infty}^{\infty}\int_{-\infty}^{\infty} G_{xx}^h(x, y \mid x', y')I_2(y')I_2(y)\, dy'\, dy. \quad (9.4.25)$$

Using (9.4.22), we can also formulate P_{xx}^h in the spectral domain:

$$P_{xx}^h(x, x') = \int_{-\infty}^{\infty}\int_{-\infty}^{\infty} Q_{xx}^h(k_x, k_y)\mid F_y(k_y)\mid^2 e^{jk_x(x-x')}\, dk_x\, dk_y, \quad (9.4.26)$$

and then a spectral representation of P^h is obtained as

$$P^h = 4\int_0^{\infty}\int_0^{\infty} Q_{xx}^h(k_x, k_y)\mid F_y(k_y)\mid^2 \Re\{F_2(k_x)F_1^*(k_x)\}\, dk_x\, dk_y. \quad (9.4.27)$$

Since the source and image singularities in G_{xx} are identical to those in G_{xx}^h, Q_{xx} and Q_{xx}^h have the same asymptotic form for large $\beta = \sqrt{k_x^2 + k_y^2}$. Thus the second term in (9.4.23) converges quickly. Either (9.4.24) or (9.4.27) can be used to compute P^h. Because of the phases of terms $e^{jk_x(x-x')}$ in (9.4.26) and $F_2(k_x)F_1^*(k_x)$ in (9.4.27), the integrands in (9.4.26) and (9.4.27) oscillate rapidly for large $\beta = \sqrt{k_x^2 + k_y^2}$, except when x is very close to x'. On the other hand, the Green's function G_{xx}^h in (9.4.25) has a singularity near point $x = x'$, but allows well-convergent integration for all other x. Therefore a scheme for evaluating P^h is designed as follows:

- If the supporting regions of $f_1(x)$ and $f_2(x)$ do not overlap, use the spatial formulations (9.4.24) and (9.4.25) to compute P^h;
- If the supporting regions of $f_1(x)$ and $f_2(x)$ overlap, rewrite $f_1(x)$ as $f_1^o(x) + f_1^r(x)$ and $f_2(x)$ as $f_2^o(x) + f_2^r(x)$, where $f_1^o(x)$ and $f_2^o(x)$ share the common supporting region, while $f_1^r(x)$ and $f_2^r(x)$ are the remaining parts whose support regions do not overlap. Then

  (1) Use the spectral formulation (9.4.27) to compute the contribution to P^h by $f_1^o(x)f_2^o(x')$.
  (2) Use the spatial formulations (9.4.24) and (9.4.25) to compute the contribution to P^h by $f_1^r(x)f_2^r(x')$, $f_1^r(x)f_2^o(x')$, and $f_1^o(x)f_2^r(x')$.

To calculate the infinite integrals related to $f_s(\cdot)$, we redefine $f_s(u)$ as

$$f_s(u) = \begin{cases} \sin u, & -M_s\pi < u < 0 \\ 0, & \text{otherwise,} \end{cases}$$

where M_s is a large integer [10, 11]. Numerical computations have demonstrated that the convergence can be achieved by requiring that $M_s > 6$. The infinite integrations in (9.4.24) and (9.4.25) thus become finite integrations since all $f_1(x)$, $f_2(x)$, and $I_y(x)$ now are of finite support.

In order to use the spectral formulation, the Fourier transform of the wavelet bases must be evaluated. An iterative formulation of the Fourier transform $\hat{\varphi}(\xi)$ of $\varphi(x)$ was obtained in (3.2.15) as

$$\hat{\varphi}(\xi) = \hat{h}\left(\frac{\xi}{2}\right)\hat{\varphi}\left(\frac{\xi}{2}\right),$$

where $\hat{\varphi}(0) = \hat{h}(0) = 1$ (noting that there is a difference of a factor $1/\sqrt{2\pi}$ between the Fourier transform defined in [43]), and $\hat{h}(\xi) = (1/\sqrt{2})\sum_k h_k e^{-jk\xi}$. Making in (3.3.9) use of dilation equations gives the Fourier transform $\hat{\psi}(\xi)$ of $\psi(x)$ in terms of $\hat{\varphi}(\xi)$:

$$\hat{\psi}(\xi) = -e^{-i\xi/2}\hat{h}^*\left(\frac{\xi}{2} + \pi\right)\hat{\varphi}\left(\frac{\xi}{2}\right).$$

Moreover the Fourier transforms $\hat{\varphi}_{m,n}(\xi)$ and $\hat{\psi}_{m,n}(\xi)$ of $\varphi_{m,n}(x)$ and $\psi_{m,n}(x)$ can easily be shown to have the forms

$$\hat{\varphi}_{m,n}(\xi) = 2^{-m/2} e^{-j\xi 2^{-m} n} \hat{\varphi}(2^{-m}\xi),$$
$$\hat{\psi}_{m,n}(\xi) = 2^{-m/2} e^{-j\xi 2^{-m} n} \hat{\psi}(2^{-m}\xi).$$

Generally, the infinite integrations in (9.4.23) and (9.4.27) can be truncated at $k_x = k_y \simeq 200k_0$ with sufficient accuracy.

9.4.4 Numerical Results and Sparsity of Impedance Matrix

A FORTRAN program was written implementing the procedure developed in the preceding sections. Daubechies's wavelets [30, 43] are employed for our calculations. It has been found that the convergence may be sped up by adding one edge basis near each end of the conductors into the conventional wavelet bases, since the edge basis provides a better representation in matching the edge current distribution. Numerical results obtained from the wavelet expansion method are compared with measurements and computational results of the PWS basis functions. The improvement in sparsity of the impedance matrix achieved by using the wavelets rather than the PWS basis is also illustrated. All of the following examples were executed on the IBM RS-6000/530, and roughly a factor of 2 in CPU time savings were recorded in the use of the wavelet rather than the PWS basis. We believe that as the number of unknowns increases and the matrix size grows, the advantages to using wavelet basis will prove more significant.

Numerical Results

Example 1 Open-Ended Microstrip Transmission Line. For the first example, let us consider an open-ended microstrip transmission line with $\epsilon_r = 9.90$, $w = 0.6$ mm and $d = 0.635$ mm. The magnitude and phase of the reflection coefficient are calculated and compared with those of the spectral domain method and measurement [46] in Fig. 9.16. Good agreement between our results and the measured values can be observed.

Example 2 Microstrip Floating Line Resonator. A microstrip floating line (see Fig. 9.15) with parameters $\epsilon_r = 8.875$, $\ell = 3.653$ mm, $g_1 = g_2 = 0.08$ mm and $d = a = w = 0.508$ mm is investigated. To search for the resonant frequency, the reflection and transmission coefficients are computed at different frequencies. Figure 9.17 depicts the magnitude of the reflection coefficient R and transmission coefficient T versus frequency as computed by this method in contrast to the calculations of the PWS basis functions [11]. The results obtained from this method agree well with those from [11]. At the resonant frequency, the magnitude of the standing wave current on the floating line and the local modes on both sides of the floating line are illustrated in Fig. 9.17. These quantities were obtained by our technique and here are shown against the curve obtained by using the PWS basis functions [12]. Again, very good agreement between the two sets of results is demonstrated. In this example the CPU time is about 4 hours for the PWS basis and 2 hours for the wavelet basis.

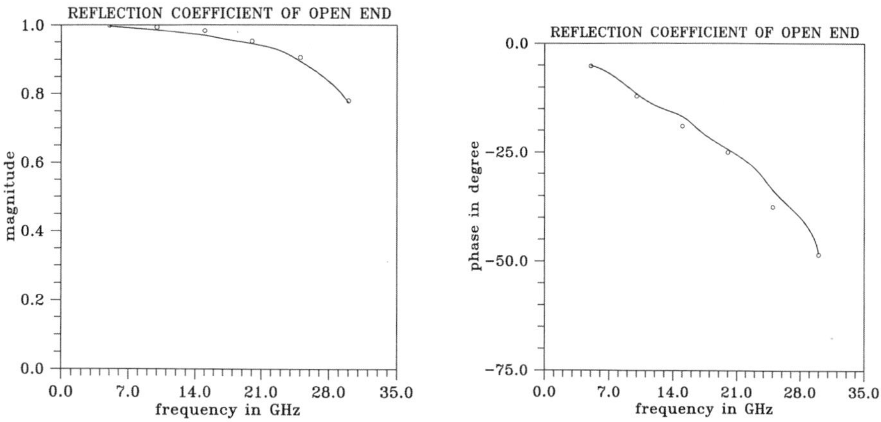

(a) (b)

FIGURE 9.16 Comparison of results of open ended microstrip transmission line using the wavelet expansion method and spectral domain method and measurement. (*a*) Magnitude of the reflection coefficient (*solid line:* wavelet; ooo: SDM). (*b*) Phase of the reflection coefficient (*solid line:* wavelet; ooo: measurement). (Source: G. Wang and G. Pan, *IEEE Trans. Microw. Theory Tech.*, **43**(1), 131–142, Jan. 1995; ©1995 IEEE.)

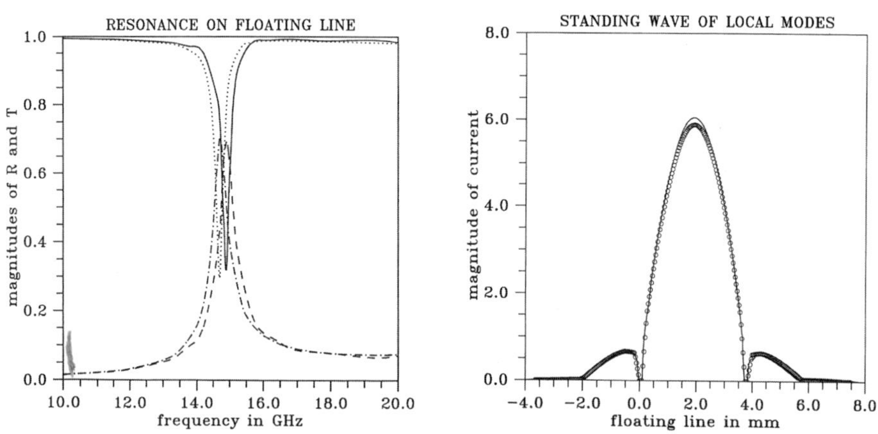

(a) (b)

FIGURE 9.17 Comparison of results of microstrip floating line resonator using this method and PWS basis functions. (*a*) Magnitude of the reflection and transmission coefficients versus frequency (*solid line:* $|R|$ from wavelet; *dashed line:* $|T|$ from wavelet; *dotted line:* $|R|$ from PWS; *dash-dot line:* $|T|$ from PWS). (*b*) Magnitude of the standing wave current on the floating line as well as the local modes on both sides of the floating line (*solid line*; wavelet; ooo: PWS). (Source: G. Wang and G. Pan, *IEEE Trans. Microw. Theory Tech.*, **43**(1), 131–142, Jan. 1995; ©1995 IEEE.)

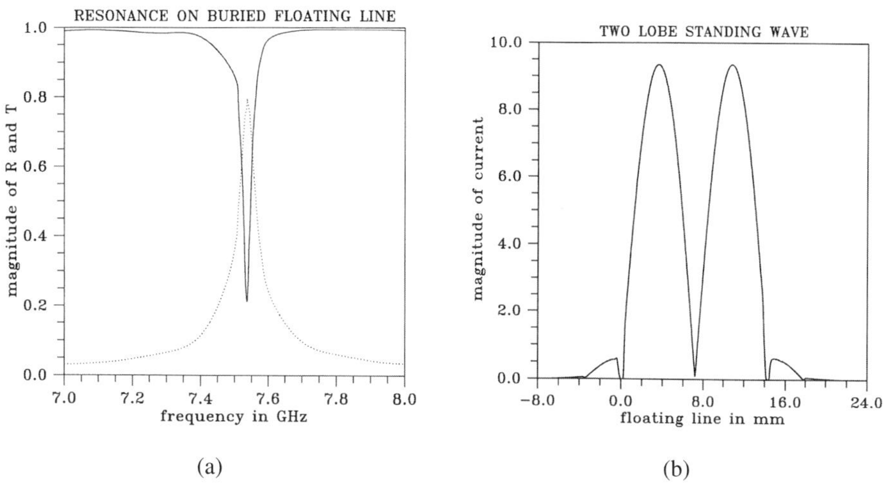

FIGURE 9.18 Results of embedded microstrip floating line resonator using wavelet expansion method. (*a*) Magnitude of the reflection and transmission coefficients versus frequency. (*b*) Magnitude of standing wave current and local modes on both sides of the embedded floating line.

Example 3 Embedded Microstrip Floating Line Resonator. The wavelet expansion method is also applied to a buried microstrip floating line. We search for the resonant frequency given parameters in Fig. 9.15 as $\epsilon_r = 10.0$, $\ell = 14.00$ mm, $g_1 = g_2 = 0.2$ mm, $d = 0.660$ mm and $a = w = 0.560$ mm. The magnitudes of the reflection coefficient R and transmission coefficient T versus frequency obtained by the wavelet expansion method are plotted in Fig. 9.18*a*. The resonant frequency obtained is about 7.54 GHz. At the resonant frequency, the magnitude of the standing wave current on the buried floating line and the local modes on both sides of the buried floating line are depicted in Fig. 9.18*b*. Comparing this example with Example 4 in [11], all parameters are the same except that the floating line in this example is about half of that in [11], and all conductor lines are 0.10 mm narrower and embedded down 0.10 mm in the substrate. There are two current lobes on the floating line here instead of four current lobes as in [11] at the resonant frequency. The resonant frequency decreases slightly from 8 GHz in [11] to 7.54 GHz here.

Sparsity of Impedance Matrices As expected the wavelet expansion method yields a sparse impedance matrix $[P_{q,l}]$. Figure 9.19*a* and *b* illustrates the 3D logarithmic plots of typical normalized impedance matrices generated in Example 2 by the wavelet expansion method (with $m_h = 15$ and $m_l = 13$) and by the PWS basis functions [11]. It can be observed that the wavelet impedance matrix is nearly diagonal or block diagonal. Although the size $N = 264$ of the impedance matrix from wavelets is larger than the size $N = 92$ from PWS basis functions, the effective size from wavelets is still smaller than that from PWS basis functions due to the

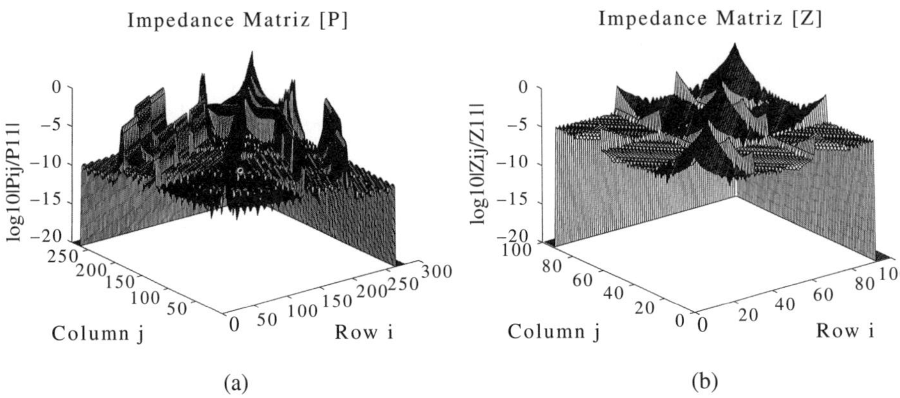

FIGURE 9.19 Comparison of impedance matrices in the computation of the current distribution on microstrip floating line resonator using this method and PWS basis functions. 3D logarithmic plots of typical impedance matrix by using (*a*) wavelets, and (*b*) PWS basis functions.

sparsity of the impedance matrix from wavelets. The sparseness of the matrix has even more profound significance for problems in which large matrices are generated.

In order to give a measure of sparsity in an impedance matrix, we replace each entry of a matrix by its magnitude normalized by the magnitude of the largest matrix element. Now the entries below a threshold, say 10^{-6}, are set to zero, and the remaining entries are considered to be the significant, nonzero, elements. The ratio of the significant entries to the total entries in the matrix measures the sparseness of the matrix. In Fig. 9.20*a* the heavy black-inked line depicts the nonzero elements of an impedance matrix in Example 3 with $m_h = 14$ and $m_l = 13$. A similar result of the same problem in Example 3 with a greater number of resolution levels ($m_h = 14$ and $m_l = 12$) is depicted in Fig. 9.20*b*. In contrast, a full black square, representing a full matrix, will be plotted if the PWS basis functions are used. From Fig. 9.20*a*

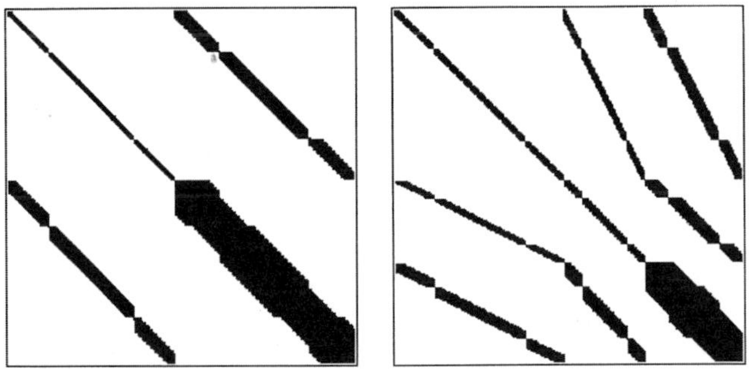

FIGURE 9.20 Sparsity of impedance matrices in the computation of the current distribution on buried microstrip floating line resonator using this method. Sparsity of typical impedance matrix with $m_h = 14$, (*a*) $m_l = 13$ and (*b*) $m_l = 12$.

of the matrix 416×416 and Fig. 9.20*b* of the matrix 351×351, it can be seen that as more resolution levels are used, the dense "plateau" area of the impedance matrix shrinks. This is not surprising. As the decomposition reaches a higher number of levels, more wavelets and fewer scalets are used. The wavelets possess cancellation and localization properties in addition to orthogonality, which is the only thing that the scalets can provide.

9.4.5 Further Improvements

Recently many articles have been published either to expedite the Sommerfeld integrals by integral transform method [54] or to approximate layered Green's functions in closed forms by the complex image method [55], Lipschitz transform method [56], and model order reduction method [57]. While wavelets cannot be applied to the integral transform method, which requires very special functional forms to facilitate the table of integrals [58], they are applicable to all other approaches. The effectiveness of wavelets should be the same as it appears here in this section.

9.5 FULL-WAVE EDGE ELEMENT METHOD FOR 3D LOSSY STRUCTURES

Various analysis methods have been proposed and derived for dealing with lossy guided wave structures, including the mixed potential integral equation method [59], the volume integral equation [60], and the modified spectral domain method [61] among others. The finite element method (FEM) is perhaps the most versatile numerical approach because of its unique ability to manage complicated geometries and boundary value problems, as well as for its ability to support highly structured and systematic procedures to achieve the solution of complex systems [62].

Three-dimensional structures consisting of combinations of metal and dielectric materials will be analyzed by means of an improved finite edge element formulation, which incorporates a newly identified term in the standard boundary conditions of the third kind (Cauchy). These conditions take into account both the transverse and longitudinal field components of the propagating signals in the planes of incidence and transmittance. Employing these boundary conditions, in conjunction with the absorbing boundary conditions (ABC) and/or the boundary conditions of the first kind (Dirichlet) and third kind, a 3D asymmetrical functional is implemented as a hybrid vector edge element method. Numerical examples are presented for air bridges, lossy transmission lines connected by a through-hole via, and a spiral inductor. The equivalent frequency-dependent circuit parameters are then extracted from the field solutions. Laboratory measurements and data comparison with previous published results strongly support the newly developed theoretical work.

In this section we will develop a new functional for general 3D guided wave structures, that need not have completely closed metallic walls. We will then derive the termination conditions at the plane of incidence and the plane of transmittance. An asymmetrical functional is formulated, using the incident and transmitted boundary conditions, in conjunction with the boundary conditions of the first and third kind

[63]. Although this functional can be further modified to a symmetrical one under certain conditions, we will elect not to do so. This 3D asymmetrical functional is then implemented by using hybrid vector finite elements. Utilizing prior information of the eigenmodes resulting from the evaluation of the $2\frac{1}{2}$D edge element solver [64], the 3D field solutions are obtained. It should be mentioned that there is no formal definition of $2\frac{1}{2}$D. Sometimes it refers to circuit with planar dielectric. Sometimes it refers to circuits for which all fields have all three dimensions, but no current is allowed in one dimension, e.g. the z direction, where z is perpendicular to the substrate surface. Here $2\frac{1}{2}$D implies 2D structures with losses along the direction of signal propagation.

After extracting the scattering parameters, the frequency-dependent circuit parameters, such as L, C, R, and G, are converted according to relevant equivalent circuits of the structures. These parameters are readily for SPICE-compatible software packages, and therefore are very useful to digital design engineers. The brief derivation of the boundary conditions at the planes of incidence and transmittance was published in recent papers [65, 66].

9.5.1 Formulation of Asymmetric Functionals with Truncation Conditions

In the finite element implementation, the basic EM equation, which is to be solved for the $3D$ structures in a full-wave analysis, is the vector wave equation

$$\nabla \times \frac{1}{\mu_r} \nabla \times \mathbf{E} - \vec{\vec{\epsilon}}_r k_0^2 \cdot \mathbf{E} = -j\omega\mu_0 \mathbf{J} \qquad \text{in } V. \tag{9.5.1}$$

The boundary conditions for (9.5.1) are

$$\begin{cases} \hat{n} \times \mathbf{E} = \mathbf{P} & \text{on } S_1 \\ \dfrac{1}{\mu_r} \hat{n} \times \nabla \times \mathbf{E} + \gamma_v \hat{n} \times \hat{n} \times \mathbf{E} = \mathbf{V} & \text{on } S_2. \end{cases} \tag{9.5.2}$$

In the previous equations, S_1 is the surface where the boundary condition of the first kind applies, and S_2 the surface where the boundary condition of the third kind applies, $\mathbf{P}$ and $\mathbf{V}$ are known functions, and $\gamma_v = jk_0\sqrt{[\epsilon_{rc} - j(\sigma/\omega\epsilon_0)]/\mu_{rc}}$, as defined in [37] with ϵ_{rc} and μ_{rc} being relative permittivity and permeability of the conductor. The detailed derivation of boundary condition of the third kind can be found in [67]. For nonzero $\mathbf{P}$, the first equation in (9.5.2) represents imperfect electric conductor. For nonzero $\mathbf{V}$ and very small γ_v, the second equation in (9.5.2) stands for imperfect magnetic conductor. When the known functions $\mathbf{P}$ and $\mathbf{V}$ are zero (i.e., the homogeneous cases), the first equation in (9.5.2) reduces to the boundary condition for perfect electric conductor (PEC), while the second equation in (9.5.2) becomes the impedance boundary condition. If we assume, in addition of $\mathbf{V} = 0$, that $\gamma_v = 0$, then the second equation in (9.5.2) becomes that for perfect magnetic conductor (PMC). Different $\mathbf{P}$ and $\mathbf{V}$ may change the functional, but the variational remains unchanged as long as $\mathbf{P}$ and $\mathbf{V}$ are known functions. Usually S_1 and S_2 are disjoint; for instance, S_1 represents the portions of the boundary in which the materials are of

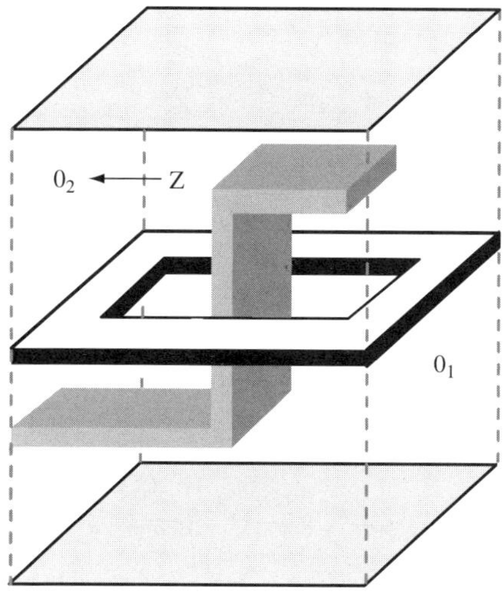

FIGURE 9.21 Via configuration.

perfect conductors, while S_2 describes the regions of the impedance boundary where energy transfers through the boundary.

The boundary condition of the second kind (Neumann) can be included within the third kind. In the application of this theory to transmission line structures and their discontinuities, the field component in the signal propagation direction is generally nonzero, and the aforementioned boundary conditions are insufficient. On both the incident and transmitted planes, the longitudinal component needs to be treated with caution [68]. Without losing generality, we will employ a typical via structure, depicted in Fig. 9.21, as an example. On the incident plane O_1 and transmitted plane O_2, the suitable termination condition is found to be

$$\hat{n} \times \nabla \times \mathbf{E} + \gamma \hat{n} \times \hat{n} \times \left[\mathbf{E} + \frac{\nabla_t E_n}{\gamma} \right] = \mathbf{U}, \tag{9.5.3}$$

where E_n is the electric field component normal to the surface, and γ is the complex propagation constant. Generally, the functional is no longer symmetric because of (9.5.3). Furthermore, to be consistent with the treatment in the $2\frac{1}{2}$D case, and with the expressions that we proposed in [64], the adjoint field should be the field that is incident upon plane O_2 and transmitted through O_1. This adjoint system satisfies

$$\nabla \times \frac{1}{\mu_r} \nabla \times \mathbf{E}^\dagger - \vec{\vec{\epsilon}}_r k_0^2 \cdot \mathbf{E}^\dagger = -j\omega\mu_0 \mathbf{J}^\dagger \qquad \text{in } V \tag{9.5.4}$$

under the associated boundary conditions

$$\begin{cases} \hat{n} \times \mathbf{E}^\dagger = \mathbf{P}^\dagger & \text{on } S_1 \\ \dfrac{1}{\mu_r}\hat{n} \times \nabla \times \mathbf{E}^\dagger + \gamma_v^\dagger \hat{n} \times \hat{n} \times \mathbf{E}^\dagger = \mathbf{V}^\dagger & \text{on } S_2 \\ \hat{n} \times \nabla \times \mathbf{E}^\dagger + \gamma^\dagger \hat{n} \times \hat{n} \times \left[\mathbf{E}^\dagger + \dfrac{\nabla_t E_n^\dagger}{\gamma^\dagger} \right] = \mathbf{U}^\dagger & \text{on } O_1 \cup O_2. \end{cases} \tag{9.5.5}$$

Mathematicians have proved that for a lossy system, the complex conjugate fields $\mathbf{E}^*$ (or $\mathbf{H}^*$) cannot be used to obtain the functional via the Rayleigh-Ritz procedure.

Suppose that we can find an $\mathbf{E}_0$ or $\mathbf{E}_0^\dagger$ that satisfies the aforementioned boundary conditions. Let us define $\mathbf{e} = \mathbf{E} - \mathbf{E}_0$, $\mathbf{e}^\dagger = \mathbf{E}^\dagger - \mathbf{E}_0^\dagger$, $\mathbf{j} = -j\omega\mu_0\mathbf{J}$, and $\mathbf{j}^\dagger = -j\omega\mu_0\mathbf{J}^\dagger$. Then the functional [23]

$$I = \langle \mathbf{e}^\dagger, \hat{L}\mathbf{e} \rangle - \langle \mathbf{e}^\dagger, \mathbf{j} \rangle - \langle \mathbf{e}, \mathbf{j}^\dagger \rangle$$

can still apply provided that the assumed known vector $\mathbf{U}$ is modified to $\mathbf{U} - \hat{n} \times \hat{n} \times \nabla_t E_n$ and the local potential method is employed. Following similar procedures presented in [69], we may further simplify the functional, yielding

$$\begin{aligned} I = &\int_V \left[\frac{1}{\mu_r} (\nabla \times \mathbf{E}^\dagger) \cdot (\nabla \times \mathbf{E}) - k_0^2 \mathbf{E}^\dagger \cdot \vec{\vec{\epsilon}}_r \cdot \mathbf{E} \right] dv \\ &+ \int_{S_2} \gamma_v (\hat{n} \times \mathbf{E}^\dagger) \cdot (\hat{n} \times \mathbf{E})\, ds + \int_{O_1+O_2} \frac{\gamma}{\mu_r} (\hat{n} \times \mathbf{E}^\dagger) \cdot (\hat{n} \times \mathbf{E})\, ds \\ &+ \int_{O_1+O_2} \frac{1}{\mu_r} (\mathbf{U}_0 \cdot \mathbf{E}^\dagger + \mathbf{U}_0^\dagger \cdot \mathbf{E})\, ds + \int_{S_2} (\mathbf{V} \cdot \mathbf{E}^\dagger + \mathbf{V}^\dagger \cdot \mathbf{E})\, ds \\ &+ j\omega\mu_0 \int_V (\mathbf{j} \cdot \mathbf{E}^\dagger + \mathbf{j}^\dagger \cdot \mathbf{E})\, dv. \end{aligned} \tag{9.5.6}$$

Equation (9.5.6) can be verified by using Galerkin's procedure to transform the vector wave equation into the weak integral form [68]. For the via structure the incident field can be expressed as $\mathbf{E}^{\text{in}}(x, y, z) = \mathbf{E}_1^{2\text{D}}(x, y, z) = \mathbf{E}_1^0(x, y)e^{-\gamma(z-z_1)}$ on the plane of incidence O_1, where $\mathbf{E}_1^{2\text{D}}(x, y, z)$ is the 2D solution obtained in [64]. Thus

$$\begin{aligned} \mathbf{E} &= \mathbf{E}_{1t}^0(x, y)e^{-\gamma_1(z-z_1)} + \Gamma \mathbf{E}_{1t}^0(x, y)e^{\gamma_1(z-z_1)} \\ &\quad + \mathbf{E}_{1z}^0(x, y)e^{-\gamma_1(z-z_1)} - \Gamma \mathbf{E}_{1z}^0(x, y)e^{\gamma_1(z-z_1)} \\ &= \mathbf{E}^{\text{in}} + \mathbf{E}^{\text{re}}, \end{aligned} \tag{9.5.7}$$

where Γ is the reflection coefficient. Consequently on this surface we obtain

$$\begin{aligned} &\hat{n} \times \nabla \times \mathbf{E} + \gamma_1 \hat{n} \times \hat{n} \times \mathbf{E} \\ &= 2\gamma_1 \hat{n} \times \hat{n} \times \mathbf{E}^{\text{in}} - \hat{n} \times \hat{n} \times \nabla_t E_n \\ &= \mathbf{U}_0, \end{aligned} \tag{9.5.8}$$

where $\hat{n}$ is the outgoing normal to O_1, meaning that $\hat{n} = -\hat{z}$. Note that γ_1 in (9.5.7) is the complex propagation constant for the $2\frac{1}{2}$D uniform line case, which has been obtained from the precomputation of the $2\frac{1}{2}$D edge element codes. Comparing (9.5.8) with (9.5.3), we have

$$\begin{cases} \gamma = \gamma_1 \\ \mathbf{U} = 2\gamma_1 \hat{n} \times (\hat{n} \times \mathbf{E}^{\text{in}}). \end{cases}$$

On O_2, the surface through which the wavefront propagates out of the via structure, we have

$$\begin{aligned} \mathbf{E} &= T\mathbf{E}_2^0(x, y)e^{-\gamma_2(z-z_2)} \\ &= \mathbf{E}^{\text{tr}}, \end{aligned} \tag{9.5.9}$$

where T is the transmission coefficient, and $\mathbf{E}_2^0(x, y)e^{-\gamma_2(z-z_2)}$ is the 2D solution at $z = z_2$. On O_2, we also have

$$\begin{aligned} \hat{n} \times \nabla \times \mathbf{E} + \gamma_2 \hat{n} \times \hat{n} \times \mathbf{E} &= -\hat{n} \times \hat{n} \times \nabla_t E_n \\ &= \mathbf{U}_0, \end{aligned} \tag{9.5.10}$$

where the outgoing normal to O_2 is $\hat{n} = \hat{z}$.

Comparing (9.5.3) with (9.5.10), we obtain $\gamma = \gamma_2$, and $\mathbf{U} = 0$. On other boundary surfaces, either the boundary conditions of the first or third kind or the radiation boundary condition apply. The adjoint field satisfies (9.5.4) and (9.5.5) on O_2 and has the form of $\mathbf{E}^{\text{in}\dagger} = \mathbf{E}_2^{0\dagger} e^{\gamma(z-z_2)}$. At port 2 we have

$$\begin{aligned} \mathbf{E}^\dagger &= \mathbf{E}_{2t}^0(x, y)e^{\gamma_2(z-z_2)} + R\mathbf{E}_{2t}^0(x, y)e^{-\gamma_2(z-z_2)} \\ &\quad - \mathbf{E}_{2z}^0(x, y)e^{\gamma_2(z-z_2)} + R\mathbf{E}_{2z}^0(x, y)e^{-\gamma_2(z-z_2)} \\ &= \mathbf{E}^{\text{in}\dagger} + \mathbf{E}^{\text{re}\dagger}. \end{aligned}$$

Similar to (9.5.8), we find that

$$\begin{aligned} \hat{n} \times \nabla \times \mathbf{E}^\dagger + \gamma_2 \hat{n} \times \hat{n} \times \mathbf{E}^\dagger &= 2\gamma_2 \hat{n} \times \hat{n} \times \mathbf{E}^{\text{in}\dagger} - \hat{n} \times \hat{n} \times \nabla_t E_n^\dagger \\ &= \mathbf{U}_0^\dagger. \end{aligned}$$

On O_1, the adjoint field is governed by

$$\begin{aligned} &\hat{n} \times \nabla \times \mathbf{E}^\dagger + \gamma_1 \hat{n} \times \hat{n} \times \mathbf{E}^\dagger \\ &= -\hat{n} \times \hat{n} \times \nabla_t E_n^\dagger \\ &= \mathbf{U}_0^\dagger. \end{aligned}$$

Two issues related to the functional (9.5.6) should be emphasized:

(1) The functional (9.5.6) will reduce to the form proposed in [64] for uniform $2\frac{1}{2}$D structures.

(2) The $\nabla_t E_n$ term is assumed to be a known function. We will maintain this dual characteristic until the optimization of the functional has been completed.

Based on the formulation developed in this section, a finite element procedure is performed. A detailed description of this procedure is provided in the next subsection. Employing the Ritz procedure and grouping together all of the local elements into the global coordinate system, we arrive at

$$\left\{[\mathcal{Z}_v^e] + \gamma\left[\frac{1}{\mu_r^s}\mathcal{Z}_z^s\right] + [\mathcal{Z}_{tz}^s] - [\mathcal{Z}_{-tz}^s] + \gamma\left[\frac{1}{\mu_r^s}\mathcal{Z}_{(-z)}^s\right] + \gamma_v[\mathcal{Z}_i^g]\right\}[E]$$
$$= 2\gamma\left[\frac{1}{\mu_r^u}\mathcal{Z}_z^u\right][E_t^{\text{in}}]. \qquad (9.5.11)$$

Once (9.5.11) has been solved, the distribution of the electrical fields will be obtained. The S parameters, including the reflection and transmission coefficients, can be evaluated and the desired circuit parameters, C, L, R, and G, can then be found from network theory.

9.5.2 Edge Element Procedure

The edge element procedure is standard, and can be found in many sources [e.g., 69]. For ease of reference, we summarize the major steps here for the via problem. For reasons of simplicity, only the isotropic case will be considered. We will assume that the same shape functions employed for the primary fields can also be used for the adjoint fields. For the edge element with a basic building block, we may express the electrical fields in each small cell [69] as

$$\mathbf{E}^e = \sum_{i=1}^{12} \mathbf{N}_i^e E_i^e$$

where, with the volume element edge numbering shown in Figure 9.22.

$$N_{x1}^e = \frac{1}{l_y^e l_z^e}\left(y_c^e - y + \frac{l_y^e}{2}\right)\left(z_c^e - z + \frac{l_z^e}{2}\right)$$

$$N_{x2}^e = \frac{1}{l_y^e l_z^e}\left(y - y_c^e + \frac{l_y^e}{2}\right)\left(z_c^e - z + \frac{l_z^e}{2}\right)$$

$$N_{x3}^e = \frac{1}{l_y^e l_z^e}\left(y_c^e - y + \frac{l_y^e}{2}\right)\left(z - z_c^e + \frac{l_z^e}{2}\right)$$

$$N_{x4}^e = \frac{1}{l_y^e l_z^e}\left(y - y_c^e + \frac{l_y^e}{2}\right)\left(z - z_c^e + \frac{l_z^e}{2}\right)$$

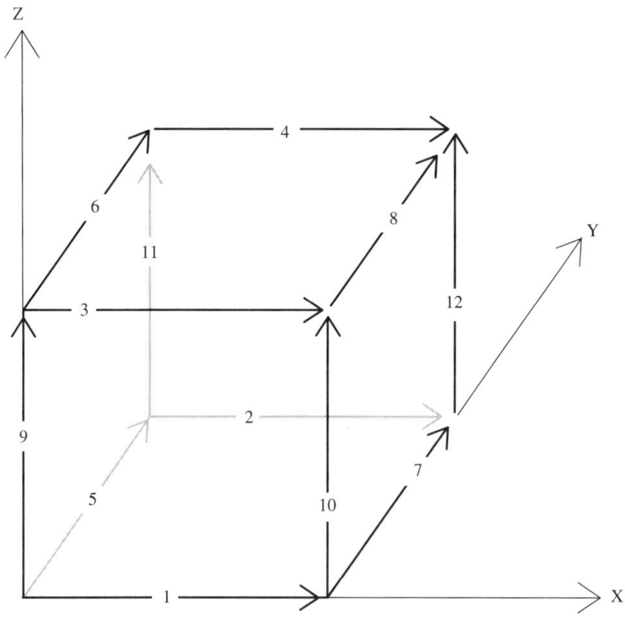

FIGURE 9.22 Volume element edge numbering arrangement.

$$N_{y1}^e = \frac{1}{l_x^e l_z^e}\left(x_c^e - x + \frac{l_x^e}{2}\right)\left(z_c^e - z + \frac{l_z^e}{2}\right)$$

$$N_{y2}^e = \frac{1}{l_x^e l_z^e}\left(x_c^e - x + \frac{l_x^e}{2}\right)\left(z - z_c^e + \frac{l_z^e}{2}\right)$$

$$N_{y3}^e = \frac{1}{l_x^e l_z^e}\left(x - x_c^e + \frac{l_x^e}{2}\right)\left(z_c^e - z + \frac{l_z^e}{2}\right)$$

$$N_{y4}^e = \frac{1}{l_x^e l_z^e}\left(x - x_c^e + \frac{l_x^e}{2}\right)\left(z - z_c^e + \frac{l_z^e}{2}\right)$$

$$N_{z1}^e = \frac{1}{l_x^e l_y^e}\left(x_c^e - x + \frac{l_x^e}{2}\right)\left(y_c^e - y + \frac{l_y^e}{2}\right)$$

$$N_{z2}^e = \frac{1}{l_x^e l_y^e}\left(x - x_c^e + \frac{l_x^e}{2}\right)\left(y_c^e - y + \frac{l_y^e}{2}\right)$$

$$N_{z3}^e = \frac{1}{l_x^e l_y^e}\left(x_c^e - x + \frac{l_x^e}{2}\right)\left(y - y_c^e + \frac{l_l y^e}{2}\right)$$

$$N_{z4}^e = \frac{1}{l_x^e l_y^e}\left(x - x_c^e + \frac{l_x^e}{2}\right)\left(y - y_c^e + \frac{l_y^e}{2}\right).$$

These equations can also be written as

$$\left.\begin{aligned} \mathbf{N}_i^e &= N_{xi}^e \hat{x} \\ \mathbf{N}_{i+4}^e &= N_{yi}^e \hat{y} \\ \mathbf{N}_{i+8}^e &= N_{zi}^e \hat{z} \end{aligned}\right\} i = 1, 2, 3, 4.$$

Suppose that in the entire mesh region we have M_v volume elements, M_{O_1} surface elements on surface O_1, M_{O_2} on surface M_{O_2}, and M_g on the side wall and on the ground plane. Then (9.5.6) can be written as

$$I = \sum_{e=1}^{M_v} I_1^e + \sum_{s=1}^{M_{O_1}} I_1^s + \sum_{s=1}^{M_{O_2}} I_2^s + \sum_{g=1}^{M_g} I^g + \sum_{u=1}^{M_{O_1}} I^u + \sum_{s=1}^{M_{O_1}} I_1^{as} + \sum_{s=1}^{M_{O_2}} I_2^{as}. \quad (9.5.12)$$

Each term in (9.5.12) can be expressed as

$$\begin{aligned} I_1^e &= [E^e]^t[Z_v^e][E^e] \\ I_1^s &= \gamma[E^s]^t[Z_z^s][E^s] \\ I_2^s &= \gamma[E^s]^t[Z_{-z}^s][E^s] \\ I^g &= \gamma_v[E^g]^t[Z_i^g][E^g] \\ I^u &= -2\gamma[E^u]^t[Z_z^u][E_t^{\text{in}}] \end{aligned}$$

with asymmetrical terms

$$\begin{aligned} I_1^{as} &= -[E^s]^t[Z_{-tz}^s][E^z] \\ I_2^{as} &= [E^s]^t[Z_{tz}^s][E^z]. \end{aligned}$$

In the previous equations

$$Z_v^e = \int_V \left[\frac{1}{\mu_r^e}(\nabla \times \mathbf{N}^e) \cdot [\nabla \times (\mathbf{N}^e)^t] - \epsilon_r k_0^2 \mathbf{N}^e \cdot (\mathbf{N}^e)^t\right] dv,$$

and noticing that

$$\int \left(x - x_c^e + \frac{l_x^e}{2}\right)^2 dx = \frac{(l_x^e)^3}{3}$$

$$\int \left(x - x_c^e + \frac{l_x^e}{2}\right)\left(x_c^e - x + \frac{l_x^e}{2}\right) dx = \frac{(l_x^e)^3}{6},$$

we may show that

$$Z_v^e = \frac{1}{\mu_r^e} \begin{Bmatrix} \frac{l_x^e l_z^e}{6l_y^e} K_1 + \frac{l_x^e l_y^e}{6l_z^e} K_2 & -\frac{l_z^e}{6} K_3 & -\frac{l_y^e}{6} K_5 \\ -\frac{l_z^e}{6} K_3^t & \frac{l_y^e l_x^e}{6l_z^e} K_1 + \frac{l_y^e l_z^e}{6l_x^e} K_2 & -\frac{l_x^e}{6} K_3 \\ -\frac{l_y^e}{6} K_5^t & -\frac{l_x^e}{6} K_3^t & \frac{l_y^e l_z^e}{6l_x^e} K_1 + \frac{l_x^e l_z^e}{6l_y^e} K_2 \end{Bmatrix}$$

$$- \epsilon_r^e k_0^2 \begin{Bmatrix} \frac{l_x^e l_y^e l_z^e}{36} K_4 & 0 & 0 \\ 0 & \frac{l_x^e l_y^e l_z^e}{36} K_4 & 0 \\ 0 & 0 & \frac{l_x^e l_y^e l_z^e}{36} K_4 \end{Bmatrix}.$$

In the previous equation

$$K_1 = \begin{bmatrix} 2 & -2 & 1 & -1 \\ -2 & 2 & -1 & 1 \\ 1 & -1 & 2 & -2 \\ -1 & 1 & -2 & 2 \end{bmatrix}$$

$$K_2 = \begin{bmatrix} 2 & 1 & -2 & -1 \\ 1 & 2 & -1 & -2 \\ -2 & -1 & 2 & 1 \\ -1 & -2 & 1 & 2 \end{bmatrix}$$

$$K_3 = \begin{bmatrix} 2 & 1 & -2 & -1 \\ -2 & -1 & 2 & 1 \\ 1 & 2 & -1 & -2 \\ -1 & -2 & 1 & 2 \end{bmatrix}$$

$$K_4 = \begin{bmatrix} 4 & 2 & 2 & 1 \\ 2 & 4 & 1 & 2 \\ 2 & 1 & 4 & 2 \\ 1 & 2 & 2 & 4 \end{bmatrix}$$

$$K_5 = \begin{bmatrix} -2 & 2 & -1 & 1 \\ -1 & 1 & -2 & 2 \\ 2 & -2 & 1 & -1 \\ 1 & -1 & 2 & -2. \end{bmatrix}.$$

Other matrices in (9.5.12) are $Z_i^s, i = x, y, z$ with x, y, z being the normal to the surface, $Z_z^u = Z_z^s$, and $Z_{-z}^s = Z_z^s$. In general, the surface matrices have the form

$$Z_i^s = \int ds(\mathbf{n} \times \mathbf{N}_{n\pm}^s) \cdot (\mathbf{n} \times \mathbf{N}_{n\pm}^s)^t.$$

We may categorize the surface integrals into three groups:

CASE 1. FOR $\mathbf{n} = \pm\hat{x}$

$$Z_x^s = \frac{l_y^e l_z^e}{6}\begin{bmatrix} 2 & 1 & 0 & 0 \\ 1 & 2 & 0 & 0 \\ 0 & 0 & 2 & 1 \\ 0 & 0 & 1 & 2 \end{bmatrix}.$$

CASE 2. FOR $\mathbf{n} = \pm\hat{y}$

$$Z_y^s = \frac{l_z^e l_x^e}{6}\begin{bmatrix} 2 & 1 & 0 & 0 \\ 1 & 2 & 0 & 0 \\ 0 & 0 & 2 & 1 \\ 0 & 0 & 1 & 2 \end{bmatrix}.$$

CASE 3. FOR $\mathbf{n} = \pm\hat{z}$

$$Z_z^s = \frac{l_y^e l_x^e}{6}\begin{bmatrix} 2 & 1 & 0 & 0 \\ 1 & 2 & 0 & 0 \\ 0 & 0 & 2 & 1 \\ 0 & 0 & 1 & 2 \end{bmatrix}.$$

Finally, for the surfaces perpendicular to the propagation direction, z, we have

$$Z_{\mp tz}^s = \frac{1}{6}\begin{bmatrix} -2l_y & 2l_y & -l_y & l_y \\ -l_y & l_y & -2l_y & 2l_y \\ -2l_x & -l_x & 2l_x & l_x \\ -l_x & -2l_x & l_x & 2l_x \end{bmatrix}.$$

After using the Ritz procedure and grouping together all of the relationships in the global coordinate system, we arrive at

$$\left\{[\mathcal{Z}_v^e] + \gamma\left[\frac{1}{\mu_r^s}\mathcal{Z}_z^s\right] + [\mathcal{Z}_{tz}^s] - [\mathcal{Z}_{-tz}^s] + \gamma\left[\frac{1}{\mu_r^s}\mathcal{Z}_{(-z)}^s\right] + \gamma_v[\mathcal{Z}_i^g]\right\}[E]$$
$$= 2\gamma\left[\frac{1}{\mu_r^u}\mathcal{Z}_z^u\right][E_t^{\text{in}}]. \tag{9.5.13}$$

Equation (9.5.13) is, in fact, the last equation of Section 9.5.1, namely (9.5.11).

9.5.3 Excess Capacitance and Inductance

Once the distribution of the 3D electrical field has been determined, the reflection coefficient can be evaluated from (9.5.7). For example, when the system is excited from port 1, we have

$$\Gamma = \frac{\int_{O_1} ds[\mathbf{E} \cdot \mathbf{E}^{2D} - \mathbf{E}^{2D} \cdot \mathbf{E}^{2D}]}{\int_{O_1} ds\mathbf{E}^{2D} \cdot \mathbf{E}^{2D}} |_{z=z_1}$$

Similarly, from (9.5.9), the transmission coefficient is

$$T = \frac{\int_{O_2} ds\mathbf{E} \cdot \mathbf{E}^{2D}}{\int_{O_2} ds\mathbf{E}^{2D} \cdot \mathbf{E}^{2D}} |_{z=z_2}$$

provided that the excitation field is properly normalized [7].

Let us explain the physical meaning of the preceding equations for the reflection coefficient Γ and transmission coefficient T. In the transmission line theory, these coefficients are defined in terms of voltages and currents, which are integrated quantities in nature. However, our field solutions are differential values. Therefore the integration over the plane of incidence was performed in the numerator. The dot product of $(\mathbf{E} - \mathbf{E}^{2D})$ with $\mathbf{E}^{2D}$ provides the scalar contributions to the Γ. The subtraction of the 3D fields from the 2D fields indicates the contributions from the discontinuities. This subtraction is a very standard procedure in the FDTD method to find the reflection coefficients. Finally, the integration on the denominator provides a normalization factor. The scattering parameters of a two-port system are well known as

$$\begin{cases} S_{11} = \Gamma \\ S_{21} = T \end{cases},$$

while S_{22} and S_{12} are obtained when port 2 is excited.

The normalized Y parameters, from two-port network theory, are [70]

$$Y_{11} = \frac{(1 - S_{11}) \cdot (1 + S_{22}) + S_{12} \cdot S_{21}}{(1 + S_{11}) \cdot (1 + S_{22}) - S_{12} \cdot S_{21}}$$

$$Y_{12} = \frac{-2S_{12}}{(1 + S_{11}) \cdot (1 + S_{22}) - S_{12} \cdot S_{21}}$$

$$Y_{21} = \frac{-2S_{21}}{(1 + S_{11}) \cdot (1 + S_{22}) - S_{12} \cdot S_{21}}$$

$$Y_{22} = \frac{(1 - S_{22}) \cdot (1 + S_{11}) + S_{12} \cdot S_{21}}{(1 + S_{11}) \cdot (1 + S_{22}) - S_{12} \cdot S_{21}}.$$

For reciprocal structures with symmetry, $S_{11} = S_{22}$ and $S_{21} = S_{12}$. Then, based on the type Π equivalent circuits, we have the equivalent capacitance and inductance

$$\begin{cases} C = \dfrac{\Im Y_{12}}{\omega} \\ L = \dfrac{1}{\omega \Im (Y_{11} - Y_{12})} \end{cases}, \tag{9.5.14}$$

where $\Im$ stands for imaginary. The resistance and the conductance can be found in the same way. The resistance can be ignored in this equivalent circuit becausc, while the vias are of fairly small cross section, their vertical height between layers is also quite small (typical via diameters and heights are in the range of 30 to 90 μm).

The excess capacitance and inductance can be obtained from (9.5.14) by subtracting the 2D uniform line parameters multiplied by the distance between O_1 and O_2.

9.5.4 Numerical Examples

The nonsymmetric complex sparse system equation (9.5.13) was solved using the Harwell subroutines. Only a few minutes are required on an IBM 6000 computer for each frequency point, while the total number of unknowns is approximately 3000.

Example 1 An Air Bridge. Air bridges, as depicted in Fig. 9.23, are employed in several high-performance integrated circuit technologies to ensure minimum interconnect capacitance and maximum signal propagation velocity along the interconnect. The dimension in micrometers (μm) are: $a = 212$, $h = 106$, $w_1 = 212$, $h_3 = 200$, $h_2 = 60$, $h_1 = 635$, $g = 635$, and $w = 635$. Using the newly developed three-dimensional simulation codes, the air bridge connection problem is successfully solved. The conducting line and the ground plane are assumed in this example to be copper, although in practice aluminum or gold are typically employed. Our assumption of copper metallurgy in the example allows us to compare our numerical

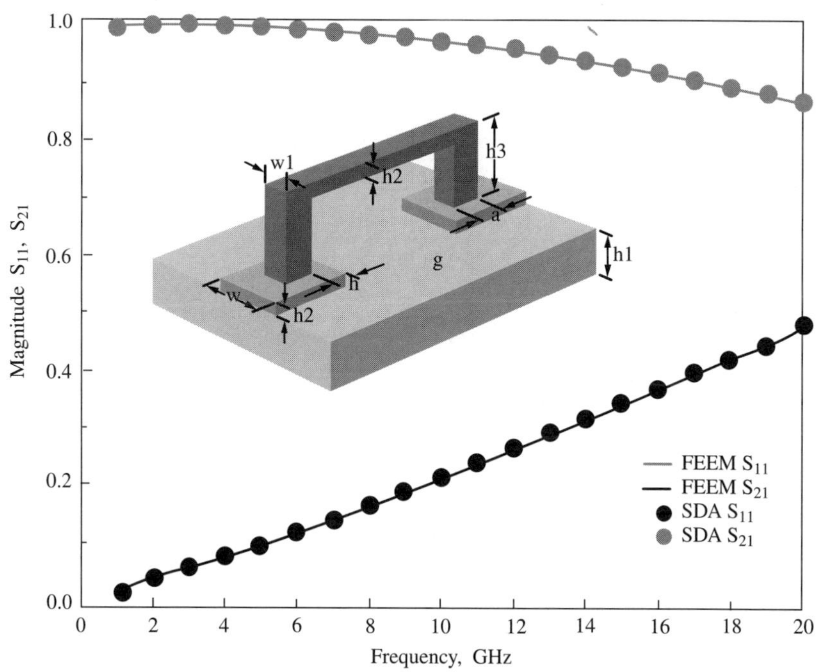

FIGURE 9.23 Geometry and *S* parameters for the air bridge. (Source: G. Pan, J. Tan and B. Gilbert, *IEE Proc. Microw. Ant. Propg.*, **147**(5), 391–397, Oct. 2000; ©2000 IEE.)

TABLE 9.12. Reflection, Transmission Coefficients S_{11} and S_{21}, and Equivalent Circuit Parameters Generated by the EEM for Through-hole Via Structure

Frequency (GHz)	Reflection S_{11}	Transmission S_{21}	Inductance (nH)	Capacitance (pF)
5	2.56E-3	0.99998	55.767E-3	21.274E-3
10	1.29E-2	0.99986	55.568E-3	24.602E-3
15	2.07E-2	0.99968	55.406E-3	24.965E-3
20	2.77E-2	0.99944	55.204E-3	25.046E-3
25	3.46E-2	0.99910	54.950E-3	25.122E-3
30	4.19E-2	0.99876	54.650E-3	25.254E-3

results with results already published in the literature, which also postulate the use of copper. However, the ohmic loss in this structure is found not to be significant after analyzing and comparing the real-world interconnect with an ideal lossless structure of equivalent geometry. Figure 9.23 also depicts the scattering parameters evaluated from this method, and from the spectral domain analysis (SDA) method [71]. Results from the two methods show excellent agreement.

Example 2 Via Structure. Through-hole vias are typically used to connect signal lines residing on different metal layers in most printed circuit board technologies and in some multichip module (MCM) technologies. From the point of view of electromagnetic fields, we would like to know, for the via structure, the transmitted power and reflected power at specific frequencies or over specific frequency ranges. For the circuit design, engineers are concerned about overall signal integrity interconnects carrying wideband signals, and thus wish to understand the magnitude of the excess inductance and excess capacitance caused by this via discontinuity. The method described herein provides the needed parametric values. Note in Table 9.12 that the two reference planes incorporated into the via structure, shown in Fig. 9.21, are placed at locations $z_1 = -0.07$ and $z_2 = 0.07$ (mm) respectively. Figure 9.24 depicts the side

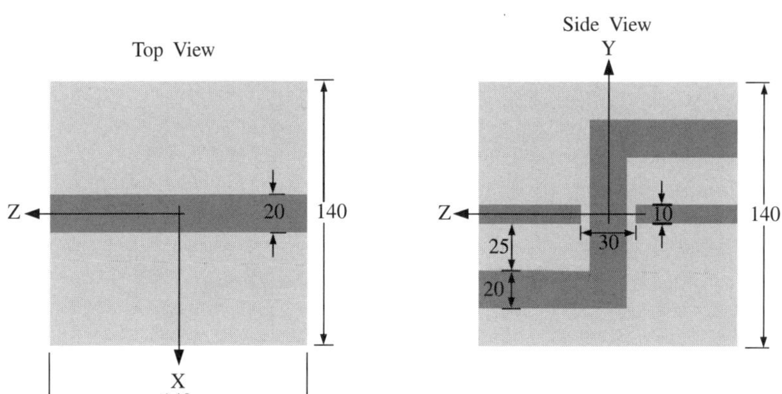

FIGURE 9.24 Top and side views of through-hole via.

and top views of the structure, with all dimensions (in μm) included. The resulting frequency dependent S parameters are listed in Table 9.12. It can be seen from the table that the signal integrity effect of the via in Figure 9.24 is minimal. Laboratory measurements support this conclusion. The capacitance values are compared with the FDTD results, with a discrepancy $\leq 7\%$.

It appears from this set of data that the new method allows the use of a minimum number of brick edge elements (2000), while nonetheless obtaining numerically acceptable results.

Example 3 A Spiral Inductor. Spiral inductors play a significant role in recent microwave monolithic integrated circuit (MMIC) and some multichip module technologies. Depicted in Fig. 9.25 are the top and side views of a square spiral inductor. The dimensions in millimeters (mm) are $a = s = 0.3125$, $w = 0.625$, $h = 0.3175$, $d = 0.635$, and $t = 0.05$, and the material parameters are $\epsilon_r = 9.8$ and $\tan\,\delta = 0$. This inductor has been analyzed by the spectral domain mixed potential equation method (MPIE) in [59]. Shown in Figure 9.26 are the magnitudes of S_{21} obtained from measurements in [59] and from this simulation method, using a tetrahedral mesh. Good agreement has been observed between our results and the reference values.

All examples presented here can be implemented by the multiscalet-based FEM (MWFEM) of Section 6.10, provided the linear shape functions in Section 9.5.2 are replaced by the multiwavelet interpolating functions and the truncation boundary

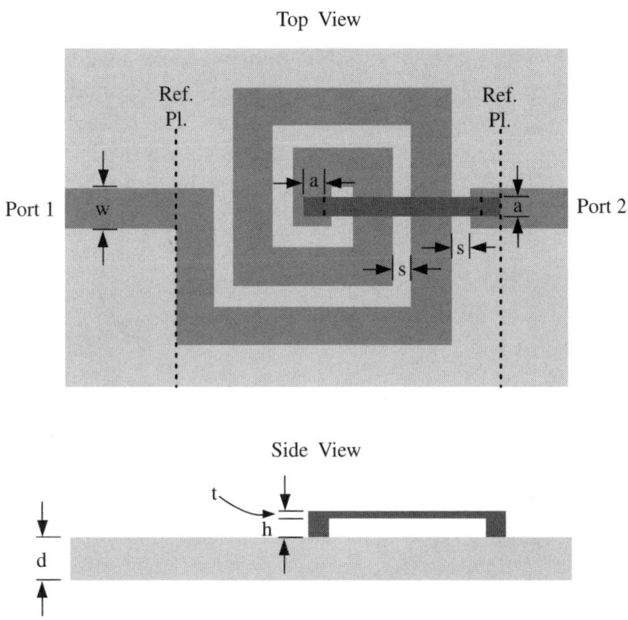

FIGURE 9.25 Top and side views of spiral inductor. (Source: G. Pan, J. Tan, and B. Gilbert, *IEE Proc. Microw. Ant. Propg.*, **147**(5), 391–397, Oct. 2000; ©2000 IEE.)

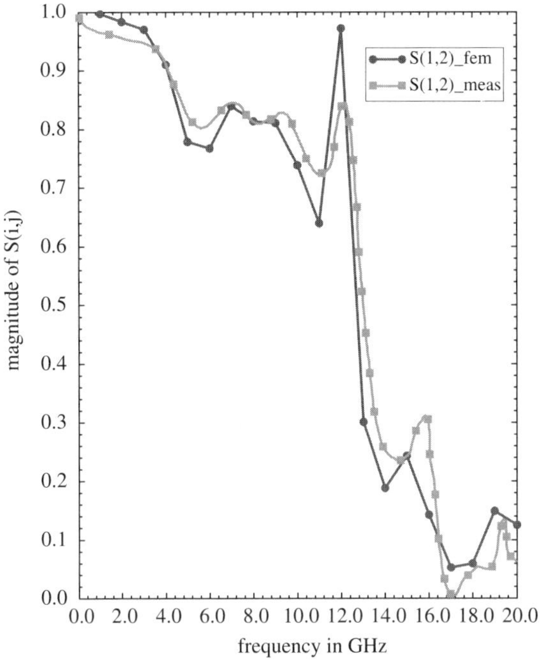

FIGURE 9.26 Comparison of *S* parameters of spiral inductor generated by EEM and MPIE. (Source: G. Pan, J. Tan, and B. Gilbert, *IEE Proc. Microw. Ant. Propg.*, **147**(5), 391–397, Oct. 2000; ©2000 IEE.)

conditions are properly treated. Currently we are extending our 2D MWFEM work [65] to general purpose 3D algorithms. The examples here will serve as the test cases to verify the new computer codes.

BIBLIOGRAPHY

[1] C. Wei, R. F. Harrington, J. R. Mautz, and T. K. Sarkar, "Multiconductor transmission lines in multilayered dielectric media," *IEEE Trans. Microw. Theory Tech.*, **32**, 439-449, Apr. 1984.

[2] G. Pan, K. Olson, and B. Gilbert, "Improved algorithmic method for the prediction of wavefront propagation behavior in multiconductor transmission lines in multilayered dielectric media," *IEEE Trans. Comp.-Aided Design*, **8** 608–621, Junc 1989.

[3] G. Pan, G. Wang, and B. Gilbert, "Edge effect enforced boundary element analysis of multilayered transmission lines," *IEEE Trans. Circ. Syst. I: Fundamental Theory and Applications*, **39**, 955–963, Nov. 1992.

[4] G. Pan, G. Wunsch, and B. Gilbert, "Frequency-domain analysis of coupled nonuniform transmission lines using Chebyshev pseudo-spatial techniques," *IEEE Trans. Microw. Theory Tech.*, **40**, 2025–2033, Nov. 1992.

[5] G. Lei, G. Pan and B. Gilbert, “Examination, clarification, and simplification of modal decoupling method for multiconductor transmission lines,” *IEEE Trans. Microw. Theory Tech.*, **43**, 2090–2100, Sept. 1995.

[6] K. Yee, “Numerical solution of initial boundary value problems involving Maxwell's equation in isotropic media,” *IEEE Trans. Ant. Propg.*, **14**, 302–307, May 1966.

[7] R. Collin, *Field Theory of Guided Waves*, 2nd ed., IEEE Press, New York, 1990.

[8] M. Tsuk and J. Kong, “A hybrid method for the calculation of the resistance and inductance of transmission lines with arbitrary cross sections,” *IEEE Trans. Microwave Theory Tech.*, **39**, 1338–1347, Aug. 1991.

[9] X. Zhu, G. Lei, and G. Pan, “On application of fast and adaptive periodic Battle–Lemarie wavelets to modeling of multiple lossy transmission lines,” *J. Comput. Phys.*, **132**, 299–311, 1997.

[10] R. Jackson and D. Pozar, “Full-wave analysis of microstrip open-end and gap discontinuities,” *IEEE Trans. Microwave Theory Tech.*, **33**, 1036–1042, Oct. 1985.

[11] G. Pan, J. Tan, and J. Murphy, “Full-wave analysis of microstrip floating line discontinuities,” *IEEE Trans. Electromagn. Compatibility*, **36**(1), 49–59, Feb. 1994.

[12] G. Pan and J. Tan, “Full-wave analysis of radiation effect of microstrip transmission lines,” *Int. J. Analog Integ. Circ. Signal Process.*, **5**(1), 77–85, 1993.

[13] N. Fache, F. Olyslager, and D. DeZutter, *Electromagnetic and Circuit Modeling of Multiconductor Transmission Lines*, Clarendon Press, Oxford, 1993.

[14] G. Wang and G. Pan, “Full wave analysis of microstrip floating line structures by wavelet expansion method,” *IEEE Trans. Microw. Theory Tech.*, **43**, 131–142, Jan. 1995.

[15] J. Tan, G. Pan, G. Lei, and B. Gilbert, “Full wave analysis of transmission lines in a multilayer substrate with heavy dielectric losses,” *IEEE Trans. Comp., Packag. Manuf. Tech.*, **19**, 621–627, August 1996.

[16] G. Pan and J. Tan, “General edge element approach to lossy and dispersive structures in anisotropic media,” *IEE Proc. Microw. Ant. Propg.*, **144**(2), 81–90, Apr. 1997.

[17] Ansoft, “Maxwell QP3D,“ Ansoft, Pittsburgh, PA, 1999.

[18] G. Pan, P. Piel, and B. Gilbert, “A hybrid integral equation and neural network approach for fast extraction of frequency dependent parameters of multiconductor transmission lines,” *Int. J. of RF Microw. Comp. Aided Eng.*, **12**(1), 37–50, Jan. 2002.

[19] N. Soveiko and M. Nakhla, “Efficient capacitance extraction computations in wavelet domain,” *IEEE Trans. Circ. and Syst. I,* **47**(5), 684–701, May 2000.

[20] G. Wang, G. Pan, and B. Gilbert, “A hybrid wavelet expansion and boundary element analysis for multiconductor transmission lines in multilayered dielectric media,” *IEEE Trans. Microwave Theory Tech.,* **43**, 664–675, March 1995.

[21] C. Brebbia, *The Boundary Element Method for Engineers*, Pentech Press, London, 1978.

[22] C. Brebbia, J. Telles, and L. Wrobel, *Boundary Element Techniques*, Springer-Verlag, Berlin, 1984.

[23] W. Chew, *Waves and Fields in Inhomogeneous Media*, VNR, New York, 1990.

[24] C. Chui, ed., *Wavelets—A Tutorial in Theory and Applications*, Academic Press, San Diego, CA, 1992.

[25] W. Press, S. Teukolsky, W. Vetterling, and B. Flannery, *Numerical Recipes in C: The Art of Scientific Computing*, 2nd ed., Cambridge University Press, Cambridge, 1992.

[26] N. Fache and D. DeZutter, “Rigorous full-wave space-domain solution for dispersive microstrip lines,” *IEEE Trans. Microw. Theory Tech.*, **36**, 731–737, Apr. 1988.

[27] G. Pan, X. Zhu, and B. Gilbert, "Analysis of transient behavior of vertical interconnects in stacked circuit board layers using quasi-static techniques," *IEEE Trans. Comp. Packag. Manuf. Tech.*, **18**, 521–531, Aug. 1996.

[28] G. Pan, M. Toupikov and B. Gilbert, "A combined finite difference and analytic expression approach to crossover capacitance in a multilayer dielectric environment," *IEEE Trans. Comp., Packag. Manuf. Tech.*, **19**, 615–620, August 1995.

[29] E. Hammerstad and O. Jensen, "Accurate models for microstrip computer-aided design," in *IEEE MTT-S Int. Microw. Symp. Dig.*, Washington DC, 1980, pp. 407–409.

[30] I. Daubechies, "Orthonormal bases of compactly supported wavelets," *Commun. Pure Appl. Math.*, **41**, 909–996, Nov. 1988.

[31] G. Walter, "Pointwise convergence of wavelet expansions," Preprint, 1992.

[32] Oh, D. Kuznetsov and J. Schutt-Aine, "Capacitance computations in a multilayered dielectric medium using closed-form spatial Green's functions," *IEEE Trans. Microwave Theory Tech.,* **42**(8), 1443–1453, Aug. 1994.

[33] A. Oppenheim and R. Schafer, *Digital Signal Processing,* Prentice-Hall, Englewood Cliffs, NJ, 1975.

[34] M. Wei and G. Majda, "A new theoretical approach for Prony's method," *Linear Algebra and Its Applications*, **136**, 119–132, 1990.

[35] Y. Chow, J. Yang, and G. Howard, "Complex images for electrostatic field computation in multilayered media," *IEEE Trans. Microw. Theory Tech.,* **39**(7), 1120–1125, July 1991.

[36] Y. Pan, University of Illinois, Personal communication.

[37] J. Tan, G. Pan, and B. Gilbert, "Edge-Element Formulation of 3D Structures," *IEEE Trans. Microw. Theory Tech.*, **46**(11), 1809–1812, Nov. 1998.

[38] G. Pan, X. Zhu, and B. Gilbert, "Analysis of transmission lines of finite thickness above a periodically perforated ground plane at oblique orientations," *IEEE Trans. Microw. Theory Tech.*, **43**(2), 383–393, Feb. 1995.

[39] P. Watson and K. Gupta, "EM-ANN models for via interconnects in microstrip circuits, *IEEE MTT-S Int. Microw. Symp. Dig. III*, IEEE Press, 1996, pp. 1819–1822.

[40] R. Wu and J. Yang, "Boundary integral equation formulation of skin effect problems in multiconductor transmission lines," *IEEE Trans. Magn.*, **25**, 3013–3015, July 1989.

[41] W. Weeks, L. Wu, M. McAllister and A. Singh, "Resistive and Iinductive skin effect in rectangular conductors," *IBM J. Res. Develop.*, **23**(6), 652–660, Nov. 1979.

[42] B. Gilbert and G. Pan, "The application of gallium arsenide integrated circuit technology to design and fabrication of future generation digital signal processors: promises and problems," *Proc. IEEE*, **76**(7), 816–834, July 1988.

[43] I. Daubechies, *Ten Lectures on Wavelet*, SIAM, Philadelphia, 1992.

[44] B. Gilbert and L. Salmon, "Interconnection design and process characteristics for digital multichip modules operating at high system clock rates," in *Thin Film Multichip Modules*, G. Messner, I. Turlik, J. Balde, and P. Garrou, Eds., ISHM Press, Reston, VA, 1992, ch. 5.

[45] T. Itoh and R. Mittra, "Spectral-domain approach for calculating the dispersion characteristics of microstrip lines," *IEEE Trans. Microw. Theory Tech.*, **21**, 496–499, July 1973.

[46] J. Mclean, H. Ling, and T. Itoh, "Full wave modeling of electrically wide microstrip open discontinuities via a deterministic spectral domain method," *MTT-S Digest*, **2**(4), 1155–1158, 1990.

[47] X. Zhang and K. Mei, "Time-domain finite difference approach to the calculation of the frequency-dependent characteristics of the microstrip discontinuities," *IEEE Trans. Microwave Theory Tech.*, **36**, 1775–1787, Dec. 1988.

[48] D. Sheen, S. Ali, M. Abouzahra, and J. Kong, "Application of the three-dimensional finite difference time-domain method to the analysis of planar microstrip circuits," *IEEE Trans. Microwave Theory Tech.*, **38**, 849–857, July 1990.

[49] H. Yang, N. Alexopoulus, and D. Jackson, "Microstrip open-end and gap discontinuities in a substrate structure," *IEEE Trans. Microwave Theory Tech.*, **37**, 1542–1546, Oct. 1989.

[50] J. Mosig and F. Gardiol, "A dynamical radiation model for microstrip structures," *Advances in Electronics and Electron Physics*, **59**, Academic Press, New York, 1982, 138–236.

[51] J. Kong, *Electromagnetic Wave Theory*, John Wiley, New York, 1990.

[52] D. Pozar, "Improved computational efficiency for the moment method solution of printed dipoles and patches," *Electromagn.*, **3**, 299–309, 1983.

[53] B. Alpert, "Wavelets and other bases for fast numerical linear algebra," in *Wavelets: A Tutorial in Theory and Applications*, C. K. Chui, Ed., Academic Press, San Diego, CA, 1992.

[54] S. Park and C. Balanis, "Analytic technique to evaluate the asymptotic part of the impedance matrix of Sommerfeld-type integrals," *IEEE Trans. Ant. Propg.*, **45**(5), 798–805, May 1997.

[55] B. Popovdki, V. Arnautovski, L. Greev, and B. Spasenovski, "A closed-form spatial electric Green's function for the anisotropic grounded layer," *IEEE Conf. Electromagnetic Field Computation*, June, 1998, Tucson, AZ.

[56] D. Hechmann and S. Dvorak, "Rapid electromagnetic analysis of shielded structures," *IEEE Conf. Electromagnetic Field Computation,* June 1998, Tucson, AZ.

[57] A. Cangellaris, "A new methodology for the direct generation of closed-form electrostatic Green's functions in layered dielectrics," in *Proc. 16th Ann. Rev. Prog. App. Comput. Electromagn.*, **1**,108–114, Monterey, CA, Mar. 2000.

[58] I. Grandshiteyn and I. Ryzhik, *Table of Integrals, Series and Products*, Academic Press, New York, Academic, 1980.

[59] R. Bunger and F. Arndt, "Efficient MPIE approach for the analysis of 3D microstrip structures in layered media," *IEEE Trans. Microw. Theory Tech.*, **45**(8), 1141–1153, 1997.

[60] J. Kiang, S. Ali, and J. Kong, "Propagation properties of striplines periodically loaded with crossing strips," *IEEE Trans. Microw. Theory Tech.*, **37**(4), 776–786, 1989.

[61] J. Ke and C. Chen, "Dispersion and attenuation characteristics of coplanar waveguides with finite metalization thickness and conductivity," *IEEE Trans. Microw. Theory Tech.*, **43**(5), 1128–1134, 1995.

[62] W. Chew, J. Jin, C. Lu, E. Michielssen, and J. Song, "Fast solution methods in electromagnetics," *IEEE Trans. Ant. Propg.*, **45**(3), 533–543, 1997.

[63] G. Arfken, *Mathematical Methods for Physicists*, 2nd ed., Academic Press, New York, 1970, at pp. 428–430.

[64] J. Tan and G. Pan, "A new edge element analysis of dispersive waveguiding structures," *IEEE Trans. Microw. Theory Tech.*, **43**(11), 2600–2607, 1995.

[65] G. Pan, K. Wang, and B. Gilbert, "Multiwavelet based finite element method," *IEEE Trans. Microwave Theory Tech.*, to appear in Jan. 2003.

[66] G. Pan, J. Tan and B. Gilbert, "Full wave edge element based analysis of 3D metal-dielectric structures for high clock rate digital and microwave applications," *IEE Proc. Microw. Ant. Propg.*, **147**(5), 391–397, Oct. 2000.

[67] T. Senior, "Impedance boundary conditions for imperfectly conducting surfaces," *Appl. Sci. Res.*, B(8), 418–429, 1960.

[68] J. Wang and R. Mittra, "Finite element analysis of MMIC structures and electronic packages using absorbing boundary conditions," *IEEE Trans. Microw. Theory Tech.*, **42**(3), 441–449, 1994.

[69] J. Jin, *The Finite Element Methods in Electromagnetics*, John Wiley , New York, 1993.

[70] D. Pozar, *Microwave Engineering*, Addison-Wesley, Reading, MA, 1990.

[71] T. Becks and I. Wolff, "Analysis of 3-D metalization structures by a full-wave spectral domain technique," *IEEE Trans. Microwave Theory Tech.*, **40**(12), 2219–2227, 1992.

CHAPTER TEN

Wavelets in Nonlinear Semiconductor Devices

Semiconductor device behavior can be described by a system of coupled partial differential equations (PDEs) with associated boundary conditions, requiring the conservation of charge and energy. In physics one is more interested in the quantities of charge concentration, average velocity, and mean energy, for example. From an engineering standpoint, potential, fields, current, and I-V curves are the desired parameters. In this chapter we will study the drift-diffusion (DD) model, which is the simplest version of the Boltzmann transport equation (BTE) coupled with Poisson's equation. The DD model has handled most engineering problems to date reasonably well. Having studied the DD model, we will use spherical expansion and Galerkin's method to solve the 1D BTE, obtaining more advanced information of hot carrier effects and ballistic transport for deep-submicron SMOS devices, or high-frequency compound semiconductor devices. Interpolating wavelets will be employed to derive the sparse point representation (SPR) that reduces the computation burden in nonlinear modeling. Multiwavelets are used for the first time to replace the ad hoc upwind algorithms.

10.1 PHYSICAL MODELS AND COMPUTATIONAL EFFORTS

The Boltzmann transport equations (BTE), and Maxwell's equations establish a relationship between charge distribution and electric potential. Under most operating conditions, the quasi-static approximation holds for the electric field inside semiconductor devices, and it is appropriate to use Poisson's equation instead of Maxwell's equations. The electron distribution f is governed by the BTE:

$$\mathbf{v}_g(\mathbf{k}) \cdot \nabla_r f(\mathbf{r}, \mathbf{k}) - \frac{q}{\hbar}\boldsymbol{\varepsilon}(\mathbf{r}) \cdot \nabla_k f(\mathbf{r}, \mathbf{k}) = \left(\frac{\partial f}{\partial t}\right)_c = \int S(\mathbf{k}', \mathbf{k}) f(\mathbf{r}, \mathbf{k}')\, d^3\mathbf{k}' - f(\mathbf{r}, \mathbf{k}) \int S(\mathbf{k}, \mathbf{k}')\, d^3\mathbf{k}', \quad (10.1.1)$$

where the involved quantities are as follows: $\mathbf{v}_g$ group velocity, $\boldsymbol{\varepsilon}$ electric field, $S(\mathbf{k}, \mathbf{k}')$ differential electron scattering probability per unit time from state $\mathbf{k}$ to state $\mathbf{k}'$, $f(\mathbf{r}, \mathbf{k})$ distribution function, $\hbar$ normalized Planck's constant, $\hbar = h/2\pi = 1.0545 \times 10^{-37}$erg-s, and $(\cdot)_c$ denotes collision.

The steady-state BTE is a semiclassical model and is a six dimensional equation in (x, y, z, k_x, k_y, k_z), which is more challenging than the electromagnetic equations of three dimensions (x, y, z). The direct solution of the BTE is highly desirable, but obtaining such a solution is a difficult task. The direct solver that is based on pseudo-random solutions is referred to as the Monte Carlo model, which can be very computationally expensive and time-consuming [1]. Approximations are usually taken in order to simplify the BTE with a reasonable trade-off between physical accuracy and computational demand. The diffusion-drifting (DD) model, hydrodynamic model (HD), and energy transport model (ET) are among the most popular approximations used.

The HD equations are derived by multiplying the BTE by powers of the momentum and integrating it over the momentum space. The HD equations solve for particle number, momentum and energy. Since the HD equations take electron energy into account, they can produce better results in high-field conditions, although the basic ideas are nearly 100 years old. However, the solutions are only given in average. No distribution information is available. The difficulties in the HD model are from the numerical nature of the equations. When the average carrier velocity exceeds certain limiting values, the conservation laws become hyperbolic in nature, which can form numerical shock waves. Similar problem arises if space charge domains arise due to, for example, the Gunn effect [2].

Although the number of existing device–simulation programs seems to indicate that from a computational point of view most problems have been solved, this not the case. Even for the physically simplest DD model, there are major problems that remain to be solved, such as the discretization of the current-continuity equations and grid aspects (generation and adaptation). These problems will be addressed in this chapter. Also in this chapter we will apply multiresolution analysis to the modeling of semiconductor devices. The use of scalets and wavelets as a complete set of basis functions is called multiresolution analysis [3]. To derive a new algorithm, the potential distribution inside the semiconductor and the electron and hole current densities are represented by a twofold expansion in scalets and wavelets. Using only scalets allows the correct modeling of smoothly varying electromagnetic fields and material parameters. In regions with strong field variations, additional basis functions (wavelets) are introduced. In our derivations we use a special class of wavelets, namely interpolating wavelets that have already been applied to the solution of boundary problems for partial differential equations (PDE). For this type of wavelets, the evaluation of differential operators is simplified due to their simple representation in terms of cubic polynomial functions in the spatial domain.

In modeling nonlinear semiconductor devices such as transistors or diodes, we deal with functions describing carrier concentrations and potential distribution that are smooth almost everywhere in the domain except at a small interval of sharp variation near the p–n junction. We apply the SPR to generate a nonuniform grid, which is fine around the sharp variation and coarse in areas where the solution is smooth. Such a grid is a dynamic object that is fully integrated into the solution. A nonuniform grid becomes so fully adaptive that changes in the grid correspond to changes in the solution at each time step.

DD-based device simulation solves Poisson's equation and carrier continuity equations for certain specified structures, physical models, and bias conditions. Most of commercial software uses variants of finite element methods (FEM) to solve the appropriate equations. Despite its simplicity and versatility, the technique suffers from serious limitations due to the substantial computer resources required to model problems with medium or large computational volumes.

A fine computational grid is necessary only when material parameters undergo rapid changes. Some sort of mesh adaptability is needed as well. We apply the multiresolution analysis described in [3]. Several different approaches for solving hyperbolic PDEs using wavelets have been considered. Jameson [4] used wavelets to determine the areas where it was necessary refining the grid in the finite difference method. It has been noted by several authors that nonlinear operators such as multiplication are too expensive computationally to be done directly in a wavelet basis. There have been several attempts to deal with this problem. Keiser [5] has used the Coifman wavelets to obtain approximations for point values in a wavelet method, thus simplifying the treatment of nonlinearities. Here we will follow the idea of Holmstrom [6] in dealing with nonlinearities employing the sparse point representation (SPR).

10.2 AN INTERPOLATING SUBDIVISION SCHEME

Introduced by Deslauriers and Dubuc [7], the dyadic grids on the real line (or the subspace of the scaling functions),

$$V_j = \{x_{j,k} \in \mathcal{R} \mid x_{j,k} = 2^{-j}k,\, k \in \mathcal{Z}\}, \qquad j \in \mathcal{Z}, \tag{10.2.1}$$

namely the grid points, are the integers in V_0 and half-integers for V_1. In general, the dyadic grid V_{j+1} contains all the grid points in V_j, as well as additional points inserted half-way in between each of the points in V_j. More information and additional references describing interpolating subdivision schemes can be found in [8].

Given function values on V_j, $\{f_{j,k}\}_{k\in\mathcal{Z}}$, where $f_{j,k} = f(x_{j,k})$ is a function defined on the grid points in V_j, the interpolating subdivision scheme defines $f_{j+1,k}$ in V_{j+1}. The even numbered grid points $x_{j+1,2k}$ already exist in V_j, and the corresponding function values are left unchanged. Values at the odd grid points $x_{j+1,2k+1}$ are computed by polynomial interpolation from the values at the even grid points. We denote this interpolating polynomial by $P_{j+1,2k+1}$. The degree, $p-1$, of this polynomial is odd to make the scheme symmetric; that is to say, we interpolate from an even number of function values. It will become clear as we proceed to the end of this section. Formally, we define one step of the subdivision scheme as

$$\begin{cases} f_{j+1,2k} = f_{j,k} \\ f_{j+1,2k+1} = P_{j+1,2k+1}(x_{j+1,2k+1}), \end{cases} \qquad \forall k \in \mathcal{Z}, \tag{10.2.2}$$

where $P_{j+1,2k+1}(x)$ is chosen such that

$$P_{j+1,2k+1}(x_{j,k+l}) = f_{j,k+l} \qquad \text{for } -\frac{p}{2} < l < \frac{p}{2}. \tag{10.2.3}$$

Thus we use p symmetric points on the coarser grid V_j to interpolate one new function value on the finer grid V_{j+1}. For dyadic grids we can explicitly define the interpolating polynomial. For the case $p = 4$, a cubic polynomial, the computed values at odd grid points are

$$f_{j+1,2k+1} = \frac{-f_{j,k-1} + 9f_{j,k} + 9f_{j,k+1} - f_{j,k+2}}{16}$$

Repeating the aforementioned subdivision recursively we obtain representations on successively finer grids V_j as j increases, and in the limit $j \to \infty$, we have a representation of the function $f(x)$ at all dyadic rational points.

If the subdivision starts with the Kronecker delta sequence $\{\delta_{0,k}\}_{k\in\mathcal{Z}}$ on V_0 and is then refined to V_j, in the limit $j \to \infty$, we will obtain the scaling function of the interpolating wavelets $\varphi(x)$. From the construction it follows that $\varphi(x)$ has a compact support $[-p+1, p+1]$ and is symmetric around $x = 0$. If we make one step in the subdivision scheme for the sequence $\{\delta_{0,k}\}$, we obtain the two-scale relation

$$\varphi(x) = \sum_{k=-p+1}^{p-1} \varphi\left(\frac{k}{2}\right)\varphi(2x - k). \tag{10.2.4}$$

Using an integer translation of $\varphi(x)$, we have a basis in V_0, and the interpolant of any continuous function $f(x)$ in V_0 can be defined as

$$\mathcal{P}f(x) = \sum_k f_{0,k}\varphi(x - k).$$

The interpolant of any continuous function $f(x)$ in V_j can be defined as

$$\mathcal{P}_j f(x) = \sum_k f_{j,k}\varphi_{j,k}(x),$$

where $\varphi_{j,k}(x) = \varphi(2^j x - k)$, $k \in \mathcal{Z}$ is a basis in V_j. Here notation V_j is used as a function space and as a grid. Since the basis functions are cardinal, $\varphi_{j,k}(x_{j,l}) = \delta_{k,l}$, $j, k, l \in \mathcal{Z}$, there is a one-to-one correspondence between grid points and basis functions.

The scaling function spaces introduced above generate a ladder of spaces

$$\cdots \subset V_{j-1} \subset V_j \subset V_{j+1} \subset \cdots,$$

and the interpolating scheme enables us to move through these spaces (i.e., to achieve either refinement or coarsening). Additional spaces W_j can be introduced to encode the difference between V_j and V_{j+1},

$$V_{j+1} = V_j \bigoplus W_j.$$

Introducing a basis $\{\psi_{j,k}\}_{k\in\mathcal{Z}}$ in W_j, we can write

$$\mathcal{P}_{j+1}f(x) - \mathcal{P}_j f(x) = \sum_k d_{j,k}\psi_{j,k}(x),$$

where $\psi_{j,k}(x) = \psi(2^j - k)$. The function $\psi(x)$ is a wavelet and $d_{j,k}$ are wavelet coefficients. One of the simplest possible choices is to define $\psi(x)$ as

$$\psi(x) = \varphi(2x - 1).$$

This wavelet was introduced by Donoho [9]. Given a representation of a function in the space V_{j+1}, one can decompose it into a coarser scale representation in V_j and a correction in W_j. Starting with a representation in V_J, this decomposition can be repeated $J - j_0$ times:

$$\sum_k f_{J,k}\varphi_{J,k}(x) = \sum_k f_{j_0,k}\varphi_{j_0,k}(x) + \sum_{j_0\le j<J}\sum_k d_{j,k}\psi_{j,k}(x).$$

On the right-hand side, our function is decomposed into the scaling function representation on a coarse grid V_{j_0} and wavelets on successively finer scales.

10.3 THE SPARSE POINT REPRESENTATION (SPR)

The idea behind the use of a wavelet basis is that certain functions are well compressed in such a basis. As a result only a few basis functions are needed to represent the function with a small error. Assume that a function is represented by N points on a uniform grid, and the same function is represented, with an error ε, by N_s wavelet coefficients, where $N_s \ll N$. We would like to be able to compute derivatives and multiply functions in this wavelets basis in $O(N_s)$ time. The interpolating wavelet transform provides the means to achieve this goal. The chosen basis has the property that each wavelet coefficient corresponds to a function value at a grid point.

Assume that we have the wavelet representation

$$\mathcal{P}_J f(x) = \sum_k f_{j_0,k}\varphi_{j_0,k}(x) + \sum_{j=j_0}^{J-1}\sum_k d_{j,k}\psi_{j,k}(x).$$

Operations such as differentiation and multiplication can be costly when performed in a wavelet basis because of interactions between scales in a wavelet representation. It would be ideal to transform the N_s wavelet coefficients to N_s point values. Such a transform does exist for the interpolating wavelets due to the one-to-one correspondence between wavelet coefficients and point values. To obtain a sparse wavelet representation, we remove all wavelet coefficients with magnitude less than some threshold value ε. Then we have the threshold expansion

$$\mathcal{P}_J f(x) = \sum_k f_{j_0,k}\varphi_{j_0,k}(x) + \sum_{(j,k)\in I(\varepsilon)}\sum_k d_{j,k}\psi_{j,k}(x), \tag{10.3.1}$$

where the set $I(\varepsilon)$ contains indices of all significant coefficients. The inverse transform can be performed, but only for those points that correspond to the significant wavelet coefficients in $I(\varepsilon)$. If any point value is needed that does not exist, it will be interpolated from the coarser scale recursively. The algorithm will terminate since we have all values on the coarsest grid V_{j_0}.

This inverse transform leads us to a sparse point representation (SPR). Note that the SPR is not a representation in a basis; rather, it is simply a collection of point values $\{f_{j,k}\}_{(j,k)\in I(\varepsilon)}$. The SPR can be computed without explicitly forming a sparse wavelet representation; that is, it is possible to store the point values in the SPR, instead of the wavelet coefficients. The wavelet coefficients are only computed to decide if the corresponding point value is to be included in the SPR or not.

To examine the approximation error arising from using the threshold expansion (10.3.1), we need the maximum norm

$$| g |_{\infty} = \max_{0\leq x\leq 1} | g(x) |.$$

We are interested in the dependence of the error on the threshold parameter ε. Donoho [9] and Holmstrom [6] have shown that the estimation

$$| f(x) - \mathcal{P}_J f(x) |_{\infty} \leq c_1\varepsilon$$

holds for a sufficiently smooth function $f(x)$ and for a large enough level J. Further the number of significant coefficients, N_s, depends on ε as

$$N_s \leq c_2\varepsilon^{-1/p},$$

or equivalently

$$\varepsilon \leq c_2^p N_s^{-p}.$$

Combining the last three inequalities we can achieve a bound on the error versus N_s as

$$| f(x) - \mathcal{P}_J f(x) |_{\infty} \leq c_3 N_s^{-p}, \tag{10.3.2}$$

where c_i $(i = 1, 2, 3)$ denote constants for a given function $f(x)$. This result indicates that the sparse interpolating wavelet approximation is of order p in the number of significant coefficients N_s.

To perform the multiplication in $O(N_s)$ time, we need to specify the SPR pattern of the product. The SPR of the product can be chosen as the union of the two operand representations. If a point value is missing, it is again interpolated from the coarser scale in the SPR.

10.4 INTERPOLATION WAVELETS IN THE FDM

Interpolation wavelets can be applied to the finite difference methods (FDM), and differentiation can be applied to the SPR of the function. For each point for which we wish to approximate the derivative, we locate the closest point in the SPR and choose

TABLE 10.1. Filter Coefficients g_i' for the First Derivative Approximation

n	−2	−1	0	1	2	3	4
$0 \le x < h$			−25/12	4	−3	4/3	−1/4
$h \le x < 2h$		−1/4	−5/6	3/2	−1/2	1/12	
$x \ge$	1/12	−2/3	0	2/3	−1/12		

TABLE 10.2. Filter Coefficients g_i'' for the Second Derivative Approximation

n	−2	−1	0	1	2	3	4	5
$0 \le x < h$			15/4	−77/6	−107/6	−13	61/12	−5/6
$h \le x < 2h$		5/6	−5/4	−1/3	7/6	−1/2	1/12	
$x \ge$	−1/12	4/3	−5/2	4/3	−1/12			

the distance to that point as the step length h. Then a centered finite difference stencil of order p can be applied, where p is the order of the interpolating wavelets in the SPR. If any point is missing, it can be interpolated from a coarser scale. If any point in the stencil is located outside the boundary, a one-sided stencil of the same order is employed. The finite difference approximations of the first and second derivatives are, respectively,

$$f'(x) \approx \frac{1}{h} \sum_i g_i' f(x + ih),$$

and

$$f''(x) \approx \frac{1}{h^2} \sum_i g_i'' f(x + ih).$$

On an interval the filter coefficients g_i' and g_i'' depend on x since a one-sided approximation near the boundaries is used; their values for the case $p = 4$ are presented in Table 10.1 and Table 10.2. In these tables the filter coefficients for the first and for the second derivatives are shown at the left boundary. The coefficients at the right boundary are reversed in order, with opposite signs. When the threshold parameter $\varepsilon \to 0$, the finite difference approximations above become ordinary finite difference approximations on a uniform grid.

In the case of two dimensions, partial derivatives in each direction are evaluated using the 1D approximation. The step length h is chosen as the distance to the closest point in the SPR, as measured along any of the coordinate directions.

10.4.1 1D Example of the SPR Application

To examine the performance of the SPR, let us consider the solution to a linear advection equation on the unit interval with initial and boundary conditions

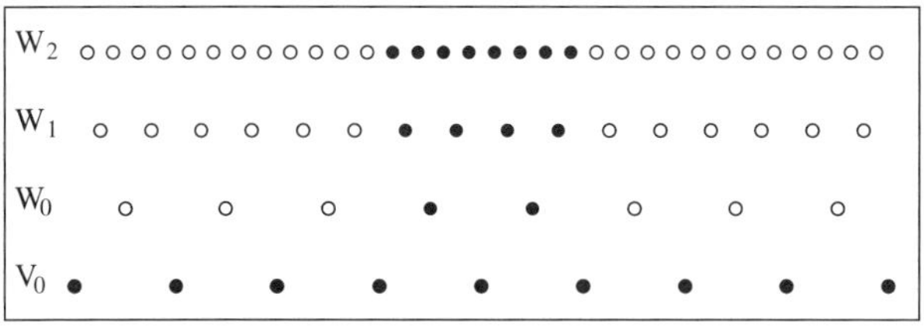

FIGURE 10.1 Example of a sparse wavelet representation.

$$\begin{cases} u_t = u_x, & 0 \leq x, t > 0, \\ u(x,0) = u_0(x), & u = u(x,t), u(1,t) = u_0(t). \end{cases}$$

The left boundary is an outflow boundary, and the right boundary is an inflow boundary. The exact solution of this problem is a periodic translation of the initial function,

$$u(x,t) = u_0[(x+t) \bmod 1].$$

As an initial function we choose

$$u_0(x) = \sin(2\pi x) + e^{-\alpha(x-1/2)^2}.$$

This is a smooth function with a sharp peak at $x = 1/2$. In this case the grid refinement must follow the solution that moves in time.

After a space discretization, forming the SPR, we have a system of ordinary differential equation with respect to time. The classical fourth-order Runge–Kutta method is used. The time step Δt is chosen as $\Delta t = sh_{\min}$, where $h_{\min}$ is the smallest distance between points in the current SPR. We have chosen $s = 0.5$ to ensure the stability of the solution. Figure 10.1 shows all significant wavelet coefficients for the function $u_0(x)$.

The solution at different time steps is shown in Fig. 10.2. The solution maintains the shape of its peak in those regions where the initial function is smooth. This demonstrates that the refined grid is moving with the solution. When using a uniform grid, we would have to work with 1025 grid points. In a SPR we retain only 159 grid points with threshold value $\varepsilon = 10^{-5}$ without loosing any accuracy. Grid refinement is performed adaptively, and grid points are updated after each time step.

10.4.2 2D Example of the SPR Application

As an example we use a function that is smooth and slowly varying, except for a small region around the point with coordinates $(\frac{1}{2}, \frac{1}{2})$:

$$u_0(x,y) = e^{-\alpha[(x-1/2)^2+(y-1/2)^2]} - 0.2 \cdot \sin(2\pi x)\sin(2\pi y),$$

where the peak slope is controlled by the parameter α. Figure 10.3 shows the graph of this function when $\alpha = 10^3$.

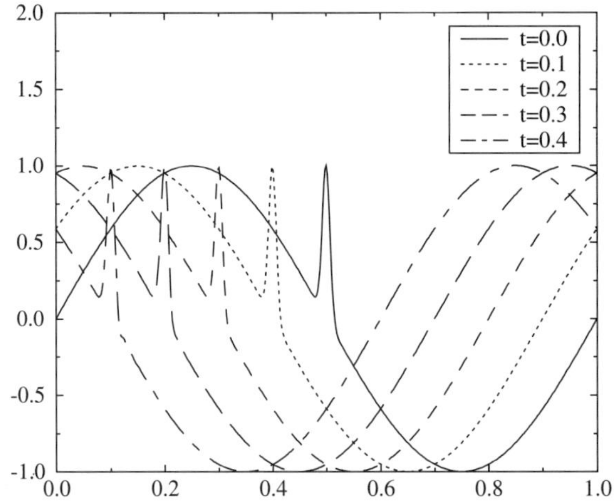

FIGURE 10.2 The solution at different time steps.

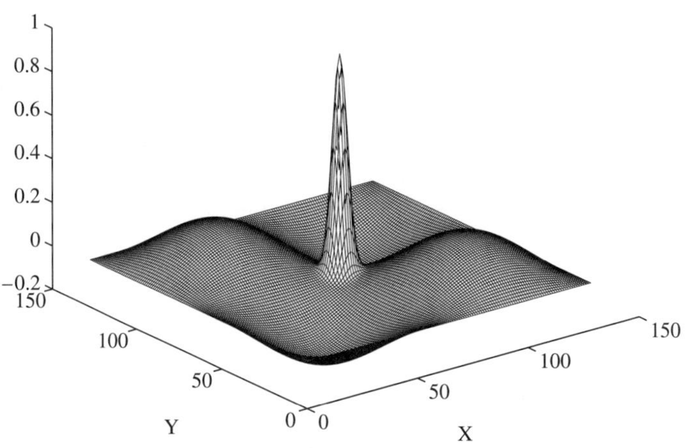

FIGURE 10.3 2D example of application of the SPR.

At the conclusion of this procedure we will obtain pictures similar to those of Figs. 10.4 and 10.5, in which each black point refers to the grid point with the assigned function value. These points correspond to significant coefficients of the test function for different values of the threshold parameter ε. Additional grid points are placed in the regions where sharp variations of the function occur.

Table 10.3 illustrates a variation in the number of significant coefficients of the test function versus the threshold parameter. The finest level of interpolating wavelets is $J = 3$. The smaller values of ε correspond to the finer mesh, until the refinement limit $\varepsilon = 0$ is reached.

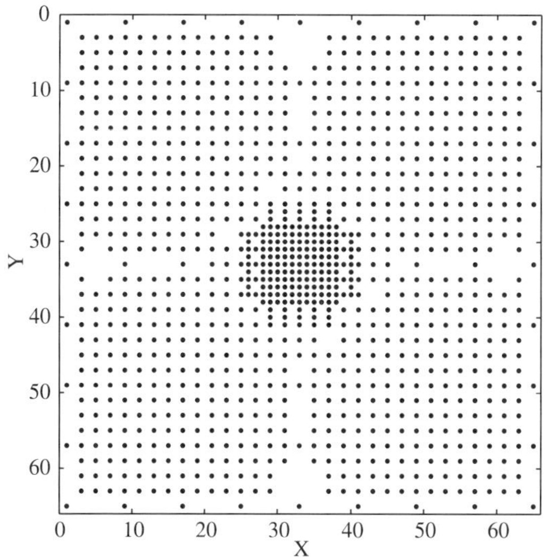

FIGURE 10.4 Significant coefficients of the test function for $\varepsilon = 10^{-5}$.

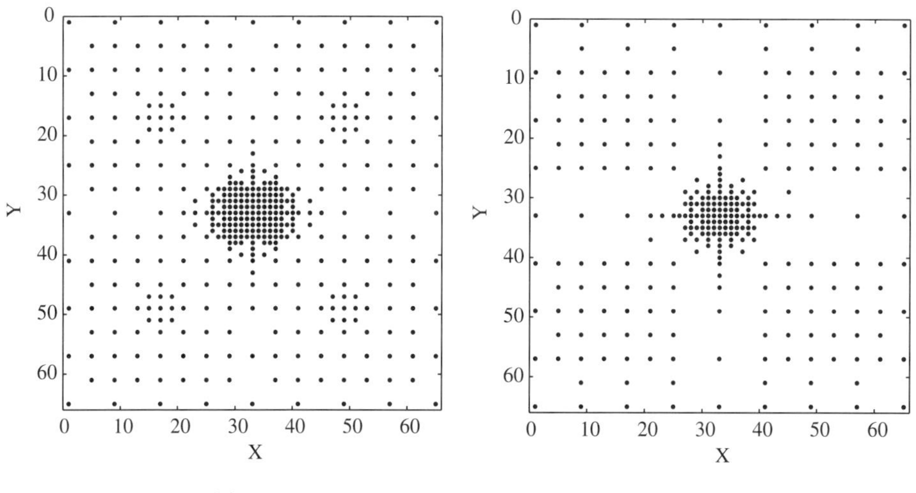

(a) (b)

FIGURE 10.5 Significant coefficients of the test function for (*a*) $\varepsilon = 10^{-4}$, (*b*) $\varepsilon = 10^{-3}$.

TABLE 10.3. Number of Significant Coefficients of the Test Function for Different Values of the Threshold Parameter

Threshold Parameter ε	Significant Coefficients
0.0	4225
0.00001	1073
0.0001	429
0.001	291

Theoretically the different meshes may cause problems when we have, for example, to add or multiply two different solution components. In such a situation it is necessary to interpolate the missing points.

To account for possible changes in the solution during a time step, or to account for abrupt discontinuities or rapidly occurring transients in the nonlinear case, it is helpful to include the neighboring values, that is, always to retain additional wavelet points in the SPR. After one or several time steps we extend all SPR to the complete solution on the finest mesh, and then using this mesh as an exact solution, we again form its SPR. If the solution changes in time rapidly, the new SPR will differ from the old one.

10.5 THE DRIFT-DIFFUSION MODEL

Under the assumption that the response of carriers to a change in the electric field is much faster than the rate of change in the field itself, we can write the set of basic equations for semiconductor transport in the form most commonly used in numerical device simulations [10, 11]. The basic physical model consists of five coupled partial differential equations: the Poisson equation for the electric field, two continuity equations for electrons and holes, and the drift-diffusion (DD) approximation for the electron and hole current densities. If a nonuniform temperature is considered, the set of equations is extended by a heat transport equation and an additional temperature diffusion term is added to the current relations. The model consists of the Poisson equation

$$\nabla^2 U = -\frac{q}{\epsilon}(N_d - N_a + p - n), \tag{10.5.1}$$

the continuity equations of the electron and hole carrier concentrations,

$$\frac{\partial n}{\partial t} - \frac{1}{q}\nabla \cdot \mathbf{J}_n = 0, \tag{10.5.2}$$

$$\frac{\partial p}{\partial t} + \frac{1}{q}\nabla \cdot \mathbf{J}_p = 0, \tag{10.5.3}$$

the DD equations of the electron and hole current densities,

$$\mathbf{J}_n = q\mu_n(\mathbf{E})n\mathbf{E} + qD_n\nabla n, \tag{10.5.4}$$

$$\mathbf{J}_p = q\mu_p(\mathbf{E})p\mathbf{E} - qD_p\nabla p, \tag{10.5.5}$$

where

U = electrical potential,

n, p = electron and hole carrier concentrations,

ϵ = dielectric permittivity of the semiconductor,

$$q = \text{magnitude of the electron charge (positive)},$$
$$N_d, N_a = \text{donor and acceptor concentrations},$$
$$\mu_n, \mu_p = \text{mobilities},$$
$$D_n, D_p = \text{diffusion coefficients},$$
$$\mathbf{J}_n, \mathbf{J}_p = \text{electron and hole electrical currents}.$$

The mobility and diffusion coefficients μ_i and D_i, $i = n, p$, respectively, may be field dependent as well as spatially dependent. The first term in (10.5.4) and (10.5.5) represents the conductivity current due to the electric field, and the second term represents a current flow due to diffusion.

Assuming the Einstein relation [12, 13] for both electrons and holes, we have

$$D_n = \mu_n(\mathbf{E})\frac{k_B T}{q},$$
$$D_p = \mu_p(\mathbf{E})\frac{k_B T}{q} \qquad \text{with } \mathbf{E} = -\nabla U,$$

where k_B is the Boltzmann's constant and $k_B = 1.3805 \times 10^{-23} JK^{-1}$. For our derivations we consider the mobility of the carriers μ_n and μ_p as constant and field-independent. Formally, we have seven unknowns $\{U, p, n, J_n^x, J_n^y, J_p^x, J_p^y\}$ and seven equations. The five equations are nonlinear namely two in (10.5.4), two in (10.5.5), and (10.5.1), where n on the RHS is a function of U. Equations (10.5.1) to (10.5.5) summarize the coupled system of partial differential equations describing the semiconductor device. It remains to specify boundary conditions for a particular geometry.

Figure 10.6 shows a representative example of a 2D cross section of a silicon abrupt diode. The potential, electron, and hole carrier concentrations sat-

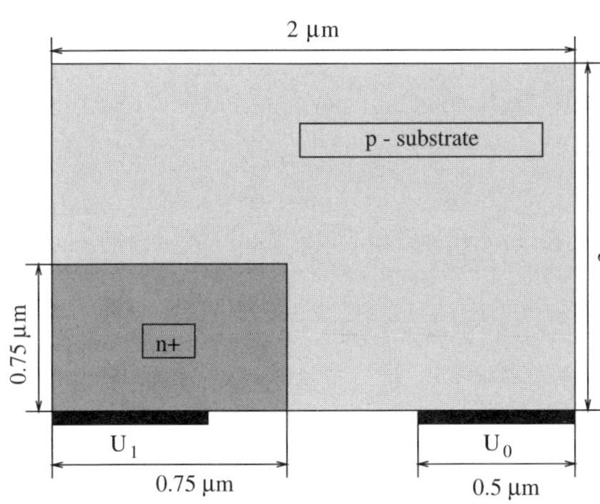

FIGURE 10.6 Idealized cross section of silicon abrupt diode.

isfy appropriate initial, boundary, and interface conditions. In general, there are semiconductor–conductor interfaces (contacts), semiconductor–isolator interfaces, and external boundaries. A semiconductor is typically attached to contacts at either metallic or polysilicon regions. We assume thermal equilibrium and charge neutrality on ohmic contacts, namely

$$np = n_i^2,$$

$$p - n + D = 0,$$

where n_i is the intrinsic carrier concentration and D is a given dope function. These conditions are supplemented by a condition on the electric potential, which is given by the built-in voltage U_{bi} and the applied potential U_a:

$$U = U_a + U_{bi}.$$

At the outside boundaries we always assume a vanishing outward electric field and vanishing outward current densities

$$\nabla U \cdot \mathbf{n} = \mathbf{J}_n \cdot \mathbf{n} = \mathbf{J}_p \cdot \mathbf{n} = 0.$$

10.5.1 Scaling

In order to solve DD equations, it is convenient to express all variables (potential, electron and hole densities, current densities, electron and hole mobilities) in terms of scaled quantities that are dimensionless. Several different scaling approaches are possible [14, 15]. Here we follow the approach proposed by De Mari [14]. Physical quantities and their scaling factors are listed in Table 10.4 where

$$L_i = \sqrt{\frac{\epsilon\epsilon_0 k_B T}{q^2 n_i}}$$

is the Debye length for the intrinsic silicon, ϵ_0 and ϵ are the permittivity in vacuum and in Si, k_B is the Boltzmann constant, and q is electronic charge. T is the temperature, which we always assume is $T = 300K$ unless otherwise specified;

TABLE 10.4. Scaling Factors after De Mari at $T300K$

Physical Quantity	Scaling Factor	Units
Spatial dimension $\mathbf{x}$	L_i	cm
Voltage potential U	U_T	V
Current density $\mathbf{J_n}$, $\mathbf{J_p}$	$qn_iL_i^{-1}$	$\mathrm{C/s\,cm^2}$
Mobility μ_n,μ_p	U_T^{-1}	1/V
Time t	L_i^2	s
Density n, p, N_a, N_d	n_i	$1/\mathrm{cm}^3$

$$U_T = \frac{k_B T}{q}$$

is called the thermal voltage. Its value at $T = 300\text{K}$ is

$$U_T = 0.0259V.$$

The scaling procedure can be seen as

$$\tilde{p} = \frac{p}{n_i}, \qquad \tilde{N_d} = \frac{N_d}{n_i}, \qquad \tilde{U} = \frac{U}{U_T},$$

$$\tilde{n} = \frac{n}{n_i}, \qquad \tilde{N_a} = \frac{N_a}{n_i}, \qquad \tilde{\mathbf{x}} = \frac{\mathbf{x}}{L_i},$$

$$\tilde{\mu_n} = \mu_n U_T, \quad \tilde{\mu_p} = \mu_p U_T, \quad \tilde{\mathbf{J}} = \frac{\mathbf{J}}{qn_i/L_i}.$$

Equations (10.5.1) through (10.5.5) can be rewritten in scaled form as

$$\nabla^2 U = -(N_d - N_a + p - n), \tag{10.5.6}$$

$$\frac{\partial n}{\partial t} - \nabla \cdot \mathbf{J}_n = 0, \tag{10.5.7}$$

$$\frac{\partial p}{\partial t} + \nabla \cdot \mathbf{J}_p = 0, \tag{10.5.8}$$

$$\mathbf{J}_n = \mu_n(-n\nabla U + \nabla n),$$

$$\mathbf{J}_p = \mu_p(-p\nabla U - \nabla p). \tag{10.5.9}$$

In the rest of the section we will use the same notation for scaled and the unscaled quantities. Without explicit declaration we always assume that the quantities are scaled.

10.5.2 Discretization

The spatial discretization techniques that are used for the analysis of semiconductor devices can be divided into three families [16]: discretization techniques based on the finite difference (FD) method, the finite volume (box) method, and the finite element method (FEM). Although these three approaches are different in origin, they often lead to equivalent systems of discrete equations. In all cases measurements should be taken, so as to prevent the domination of convection terms. This is usually done by a scheme of Scharfetter–Gummel type [17].

Within the FD scheme, the physical domain is mapped onto a topologically regular grid that is obtained by partitioning each spatial dimension of the simulation domain into a finite number of intervals. We assume the vector quantities in the equations to be constant over the grid edge. We therefore denote the edge vector components with indexes of the edge midpoint. For example, the electron current flowing from node (i, j) to node $(i+1, j)$ will be labeled as $J_n(i+1, j)$. The scalar quantities in our equations are labeled according to the node where they are defined. For example, the electron density at node (i, j) will be labeled $n(i, j)$.

At node (i, j) the FD discrete approximation of the 2D Poisson equation is

$$\frac{U(i+1,j)-2U(i,j)+U(i-1,j)}{\Delta x^2}+\frac{U(i,j+1)-2U(i,j)+U(i,j-1)}{\Delta y^2}$$
$$= -(N_d(i,j)-N_a(i,j)+p(i,j)-n(i,j)). \tag{10.5.10}$$

Similar expressions are derived from the current continuity equations

$$\frac{\partial n(i,j)}{\partial t} = \frac{J_n(i+\frac{1}{2},j)-J_n(i-\frac{1}{2},j)}{\Delta x} + \frac{J_n(i,j+\frac{1}{2})-J_n(i,j-\frac{1}{2})}{\Delta y}, \tag{10.5.11}$$

$$\frac{\partial p(i,j)}{\partial t} = -\frac{J_p(i+\frac{1}{2},j)-J_p(i-\frac{1}{2},j)}{\Delta x} - \frac{J_p(i,j+\frac{1}{2})-J_p(i,j-\frac{1}{2})}{\Delta y}. \tag{10.5.12}$$

We cannot follow the same procedure for Eqs. (10.5.9). A piecewise linear electric potential and currents were implicitly assumed in approximations (10.5.10) to (10.5.12). This assumption is too crude for carrier densities. Unless a prohibitively dense grid is used, a more accurate fitting expression for the carrier densities along the edges is needed. In 1969 Scharfetter and Gummel published a robust method [17] for the discretization of the current equations in 1D. By integrating these equations over each interval, and accounting for a linear potential change, the edge currents that appear in (10.5.9) can be written as

$$J_n\left(i+\frac{1}{2},j\right) = \frac{n(i+1,j)B(U(i+1,j)-U(i,j))-n(i,j)B(U(i,j)-U(i+1,j))}{\Delta x},$$

$$J_p\left(i+\frac{1}{2},j\right) = -\frac{p(i+1,j)B(U(i,j)-U(i+1,j))-p(i,j)B(U(i+1,j)-U(i,j))}{\Delta x},$$

where $B(x)$ is the Bernoulli function

$$B(x) = \frac{x}{e^x-1}.$$

The derivation of these equations can be found in [17]. The discrete model therefore consists of a system of two nonlinear ordinary differential equations (ODE) of the first order (10.5.11) and (10.5.12) and the algebraic equation (10.5.10). The two ordinary differential equations are coupled through the algebraic equation.

10.5.3 Transient Solution

In scaled form, the basic semiconductor equations (10.5.10), (10.5.11), and (10.5.12) can be written as

$$\begin{aligned} g_1(\mathbf{U}, \mathbf{n}, \mathbf{p}) &= 0, \\ g_2(\mathbf{U}, \mathbf{n}, \mathbf{p}) &= 0, \\ g_3(\mathbf{U}, \mathbf{n}, \mathbf{p}) &= 0, \end{aligned} \tag{10.5.13}$$

where $\mathbf{U}$, $\mathbf{n}$, and $\mathbf{p}$ are now the SPR of the normalized electrostatic potential and carrier concentrations.

Essentially, all attempts have followed two different approaches in solving the system (10.5.10), (10.5.11), and (10.5.12) numerically. The first approach is known as the Gummel iteration [18] or as the nonlinear Gauss–Seidel–Jacobi iteration [19]. The second popular approach is based upon some form of the Newton-like iteration [20]. In our calculation we will follow the modified first approach.

The resulting system of ODEs can be solved by an ODE solver. We use the fourth-order Runge–Kutta method for the solution [21].

The iterative procedure to solve the problem (10.5.13) can be presented in seven steps:

Step 1. Set initial values for the function p and n, and fix a threshold value ε.
Step 2. Obtain SPR for p, n, and U.
Step 3. Solve Poisson's equation g_1 and obtain a SPR for the potential.
Step 4. Make one step in the continuity equation g_2 for n.
Step 5. Repeat previous step in the continuity equation g_3 for p.
Step 6. Update all SPRs.
Step 7. Go to step 2.

If we are interested in a steady-state solution, iterations will continue until the solution does not change with time. The basic difficulty in the solution of the transient system is the requirement that the numerical method must be unconditionally stable. For the solution of the linear algebraic system (10.5.10) we use a fast variant of the Bi-CG iterative solver, named Bi-CGSTAB [22].

x_0 is an initial guess; $r_0 = b - Ax_0$;
$\hat{r}_0$ is an arbitrary vector, such that
$(\hat{r}_0, r_0) \neq 0$, e.g., $\hat{r}_0 = r_0$;
$\rho = \alpha = \omega_0 = 1$;
$v_0 = p_0 = 0$;
for $i = 1, 2, 3, \ldots,$
    $\rho_i = (\hat{r}_0, r_{i-1})$; $\beta = (\rho_i/\rho_{i-1})(\alpha/\omega_{i-1})$;
    $p_i = r_{i-1} + \beta(p_{i-1} - \omega_{i-1} v_{i-1})$;
    $v_i = Ap_i$;

$\alpha = \rho_i/(\hat{r}_0, v_i);$
$s = r_{i-1} - \alpha v_i;$
$t = As;$
$\omega_i = (t, s)/(t, t);$
$x_i = x_{i-1} + \alpha p_i + \omega_i s;$
if x_i is accurate enough then quit;
$r_i = s - \omega_i t;$

end

Bi-CGSTAB Algorithm. It is important that the Bi-CGSTAB method not involve any use of the transpose matrix A^T. Because we use interpolation to obtain the missing points in the finite difference procedure, it is difficult to explicitly assemble the system matrix A. Instead, we calculate all matrix vector products directly, without forming the matrix A. The matrix vector product is treated as an operator acting on the SPR of the unknown function.

10.5.4 Grid Adaptation and Interpolating Wavelets

Accuracy and efficiency are strongly related to the discretization of the equations and thus to the chosen mesh. In a standard finite difference algorithm, a tensor product mesh is usually selected. Figure 10.7 depicts a tensor product mesh for an abrupt $n - p$ diode. The mesh lines continue in the regions far from the junction, where the potential is a slowly varying function and there is no need for a fine mesh.

Methods of mesh line termination are introduced in [23, 24]. The Laplacian discretization in any node (i, j) at the end of a terminated line is obtained as a linear combination of potential values in the surrounding nodes. The local truncation error is of the order of a third derivative of the potential, and it is used as a refinement criterion. Problems of discretization and grid adaptation are addressed in [25].

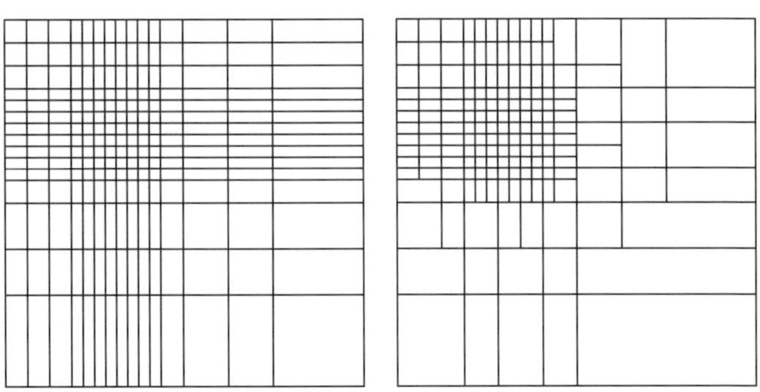

(a) (b)

FIGURE 10.7 Tensor product mesh for (*a*) an abrupt diode, (*b*) an abrupt diode with line termination.

Figure 10.7 illustrates the same grid without termination and with terminated mesh lines. Several different criteria for the grid refinement process are available. One can refine the mesh by taking into account the error in the Poisson equation, or the electron/hole continuity equations, or through the use of doping concentrations. Unfortunately, there is a high likelihood that a mesh that is optimal for the potential will be insufficient for a carrier concentration, and vice versa. Attempts to satisfy all criteria will lead to a very dense mesh. However, the interpolating wavelets provide a unique opportunity to overcome these difficulties. The SPR of a function contains the only points that correspond to significant wavelet coefficients. In the area where a function has slow variation, the SPR contains only a few function values. All other function values can be obtained through interpolation. In the area where the function varies sharply, the interpolation error may exceed the given threshold level ε. The function values there cannot be obtained through interpolation with the required precision, and additional points must be added to the SPR set until the interpolation error falls below the threshold.

Assume that we have a continuous function which is the initial condition value. The construction of the mesh can be demonstrated for a rectangular domain as follows:

Step 1. Set the smallest discretization value h, and calculate all mesh points equally spaced by the distance h in the given domain, as illustrated in Fig. 10.8. This is the refinement limit.

Step 2. Set the largest discretization value H, and calculate all mesh points equally spaced by the distance H in the given domain. This is the coarsest limit. These mesh points will define interpolating scaling functions. Let us assign the initial function values to coarse mesh points. Function values corresponding to coarse mesh points will always be present in the SPR. In Fig. 10.8 these mesh points with function values are denoted by black dots. There is a limitation on the choice of H. The condition

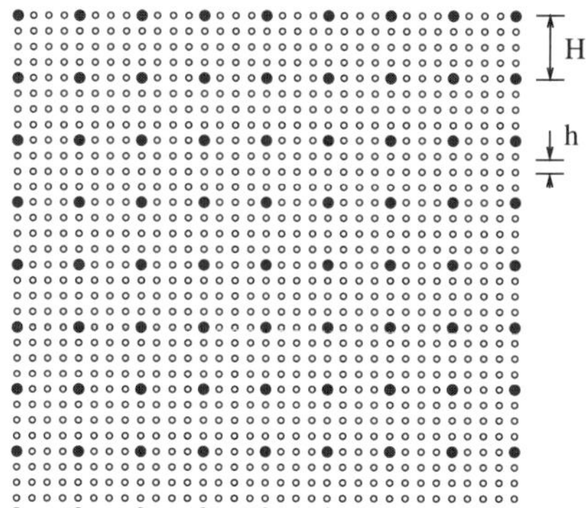

FIGURE 10.8 Lowest resolution size H and highest resolution size h in the interpolating wavelet method.

$H = 2^J h$ for a fix value of J must hold. J is then called an interpolating wavelet level.

Step 3. For all intermediate mesh points equally spaced by the distance $H/2$ and different from points defined during the previous step, we calculate their interpolating values. We then compare the interpolated value with the real value of the function at this point. If the difference is less than the given threshold value ε, the corresponding node is excluded from consideration; otherwise, it is added to the SPR. Step 3 is repeated with the half-space distance until the refinement limit is reached.

At the completion of this procedure we will get pictures similar to those of Figs. 10.4 and 10.5, where each black point, referred to the grid point, has an assigned function value. It is expected that more grid points are present in the area where the function undergo a sharp variation.

As will be seen in the numerical examples, the SPR of the potential will be different from the SPR of the carrier concentrations; even the SPR of the electron concentration will differ from the the SPR of the hole concentration. Because interpolating wavelets have an one-to-one correspondence with grid nodes, it is possible to say that by forming the SPR of the function, we create the corresponding mesh. Each quantity to be found in the solution has its own mesh which is optimally adapted to the behavior of that quantity. By choosing different values of the threshold ε, we can control the mesh. The smaller value of ε corresponds to a finer mesh, until the refinement limit is reached. Theoretically different meshes may cause problems when we add or multiply two different solution quantities. In such a case we would have to interpolate the missing points. There is a certain freedom in choosing the SPR of the result. For example, the resulting SPR structure may resemble the combined structure of the operands.

To account for possible changes in the solution during a time step, we always keep a few more wavelet points in the SPR. After a small number of time steps we extend all SPR to the complete solution on the finest mesh. Using this as the final solution, we again form its SPR. If the solution changes in time rapidly, the updated SPR will differ from the previous one.

10.5.5 Numerical Results

After a new algorithm is developed it is always necessary to verify its correctness and to test its numerical accuracy. To this end we have tested our computer programs with input parameters either from previously published papers or from well-known textbooks.

Example 1 Consider a 1D silicon p–n junction in Fig. 10.9. The volume concentration of the implanted acceptors is $N_a = 5 \times 10^{15}\ \text{cm}^{-3}$ and the volume concentration of the implanted donors $N_d = 1 \times 10^{15}\ \text{cm}^{-3}$. The 1D problem has been discretized using the SPR with interpolating wavelets. Unlike analytic solutions, in which as-

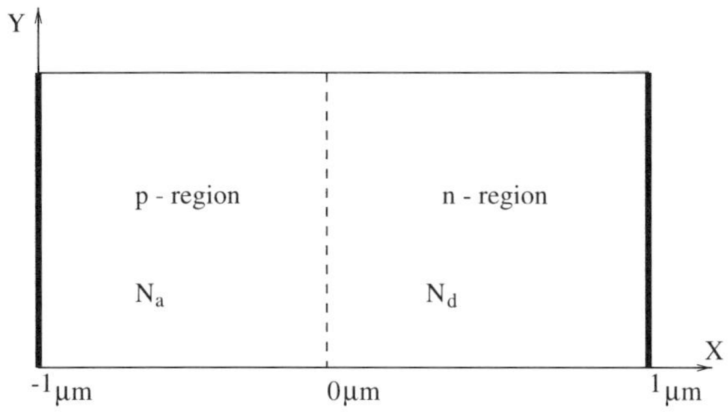

FIGURE 10.9 Diagram of a 1D silicon p–n junction.

sumptions are made to simplify the mathematics, we have faithfully followed the tedious numerical procedures outlined in the previous sections.

The resulting electron and hole concentrations for an abrupt silicon p-n junction with zero external bias are presented in Fig. 10.10*a* and potential distribution appears in Fig. 10.10*b*. Markers on the curves show corresponding mesh points. It is apparent from the figures that all components of the solution have their own meshes. In the case of the potential distribution this mesh is quite coarse. Numerical calculations show that several hundred iterations (250–400) are required to achieve a steady state solution of the equations. The number of iterations depends on the value of the threshold parameter ε. As we noted earlier, the smaller parameter leads to the finer mesh and more iteration steps to achieve the converged solution. As expected, these two figures are in excellent agreement with Figs. 2-2-2 and 2-2-7 of [26], indicating the superior numerical precision of this new method.

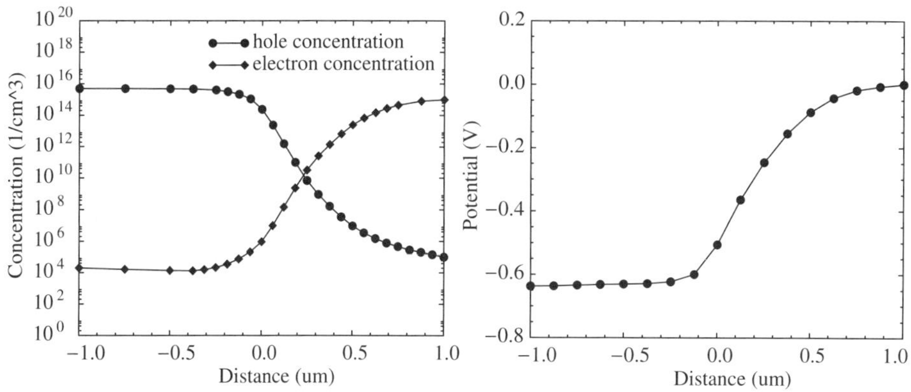

FIGURE 10.10 1D abrupt silicon p–n junction at zero external bias: (*a*) electron and hole carrier concentration, (*b*) potential distribution.

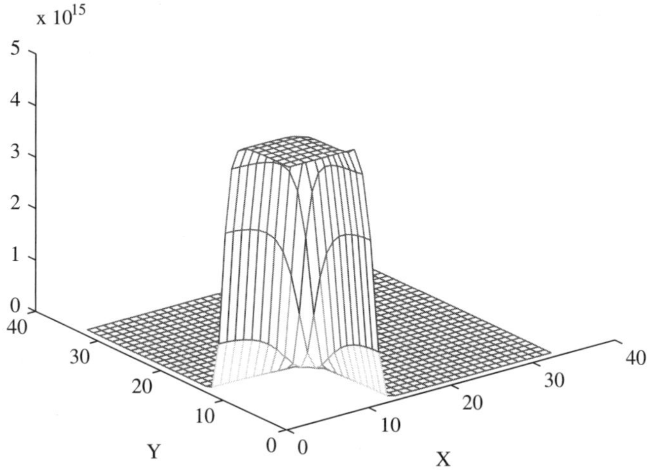

FIGURE 10.11 Electron concentration for a 2D abrupt silicon p–n junction with zero external bias.

Example 2 Consider an abrupt $n^+ - p$ diode in 2D as depicted in Fig. 10.6. As reported in [27], the doping concentration under the left contact is $N_d = 5.0 \times 10^{15}\,\text{cm}^{-3}$. In the substrate $N_a = 1.0 \times 10^{15}\,\text{cm}^{-3}$ (p-type). Figures 10.11 and 10.12 illustrate the distribution of the electron concentration and its corresponding mesh. The number of nodes in the mesh is 325. Figures 10.13 and 10.14 illustrate

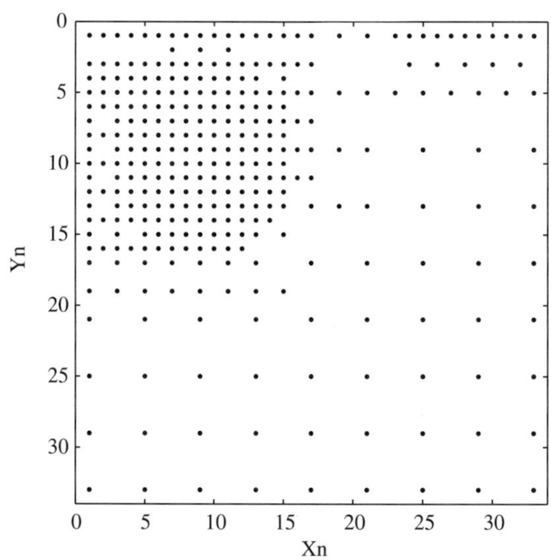

FIGURE 10.12 Grid points of electron concentration for a 2D abrupt silicon p–n junction with zero external bias.

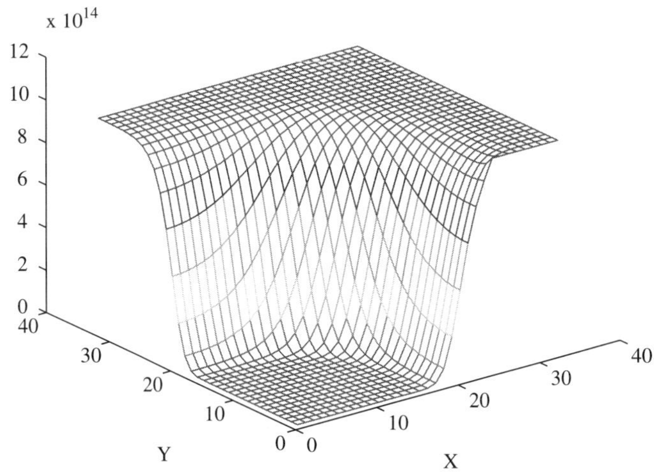

FIGURE 10.13 Hole concentration for a 2D abrupt silicon p–n junction with zero external bias.

the distribution of the hole concentration and its corresponding mesh. The number of nodes in the mesh is 613. It is apparent that the SPR of the electron concentration differs from the SPR of the hole concentration. Further, each component of the solution has its own mesh, which is optimally adapted to the behavior of that component. The 2D electron and hole concentrations have been compared with those from

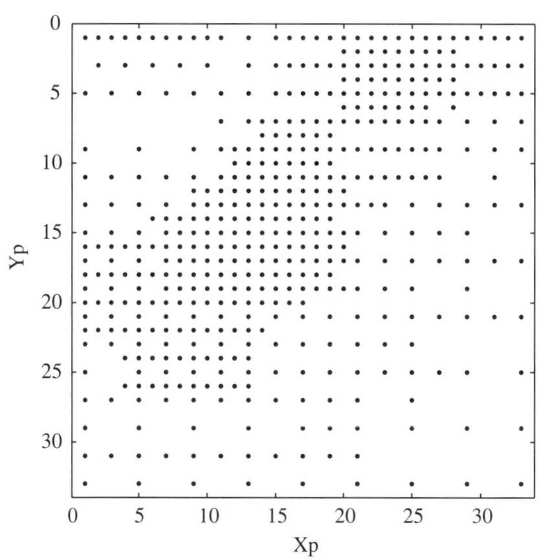

FIGURE 10.14 Grid points of hole concentration in a 2D abrupt silicon p–n junction with zero external bias.

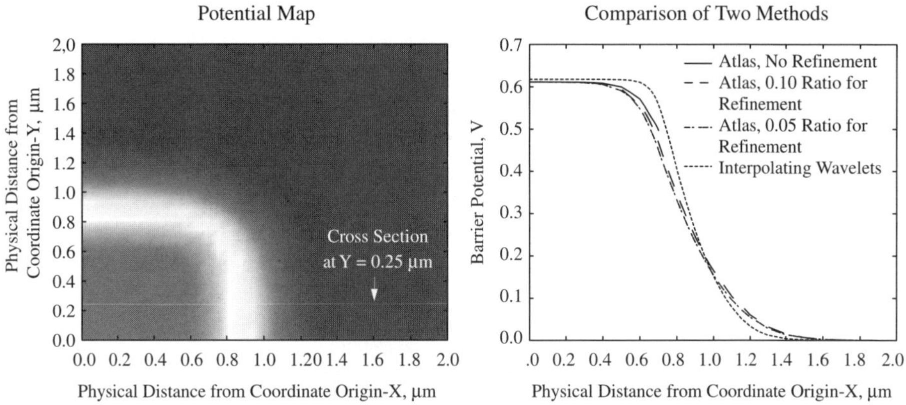

FIGURE 10.15 Equipotential map and voltage profile for a 2D abrupt silicon p–n junction with zero external bias. (Courtesy: M. Toupikov, G. Pan, and B. Gilbert, *IEEE Trans. Microw. Theory Tech.*, **48**, 500–509, Apr. 2000.)

a commercial package, ATLAS. The two sets of results are compatible, though not exactly identical. To extend the investigation, we plotted the 2D potential distribution of our results as an equipotential map in the lefthand panel of Fig. 10.15. A potential profile at $y = 0.25\ \mu$m from the map was taken and appears as the right-hand panel of Fig. 10.15. Again, our results and those from ATLAS are in good agreement.

It is often difficult to judge the precision of two numerical solutions when they show slight differences from one another. Detailed laboratory measurements would appear to be the only way to resolve these differences. However, it is extremely difficult to measure the potential profiles in such a tiny region as inside the diode. The literature does document a few so-called hero experiments in which specially designed diodes have been fabricated and passivated, and then probed with a scanning tunneling microscope to create approximate measurements of the field distributions. However, it is believed by practitioners in this field that these measurements are sufficiently indirect that the simulation results are probably closer representations of the actual device behavior than the reported measurements. Thus, we believe that we were justified in integrating the simulated current densities to obtain the device circuit parameter as the I-V curve, as depicted in Fig. 10.16. The two curves in Fig. 10.16, calculated with ATLAS and with our method, do exhibit small differences when the device bias voltage exceeds 0.6 V and the relative error reaches 19% at 0.8 V. The discrepancy between the two curves is due to slight differences in material parameters, including mobility, intrinsic density, among these parameters. If we account for the exponential behavior of the I-V curves, the small discrepancy is quite satisfactory; in this case (unlike for the potential distribution discussed above) detailed laboratory tests might help resolve the differences between these two methods.

It is worth noting the numerical efficiency of the new approach. In the solution of the potential distribution, 423 nodes were needed to achieve a precision of 1.6% for

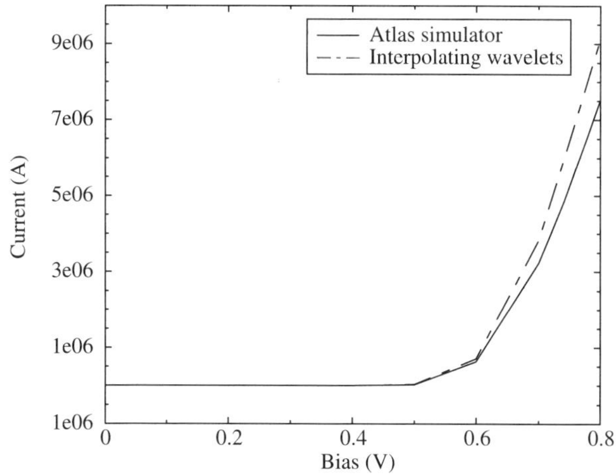

FIGURE 10.16 Comparison of I-V curves between ATLAS and wavelet results for a 2D abrupt silicon p–n junction.

the wavelets, while for a 5% precision, the Silvaco ATLAS simulator required 1756 triangles.

In Fig. 10.17 we have plotted the normalized electron current (the majority carrier current) and the hole current (the minority carrier current), both under a forward bias of 0.4 V. Figure 10.18 illustrates the electric field distributions for zero bias and the

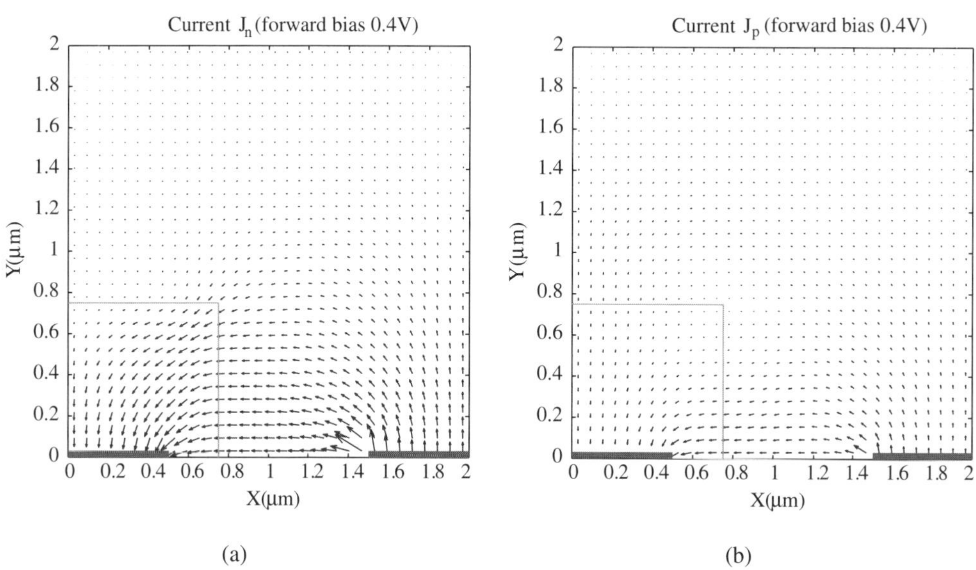

FIGURE 10.17 Current distributions for an idealized abrupt silicon p–n junction under a 0.4 V forward bias: (*a*) Electron current; (*b*) Hole current. (Courtesy: M. Toupikov, G. Pan, and B. Gilbert, *IEEE Trans. Microw. Theory Tech.*, **48**, 500–509, Apr. 2000.)

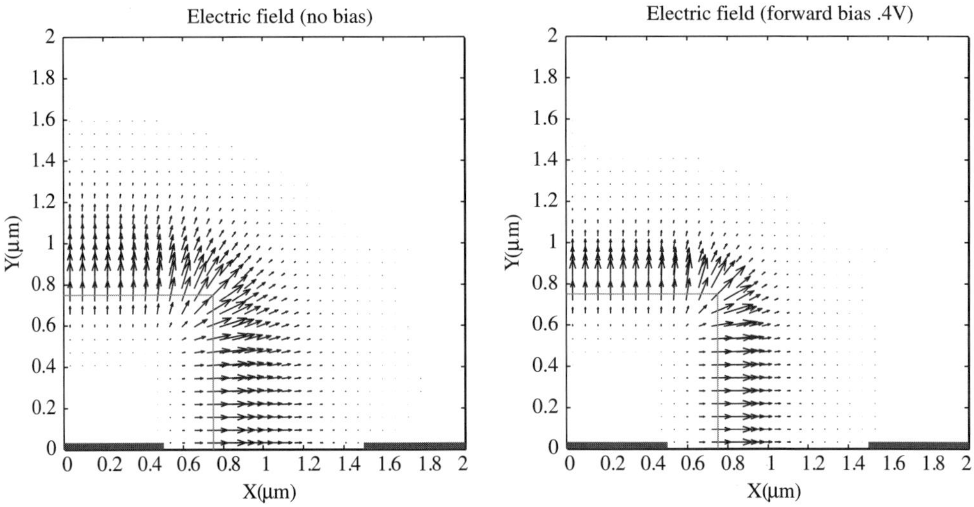

(a) (b)

FIGURE 10.18 Electric field distribution for an idealized abrupt silicon p–n junction with zero bias: (*a*) zero bias, (*b*) 0.4 V forward bias. (Courtesy: M. Toupikov, G. Pan, and B. Gilbert, *IEEE Trans. Microw. Theory Tech.*, **48**, 500–509, Apr. 2000.)

0.4 V bias. It can be seen clearly that the depletion region shrinks as the forward bias is applied. The number of iterations required in the computation was approximately 1000. The meshes developed with the computationally efficient nonuniform wavelet method described in the previous paragraphs were generated with the threshold parameter $\varepsilon = 0.1$ for the electron and hole concentrations, and $\varepsilon = 0.01$ for the potential. The full mesh of the uniform grid consisted of 1089 nodes, in contrast to the 423 nodes for the nonuniform wavelet approach. The computational results associated with the nonuniform meshes compared favorably with simulation results obtained from the full mesh in its refinement limit of size h. The number of iterations in the numerical examples presented above were in the range of 1000 to 5000 for different values of the threshold parameter ε.

10.6 MULTIWAVELET BASED DRIFT-DIFFUSION MODEL

In the previous section the drift-diffusion (DD) model was solved by the finite difference (FD) method, based upon the Scharfetter–Gummel discretization (SGD) and incorporated with interpolating wavelets.

Because of the high nonlinearity and mixed (elliptic/parabolic/hyperbolic) nature of the PDE systems, spatial discretization becomes crucial in the device simulation [28]. Some early attempts were made by using the conventional FEM and FD schemes [29, 30], but they are impractical for semiconductor simulation because of their instability. A fundamental step in the development of stable numerical schemes was made by Scharfetter and Gummel for 1D current continuity equation, which was later extended to 2D and 3D cases by the so-called generalized FD (finite box) approach [31]. As presented in Section 10.5 under linearly verying voltage Scharfetter and Gummel actually solved the 1D differential equation for current along a segment, instead of just using a two-point discretization to evaluate the current in the mesh. The SGD was extensively exploited in well-known device simulators. However, it suffers from the crosswinding effect and cannot be extended to higher orders. More recently, the canonical upwinding methods emerged as a more general and effective alternative to the ad hoc SGD and they ensure spatial stability. These methods include the Petrov–Galerkin FEM and its equivalance that has an artificial diffusivity term. Nonetheless the Petrov–Galerkin method is highly empirical without a solid mathematic underpinning; the relative weights of its symmetric and antisymmetric bases are critical, yet difficult to determine a priori. While upwind scheme is probably the most adaptive among these ad hoc approaches its numerical error is also the largest.

In this section, we apply the multiwavelet-based finite element method (MWFEM), discussed in Chapter 6, which has been shown to be effective in solving electromagnetic problems [32]. Unlike the conventional FEM, the MWFEM tracks the unknown function as well as its first derivative to guarantee a stable solution for a stiff, highly nonlinear system.

10.6.1 Precision and Stability versus Reynolds

We now implement the MWFEM for a generalized continuity equation and compare the results in terms of precision and stability against the conventional FEM, upwind FEM, SGD, and analytic solution. Let us consider the 1D continuity equation for the electrons with constant drift velocity and zero recombination,

$$\frac{d}{dx}\left(n - \frac{1}{R}\frac{dn}{dx}\right) = 0 \tag{10.6.1}$$

where $R = v\Delta x / D_n$ is the Reynolds cell number [28], which describes the nonlinearity of the system. The conventional FEM usually produces spurious spatial oscillations when $R > 2$. An approach to overcome this drawback is the Petrov–Galerkin method [33]. In the Petrov–Galerkin method, an upwind basis function q_i is superimposed on the conventional FEM basis function w_i. Under one spatial dimension, the upwind basis function can be defined as

$$q_i = \text{sign}(v) a \frac{dw_i}{dx} \tag{10.6.2}$$

where v is the local drift velocity, and a is the weighting to determine the amount of upwinding. For $a = 0.5\Delta x$, where Δx is the discretization step, we obtain the full upwind scheme, and for $a = 0$, the conventional FEM is recovered. It is interesting that the Scharfetter–Gummel scheme can be obtained by setting

$$a = \Delta x \left[\frac{1}{2} \coth \left(\frac{R}{2} \right) - \frac{1}{R} \right]. \tag{10.6.3}$$

Because of the empirical nature of the Petrov–Galerkin method, it is not easy to determine the optimum a under complicated situations. Here we introduce the multiwavelet finite element method. By using the interpolating property of the multiscalets, the electron concentration on one element can be written in terms of the interpolating functions as

$$\begin{aligned} n(x) &= \sum_{j=1}^{4} n_j N_j^e(x), \\ n'(x) &= \sum_{j=1}^{4} n_j N_j'^e(x), \end{aligned} \tag{10.6.4}$$

where N_j^e are multiwavelet basis functions

$$\begin{aligned} N_1^e &= 3\left(\frac{x_2 - x}{\Delta x}\right)^2 - 2\left(\frac{x_2 - x}{\Delta x}\right)^3, \\ N_2^e &= 3\left(\frac{x - x_1}{\Delta x}\right)^2 - 2\left(\frac{x - x_1}{\Delta x}\right)^3, \\ N_3^e &= -\left[\left(\frac{x_2 - x}{\Delta x}\right)^3 - \left(\frac{x_2 - x}{\Delta x}\right)^2\right] \Delta x, \\ N_4^e &= \left[\left(\frac{x - x_1}{\Delta x}\right)^3 - \left(\frac{x - x_1}{\Delta x}\right)^2\right] \Delta x, \end{aligned} \tag{10.6.5}$$

and the unknown coefficients are solved:

$$\begin{aligned} n_1 &= n(x_1), \\ n_2 &= n(x_2), \\ n_3 &= n'(x_1), \\ n_4 &= n'(x_2). \end{aligned} \tag{10.6.6}$$

Galerkin's procedure yields the following system equation

$$\left([A] + \frac{1}{R}[B] \right) = 0, \tag{10.6.7}$$

where

$$A = \begin{bmatrix} -\frac{1}{2} & \frac{1}{2} & \frac{\Delta x}{10} & -\frac{\Delta x}{10} \\ -\frac{1}{2} & \frac{1}{2} & -\frac{\Delta x}{10} & \frac{\Delta x}{10} \\ -\frac{\Delta x}{10} & \frac{\Delta x}{10} & 0 & -\frac{\Delta x^2}{60} \\ \frac{\Delta x}{10} & -\frac{\Delta x}{10} & \frac{\Delta x^2}{60} & 0 \end{bmatrix},$$

$$B = \begin{bmatrix} \frac{6}{5\Delta x} & -\frac{6}{5\Delta x} & \frac{1}{10} & \frac{1}{10} \\ -\frac{6}{5\Delta x} & \frac{6}{5\Delta x} & -\frac{1}{10} & -\frac{1}{10} \\ \frac{1}{10} & -\frac{1}{10} & \frac{2\Delta x}{15} & -\frac{\Delta x}{30} \\ \frac{1}{10} & -\frac{1}{10} & -\frac{\Delta x}{30} & \frac{2\Delta x}{15} \end{bmatrix}. \tag{10.6.8}$$

The numerical solution of the normalized electron concentration as a function of the normalized distance is shown in Figs. 10.19 and 10.20 for $R = 0.5$ and $R = 5$; the boundary conditions are $n(0) = 1$ and $n(1) = 0$, and the mesh size is $\Delta x = 0.1$. The SGD yields the exact values at the discretization nodes, which only holds in one dimension with constant velocity. The conventional FEM is more accurate for low Reynolds numbers but becomes oscillating for larger Reynolds numbers. In opposition, the upwind scheme is not accurate for small Reynolds numbers but remains

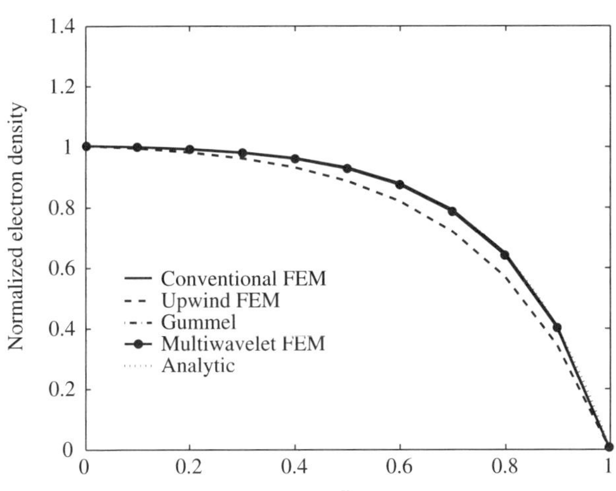

FIGURE 10.19 Comparison of different methods ($R = 0.5$).

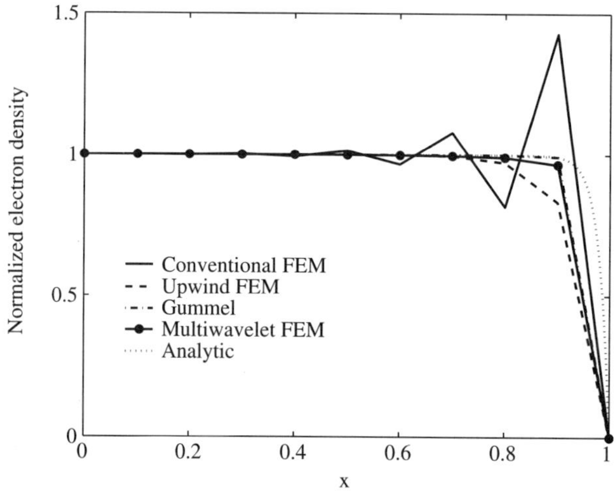

FIGURE 10.20 Comparison of different methods ($R = 5$).

stable. On the other hand, our MWFEM scheme keeps accurate and stable for either small or large Reynolds numbers. It is worth mentioning that the MWFEM is as versatile as the conventional FEM, which along with its stability makes the MWFEM suitable for complicated systems.

10.6.2 MWFEM-Based 1D Simulation

The continuity equations of the drift-diffusion model for electrons and holes under steady-state condition can be written as

$$\begin{aligned} \frac{1}{q}\nabla \cdot \mathbf{J}_n &= R, \\ \frac{1}{q}\nabla \cdot \mathbf{J}_p &= -R \end{aligned} \tag{10.6.9}$$

and

$$\begin{aligned} \mathbf{J}_n &= qu_n n\mathbf{E} + qD_n\nabla n, \\ \mathbf{J}_p &= qu_p p\mathbf{E} - qD_p\nabla p. \end{aligned} \tag{10.6.10}$$

Discretizing the equations by using MWFEM under one dimensional condition, we obtain

$$\begin{aligned} \{-u_n E[D] + u_n E[A] - D_n[B]\}\{n\} &= R, \\ \{-u_p E[D] + u_p E[A] + D_p[B]\}\{p\} &= -R, \end{aligned} \tag{10.6.11}$$

where [A] and [B] are shown in (10.6.8), and

$$D = \begin{bmatrix} -1 & 0 & 0 & 0 \\ 0 & 1 & 0 & 0 \\ 0 & 0 & 0 & 0 \\ 0 & 0 & 0 & 0 \end{bmatrix}. \tag{10.6.12}$$

To correctly implement the matrix equation, a proper normalization is necessary. Here we normalize Δx to unity, which results in a properly conditioned system.

A PIN diode structure as shown in Fig. 10.21 is simulated using the MWFEM formulation. Arora's model for field and concentration dependent mobility is adopted in this simulation. The SRH generation-recombination mechanism is considered.

The resultant electron and hole concentrations, potential profile, and electrical field under 0.2 V bias condition are obtained and plotted in Figs. 10.22 through 10.24.

In conclusion the MWFEM that was derived in Chapter 6 is applied to 1D drift-diffusion device simulation. The MWFEM basis functions make it possible for this new algorithm to be both accurate and flexible for complex nonlinear systems. Future research will extend the MWFEM to the Boltzmann–Poisson system for modeling deep submicron devices.

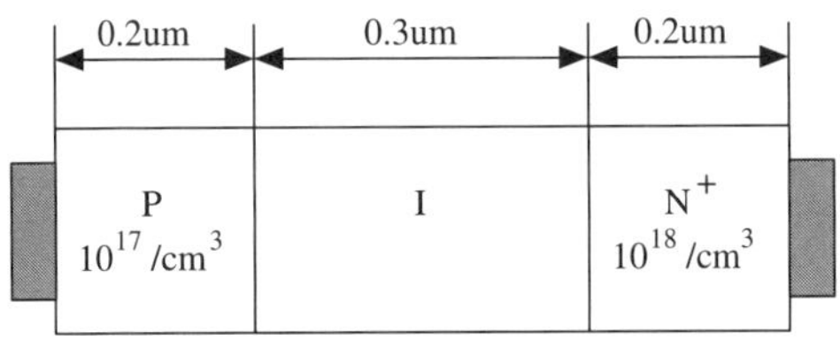

FIGURE 10.21 Configuration of the PIN diode.

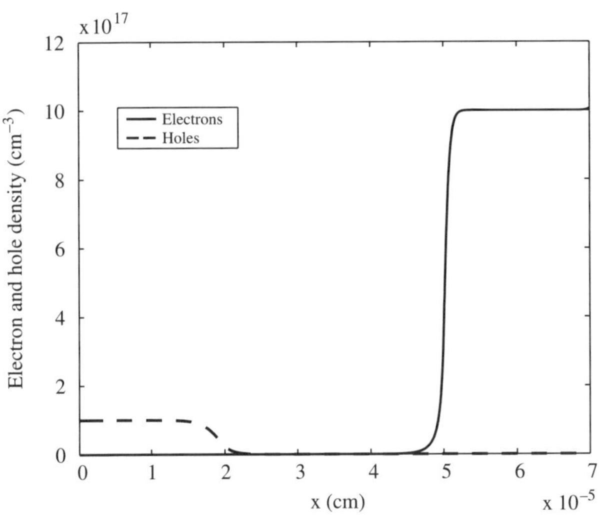

FIGURE 10.22 Electron and hole density.

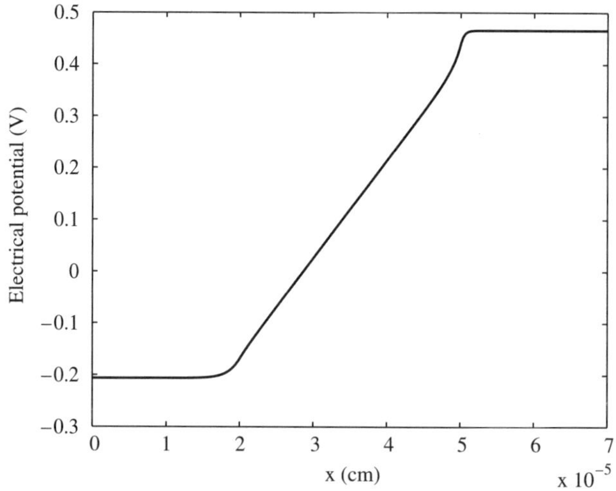

FIGURE 10.23 Electrical potential inside the PIN diode.

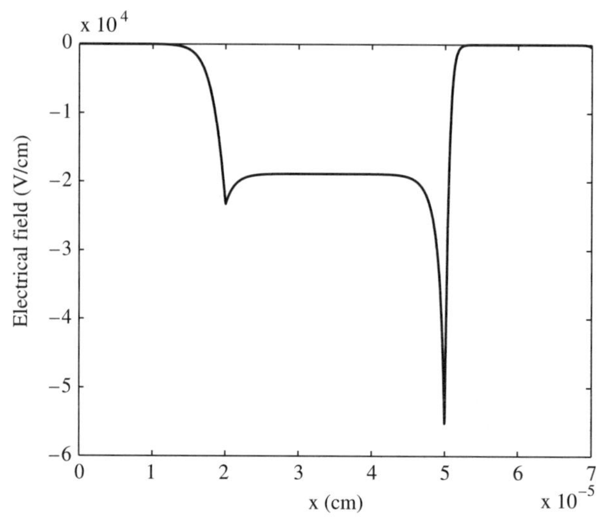

FIGURE 10.24 Electrical field inside the PIN diode.

10.7 THE BOLTZMANN TRANSPORT EQUATION (BTE) MODEL

The BTE is the most accurate semiclassic model, and it solves the distribution function. From the distribution function one can obtain the potential, velocity, energy, current information precisely, and can model the hot carrier, ballistic transport accurately. The two major approaches to solve the BTE are the Monte Carlo method [34] and orthogonal expansion method [35–37].

10.7.1 Why BTE?

The drift-diffusion (DD) and hydrodynamic (HD) models are called the continuum models. The limitation of the continuum models lie in the fact that the distribution function is not known a priori. As a result the computations that rely on the distribution function, such as the electron-electron interactions, will produce erroneous results based on the actual form of the distribution function used. Another major drawback of the continuum models is that electron–electron and electron–ion interactions can only be treated as an additional scattering mechanism in the mobility model, or the momentum and energy relaxition times. Hence the continuum models cannot fully account for carrier–ion interactions that cause multiple scattering events and introduce local variations in electron and ion densities. Because of these difficiencies, advanced device modeling turns to the Boltzmann transport equation (BTE) [38].

10.7.2 Spherical Harmonic Expansion of the BTE

The probability density function can be expanded in term of spherical harmonics

$$f(\mathbf{r}, \mathbf{k}) = \sum_{nm} f_{nm}(\mathbf{r}, k) Y_{nm}(\theta, \varphi) \tag{10.7.1}$$

for the BTE

$$\begin{aligned}\mathbf{v}(\mathbf{k}) \cdot \nabla_{\mathbf{r}} f(\mathbf{r}, \mathbf{k}) - \frac{q\boldsymbol{\varepsilon}(\mathbf{r})}{\hbar} \cdot \nabla_{\mathbf{k}} f(\mathbf{r}, \mathbf{k}) &= \int S(\mathbf{r}, \mathbf{k}', \mathbf{k}) f(\mathbf{r}, \mathbf{k}')\, d^3\mathbf{k}' \\ &\quad - f(\mathbf{r}, \mathbf{k}) \int S(\mathbf{r}, \mathbf{k}, \mathbf{k}')\, d^3\mathbf{k}', \end{aligned} \tag{10.7.2}$$

where $\mathbf{v}$ is the group velocity, and we have dropped the subscript g in (10.1.1).

The spherical harmonics are defined in [39] as

$$Y_{nm}(\theta, \varphi) = (-1)^m \sqrt{\frac{2n+1}{4\pi} \frac{(n-m)!}{(n+m)!}}\, P_n^m(\cos\theta) e^{im\varphi},$$

where $P_n^m(x)$ is the associated Legendre function, and

$$P_n^m(x) = (1-x^2)^{m/2} d^m P_n(x)/dx^m,$$

with P_n being the Legendre polynomial of order n.

First a few spherical harmonics are

$$Y_{00}(\theta, \varphi) = \frac{1}{\sqrt{4\pi}}$$

$$Y_{11}(\theta, \varphi) = -\sqrt{\frac{3}{8\pi}}\, \sin\theta e^{i\varphi}$$

$$Y_{10}(\theta, \varphi) = \sqrt{\frac{3}{4\pi}} \cos\theta$$

$$Y_{-1,1}(\theta, \varphi) = \sqrt{\frac{3}{8\pi}} \sin\theta e^{-i\varphi}$$

$$Y_{22}(\theta, \varphi) = \sqrt{\frac{5}{96\pi}} 3\sin\theta e^{2i\varphi}$$

$$Y_{21}(\theta, \varphi) = -\sqrt{\frac{5}{24\pi}} 3\sin\theta e^{i\varphi}$$

$$Y_{20}(\theta, \varphi) = \sqrt{\frac{5}{4\pi}} \frac{3\cos 2\theta + 1}{4}$$

$$\vdots$$

For the real 1D (z-directed electric fields) case there is no φ dependent, meaning that $m = 0$. Hence

$$f(z, \mathbf{k}) = \sum_n f_n(z, k)\sqrt{\frac{2n+1}{4\pi}} P_n(\cos\theta). \tag{10.7.3}$$

***Diffusion Term:* $\mathbf{v}(\mathbf{k}) \cdot \nabla_{\mathbf{r}} f(\mathbf{r}, \mathbf{k})$.** Velocity $\mathbf{v}(\mathbf{k})$ is determined from the band structure

$$\mathbf{v}(\mathbf{k}) = \frac{1}{\hbar} \cdot \frac{\partial E}{\partial \mathbf{k}} \tag{10.7.4}$$

under spherical symmetry and

$$\mathbf{v}(\mathbf{k}) = \left(\frac{2\gamma}{m^*}\right)^{1/2} \frac{dE}{d\gamma}\hat{k} \tag{10.7.5}$$

for a nonparabolic band structure, and m^* is the effective mass. In the equation above

$$\gamma(E) = \frac{(\hbar k)^2}{2m^*}$$

is for general nonparabolic band structure, $\gamma = E(1 + \alpha E)$, where α is the nonparabolic factor. When $\gamma(E) = E$, (10.7.5) reduces to parabolic band model of (10.7.4).

Hence the diffusion term

$$\begin{aligned} &\mathbf{v}(\mathbf{k}) \cdot \nabla_{\mathbf{r}} f(\mathbf{r}, \mathbf{k}) \\ &= v(k)\underbrace{\hat{k} \cdot \hat{z}}_{\cos\theta} \frac{\partial}{\partial z} \sum_n \sqrt{\frac{2n+1}{4\pi}} f_n(z, k) P_n(\cos\theta) \end{aligned}$$

$$
\begin{aligned}
&= v(k)\cos\theta \sum_n \sqrt{\frac{2n+1}{4\pi}} \frac{\partial f_n(z,k)}{\partial z} P_n(\cos\theta) \\
&= v(k) \sum_n \frac{\partial f_n(z,k)}{\partial z} \cdot \frac{(n+1)P_{n+1}(\cos\theta) + nP_{n-1}(\cos\theta)}{\sqrt{4\pi(2n+1)}},
\end{aligned}
\tag{10.7.6}
$$

where we have used the recursive relation of the Legendre polynomials

$$
xP_n(x) = \frac{(n+1)P_{n+1}(x) + nP_{n-1}(x)}{2n+1}.
$$

***The Drift Term:* $[q\boldsymbol{\varepsilon}(\mathbf{r})/\hbar] \cdot \nabla_{\mathbf{k}} f(\mathbf{r}, \mathbf{k})$** For a continuous function $u(\mathbf{k})$

$$
\nabla_{\mathbf{k}} u(\mathbf{k}) = \frac{\partial u(k,\theta)}{\partial k}\hat{k} + \frac{1}{k}\frac{\partial u(k,\theta)}{\partial \theta}\hat{\theta}.
$$

Thus from (10.7.3) we have

$$
\begin{aligned}
\nabla_{\mathbf{k}} f(\mathbf{v},\mathbf{k}) &= \sum_n \sqrt{\frac{2n+1}{4\pi}} \frac{\partial f_n(z,k)}{\partial k} P_n(\cos\theta)\hat{k} \\
&\quad + \sum_n \sqrt{\frac{2n+1}{4\pi}} \frac{f_n(z,k)}{k} \frac{\partial P_n(\cos\theta)}{\partial \cos\theta} (-\sin\theta)\hat{\theta}.
\end{aligned}
$$

Noticing that $\hat{z} = \hat{k}\cos\theta - \hat{\theta}\sin\theta$, we obtain

$$
\begin{aligned}
\frac{q\boldsymbol{\varepsilon}(\mathbf{r})}{\hbar} \cdot \nabla_{\mathbf{k}} f(\mathbf{r},k) &= \frac{q\boldsymbol{\varepsilon}(z)}{\hbar}\hat{z} \\
&\quad \times \left\{ \sum_n \sqrt{\frac{2n+1}{4\pi}} \frac{\partial f_n(z,k)}{\partial z} P_n(\cos\theta)\hat{k} \right. \\
&\quad \left. - \sum_n \sqrt{\frac{2n+1}{4\pi}} \frac{f_n(z,k)}{k} \frac{\partial P_n(\cos\theta)}{\partial \cos\theta} (\sin\theta)\hat{\theta} \right\} \\
&= \frac{q\boldsymbol{\varepsilon}(z)}{\hbar} \\
&\quad \times \left\{ \cos\theta \sum_n \sqrt{\frac{2n+1}{4\pi}} \frac{\partial f_n(z,k)}{\partial k} P_n(\cos\theta) \right. \\
&\quad \left. + \sin^2\theta \sum_n \sqrt{\frac{2n+1}{4\pi}} \frac{\partial f_n(z,k)}{\partial \cos\theta} P_n(\cos\theta) \right\}.
\end{aligned}
\tag{10.7.7}
$$

After some algebra, we arrive at

$$\frac{q\boldsymbol{\varepsilon}(z)}{\hbar}\cdot\nabla_{\mathbf{k}}f(\mathbf{r},\mathbf{k})$$
$$=\frac{q\boldsymbol{\varepsilon}(z)}{\hbar}\left\{\sum_n\frac{1}{\sqrt{4\pi(2n+1)}}\left[n\frac{\partial f_n(z,k)}{\partial k}+n(n+1)\frac{f_n(z,k)}{k}\right]P_{n-1}(\cos\theta)\right.$$
$$\left.+\sum_n\frac{1}{\sqrt{4\pi(2n+1)}}\left[(n+1)\frac{\partial f_n(z,k)}{\partial k}-n(n+1)\frac{f_n(z,k)}{k}\right]P_{n+1}(\cos\theta)\right\}. \tag{10.7.8}$$

Complete BTE in Legendre Expansion Using parabolic band relations

$$\gamma(E)=\frac{(\hbar k)^2}{2m^*},$$
$$v=\frac{1}{\hbar}\cdot\frac{\partial E}{\partial k},$$
$$\frac{1}{\hbar}\cdot\frac{\partial}{\partial k}=v\frac{\partial}{\partial E},$$

we obtain

$$\frac{1}{\hbar}\frac{\partial f_n(z,k)}{\partial k}=\frac{\partial f_n(z,E)}{\partial E}v(E). \tag{10.7.9}$$

In the same fashion

$$\frac{1}{v}\frac{1}{\hbar}\frac{f_n(z,k)}{k}=f_n(z,E)\frac{\gamma'}{2\gamma}, \tag{10.7.10}$$

where $\gamma':=\partial\gamma/\partial E$. Using (10.7.9) and (10.7.10), we convert (10.7.8) into

$$\frac{q\boldsymbol{\varepsilon}(z)}{\hbar}\cdot\nabla_{\mathbf{k}}f(\mathbf{r},\mathbf{k})=q\boldsymbol{\varepsilon}(z)v(E)\sum_n\frac{1}{\sqrt{4\pi(2n+1)}}\left\{\left[n\frac{\partial f_n}{\partial k}+n(n+1)\frac{f_n}{k}\right]P_{n-1}\right.$$
$$\left.+\left[(n+1)\frac{\partial f_n}{\partial k}-n(n+1)\frac{f_n}{k}\right]P_{n+1}\right\}. \tag{10.7.11}$$

The Boltzmann equation can therefore be rewritten as an infinite set of coupled partial differential equations for the spherical harmonic expansion coefficients, one partial differential equation for each order. For example, the partial differential equations generated by the two lowest-order expansions of $n=0$ and $n=1$ are as follows:

CASE 1. FOR $n=0$

$$\frac{v(E)}{\sqrt{4\pi}}\left[\frac{1}{\sqrt{3}}\frac{\partial f_1}{\partial z}-q\varepsilon\left(\frac{1}{\sqrt{3}}\frac{\partial f_1}{\partial E}+\frac{1}{\sqrt{3}}\frac{\gamma'}{\gamma}f_1\right)\right]=\frac{1}{\sqrt{4\pi}}\left(\frac{\partial f_0}{\partial t}\right)_c,$$

namely

$$\frac{\partial f_1}{\partial z} - q\varepsilon\left(\frac{\partial f_1}{\partial E} + \frac{\gamma'}{\gamma} f_1\right) = \frac{\sqrt{3}}{v(E)}\left(\frac{\partial f_0}{\partial t}\right)_c. \qquad (10.7.12)$$

where $(\cdot)_c$ denotes collision.

CASE 2. FOR $n = 1$

$$\frac{v(E)}{\sqrt{4\pi}}\left[\frac{\partial f_0}{\partial z} + \frac{2}{\sqrt{5}}\frac{\partial f_2}{\partial E} - q\varepsilon\left(\frac{\partial f_0}{\partial E} + \frac{2}{\sqrt{5}}\frac{\partial f_1}{\partial E} + \frac{6}{\sqrt{5}} f_2\right)\right] = \frac{\sqrt{3}}{\sqrt{4\pi}}\left(\frac{\partial f_1}{\partial t}\right)_c. \qquad (10.7.13)$$

In (10.7.13) all coefficients with orders higher than 1 must be set to zero to create a closed system of equations, yielding

$$\frac{\partial f_0}{\partial z} - q\varepsilon(z)\left(\frac{\partial f_0}{\partial E}\right) = \frac{\sqrt{3}}{v(E)}\left(\frac{\partial f_1}{\partial t}\right)_c. \qquad (10.7.14)$$

In [35], Eqs. (10.7.12) and (10.7.14) were discretized and, after including appropriate scattering mechanisms, solved for the zeroth-order, f_0 and first-order, f_1 coefficients. The results obtained using this technique coincide with the formulations in [35].

10.7.3 Arbitrary Order Expansion and Galerkin's Procedure

Multiplying the BTE of (10.1.1) by $\int d\Omega Y^*_{l'm'}(\theta, \varphi)$, we obtain

$$\begin{aligned}
&\int Y^*_{l'm'}(\theta,\varphi)\mathbf{v}(\mathbf{k})\cdot\frac{\partial}{\partial r}\sum_{lm} f_{lm}(\mathbf{r},k)Y_{lm}(\theta,\varphi)\,d\Omega \\
&\quad - \frac{q\varepsilon}{\hbar}\int Y^*_{l'm'}(\theta,\varphi)\,\nabla_k\sum_{lm} f_{lm}(\mathbf{r},k)Y_{lm}(\theta,\varphi)\,d\Omega \\
&\quad = \frac{\partial}{\partial t}\left(\int Y^*_{l'm'}(\theta,\varphi)\sum_{lm} f_{lm}(\mathbf{r},k)Y_{lm}(\theta,\varphi)\,d\Omega\right)_c,
\end{aligned}$$

where $d\Omega = \sin\theta\, d\theta\, d\varphi$ is the differential solid angle.

The Diffusion Term

$$\begin{aligned}
&\int Y^*_{l'm'}(\theta,\varphi)v(k)\hat{k}\cdot\hat{z}\frac{\partial}{\partial z}\sum_{lm} f_{lm}(z,k)Y_{lm}(\theta,\varphi)\,d\Omega \\
&\quad = v(k)\int Y^*_{l'm'}(\theta,\varphi)\sum_{lm}\cos\theta\frac{\partial f_{lm}(z,k)}{\partial z}Y_{lm}(\theta,\varphi)\,d\Omega. \qquad (10.7.15)
\end{aligned}$$

Now, consider a two-point finite difference approximation to the space derivative

$$\frac{\partial}{\partial z} f_{lm}(z,k) \approx \frac{f_{lm}^{i+1,j} - f_{lm}^{i,j}}{\Delta z},$$

where index i denotes discretization in space and index j in energy. We can rewrite (10.7.15) as

$$v(k) \int \cos\theta \sum_{lm} \frac{f_{lm}^{i+1,j} - f_{lm}^{i,j}}{\Delta z} Y_{l'm'}^{*}(\theta,\varphi) Y_{lm}(\theta,\varphi)\, d\Omega.$$

To simplify the notation, we use

$$R_{l'm';lm} := \int_0^{2\pi} \int_0^{\pi} Y_{l'm'}^{*}(\theta,\varphi) Y_{lm}(\theta,\varphi) \sin\theta \cos\theta \, d\theta \, d\varphi. \qquad (10.7.16)$$

Thus the discretized diffusion term of the BTE is written as

$$v^j \sum_{lm} \frac{f_{lm}^{i+1,j} - f_{lm}^{i,j}}{\Delta z} R_{l'm';lm} = \frac{v^j}{\Delta z} R(f^{i+1,j} - f^{i,j}), \qquad (10.7.17)$$

where f is the coefficient vector and R is the matrix given by (10.7.16).

The Drift Term Similar to (10.7.7), we have

$$-\frac{q\varepsilon(z)}{\hbar} \int Y_{l'm'}^{*}(\theta,\varphi) \sum_{lm} \left[\frac{\partial f_{lm}(z,k)}{\partial k} Y_{lm}(\theta,\varphi)\hat{k} + \frac{f_{lm}(z,k)}{k} \frac{\partial Y_{lm}(\theta,\varphi)}{\partial \theta}\hat{\theta} \right] \cdot \hat{z}\, d\Omega$$

$$= -\frac{q\varepsilon(z)}{\hbar} \int Y_{l'm'}^{*}(\theta,\varphi) \sum_{lm} \left[\frac{\partial f_{lm}(z,k)}{\partial k} Y_{lm}(\theta,\varphi) \cos\theta \right.$$

$$\left. - \frac{f_{lm}(z,k)}{k} \frac{\partial Y_{lm}(\theta,\varphi)}{\partial \theta} \sin\theta \right] d\Omega.$$

We approximate the derivative

$$\frac{\partial f_{lm}(z,k)}{\partial k} = \frac{\partial f_{lm}(z,E)}{\partial E} \hbar v(E)$$

$$\approx \hbar v^j \frac{f_{lm}^{i,j+1} - f_{lm}^{i,j}}{\Delta E}.$$

Eventually the drift term becomes

$$- q\varepsilon(z) v^j \sum_{lm} \frac{f_{lm}^{i,j+1} - f_{lm}^{i,j}}{\Delta E} \iint Y_{l'm'}^{*}(\theta,\varphi) Y_{lm}(\theta,\varphi) \cos\theta \, d\Omega$$

$$+ q\varepsilon(z) v^j \frac{\gamma'^j}{2\gamma^j} \sum f_{lm}^{i,j} \iint Y_{l'm'}^{*}(\theta,\varphi) \frac{\partial Y_{lm}(\theta,\varphi)}{\partial \theta} \sin\theta \, d\Omega.$$

We denote the second integral as

$$Q_{l'm';lm} := \int_0^{2\pi}\int_0^{2\pi} Y_{l'm'}(\theta,\varphi)\frac{\partial Y_{lm}(\theta,\varphi)}{\partial\theta}\sin\theta\, d\Omega. \tag{10.7.18}$$

Thus the drift term acquires a compact form as

$$-q\varepsilon^i\frac{v^j}{\Delta E}R(f^{i,j+1}-f^{i,j})+q\varepsilon\frac{v^j\gamma'^j}{2\gamma^j}Qf^{i,j}.$$

The Complete Equation We put the previous expressions in a matrix form. To simplify the form, let us introduce more notations

$$\begin{aligned}
a^{i,j} &:= \frac{q\varepsilon^i}{\Delta E}v^j \approx \frac{q(U^{i-1}-U^{i+1})}{2\Delta E\Delta z}v^j\\
b^{i,j} &:= \frac{q\varepsilon^i\gamma'^j}{2\gamma^j}v^j \approx \frac{q(U^{i-1}-U^{i+1})\gamma'^j}{4\gamma^j\Delta z}v^j\\
c^{i,j} &:= \frac{v^j}{\Delta z}\\
G_1^{i,j} &:= (-c^{i,j}+a^{i,j})R+b^{i,j}Q && \text{(self)}\\
G_2^{i,j} &:= c^{i,j}R && \text{(space coupling)}\\
G_3^{i,j} &:= -a^{i,j}R && \text{(energy coupling).}
\end{aligned}$$

Then the complete BTE can be written as

$$\begin{bmatrix} \ddots & & \ddots & & \ddots \\ G_1^{i,j} & G_2^{i,j} & 0 & G_3^{i,j} & 0 \\ 0 & G_1^{i+1,j} & G_2^{i+1,j} & 0 & G_3^{i+1,j} \\ & \ddots & & \ddots & \\ 0 & & & & 0 \end{bmatrix}\begin{pmatrix} \vdots \\ f^{i,j} \\ f^{i+1,j} \\ 0 \\ f^{i,j+1} \\ f^{i+1,j+1} \\ \vdots \end{pmatrix} = [S]\begin{pmatrix} \vdots \\ f^{i,j} \\ f^{i+1,j} \\ 0 \\ f^{i,j+1} \\ f^{i+1,j+1} \\ \vdots \end{pmatrix}, \tag{10.7.19}$$

where $[S]$ is the scattering matrix, which will be discussed below.

We may verify the previous matrix-vector product on the LHS by taking one row, which is

$$
\begin{aligned}
&G_1^{i,j} f^{i,j} + G_2^{i,j} f^{i+1,j} + G_3^{i,j} f^{i,j+1} \\
&\quad = [(-c^{i,j} + a^{i,j})R + b^{i,j} Q] f^{i,j} + c^{i,j} R f^{i+1,j} - a^{i,j} R f^{i,j+1} \\
&\quad = c^{i,j} R(f^{i+1,j} - f^{i,j}) + a^{i,j} R(f^{i,j} - f^{i,j+1}) + b^{i,j} Q f^{i,j} \\
&\quad = \frac{v^i}{\Delta z} R(f^{i+1,j} - f^{i,j}) - \frac{q\varepsilon^i}{\Delta E} v^j R(f^{i,j+1} - f^{i,j}) + \frac{q\varepsilon^i \gamma'^j}{2\gamma^j} v^j Q f^{i,j}.
\end{aligned} \tag{10.7.20}
$$

In (10.7.20), the first term is the diffusion term (10.7.17); the other two terms are the drift terms of (10.7.18).

The right-hand side of the matrix equation comprises the scattering terms and the distribution function boundary conditions. Up to the third order, R and Q are precomputed from (10.7.16) and (10.7.18) for the 1D case as

$$
R = \begin{bmatrix} 0 & \frac{1}{\sqrt{3}} & 0 & 0 \\ \frac{1}{\sqrt{3}} & 0 & \frac{2}{\sqrt{15}} & 0 \\ 0 & \frac{2}{\sqrt{15}} & 0 & \frac{3}{\sqrt{35}} \\ 0 & 0 & \frac{3}{\sqrt{35}} & 0 \end{bmatrix},
$$

$$
Q = \begin{bmatrix} 0 & -\frac{2}{\sqrt{3}} & 0 & 0 \\ 0 & 0 & -2\sqrt{\frac{3}{5}} & 0 \\ 0 & \frac{2}{\sqrt{15}} & 0 & -\frac{12}{\sqrt{35}} \\ 0 & 0 & \frac{6}{\sqrt{35}} & 0 \end{bmatrix}.
$$

The Scattering Term

CASE 1. ACOUSTIC PHONON SCATTERING Acoustic phonon scattering is isotropic and completely elastic [40]:

$$
\begin{aligned}
S_{ac}(\mathbf{k}, \mathbf{k}') &= \frac{2\pi k_B T_0 \varepsilon^2}{\hbar V u_\ell^2 \rho} \delta[E(\mathbf{k}') - E(\mathbf{k})] \\
&= c_{ac} \delta[E(\mathbf{k}') - E(\mathbf{k})],
\end{aligned}
$$

where ε is the deformation potential, ρ is the density of silicon, u_ℓ is the sound velocity in silicon, and $V = (2\pi)^3$ is the volume in k-space.

The net scattering term due to acoustic phonon is

$$
\int S_{ac}(\mathbf{k}', \mathbf{k}) f(\mathbf{k}')\, d^3\mathbf{k}' - f(\mathbf{k}) \int S_{ac}(\mathbf{k}, \mathbf{k}'))\, d^3\mathbf{k}'
$$

$$= c_{ac}\left\{\int\int \delta[E(\mathbf{k}) - E(\mathbf{k}')]f(\mathbf{k}')\,d^3\mathbf{k}' - f(\mathbf{k})\int \delta[E(\mathbf{k}') - E(\mathbf{k})]\,d^3\mathbf{k}'\right\}$$

$$= c_{ac}\left[f_{0,0}Y_{0,0} - \sum_{lm} f_{lm}Y_{lm}(\theta,\varphi)\right]g(E), \qquad (10.7.21)$$

where $g(E)$ is the density of states.

Using vector-matrix expression, up to the third order, we rewrite (10.7.21) as

$$c_{ac}g(E)[Y_{0,0}Y_{1,0}Y_{2,0}Y_{3,0}]\begin{bmatrix}0 & 0 & 0 & 0\\ 0 & -1 & 0 & 0\\ 0 & 0 & -1 & 0\\ 0 & 0 & 0 & -1\end{bmatrix}\begin{bmatrix}f_{0,0}\\ f_{1,0}\\ f_{2,0}\\ f_{3,0}\end{bmatrix},$$

or compactly as

$$c_{ac}g(E)Y^T S_{ac} f.$$

Finally, we apply the Galerkin procedure to obtain

$$c_{ac}g(E)\int d\Omega Y^* Y^T S_{ac} f.$$

CASE 2. OPTICAL PHONON SCATTERING The optical phonon scattering rate

$$\begin{aligned} S_{\text{op}}(\mathbf{k},\mathbf{k}') &= \frac{\pi(D_t K)^2}{(2\pi)^3\rho\omega_{\text{op}}}\left\{N_{\text{op}};\,N_{\text{op}}+1\right\}\delta[E(\mathbf{k}') - E(\mathbf{k}) \mp \hbar\omega_{\text{op}}] \\ &= c_{\text{op}}\{N_{\text{op}}\delta[E(\mathbf{k}') - E(\mathbf{k}) - \hbar\omega_{\text{op}}] \\ &\quad + (N_{\text{op}}+1)\delta[E(\mathbf{k}') - E(\mathbf{k}) + \hbar\omega_{\text{op}}]\}, \end{aligned} \qquad (10.7.22)$$

where $D_t K$ is the coupling constant, ω_{op} is the frequency of the optical phonon, ρ is the density of the material, $\hbar\omega_{\text{op}}$ is the optical phonon energy, and N_{op} is the optical phonon number.

The net scattering rate due to optical phonon is

$$\begin{aligned} &\int S_{\text{op}}(\mathbf{k}',\mathbf{k})f(\mathbf{k}')\,d^3\mathbf{k}' - f(\mathbf{k})\int S_{\text{op}}(\mathbf{k},\mathbf{k}')\,d^3\mathbf{k}' \\ &\quad = c_{\text{op}}\left\{\int\int \left(N_{\text{op}}\delta[E(\mathbf{k}) - E(\mathbf{k}') - \hbar\omega]f(\mathbf{k}')\right.\right. \\ &\quad\quad \left. + N_{\text{op}}^{+}\delta[E(\mathbf{k}) - E(\mathbf{k}') + \hbar\omega]f(\mathbf{k}')\right)d^3\mathbf{k}' \\ &\quad\quad - f(\mathbf{k})\left[\int\int \left(N_{\text{op}}\delta[E(\mathbf{k}') - E(\mathbf{k}) - \hbar\omega]\right.\right. \end{aligned}$$

$$
\begin{aligned}
&\left. \left. + N_{\text{op}}^{+} \delta[E(\mathbf{k}') - E(\mathbf{k}) + \hbar\omega] \right) d^3\mathbf{k}' \right] \Bigg\} \\
&= c_{\text{op}}[N_{\text{op}} Y_{0,0} f_{0,0}(E - \hbar\omega) g^{-} + N_{\text{op}}^{+} Y_{0,0} f_{0,0}(E + \hbar\omega) g^{+} \\
&\quad - N_{\text{op}} Y_{0,0} f_{0,0}(E) g^{+} - N_{\text{op}}^{+} Y_{0,0} f_{0,0}(E) g^{-} \\
&\quad - N_{\text{op}} \sum_{l,m \neq 0} f_{lm}(E) Y_{lm}(\theta, \varphi) g^{+} - N_{\text{op}}^{+} \sum_{l,m \neq 0} f_{lm}(E) Y_{lm}(\theta, \varphi) g^{-}],
\end{aligned}
$$

where

$$
\begin{aligned}
g^{-} &:= g(E - \hbar\omega), \\
g^{+} &:= g(E + \hbar\omega), \\
N_{\text{op}}^{+} &= N_{\text{op}} + 1.
\end{aligned}
$$

Note that we have dropped the subscripts "op" in $\hbar\omega_{\text{op}}$ for simplicity.

In vector notation, the scattering term for 1D space up to third-order is

$$
\begin{aligned}
c_{\text{op}} g[Y_{0,0} Y_{1,0} Y_{2,0} Y_{3,0}] &\left\{ \begin{bmatrix} 1 & 0 & 0 & 0 \\ 0 & 0 & 0 & 0 \\ 0 & 0 & 0 & 0 \\ 0 & 0 & 0 & 0 \end{bmatrix} \begin{bmatrix} f_{0,0}(E + \hbar\omega) \\ f_{1,0}(E + \hbar\omega) \\ f_{2,0}(E + \hbar\omega) \\ f_{3,0}(E + \hbar\omega) \end{bmatrix} (N_{\text{op}}^{+} g^{+}) \right. \\
&- \begin{bmatrix} 1 & 0 & 0 & 0 \\ 0 & 1 & 0 & 0 \\ 0 & 0 & 1 & 0 \\ 0 & 0 & 0 & 1 \end{bmatrix} \begin{bmatrix} f_{0,0}(E) \\ f_{1,0}(E) \\ f_{2,0}(E) \\ f_{3,0}(E) \end{bmatrix} (N_{\text{op}} g^{+} + N_{\text{op}}^{+} g^{-}) \\
&\left. + \begin{bmatrix} 1 & 0 & 0 & 0 \\ 0 & 0 & 0 & 0 \\ 0 & 0 & 0 & 0 \\ 0 & 0 & 0 & 0 \end{bmatrix} \begin{bmatrix} f_{0,0}(E - \hbar\omega) \\ f_{1,0}(E - \hbar\omega) \\ f_{2,0}(E - \hbar\omega) \\ f_{3,0}(E - \hbar\omega) \end{bmatrix} (N_{\text{op}} g^{-}) \right\}.
\end{aligned}
$$

If E is with index j, then $E + \hbar\omega$ will be with index $j + 1$, and $E - \hbar\omega$ with $j - 1$.

CASE 3. IONIZED IMPURITY SCATTERING The scattering operator for ionized impurity by the Brooks–Herring model is

$$
S_{BH}(\mathbf{k}, \mathbf{k}') = \frac{Z^2 q^4 N_I}{(2\pi)^3 \epsilon_{Si}^2 \hbar [(1/L_D)^2 + 2k^2(1 - \cos\theta)]^2} \delta[E(\mathbf{k}') - E(\mathbf{k})],
$$

where $\theta = \arccos(\hat{k}, \hat{k}')$, Zq is the ionized impurity charge (for silicon $Z = 1$), N_I is the impurity concentration, ϵ_{Si} is the dielectric constant of silicon, and L_D is the Debye length that depends on doping concentration.

The scattering may be expanded in spherical harmonics, or in Legendre polynomials for 1D problems as

$$S_{BH}(\mathbf{k}, \mathbf{k}') = \sum_n \sqrt{\frac{2n+1}{4\pi}} P_n(\cos\theta) I_n \delta[E(\mathbf{k}') - E(\mathbf{k})],$$

where I_n are the coefficients of the spherical expansion. Thus the scattering due to ionized impurity is

$$\begin{aligned}
&\int S_{BH}(\mathbf{k}', \mathbf{k}) f(\mathbf{k}')\, d^3\mathbf{k}' - f(\mathbf{k}) \int S_{BH}(\mathbf{k}, \mathbf{k}')\, d^3\mathbf{k}' \\
&\quad = \sum_{lm} I_{lm} \int Y_{lm}(\theta, \varphi)\delta[E(\mathbf{k}) - E(\mathbf{k}')] \sum_{l'm'} f_{l'm'} Y_{l'm'}(\theta, \varphi))\, d^3\mathbf{k}' \\
&\qquad - I_{0,0} Y_{0,0}(\theta, \varphi) \sum_{lm} f_{lm} \int Y_{lm}(\theta, \varphi)\delta[E(\mathbf{k}') - E(\mathbf{k})]\, d^3\mathbf{k}' \\
&\quad = \left\{ \sum_{lm} I_{lm} f_{lm} Y_{lm}(\theta, \varphi) - I_{0,0} \sum_{lm} f_{lm} Y_{lm}(\theta, \varphi) \right\} g(E) \\
&\quad = g(E) Y^T S_{BH} f,
\end{aligned} \tag{10.7.23}$$

where

$$S_{BH} = 4\pi \begin{bmatrix} 0 & 0 & 0 & 0 \\ 0 & \frac{1}{3}I_1 - I_0 & 0 & 0 \\ 0 & 0 & \frac{1}{5}I_2 - I_0 & 0 \\ 0 & 0 & 0 & \frac{1}{7}I_3 - I_0 \end{bmatrix}.$$

Imposing the Galerkin procedure yields

$$g(E) \int Y^* Y^T S_{BH} f\, d\Omega.$$

Note that in (10.7.23) we have applied the addition theorem of spherical harmonics in $S(\mathbf{k}', \mathbf{k})$, where the outscattering term is only the lowest order harmonic, which is isotropic.

10.7.4 The Coupled Boltzmann–Poisson System

The Poisson equation can be expressed in terms of the distribution function, which is governed by the BTE. Hence

$$
\begin{aligned}
\nabla^2 U^i &= -\frac{q}{\epsilon_s}(N_d^i - n^i) \\
&= -\frac{q}{\epsilon_s}\left(N_d^i - \sum_j g_c(E_j) f_{0,0}^{i,j} \Delta E\right),
\end{aligned} \tag{10.7.24}
$$

where ϵ_s is the permittivity of the semiconductor, N_d^i is the doping density at node i, and g_c is the density of states. Using the finite difference approximation for the Laplacian operator

$$
\nabla^2 U \approx \frac{U^{i+1} - 2U^i + U^{i-1}}{(\Delta z)^2}, \tag{10.7.25}
$$

we may rewrite (10.7.24) as

$$
\sum_j \frac{g \Delta E \Delta z^2}{\epsilon_s} g_c(E_j) f_{0,0}^{i,j} + U^{i+1} - 2U^i + U^{i-1} = -\frac{q \Delta z^2}{\epsilon_s} N_d^i. \tag{10.7.26}
$$

Combining the discretized Poisson (10.7.26) equation with the Boltzmann transport equation (10.7.19), we arrive at

$$
\left(\begin{bmatrix}
\ddots & \ddots & & \ddots & & \ddots & \ddots \\
\ddots & G_1^{i,j} & G_2^{i,j} & 0 & G_3^{i,j} & 0 & \ddots \\
\ddots & 0 & G_1^{i+1,j} & G_2^{i+1,j} & 0 & G_3^{i+1,j} & \ddots \\
\ddots & & \ddots & & \ddots & & \ddots \\
B & \cdots & \cdots & 1 & -2 & 1 & 0 \\
& B & & 0 & 1 & -2 & 1 \\
\cdots & \cdots & B & \cdots & \cdots & \cdots & \cdots
\end{bmatrix} - \begin{bmatrix} S & 0 \\ 0 & 0 \end{bmatrix}\right)
\begin{pmatrix} \vdots \\ f^{i,j} \\ f^{i+1,j} \\ \vdots \\ U^{i,j} \\ U^{i+1,j} \\ \vdots \end{pmatrix}
=
\begin{pmatrix} \vdots \\ 0 \\ 0 \\ \vdots \\ pN_d^i \\ pN_d^{i+1} \\ \vdots \end{pmatrix}. \tag{10.7.27}
$$

The coupled Boltzmann–Poisson system is nonlinear because now the matrix elements in (10.7.27) are inexplicit functions of the distribution function f.

There are two different ways to solve the coupled Boltzmann–Poisson equations:

Case 1. The Direct Solver In this approach the BTE of (10.7.19) and the Poisson equation of (10.7.26) are solved separately. From an initial guess of potential U, we obtain the submatrices of $G_1^{i,j}$, $G_2^{i,j}$, ..., in (10.7.19). We then solve (10.7.19) for distribution $f^{i,j}$, $f^{i+1,j}$, ... Employing the obtained solution of the distribution function in the Poisson equation (10.7.24), we may solve the potential U^i. This iteration procedure repeats until a convergent solution is reached. From our experience, it takes 1000 to 2000 iterations to obtain a stable result for a simple 1D problem. This method is very sensitive to physical parameters and to initial setting. On many occasions the solution may not be smooth. Depending on the initial guess, the procedure may be divergent.

CASE 2. THE NONLINEAR SOLVER Instead of solving the BTE and Poisson equation separately, we solve the coupled nonlinear system equation (10.7.27). The initial guess of the distribution is the Maxwellian. The standard Newton–Raphson method is employed. For a nonlinear system

$$F(\mathbf{x}) = 0, \tag{10.7.28}$$

the Newton–Raphson method is formally written as

$$\mathbf{x}^{(k+1)} = \mathbf{x}^{(k)} - J^{-1} F\mathbf{x}^{(k)}. \tag{10.7.29}$$

In the previous equation, the Jacobian matrix elements

$$J_{i,j} = \frac{\partial F_i}{\partial x_j}, \tag{10.7.30}$$

where F_i is the ith row of matrix F, x_j is jth component of the independent variable vector $\mathbf{x}$.

The convergence of this nonlinear solver is very rapid. Usually 10 to 20 iterations provide us with a solution that is smoother than that from the direct solver.

10.7.5 Numerical Results

We first present a uniformly doped silicon, followed by an n^+nn^+ structure. We will keep the band structure simple and yet provide real physical insight, although the algorithm allows the use of more complicated physical models, including nonparabolic bands and multiple bands. In all examples presented the discretization utilizes 50 to 100 space intervals, and the energy level ranges from 0 to 1 eV with 25 meV per step, meaning 40 levels. Each space-energy mode has four unknowns of f_0, f_1, f_2, f_3. As a result the sparse matrix has about 8000 to 16, 000 unknowns. The iteration for the Newton–Raphson is about 10 steps. In the first two examples we use the parameters in [36] so as to verify our own codes. It turns out that the results in [36] were incorrect due to inconsistent normalization between the spherical harmonics and Legendre polynomials. Finally we simulate a deep-submicron structure to show the difference between the DD and BTE models.

Example 1 Uniformly Doped Silicon with Doping of* $2 \times 10^{18}\,\mathrm{cm}^{-3}$ *and a Bias of 0.3 V. Figure 10.25 represents the coefficients, f_{00} to f_{30} of the spherical harmonic expansion, for the doped bulk silicon. Notice in the figure that clearly, the coefficients are functions of position. The curves are with fixed energy from 0.25 meV to 1 eV with an increment in 0.25 meV. The higher the energy is, the lower the curve will be.

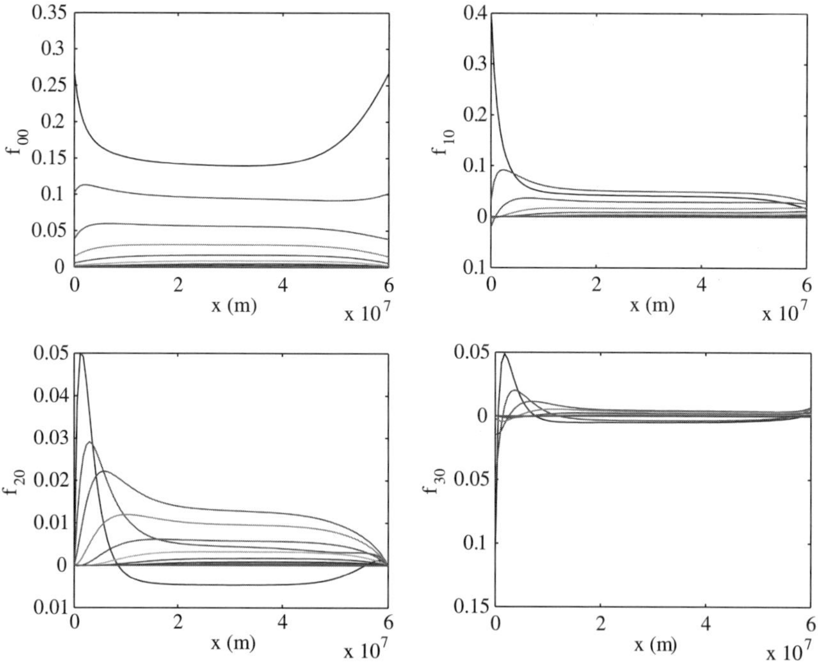

FIGURE 10.25 Spherical expansion coefficients of zeroth, first, second, and third orders for bulk silicon under a 0.3 V bias.

Figure 10.26 illustrates the potential, electric field, electron concentration and current of the bulk silicon from the BTE solution up to third order. Shown in Fig. 10.27 is the electron energy profile in the bulk silicon.

Example 2 An n^+nn^+ Structure with a Doping of $2 \times 10^{18}\,\mathrm{cm}^{-3}$ and $1 \times 10^{17}\,\mathrm{cm}^{-3}$ in the n^+ and n Regions, and a Bias of 0.6 V. The three doping regions are, respectively, [0, 0.15], [0.15, 0.45], and [0.45, 0.6] in microns. Figures 10.28 to 10.30 depict the quantities and characteristics of this n^+nn^+ structure. These three figures are arranged in the same order as in Figs. 10.25 to 10.27 so as to allow easy comparisons.

Figure 10.30 shows the electron energy distribution obtained from the BTE model. While developing the computer code of the BTE model, we prefered not to use the relaxation time, although it presents a convenient way to handle the BTE. This is because with the relaxation approach one cannot incorporate the anisotropic behavior of the scattering. Figure 10.31 shows ionized scattering from the Brooks-Herring (B-H) model and the approximation by Legendre expansion. As electron energy increases, the Legendre expansions (up to a third order) are gradually departing from the B-H model, suggesting that higher order may be needed. The individual contributions from the acoustic, optical phonon and the B-H model ionized scattering are demonstrated in Fig. 10.32. At high-energy levels, acoustic and optical

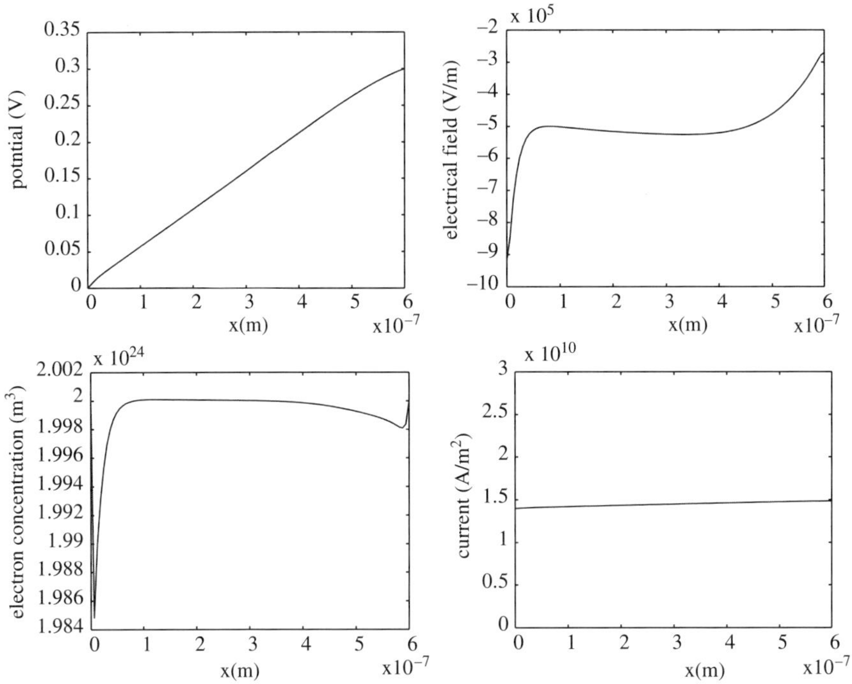

FIGURE 10.26 BTE solution up to third order for potential, electric field, electron concentration, and current in a bulk silicon under a 0.3 V bias.

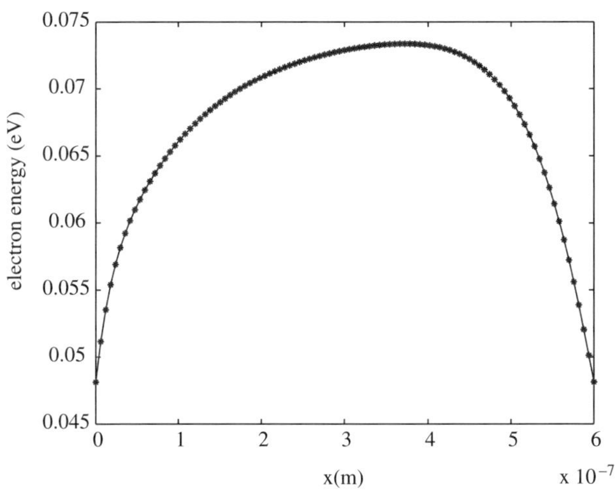

FIGURE 10.27 Electron energy distribution for bulk silicon under a 0.3 V bias.

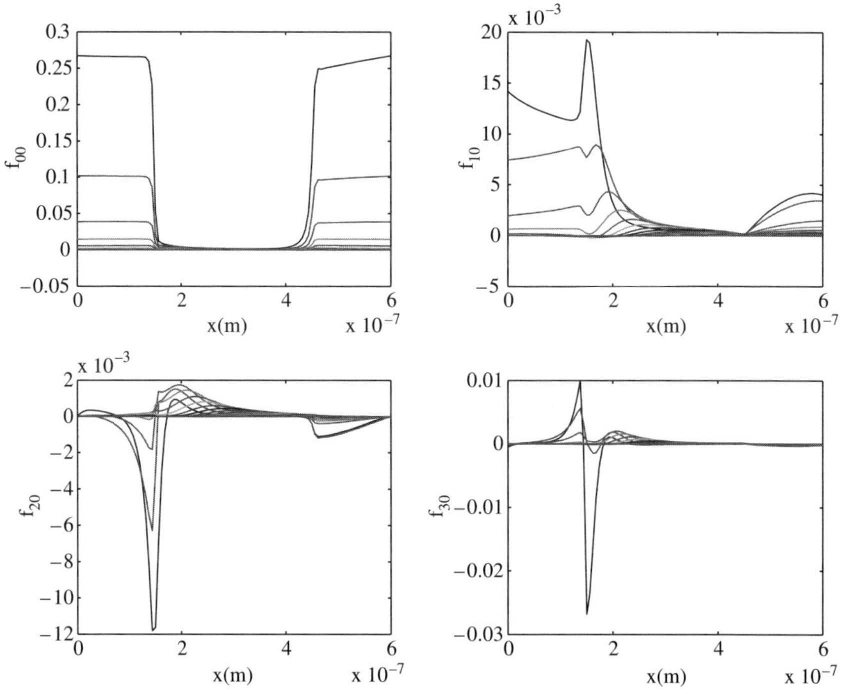

FIGURE 10.28 Spherical expansion coefficients of zeroth, first, second, and third orders for a 0.6 μm n^+nn^+ under a 0.6 V bias.

scattering are dominating, while on low-energy levels ionized scattering is the main mechanism.

Example 3 The I-V Curves of an n^+nn^+ Diode from the BTE and DD Models. This example demonstrates the necessity of using the BTE for the deep-submicron silicon device, in which the traditional drift-diffusion (DD) model introduces substantial errors in estimating the current density. The doping densities are $2 \times 10^{18}\,\text{cm}^{-3}$ in the n^+ regions and $10^{17}\,\text{cm}^{-3}$ in the n region. In the BTE model we utilized the nonparabolic band structure of $\alpha = 0.5\,\text{eV}^{-1}$ to be more realistic. The DD model is a field- and concentration-dependent program that is incorporated in the commercial software PISCES. For ease of reference, we quote the PISCES simulation codes below.

PISCES Commands of the Drift-Diffusion Model

```
title Si n+nn+diode simulation
mesh rect nx=51 ny=51
x.m n=1 loc=0.0
x.m n=51 loc=0.2
y.m n=1 loc=0
y.m n=51 loc=1
```

```
region num=1 iy.l=1 iy.h=51 ix.l=1 ix.h=51 silicon
elec num=1 ix.l=1 ix.h=1 iy.l=1 iy.h=51
elec num=2 ix.l=51 ix.h=51 iy.l=1 iy.h=51
doping uniform conc=1e17 n.type
doping uniform conc=2e18 n.type x.l=0 x.r=0.05 y.t=0 y.b=1
doping uniform conc=2e18 n.type x.l=0.15 x.r=0.2 y.t=0 y.b=1

models fldmob conmob
symb newton carr=1
method
log outf=nnn1.log
solve init
solve v1=0.1 vstep=0.1 nstep=9 elect=1
end
```

In the BTE simulation we use the following parameters taken from [41]:

$m^* = 0.26m_0$	effective mass
$u_l = 9.00 \times 10^3 \mathrm{m/s}$	sound velocity in silicon
$\rho = 2.33 \times 10^3 \mathrm{kg/m^3}$	density of silicon

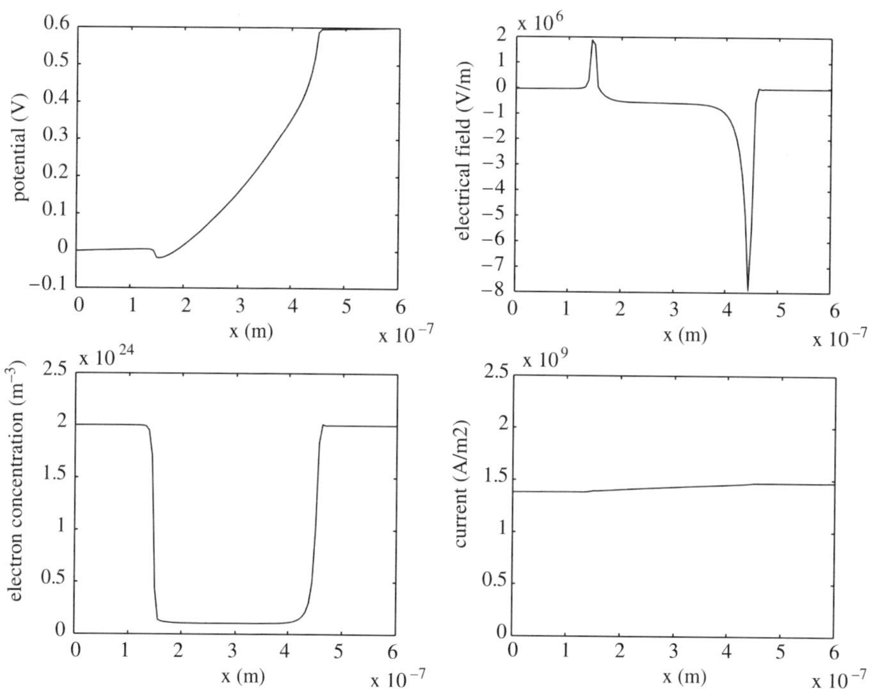

FIGURE 10.29 Average quantities for a 0.6 μm n^+nn^+ diode with a 0.6 V bias.

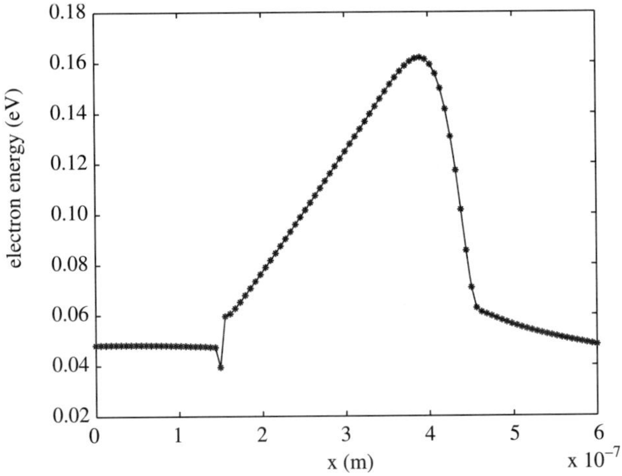

FIGURE 10.30 Electron energy for a 0.6 μm n^+nn^+ diode with a bias of 0.6 V.

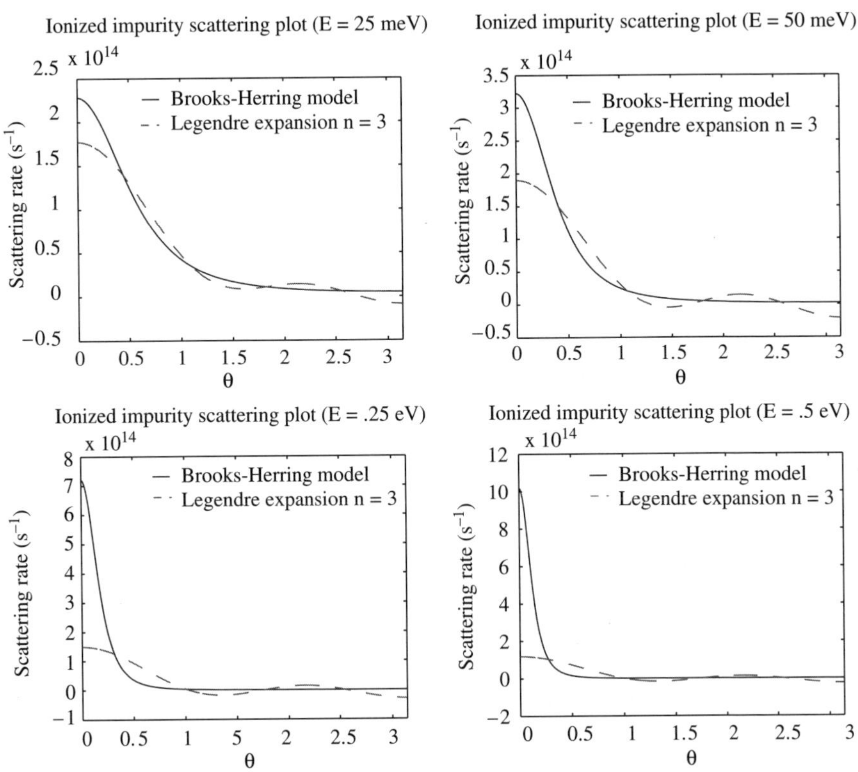

FIGURE 10.31 Spherical expansion of ionized scattering.

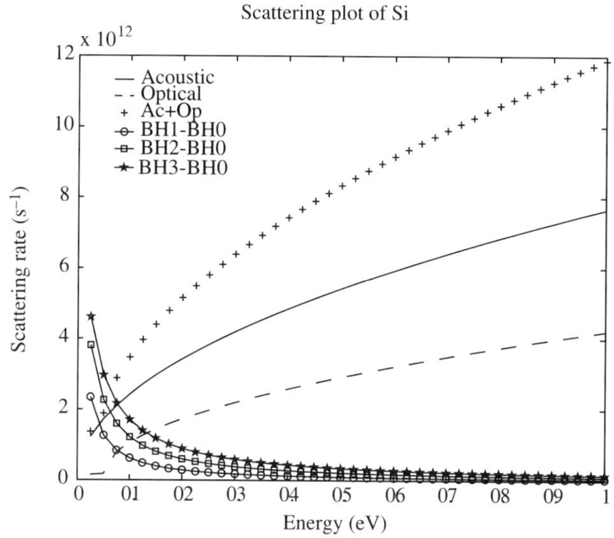

FIGURE 10.32 Contributions of optical, acoustic, and ionized scattering.

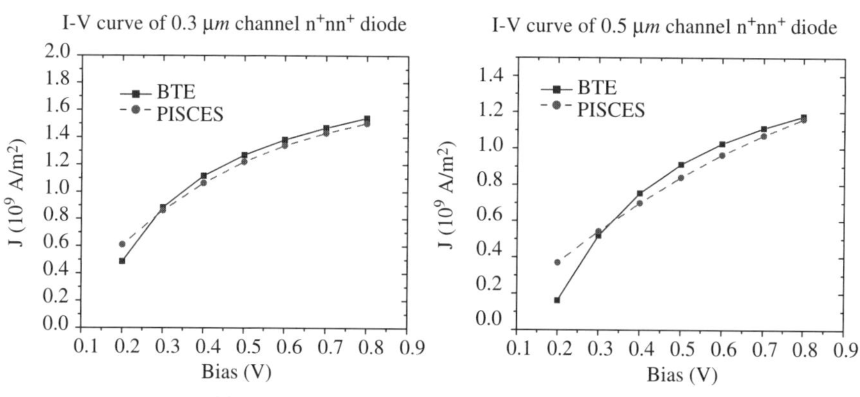

FIGURE 10.33 I-V curves of (a) 0.3 μm and (b) 0.5 μm channel in the n^+nn^+ diode.

$\varepsilon = 9.00\,\text{eV}$	acoustic coupling constant
$D_t K = 8 \times 10^8\ \text{V/cm}$	optical phonon coupling constant
$\alpha = 0.5/\text{eV}$	nonparabolic coefficient
$\hbar w_{\text{op}} = 50\,\text{meV}$	optical phonon energy

Figure 10.33a demonstrates the I-V curve, that is, the current density versus bias voltage, for a 0.3 μm channel, in which the n^+nn^+ are, respectively, in intervals [0, 0.15], [0.15, 0.45], and [0.45, 0.60] μm. The BTE model and DD model show

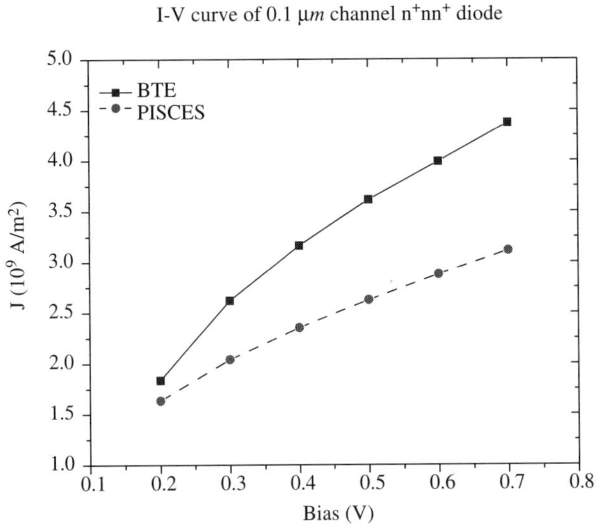

FIGURE 10.34 I-V curves of a 0.1 μm channel in the n^+nn^+ diode.

very good agreement in the bias region from 0.2 to 0.8 V, which agrees with the common belief that 0.3 μm is the limit to which the DD model can go.

Figure 10.33b is a validation of the BTE model for long channel of 0.5 μm, where the n^+nn^+ structure occupies intervals [0, 0.25], [0.25, 0.75], and [0.75, 1.00] μm. As expected, the BTE and DD models are in fairly good agreement, except for the low bias of 0.2 V where the BTE appears to have lost accuracy because of the one-side finite difference in its numerical treatment.

Figure 10.34 reveals significant discrepancies between the BTE and DD models for a short channel of 0.1 μm. The n^+nn^+ structure possesses intervals [0, 0.05], [0, 05, 0.15], and [0.15, 0.20] μm. At the 0.7 V bias, the DD model results obtained from the commercial software PISCES have introduced an error of 30%, underestimating the current. This error is from an intrinsic deficiency of the DD model, which is that it cannot address velocity overshooting nor ballistic transport in short channels.

We conclude this chapter here. Discretization is still too expensive computationally for 2D BTE models to be used. Wavelets are an essential factor in such models.

BIBLIOGRAPHY

[1] W. Gross, D. Vasileska, and D. Ferry, "Ultrasmall MOSFETs: The importance of the full coulomb interaction on device characteristics," *IEEE Trans. Electron Devices*, **47**(10), 1831–1837, Oct. 2000.

[2] W. Gross, D. Vasileska and D. Ferry, "3D simulation of ultra-small MOSFETs with real-space treatment of the electron-electron and electron-ion interactions," *VLSI Design*, **10**(4), 437–452, 2000.

[3] B. Jawerth and W. Sweldens, "An overview of wavelet based multiresolution analyses," *SIAM Rev.*, **36**(3), 377–412, Sept. 1994.

[4] L. Jameson, "On the wavelet optimized finite difference method," Technical Report ICASE 94-9, NASA Langley Research Center, Mar. 1994.

[5] J. Keiser, "Wavelet based approach to numerical solution of nonlinear partial differential equations," Ph.D. dissertation, University of Colorado, Boulder, 1995.

[6] M. Holmstrom, "Solving hyperbolic PDEs using interpolating wavelets," Technical Report No.189, 1996, Uppsala University, Sweden.

[7] G. Deslauriers and S. Dubuc, "Symmetric iterative interpolation processes," *Const. Approx.*, **5**(1), 49–68, 1989.

[8] I. Daubechies, *Ten Lectures on Wavelets*, SIAM, Philadelphia, 1992.

[9] D. Donoho, "Interpolating wavelet transform," Technical Report 408, Dept. of Statistics, Stanford University, Nov. 1992.

[10] W. Shockley, "The theory of p–n junctions in semiconductors and p–n junctions transistors," *Bell Syst. Tech. J.*, **28**, 435–489, 1949.

[11] W. Van Roosbroeck, "Theory of flow of electrons and holes in germanium and other semiconductors," *Bell Syst. Tech. J.*, **19**, 560–607, 1950.

[12] A. Marshak, "Carrier densities and emitter efficiency in degenerate materials with position dependent band structure," *Sol. State Electron.*, **21**, 429–434, 1978.

[13] R. Smith, *Semiconductors*, Cambridge University Press, Cambridge, 1978.

[14] A. De Mari, "An accurate numerical steady-state one dimensional solution of the p–n junction." *Sol. State Electron.*, **11**, 33–58, 1968.

[15] P. Markowich, *The Stationary Semiconductor Device Equations*, Springer-Verlag, New York, 1985.

[16] G. Strang and G. Fix, *An Analysis of the Finite Element Method*, Prentice-Hall, Englewood Cliffs, NJ, 1973.

[17] D. Scharfetter and K. Gummel, "Large-signal analysis of a silicon read diode oscillator," *IEEE Trans. Electron. Devices*, **16**, 64–77, 1969.

[18] H. Gummel, "A self-consistent iterative scheme for one-dimensional steady state transistor calculations," *IEEE Trans. Electron. Devices*, **11**, 455–465, 1964.

[19] J. Orgega and W. Rheinbolt, *Iterative Solution of Nonlinear Equations in Several Variables*, Academic Press, New York, 1970.

[20] R. Bank and D. Rose, "Global approximate newton methods," *Num. Math.*, **37**, 279–295, 1981.

[21] J. Ortega and W. Poole. *An Introduction to Numerical Methods for Differential Equations*. Pitman Publishing, New York, 1981.

[22] H. Vander Vorst, "Bi-CGSTAB: A fast and smoothly converging variant of Bi-CG for the solution of nonsymmetric linear systems," *SIAM J. Sci. Statist. Comput.*, **13**(2), 631–644, Mar. 1992.

[23] M. Adler, "A method for terminating mesh lines in finite difference formulations of the semiconductor device equations," *Sol. State Electron.*, **23**, 845–853, 1980.

[24] A. Franz, G. Franz, S. Selberherr, C. Ringhofer and P. Markovich, "Finite boxes: A generalization of the finite difference method suitable for semiconductor device simulation," *IEEE Trans. Electron. Devices*, **30**, 1070–1082, 1983.

[25] J. Burgler, *Discretization and Grid Adaptation in Semiconductor Device Modeling*, Konstanz: Hartung-Gorre, 1990.

[26] M. Shur, *Physics of Semiconductor Devices*, Prentice-Hall, Englewood Cliffs, NJ, 1990, Section 2-2.

[27] M. Toupikov, G. Pan, and B. Gilbert, "On nonlinear modeling of microwave devices using interpolating wavelets," *IEEE Tran. Microw. Theory Tech.*, **48**(4), 500–509, April 2000.

[28] G. Ghione, "Discretization schemes for high-frequency semiconductor device models," *IEEE Trans. Ant. Propg.*, **45**(3), 443–456, Mar. 1997.

[29] M. Reiser, "A two-dimensional numerical FET model for DC, AC, and large signal analysis," *IEEE Trans. Electron Devices*, **20**(1), 35–45, Jan. 1973.

[30] J. Slotboom, "Computer-aided two-dimensional analysis of bipolar transistors," *IEEE Trans. Electron Devices*, **ED-20**(8), 669–679, Aug. 1973.

[31] S. Selberherr, *Analysis and Simulation of Semiconductor Devices*. Springer-Verlag, New York, 1984.

[32] G. Pan, K. Wang, and B. Gilbert, "On multiwavelet based finite element method," *IEEE Trans. MTT.*, to appear in Jan. 2003.

[33] M. Sharma, "Semiconductor device modeling using flux upwind finite element," *COMPEL*, **8**(4), 219–224, 1989.

[34] C. Jacoboni and L. Reggiani, "The Monte Carlo method for the solution of charge transport in semiconductors with applications to covalent matrices," *Rev. Modern Phys.*, **55**, 645–704, 1983.

[35] A. Gnudi, D. Ventura, and G. Baccarani, "Modeling impact ionization in a BJT by means of spherical harmonics expansion of the Boltzmann transport equation," *IEEE Trans. Comp. Aided Design of ICS*, **12**(11), 1706–1713, Nov. 1993.

[36] K. Rahmat, J. White and A. Antoniadis, "Simulation of semiconductor devices using a Galerkin/spherical harmonic expansion approach to solve the coupled Poisson-Boltzmann system," *IEEE Trans. Comp. Aided Design of ICS*, **15**(10), 1181–1195, Oct. 1996.

[37] S. Reggiani, M. Vecchi and M. Rudan, "Investigation on electron and hole transport properties using the full-band spherical harmonics expansion method," *IEEE Trans. Electron. Devices*, **45**(9), 2010–2017, Sept. 1998.

[38] S. Sze, *Modern Semiconductor Device Physics*, John Wiley, New York, 1998.

[39] G. Arfken, *Mathematical Methods for Physicists*, Academic Press, New York, 1970.

[40] M. Lundstrom, *Fundamentals of Carrier Transport*, 2nd ed., Cambridge University Press, Cambridge, 2000.

[41] W. Gross, "Three-dimensional particle-based simulations of deep sub-micron MOSFET devices," Ph.D. dissertation, Arizona State University, 1999.

Index